CLINICAL CHEMISTRY

Interpretation and Techniques

ALEX KAPLAN, Ph.D., D.A.B.C.C.

Department of Laboratory Medicine
Professor Emeritus
University of Washington
Seattle, Washington

RHONA JACK, Ph.D.

Clinical Assistant Professor
Department of Laboratory Medicine
University of Washington
Clinical Chemist
Children's Hospital and Medical Center
Seattle, Washington

KENT E. OPHEIM, Ph.D., D.A.B.B.C.

Associate Professor, Clinical Chemistry Division
Department of Laboratory Medicine
University of Washington
Director, Clinical Chemistry Laboratory
Children's Hospital and Medical Center
Seattle, Washington

BERT TOIVOLA, Ph.D., N.R.C.C.(C)

Assistant Professor
Clinical Chemistry and Medical Technology Divisions
Department of Laboratory Medicine
University of Washington
Seattle, Washington

ANDREW W. LYON, Ph.D., N.R.C.C.(C), F.C.A.C.B.

Assistant Professor, Department of Pathology
University of Saskatchewan and Royal University Hospital
Saskatoon, Saskatchewan, Canada

CLINICAL CHEMISTRY

Interpretation and Techniques

FOURTH EDITION

Executive Editor: J. Matthew Harris
Development Editor: Lisa Stead
Project Editor: Denise Wilson
Production Coordinator: Mary Clare Beaulieu

Accurate indications, adverse reactions, and dosage schedules for drugs are provided in this book, but it is possible they may change. The reader is urged to review the package information data of the manufacturers of the medications mentioned.

Printed in the United States of America

First Edition, 1979
Second Edition, 1983
Third Edition, 1988

Library of Congress Cataloging-in-Publication Data

Clinical chemistry : interpretation and techniques / Alex Kaplan . . .
[et al.]. — 4th ed.
p. cm.
Rev. ed. of: Clinical chemistry / Alex Kaplan, LaVerne L. Szabo,
Kent E. Opheim. 3rd ed. 1988.
Includes bibliographical references and index.
ISBN 0-683-04560-1
1. Clinical chemistry. I. Kaplan, Alex, 1910– . II. Kaplan,
Alex, 1910– Clinical chemistry.
[DNLM: 1. Chemistry, Clinical. QY 90 C6412 1995]
RB40.K38 1995
616.07′56—dc20
DNLM/DLC
for Library of Congress

93-32472
CIP

96 97 98
3 4 5 6 7 8 9 10

To those dedicated medical technologists
who choose to serve humanity by
working assiduously and accurately
in clinical chemistry laboratories

Preface

We have added the following co-authors for the preparation of this fourth edition: Rhona Jack, Ph.D., whose specialty is molecular biology (genetics) and pediatric clinical chemistry, Bert Toivola, Ph.D. (expertise in endocrinology plus wide teaching experience in clinical chemistry), and Andrew W. Lyon, Ph.D. (expertise in laboratory instrumentation and lipid metabolism). It is with regret that we announce the retirement of Professor LaVerne Szabo after her participation in the first three editions.

The basic goals of this fourth edition are the same as those in previous editions: to prepare a good teaching manual for students who aspire to work in a clinical chemistry laboratory; to provide materials for teachers and supervisors to help them to instruct and to supervise more effectively; to create in laboratory workers a feeling of pride in and responsibility for the results of their analyses, in full awareness that they deal with sick people and not just numbers. To achieve these ends, we have updated all chapters to reflect developments and trends in our rapidly changing field. We have added two new chapters and have redrawn all illustrations that are now presented in a two-color format. We have added several instructional devices to facilitate learning: chapter objectives appear at the beginning of each chapter and pertinent review questions at the end.

All the chapters have been revised to some extent. The Basic Principles chapter has been converted and expanded into two chapters, Basic Principles and Basic Instrumentation. The chapter on lipid metabolism now presents the lipoproteins and their metabolism in much greater depth. The chapter on minerals has been expanded to include a nutrition section that deals extensively with trace elements. The chapter on immunotechniques has been enlarged to accommodate treatment of the newer technologies. The Clinical Toxicology chapter has been broadened to include recent drugs. Two new chapters, Prenatal and Perinatal Testing, and Genetic Disorders and their Diagnosis, have been added. Some deletions in the third edition were necessary to accommodate the new material. Accordingly, we have eliminated many of the detailed instructions for carrying out analyses because most of the laboratories now use commercially available kits and reagents or employ automated instruments that are programmed by the manufacturer for specific methods and reagents.

For each constituent, we have continued the practice of providing a reference to a good method that is accepted and widely used by many of the laboratories in the United States. The principles underlying each method are explained. In a book of this limited size, we are compelled to restrict the review of instruments to those that are widely used or that have potential for the future.

There is an ongoing trend in hospitals to move some of the automated tests from the central laboratory to special units or nursing stations to reduce turnaround time for the test results, whether it be for serum glucose in a diabetes clinic, for electrolytes or blood gases in intensive care units, or for other tests. Our goals have been expanded to include material that would be useful to nurses, respiratory therapists, and other health care professionals who suddenly might have the responsibility for performing such tests without having had a formal background preparation in clinical chemistry.

The material on clinical interpretation serves as a motivating link between the laboratory worker and the physician in their joint efforts for the diagnosis and treatment of disease. Work performance is always better when technologists understand the application of their results; this gives them a feeling of participation in the total medical effort. The technologists' work is extremely important because the modern health care expert relies heavily on their analyses of body fluid constituents. A sense of concern, accompanied by the initiative to ensure that the results are dependable and that they reach their proper destination in a timely manner, makes the difference between a good and a mediocre laboratory worker.

Seattle, Washington

Alex Kaplan, Ph.D.
Rhona Jack, Ph.D.
Kent E. Opheim, Ph.D.
Bert Toivola, Ph.D.
Andrew W. Lyon, Ph.D.

ACKNOWLEDGMENTS

Professor LaVerne Szabo, our colleague and a co-author of the first three editions, has retired. We miss working with her and are grateful for her permission to build on her original contributions to the preceding editions.

We also thank our colleagues in the Department of Laboratory Medicine for their helpful suggestions and comments.

We are indebted to Timothy Hengst and Associates for their assistance in preparation of the technical illustrations.

A.K.
R.J.
K.E.O.
B.T.
A.W.L.

Contents

1. INTRODUCTION TO CLINICAL CHEMISTRY

2. BASIC PRINCIPLES

3. BASIC INSTRUMENTATION

4. IMMUNOCHEMICAL TECHNIQUES

5. WATER BALANCE, OSMOLALITY, BLOOD GASES, pH, AND ELECTROLYTES

6. THE KIDNEY AND TESTS OF RENAL FUNCTION

7. CARBOHYDRATE METABOLISM

8. LIPID METABOLISM

9. PROTEINS

10. ENZYMES

11. THE LIVER AND TESTS OF HEPATIC FUNCTION

12. MINERALS AND TRACE ELEMENTS

13. ENDOCRINOLOGY

14. CLINICAL TOXICOLOGY

15. PRENATAL AND PERINATAL TESTING

16. GENETIC DISORDERS AND THEIR DIAGNOSIS

APPENDICES

INDEX

1 Introduction to Clinical Chemistry

OUTLINE

I. INTRODUCTION
II. CELLS
  A. Composition
  B. Malfunction
III. ROLE OF THE CLINICAL CHEMISTRY LABORATORY

OBJECTIVES

After reading this chapter, the student will be able to:

1. Describe the principal organelles within a cell and their usual functions; some of the organelles considered are the nucleus, mitochondria, chromosome, ribosome, endoplasmic reticulum, and plasma membrane.
2. Understand that many different factors at the cellular level can affect the concentration in plasma of a particular analyte (chemical constituent).

INTRODUCTION

Clinical chemistry is a basic science that utilizes the specialty of chemistry to study human beings in various stages of health and disease. It is an applied science when analyses are performed on body fluids or tissue specimens to provide important information for the diagnosis or treatment of disease. The dual nature of clinical chemistry requires that its practitioners be skilled in both phases. They should have an understanding of the physiologic and biochemical processes occurring in the body, as well as technical skills to perform the various tests. Clinical chemistry technologists and technicians are important members of the modern medical team because approximately half of hospital laboratory work involves the clinical chemistry laboratory.

Some elementary biology is presented for student orientation because a knowledge of the functions of the healthy, normal human body is essential for an understanding of changes that may occur in abnormal or pathologic states. An introductory text in biochemistry,[1,2] cell biology,[3] or human physiology[4] is recommended for those who have not had courses in these subjects.

CELLS

The basic unit of the body is the cell, where most of the body's chemical reactions occur. The cell is an integrated structure with many component parts, as illustrated in Figure 1.1. The cell has an outer membrane (bilayer) that separates it from other cells and from the interstitial fluids that bathe them.

Composition

Mammalian cells have a well-defined nucleus that contains the genetic material (DNA) distributed among chromosomes, and many organized structures or compartments (organelles) in the cytoplasm. Most of the cellular reactions take place in the organelles, some of which are illustrated in Figure 1.1. The organelles are specialized to perform different functions: the nucleus stores the genetic information; the main oxidative reactions for energy production occur in mitochondria; protein synthesis takes place on ribosomes; most cytoplasmic enzymatic reactions occur on endoplasmic reticulum; secretory granules store material for later release from the cell; the outer cellular (plasma) membrane controls the passage of substances into or out of the cell. Each organelle or compartment performs its functions by a series of linked, enzymatic reactions. The organization within specific cells may vary according to the structure and function of the cell. The formation, growth, and functioning of cells require the presence of appropriate raw materials and enzymes and a readily available energy supply.

The outer cell membrane is composed of a mosaic of a double layer of lipid molecules interspersed with protein and constitutes a means for selective permeability. Some ions and small molecules can pass easily into or out of the cell, others require special transport mechanisms for entry or exit, and some cannot cross the membrane of the living cell. Special molecules (glycoproteins) on the membrane act as recognition points or receptors for the attachment of

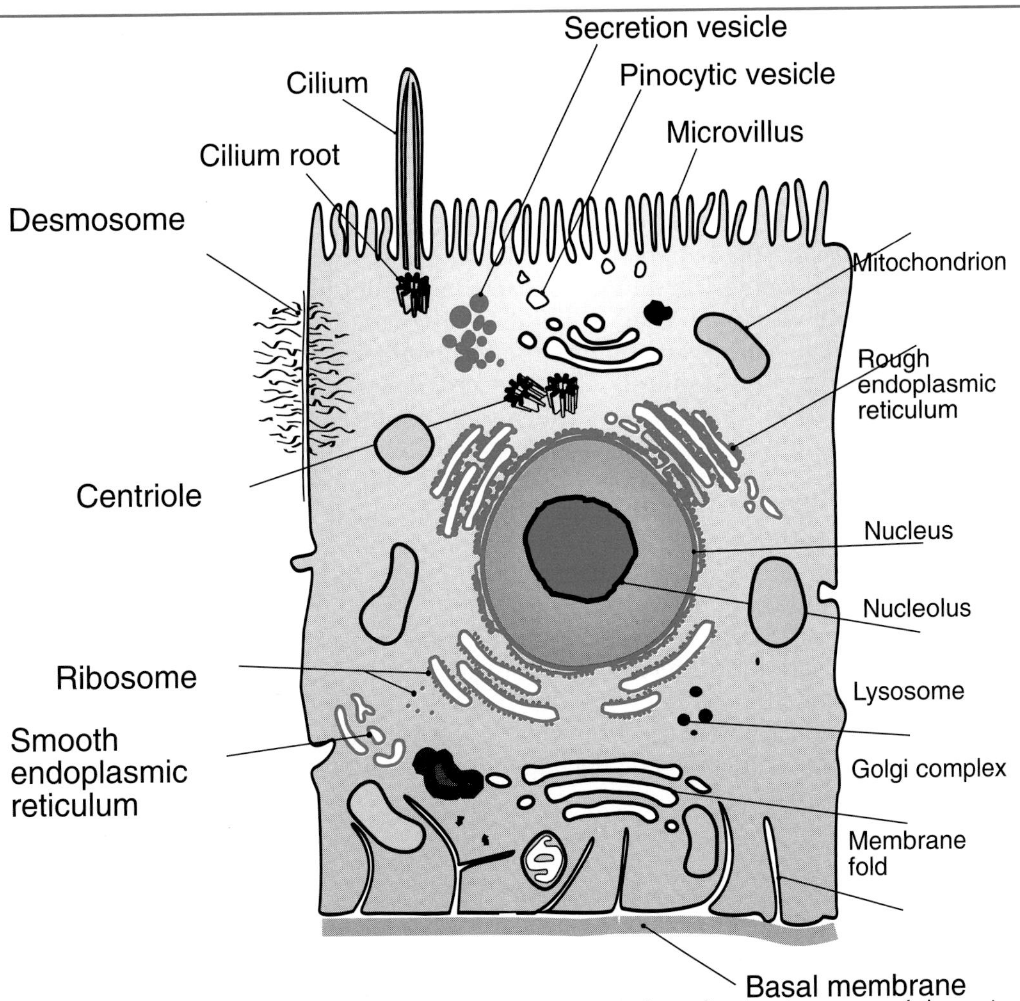

FIG. 1.1. Schematic diagram of the ultrastructure of a cell, showing some of the various organelles.

particular substances, such as hormones or antibodies, that are present in the circulating blood plasma.

Malfunction

Malfunction of a cell may be caused by a variety of factors, such as (1) destruction by trauma or by invasive agents, including pathogenic microorganisms, viruses, and toxins; (2) genetic deficiency of a vital enzyme; (3) insufficient supply of one or more essential nutrients (e.g., amino acids, vitamins, or minerals); (4) insufficient blood supply; (5) insufficient oxygen supply; (6) malignancy (uncontrolled tissue growth); (7) accumulation of waste products; (8) failure of a control system; or (9) a defect in the cellular recognition of certain signals. It is much easier to detect a malfunction than to elucidate or document the cause.

ROLE OF THE CLINICAL CHEMISTRY LABORATORY

Most clinical chemistry tests entail measuring the concentration of a particular constituent (the analyte) in body fluids, primarily blood plasma or serum. It is, however, not easy to interpret what is happening at the cellular level when the concentration of an analyte is abnormally high or low; additional information is needed. An elevated concentration could be caused by excessive intake, excessive body synthesis, deficient utilization, deficient excretion, or severe dehydration. The reverse is true for a low concentration. Information derived from the physical examination and the patient's history helps to elucidate the problem, but additional selected tests may be necessary. Gross abnormalities always indicate that something is wrong, but diagnosis is difficult to make early in a disease, when symptoms are obscure or absent and changes in concentrations are minimal. When concentration changes are small, the validity of the test result may be questioned. Thus, precision and quality control are necessary; your contributions are essential for good patient care.

The material in the following chapters is designed to prepare you for both phases of clinical chemistry: (1) a basic understanding of the physiologic and biochemical processes occurring in the body and (2) the basic information necessary for performing dependable analyses.

REFERENCES

1. Stryer, L.: Biochemistry. 3rd Edition. New York, W.H. Freeman, 1988.
2. Murray, R.K., et al. (Eds.): Harper's Biochemistry. 23rd Edition. Norwalk, Appleton and Lange, 1993.
3. Darnell, J., Lodish, H., and Baltimore, D.: Molecular Cell Biology. 2nd Edition. New York, W.H. Freeman, 1990.
4. Fox, S.I.: Human Physiology. 3rd Edition. Dubuque, IA, W.C. Brown, 1990.

REVIEW QUESTIONS

1. What are the main functions of the following cell organelles: nucleus, chromosomes, mitochondria, ribosomes, endoplasmic reticulum, and plasma membrane?
2. What are at least six possible causes of cell malfunction?
3. What are some of the possible mechanisms by which the concentration of a particular analyte may become elevated?

Basic Principles 2

OUTLINE

I. ROLE OF THE TECHNOLOGISTS

II. UNITS OF MEASURE
- A. Use of metric system
  - 1. Prefixes for units larger and smaller than the basic unit
  - 2. Système Internationale
- B. Units of temperature

III. VOLUMETRIC EQUIPMENT
- A. Pipets
  - 1. Transfer pipets
  - 2. Ostwald-Folin pipets
  - 3. Measuring pipets
  - 4. Serologic pipets
  - 5. Micropipets
  - 6. Pipettors
  - 7. Plunger-type dispensers
- B. Volumetric flasks
- C. Burets
- D. Cleaning laboratory glassware

IV. THE ANALYTIC BALANCE

V. REAGENT WATER

VI. LABORATORY CHEMICALS

VII. PRIMARY STANDARDS

VIII. LABORATORY CALCULATIONS
- A. Weight of solute per volume of solution
- B. Percent solutions
- C. Moles: molarity
- D. Gram equivalent weight
- E. Hydrated salts
- F. Dilution
- G. Conversion of units
- H. Molarity of concentrated reagents
- I. Hydrogen ion concentration
- J. Buffer solutions

IX. QUALITY ASSURANCE
- A. Quality control
  - 1. Precision and accuracy
  - 2. Standard deviation
  - 3. Coefficient of variation
  - 4. Control specimens
- B. Quality control charts
- C. Interpretation of control sera results
  - 1. Multi-rule procedure
- D. Proficiency testing
- E. Preventive maintenance
- F. Evaluation of new methodology

X. SPECIMEN COLLECTION
- A. Blood collection
  - 1. Arterial blood
  - 2. Venous blood
  - 3. "Capillary" samples
- B. Urine collection

XI. SAFETY
- A. Fire
  - 1. Precautions to take
- B. Infection
  - 1. Precautions against hepatitis and AIDS
  - 2. Universal precautions
- C. Corrosive chemicals
  - 1. Toxic fumes
- D. Broken glassware
- E. Carcinogens
- F. Radioactivity
- G. Hazardous waste disposal

OBJECTIVES

After reading this chapter, the student will be able to:

1. Use the metric system correctly for units of mass, length, volume, and temperature; understand the Système Internationale (SI) for expressing the concentration of blood and urine constituents; and convert SI units to conventional values and vice versa.
2. Express the concentration of a given reagent in several ways, such as mol/L, mg/dL, % (w/v), g/L, or meq/L.
3. Perform calculations involving the dilution of reagents or the dilution of patient's sample.
4. Calculate the pH and/or buffer concentration by using the Henderson-Hasselbalch equation.
5. Understand the necessity for the daily plotting of control serum values on a Levey-Jennings control chart and for judging whether the results of the run should be accepted.
6. Calculate the mean, standard deviation, and coefficient of variation of a control serum after 20 determinations.
7. Set up a protocol for evaluation of a new method.
8. Safely handle blood and urine specimens from patients who may be infected with hepatitis virus or HIV.

ROLE OF THE TECHNOLOGISTS

The clinical chemistry laboratory is used to measure chemical changes in the body for diagnosis, therapy, and prognosis of disease. The primary work of its technologists is the assay of various chemical constituents in blood, urine, and other fluids or tissues. Normally, the concentrations of these constituents are relatively constant, but in disease states their levels become altered. The magnitude of change usually parallels the degree of disease. In advanced disease, the abnormalities are usually large, easily detected, and present no analytical challenge to the laboratory. Early detection of organ dysfunction is much more difficult, however, because the chemical changes are usually slight and must be distinguished from possible errors in the performance of the test. Hence, test accuracy is a prerequisite for the proper interpretation of a laboratory test.

The technologist must always keep in mind that each laboratory specimen is taken from a person with a real or potential health problem, and that a technologist is an essential part of a highly skilled team that contributes to an assessment of the patient's condition.

Most test procedures are quantitative and require careful, precise measurements. An understanding of the principles of the testing method and some knowledge of the medical uses of the determinations provide the necessary background knowledge for performing the tests and understanding their rationale. A technologist, like any skilled worker, must understand thoroughly the tools of the trade: the equipment, the reagents, and the principles and calculations involved in the assays. Accordingly, this chapter summarizes some of the background information and the basic principles common to most chemistry laboratory methods and essential to good laboratory practice and technique.

Also included are a review of common laboratory calculations and an introduction to quality assurance and quality control. A brief description of medical usage and an interpretation of laboratory tests appear in the chapters dealing with the tests themselves.

The addresses of all manufacturers referred to in the text are listed in Appendix 4.

UNITS OF MEASURE

Use of Metric System

All quantitative measurements must be expressed in clearly defined units that are accepted and understood by all scientists. The metric system is used in scientific measurements; hence, the gram, meter, liter, and second are employed as the basic units for the expression of weight, length, volume, and time, respectively.

Prefixes for Units Larger and Smaller Than the Basic Unit

Identical prefixes are used with the basic units to denote larger or smaller size. Thus, one *milli*meter (mm) refers to one-thousandth part of a meter (10^{-3} meters), while *milli*gram (mg), *milli*liter (mL), and *milli*second (ms) are used to describe 10^{-3} g, 10^{-3} L, and 10^{-3} s. respectively.

Table 2.1 summarizes the more commonly used prefixes and their abbreviations. The terms marked with asterisks are correct, but are seldom used in the clinical laboratory. An additional prefix, "mega," denoting one million of the units (10^6) is used in measurements of electrical resistance, for example, one megohm (MΩ) is equal to 10^6 ohms.

Système Internationale

Although scientists have long used the metric system, over the years discrepancies in the expression of many units began to appear in different disciplines, different countries, and even different laboratories. Wavelength was designated in angstrom units (10^{-10}m) in some countries or laboratories and as millimicrons (10^{-9}m) in others; the concentrations of blood constituents were expressed as g/L in some countries and as mg/dL in others. Some

TABLE 2.1
SOME BASIC UNITS AND THEIR PREFIXES DENOTING MULTIPLE OR DECIMAL FACTORS

PREFIX BASIC UNIT	ABBREVIATION	MULTIPLE OF BASIC UNIT ($10^0 = 1$)	WEIGHT G (GRAM)	LENGTH M (METER)	VOLUME L (LITER)
tera-*	T	10^{12}	Tg	Tm	TL
giga-*	G	10^{9}	Gg	Gm	GL
mega-*	M	10^{6}	Mg	Mm	ML
kilo-	k	10^{3}	kg	km	kL
deci-	d	10^{-1}	dg*	dm*	dL
centi-	c	10^{-2}	cg*	cm	cL*
milli-	m	10^{-3}	mg	mm	mL
micro-	μ	10^{-6}	μg	μm	μL
nano-	n	10^{-9}	ng	nm	nL
pico-	p	10^{-12}	pg	pm	pL
femto-*	f	10^{-15}	fg	fm	fL
atto-*	a	10^{-18}	ag	am	aL

*Denotes prefix or term seldom used in clinical chemistry.

authors expressed the concentrations of reagents in g/100 mL, but others worked with molar concentrations. Such discrepancies led to confusion and to misunderstanding. A start in the standardized presentation of clinical chemical laboratory data was made in 1967 when the 1966 Recommendation of the Commission on Clinical Chemistry of the International Union of Pure and Applied Chemistry and of the International Federation for Clinical Chemistry[1] was published. The recommended units and mode of expression are known as the Système Internationale or the SI. In SI units, wavelength is designated as nanometers (10^{-9}m) and abbreviated nm. The Système Internationale recommends molar units rather than mass units whenever possible; e.g., for glucose, the designation 5.6 mmol/L is preferred to 100 mg/dL. Young has written an explanation of the units acceptable to the Système Internationale and how they are to be used.[2] He includes an extensive table of assays listing current reference ranges and units for the measured components, corresponding reference ranges in SI units, and conversion factors.

The units of length, weight, and volume in the metric system were designed to be interrelated. The liter was originally defined as the volume occupied by one kilogram of water at the temperature of its greatest density (4° C), but in 1964 the definition was changed by international agreement so that the liter is now defined as exactly equal to a cubic decimeter. The old liter was 1.000028 times as large as the new liter, a negligible difference. Although the terms *cubic centimeter* (cc) and *milliliter* (mL) are synonymous, only the latter term, mL, is acceptable in the Système Internationale.

Units of Temperature

The temperature scale most commonly used in the clinical laboratory is the centigrade scale (Celsius or ° C), which places the freezing point of water at 0° C and the boiling point of water at 100° C.

For calculations involving temperature, a scale based on absolute zero is needed. The Kelvin scale (° K) employs units identical to those in the Celsius system, but its zero point corresponds to absolute zero (0° K), which is equivalent to −273.15° C. Thus, for conversion from centigrade to Kelvin temperatures, we can use the following formula:

$$^\circ K = {}^\circ C + 273$$

In the SI, temperatures are expressed as ° K, but clinical laboratories still use ° C.

VOLUMETRIC EQUIPMENT

Clinical chemistry procedures require accurate measurements of specimens and reagents for valid, useful results; all equipment and instruments must be accurate and reliable. High-quality volumetric glassware should be purchased, and calibrations should be verified.

Pipets

Several types of pipets are available, each designed for a specific purpose, but a few rules for correct pipetting technique are common to all of them. The pipet must be held in a vertical position while setting the liquid level to the calibration line and during delivery. The lowest point of the meniscus should be level with the calibration line on the pipet when it is sighted at eye level. The flow of the liquid should be unrestricted when using volumetric pipets; flow may have to be slowed slightly with the finger when graduated pipets are used for fractional delivery.

One rule should *always* be observed in the laboratory: Never pipet anything by mouth. Always use a safety bulb or other pipet filler.

Transfer Pipets

The *volumetric*, or transfer, pipet is used when the greatest accuracy and precision are required. It consists of a long, narrow tube with an elongated or rounded bulb near the middle and is designed to deliver an exact volume of water at a specified drainage time. The tip is tapered to slow the rate of delivery, and a single calibration ring is etched into the tubing above the bulb. Class A pipets, conforming to narrow limits specified by the National Bureau of Standards, are necessary for the greatest accuracy. In use, the pipet tips are touched to the inclined surface of the receiving vessel until the fluid has ceased to flow and for 2 sec thereafter. The tip is then withdrawn horizontally from contact with the receiver. A volumetric pipet with a broken or chipped tip is inaccurate and must be discarded. Pipets are available in which the glass has been chemically tempered to increase strength. Because these pipets are especially resistant to chipping and breaking, they are economical in the long run.

Ostwald-Folin Pipet

This specialized version of the volumetric pipet was designed for transferring fluids, such as whole blood, to minimize drainage errors. The pipet is shorter, the bulb is near the delivery tip, and the opening is slightly larger for faster drainage. As soon as drainage has stopped, the last drop is blown out. All blowout pipets are marked with an etched single or double ring around the upper end.

Measuring Pipets

The Mohr-type measuring pipets consist of long, straight tubes with graduated markings to indicate the volume delivered from the pipet. The calibrations are made between two marks etched entirely on the straight portion of the tube. The tapered tips are not part of the calibrated portion. Measurements are made by reading the meniscus of the fluid against the pipet calibration marks before and after delivery of an aliquot.

Serologic Pipets

Serologic pipets are like measuring pipets, but are calibrated to the tip. These blowout pipets are marked with the etched ring on the upper-end mouthpiece; to deliver the total volume, the last drop must be blown out. Serologic pipets may be used for "point-to-point" measurements in the straight part of the tube, just as measuring pipets are used, or for total volume measurements, blowing out the last drop. For good precision when dispensing aliquots of a reagent (for example, 2-mL aliquots from a 10-mL serologic pipet), only the calibrations on the straight tubing should be used, not the tapered section. Thus, only four 2-mL aliquots can be obtained by using a 10-mL serologic pipet. The Mohr-type measuring pipets have smaller openings in the tip than do serologic pipets; they are therefore slower and more easily controlled for accurate measurements. Graduated pipets (Mohr or serologic) are not accurate enough to use for measuring standards or samples; only Class A volumetric pipets are suitable for this purpose.

All the described pipets are designed "to deliver" (TD) the stated volumes and are marked with "TD" near the upper end of the tube. Other pipets, designed for measuring small quantities of fluids, are calibrated "to contain," (TC) rather than to deliver the specified volume. These pipets are described in the following sections.

Micropipets

In measuring small volumes of sample (5 μL to 250 μL), variations in the liquid film adhering to the pipet wall become significant. The problem of residual liquid in micropipets calibrated to contain a given volume of liquid is eliminated by rinsing the pipet several times with the diluent into which its

contents have been dispensed; the last drop is then blown out. Other micropipets are calibrated to deliver a measured volume. The calibration of micropipets to be used for measuring standards and samples must be checked.

Pipettors

Semiautomatic pipettors are designed to pick up and to dispense a preset volume of solution as a plunger is smoothly released and then depressed; these are available commercially (Fig. 2.1). In some models, disposable plastic tips are used to eliminate carry-over from sample to sample. Another type employs a Teflon-tipped plunger that sweeps the capillary delivery tip clean as the contents are expelled. Multiple volume pipettors are available in two types: the dual- or triple-range samplers, which can be positioned to one of two or three preset volumes, and the pipettor with a continuously variable digital volume setting. Each of these is made in a wide range of sizes from 10 μL or less to 1.0 mL or more. Pipettors can deliver reproducible volumes, but the accuracy may be diminished at low volumes. It is advisable to check the accuracy of any particular pipet to be used for measuring small volumes of sample. The attainment of good precision with these pipets requires close adherence to the directions for usage, smooth manual action, and practice.

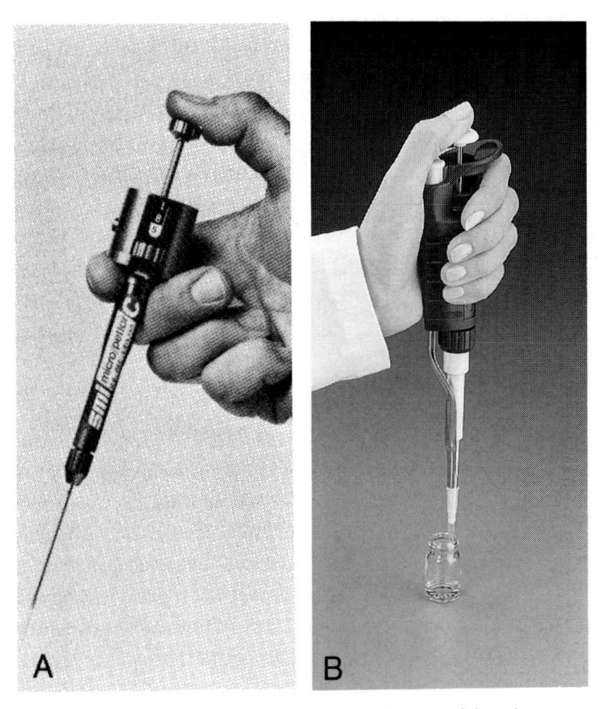

FIG. 2.1. Automatic pipets. A, With reusable glass tip. (Courtesy of SMI, a subsidiary of American Dade.) B, With a disposable plastic tip. (Courtesy of Rainin Instrument Co., Inc. "Pipetman" is a registered trademark of Gilson Medical Electronics, with exclusive license to Rainin Instrument Co., Inc.)

Micropipettors can be calibrated by weighing a solution of known density delivered by the pipet. The actual delivery volume in mL is calculated by dividing the weight in g by the density appropriate for the temperature of the liquid. The calibrated pipets are used as standards for calibrating other pipets by a simpler colorimetric procedure.*

Plunger-Type Dispensers

Reagent supply bottles may be fitted with a dispensing apparatus built into the screw top lids. These can be adjusted in the laboratory to deliver the desired volume of reagent repeatedly with good precision. Such dispenser bottles save time when repetitive volumes of a reagent are dispensed.

Large laboratories are often involved with procedures requiring many samples. For handling workloads of this magnitude, *electrically operated dispensers* and combination *sampler-dilutors* are more practical.

Volumetric Flasks

Volumetric flasks that meet Class A requirements should be used for all standard solutions and many laboratory reagents. Because changes in temperature cause variations in the volume of solutions, glassware and reagent solutions must be at room temperature at the time of final dilution in the volumetric flask. Solid reagents must be completely dissolved, and the solutions must be well mixed before final dilution to the calibration mark.

Graduated cylinders are relatively inaccurate measuring devices and are used for situations in which accuracy of a high order is not required. They are convenient for measuring 24-hr urine volumes of 500 to 2000 mL, where a low order of accuracy is sufficient. They may also be used for the preparation of noncritical solutions.

Beakers, flasks, and test tubes sometimes have graduated markings along one side indicating approximate volumes. These are only rough approximations, convenient though they may be, and are not appropriate when accurate measurements are required.

Burets

Burets may be used in the laboratory for titrations or for dispensing aliquots of a solution. They are available in a wide range of sizes, most commonly from 1-mL to 100-mL capacity. Those with a total volume of 10 mL or less are called microburets. Some microburets are fitted with fine metal tips that deliver drops of approximately 0.01 mL. For delivery and measurement of small volumes (microliters), syringe-type microburets use the forward motion of a plunger to

*For the colorimetric procedure, 10 mL water is dispensed into tubes (10 tubes for each pipet to be calibrated). An aliquot of $K_2Cr_2O_7$ (4.00 g/dL) is delivered into each tube and the absorbance is read at 450 nm against water. The actual volume of each pipet is calculated from the absorbance and dilution factor compared to that of the calibrated pipet. The aliquots can be weighed rather than measured spectrophotometrically.

displace the fluid that is dispensed. The plunger movement is measured accurately by a micrometer. Because the plunger's area of cross section is known, the volume of fluid dispensed is directly proportional to the distance the plunger moves.

Glass burets may be purchased with either Teflon or ground-glass stopcocks. The major advantage of the Teflon stopcock is that it does not require lubrication; the Teflon plug is also highly resistant to alkali and does not "freeze" to the buret when exposed to such solutions. The glass stopcocks require a thin coating of lubricant that must be applied sparingly so that an excess does not plug the stopcock channels or buret tip.

Cleaning Laboratory Glassware

Laboratory glassware must be clean and must not add substances to the solutions it contains or adsorb substances from these solutions.

Glassware should be rinsed after using and then soaked in a mild solution containing a laboratory detergent that is a wide-range cleanser. Good automatic dishwashers for use in the laboratory can prerinse, wash with hot detergent solution, then rinse thoroughly with tap water, and rinse again with distilled water. This cleaning process is sufficient for glassware used in most laboratory assays.

Some glassware requires more specialized treatment. Pipets may be soaked in a detergent solution or a dichromate acid solution ($K_2Cr_2O_7$ and concentrated H_2SO_4) before washing.

Glassware to be used for trace metal analysis is kept totally separate from other glassware. It is cleaned by first soaking it in a 50% (v/v) nitric acid solution for a minimum of 2 hr; then it is rinsed thoroughly with distilled water.

THE ANALYTIC BALANCE

Reagents for the preparation of standard solutions must be weighed accurately, with an error of less than 0.1%. Good analytical balances are accurate to at least 0.1 mg. Ultramicro balances accurate to a fraction of a μg are available, although most clinical laboratories do not need this capability.

The analytic balance found in the average laboratory is probably one of two types: a single-pan analytical balance with internal weights or an electronic analytical top-loading balance (Fig. 2.2). In the single-pan type, the weighing pan is suspended from a knife edge, and the counterbalancing internal weights are added or removed by turning a knob on the front of the instrument. To protect the delicate knife edge, the balance beam must be raised off the knife edge whenever weights, containers, or the substance to be weighed is added to, or removed from, the weighing pan.

The balance can be set to read zero with the empty container on the pan by adding "tare" weights to the system. The balance then indicates only the additional weight of the sample, because it has already compensated for the weight of the container (tare).

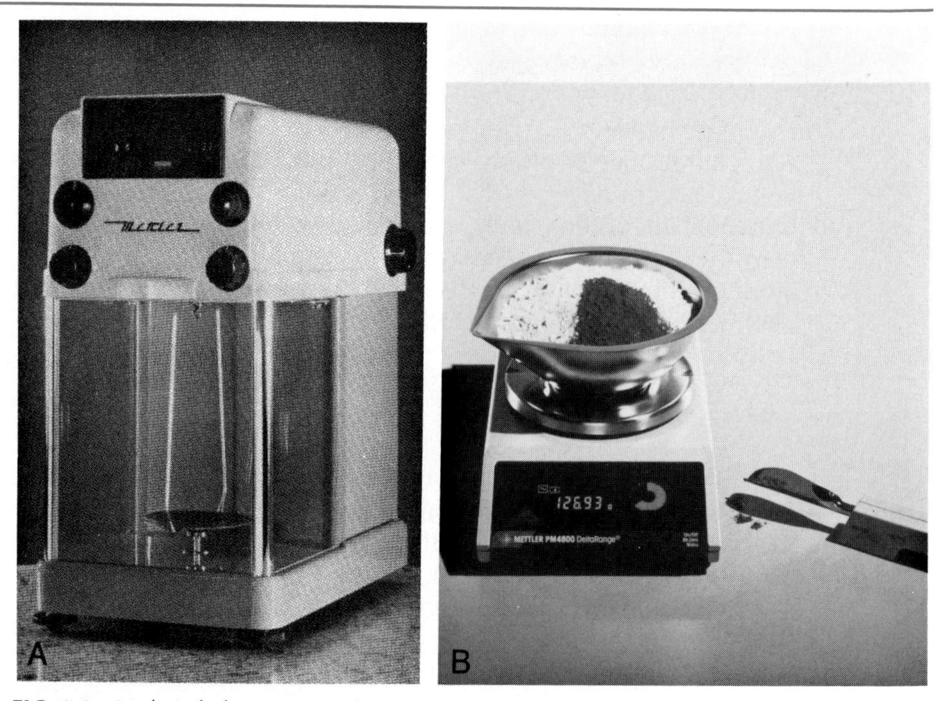

FIG. 2.2. Analytic balances. A, Balance with a single pan and internal weights dialed in by the operator. B, Top-loading electronic precision balance that automatically indicates the weight added to the pan. (Both photos courtesy of Mettler Corp.).

The top-loading balance is much simpler to use. The empty container can be added to the platform, and a button pushed to set the instrument to read zero with the tare weight. Then, as sample is added to the container, the digital readout automatically presents the weight of the sample, with no manipulation on the part of the operator. The top-loading balance uses an electromagnet to measure the weight on its platform. A wire conducting an electric current is positioned in a magnetic field. When weight is added to the pan, the wire is forced out of its position; the amount of current needed to return the wire to its original position is a function of the amount of weight added to the platform. The electronic balance is faster than those with weights to add, and has no delicate knife-edge surface to protect.

Rules for good technique, however, apply to both types of balance.

1. The balance must be level and vibration-free; marble balance tables greatly reduce vibration, but if they are not practical, vibration can be substantially reduced by inserting a cushioned marble slab between the analytical balance and the laboratory bench.
2. Air currents during the final step of weighing must be avoided by closing the doors of the weighing compartment. Because temperature differences also cause convection currents, the samples must be at room temperature.
3. The balance must be kept scrupulously clean. Weights, pans, and internal parts of the balance must never be touched with the fingers; chemicals left on the balance, either spilled or in fingerprints, may corrode the metal and

impair its accuracy. Any spills should be carefully brushed off the pans and removed from the balance compartment. Like all precision instruments, an analytical balance requires periodic maintenance and cleaning for trouble-free performance.

REAGENT WATER

Water is the most commonly used solvent in the clinical laboratory. The quality of laboratory reagents, and consequently the quality of the laboratory testing, depends on the reagent water available to the laboratory. The College of American Pathologists[3] and the National Committee for Clinical Laboratory Standards (NCCLS)[4] have both issued minimum specifications for reagent water to be used in the clinical laboratory.

The water quality is graded into three types, each appropriate for its specific use. Table 2.2 shows the specifications for each type, as defined by NCCLS.

Type I, with the most stringent requirements, is recommended for use in the assays most sensitive to impurities. These include trace element and enzyme analysis, immunoassays, and fluorescence quantitation. For high-performance liquid chromatography, Type I water may require further treatment to remove organic impurities.

Type II water is suitable for reagents in most routine laboratory assays, and Type III water should only be used for rinsing general glassware after washing. For assays requiring Type I water in its reagents, the glassware should also be rinsed with Type I water. Prepare all reagents with Type II water unless Type I is specified.

The specifications for water quality do not recommend any specific purification systems. The methods should be selected according to the quality of the available tap water. Combinations of treatments are generally needed to

TABLE 2.2
REAGENT WATER SPECIFICATIONS

	TYPE I	TYPE II	TYPE III
Bacterial content,* colony-forming units per mL (cfu/mL)—maximum	<10	10^2	NA
pH	NA	NA	5.0–8.0
Resistivity,† megohms/centimeter (MΩ/cm), 25° C	1.0 (in-line)	1.0	0.1
Silicates, mg/L SiO_2, maximum	0.05	0.1	1.0
Particulate matter‡	0.22 μm filter in system	NA	NA
Organic contaminants‡	Activated carbon filter in system	NA	NA

*Microbiologic content, evaluated by colony count after incubation for 24 hours at 37° C, followed by 24 hours at 23° C.
†Resistance measured between opposite faces of a 1-cm cube at 25° C. For Type I water, this measurement must be made in-line.
‡Specifications for particulates and organics do not require measurement in the end product; both are "process specifications," requiring that, for Type I water, the purification process includes a filter no larger than 0.22 μm and a cartridge containing activated charcoal.

produce the highest-quality (Type I) water. Distillation, filtration, and ultrafiltration remove particulates and bacteria; deionization removes ionized solids and gases; and activated carbon adsorption removes organic contaminants and ammonia.

As water is distilled or deionized, its ability to conduct an electric current is vastly reduced because of the elimination of charged particles (ions). Passage through a second deionizing cartridge may be necessary to meet the resistance specifications for Type I water.

Either distilled or deionized water may become contaminated by the container in which it is stored; soda lime glass, especially, may be a source of metal ions in the water. For this reason, water should not be stored for long periods before it is used, and any reagents that are to be kept for a long time should be stored in borosilicate glassware.

In the procedures described in the following chapters, appropriately pure water (Type I or II) is to be used in the preparation of all aqueous solutions.

LABORATORY CHEMICALS

Reagents used for clinical chemistry assays must be made from highly purified chemicals. For most chemicals, this grade is known as *Reagent Grade, Analytical Grade,* or ACS grade (meets American Chemical Society specifications). Each chemical is guaranteed to meet certain stated standards printed on the label. The label may show the actual assay for the chemical and a number of specified contaminants, or it may state that the contaminants are below certain specified concentrations.

For some laboratory procedures, such as fluorometry, gas chromatography, and trace metal analysis, reagents of exceptionally high purity are needed. In the past, the analyst had to further purify each batch of the purchased chemical or at least to check each lot against the laboratory's special requirements. To meet this need, manufacturers now offer selected chemicals that have been subjected to additional purification steps to further reduce the content of specific impurities. These reagents may be known by such names as *Ultrex Ultra pure* (J.T. Baker). Other reagents may be treated to remove impurities that might interfere in one type of laboratory procedure, such as spectrophotometry (*PHOTREX*, Baker), high-performance liquid chromatography (*HPLC*, Baker or *ChromAR*, Mallinckrodt), and trace metal analysis (*Instra-Analyzed*, Baker).

On the other hand, reagents may be labeled with descriptive terms, such as *purified* or *chemically pure*. There is no standard for such labeling, and the quality of the product varies from one brand to another. Organic compounds are often graded in this way, with the melting point range given as a further indication of purity. These chemicals are not recommended for the clinical chemistry laboratory, but they may have to be used when none of better quality is available. The analyst may have to further purify these compounds before use or to include a reagent blank with each batch of tests.

Technical, practical, or *commercial* grade chemicals are not suitable for the laboratory except for use as cleaning solutions.

A USP or NF rating on a reagent bottle indicates that the contents conform to the specifications of the *United States Pharmacopeia* or the National Formulary. Because these specifications allow no impurities at a concentration level that would be injurious to health, pharmacists can safely use them for medications. Some of these chemicals are pure enough for laboratory purposes; others are not. Chemical purity as such is not the primary goal of this classification, and such purity must not be assumed by the user of USP grade reagents.

For some of the assays described in this book, a source for a reagent or other material is given in parentheses (for example, Elvanol 70–05, DuPont). Appendix 4 gives the complete names and addresses of manufacturers so mentioned.

PRIMARY STANDARDS

A *primary standard* is a substance that can be accurately weighed or measured to produce a solution of an exactly known concentration. To qualify as a primary standard, a substance must be essentially free of impurities. Criteria for primary standards set up by the International Union of Pure and Applied Chemistry (IUPAC) require that the compound be at least 99.98% pure (for working standards, 99.95% pure).[5] This level of purification is impossible to attain with most biologic materials and is not necessary for routine clinical laboratory tests.

Biologic materials of acceptable purity, or even of *known* purity, are difficult to obtain. To help to alleviate this problem, the National Bureau of Standards (NBS) has a program of producing a series of Standard Reference Materials for Clinical Chemistry, most of which are at least 99% pure.[6,7] These include cholesterol and bilirubin, compounds that are difficult to purify but that are needed as standards for laboratory procedures. The NBS standards are much too expensive for daily use in the laboratory, but they serve as excellent reference standards. With the aid of the reference standard, a laboratory may accurately assay its commercially obtained reagent, which may then serve as a routine (secondary) standard.

LABORATORY CALCULATIONS

Weight of Solute per Volume of Solution

The simplest reagents are made by transferring a weighed amount of the desired chemical to a volumetric flask, dissolving it in solvent, and filling the flask exactly to the calibration mark. With pure chemicals and good technique, the concentration of the resulting solution is an exact quantity that may be expressed, in terms of mass per volume of solution, in several ways. As an illustration, 5 g of Na_2SO_4 in a liter of solution is equivalent to 5 g/L, 0.5 g/dL, 500 mg/dL, or 5 mg/mL.

Such expressions of mass of solute per unit volume do not indicate the functional concentration of a solution, that is, the relative number of molecules available to react with other molecules. Molar concentration, reviewed later, is based on the latter concept, but the mass per volume designation is often used in the laboratory as a convenience in identifying and making simple reagents.

Percent Solutions

Another convention consists of expressing the concentration of a solute as "percent" (per hundred parts of the total solution). The three variations of percent solutions are:

1. *Weight/volume (w/v):* A solution containing 5 g of Na_2SO_4 dissolved in water and diluted to a final volume of 100 mL of solution can be designated as 5% (w/v) solution.
2. *Volume/volume (v/v):* This designation is convenient for solutions composed of two liquids. Thus, 5 mL of glacial acetic acid diluted with water to a total volume of 100 mL can be described as a 5% (v/v) acetic acid solution.
3. *Weight/weight (w/w):* The percentage of solute can also refer to grams of solute per 100 g of final solution. For example, 5 g Na_2SO_4 dissolved in 95 g of water (approx. 95 mL) would result in a concentration of 5% (w/w), or 5 g solute in a total weight of 100 g (solute plus solvent). This designation differs significantly from the 5% (w/v) Na_2SO_4 solution previously described and is seldom used in clinical chemistry.

Percent solutions that do not designate weight or volume for solute and solution are ambiguous at best and are open to error and confusion.

A historical term, *milligrams percent* (mg%), was formerly used to represent milligrams per 100 mL of solution. This is nondefinable as percentage because mg/dL is 1 part in 100,000. *Milligrams per deciliter* (mg/dL) is the correct term implied by the mg% expression and should always be used in its place.

The SI recognizes moles per liter (mol/L) as the only mass per volume designation; the use of % or mg/dL is discouraged.

Moles: Molarity

In the previous section, the concentrations of solutions were expressed in terms of mass of solute in a given volume of solution. Except when actually weighing out the reagent, this designation is not particularly useful. A more meaningful expression is the relative number of molecules available in the solution to react with other molecules. To express this relationship, the SI recommends that the liter be used as the unit of volume and that the solute concentration be expressed as mol/L whenever the molecular structure is unequivocably known.[1] The use of mol/L allows direct comparisons of the functional concentrations of solutions.

A mole of any pure compound is the molecular weight of that compound expressed in grams. It consists of the same number of molecules (6.02×10^{23}) as 1 mole of any other compound. "Moles per liter" is sometimes expressed as *molarity*, a 1-molar (1 M) solution containing one mole of solute per liter of solution. The SI does not recommend use of the symbol M to represent molar concentration, but it is widely used, and technologists must be familiar with it.

To calculate the number of moles in a given weight of solute, divide the number of grams by the molecular weight.

Example: 5 g Na_2SO_4 is equivalent to how many moles? The molecular weight of Na_2SO_4 is 142, so there are 5/142 or 0.035 moles. If the 5 g Na_2SO_4 were dissolved in water to make 1 L of solution, the concentration would be 0.035 mol/L. To avoid cumbersome decimal fractions, small concentrations may be expressed with an appropriate prefix to the mol/L. Thus, 0.035 mol/L may also be written as 35 mmol/L (millimoles per liter). Micromoles (μmol) and nanomoles (nmol) represent 10^{-6} and 10^{-9} moles, respectively, and may also be used when appropriate.

Example: What is the concentration in SI units of a solution containing 1.20 g of Na_2CO_3 in 200 mL of solution?

$$\text{mol/L} = \frac{1.20\text{ g}}{106\text{ g/mol}} \times \frac{1000\text{ mL/L}}{200\text{ mL}} = 0.0566\text{ mol/L}$$

$$= 56.6\text{ mmol/L}$$

Gram Equivalent Weight

The gram equivalent weight (g/eq) of a substance may be defined as the weight of that substance that can combine with, or displace, 1.008 g of hydrogen.

Example: What is the gram equivalent weight of calcium (atomic weight = 40.08)?

Comparison of the formulas, $CaCO_3$ and H_2CO_3, shows that 40.08 g calcium (1 mole) displaces 2.016 g hydrogen. Setting up the ratio gives:

$$\frac{40.08}{\text{eq wt Ca}} = \frac{2.016}{1.008} = \frac{2}{1}$$

$$\text{eq wt Ca} = \frac{40.08}{2} = 20.04$$

This example illustrates that the equivalent weight of an element is equal to the atomic weight divided by the number of atoms of hydrogen with which one atom of that element can combine or can displace. For aqueous solutions of electrolytes (acids, bases, and salts), the gram equivalent weight may be determined by dividing the gram formula weight by the total positive or negative charge.

Examples: (1) In Na_2O ($2Na^{+1}$, O^{-2}) the equivalent weight of oxygen is 16.00/2 = 8.00, and the equivalent weight of the compound Na_2O is 62/2 = 31. (2) The equivalent weight of $CaCO_3$ (Ca^{+2}, $CO_3{}^{-2}$) is equal to the molecular weight (100.1) divided by 2, or 50.0.

Note: The *gram formula weight* is defined as the weight in grams of the entity represented by a formula. The gram formula weight then is equal to the gram molecular weight of a compound, to the gram atomic weight of an element, or to the weight in grams expressed by the formula for an ion, such as SO_4^{-2}.

As reviewed in the preceding section, a *molar* solution is a solution that contains one mole of solute per liter of solution. Correspondingly, a normal solution contains one gram equivalent of solute per liter of solution. *Normality* is defined as the number of equivalents of solute present per liter of solution; a 3 normal (3N) solution of $CaCO_3$ contains 3 gram equivalents, or 150.1 g of $CaCO_3$ per liter.

Example: What is the normality of a solution containing 5 g Na_2SO_4 per liter? The molecular weight is 142 and the equivalent weight is 142/2 = 71, because the 2 atoms of Na can be replaced by 2 atoms of hydrogen, as in H_2SO_4. The normality of the solution is 0.070 N (5 g/L ÷ 71 g/eq).

The terms *normality* and *equivalents* are not recommended in the SI, but both terms are widely used in clinical laboratories and must be understood by technologists. These units truly represent the functional or stoichiometric quantities of substances; one equivalent of any substance reacts with exactly one equivalent of another substance, whether the reaction is a simple combination or displacement reaction or a more complex oxidation-reduction reaction.

It may be convenient, but is by no means necessary, to use the concepts of "normality" and "equivalence" to calculate the quantities of reactants that react stoichiometrically. The calculations can be made readily using balanced equations containing molecular formulas. An acid-base titration is a simple example of this type of problem.

Example: How many mL of NaOH (1.5 mol/L) are required to neutralize 10 mL of H_2SO_4 (2.0 mol/L)?

Solution A. $2\ NaOH + H_2SO_4 \rightarrow Na_2SO_4 + 2\ H_2O$. From this equation we can see that 2 moles of NaOH are required to neutralize 1 mole of H_2SO_4.

$$\text{mmol } H_2SO_4 = \text{Volume (mL)} \times \text{Concentration} \left(\frac{\text{mmol}}{\text{mL}}\right) = 10 \times 2 = 20$$

The amount of NaOH required = 2 × 20 mmol = 40 mmol

$$\text{mL NaOH} = 40 \text{ mmol}/1.5 \text{ mmol/mL} = 26.67 \text{ mL}$$

Solution B. The problem may also be solved by using the normalities of the solutions:

$$1.5 \text{ M NaOH} = 1.5 \text{ N NaOH}$$

$$2.0 \text{ M } H_2SO_4 = 4.0 \text{ N } H_2SO_4$$

Because we are dealing with equivalents, the equation becomes $V_1C_1 = V_2C_2$.

$$1.5\ V_1 = 10 \times 4$$

$$V_1 = 26.67 \text{ mL of NaOH}$$

Either method of solving the problem is acceptable; use the one that is easier for you.

Hydrated Salts

When salts are crystallized, water molecules sometimes form an integral part of the crystal structure as water of hydration. Some salts have several different hydrated forms that contain differing amounts of water in the crystal lattice; these are always indicated in the molecular formula and contribute to the molecular weight.

For example, sodium sulfate crystals may appear in three forms:

1. Anhydrous (Na_2SO_4); Mol wt. 142
2. Heptahydrate ($Na_2SO_4 \cdot 7H_2O$); Mol wt. 268
3. Decahydrate ($Na_2SO_4 \cdot 10H_2O$); Mol wt. 322

A 1 mol/L solution of each of these salt forms contains 142, 268, and 322 g/L, respectively; each solution contains exactly the same concentration of sodium sulfate.

When making up reagents, the technologist must always be aware of the hydration state of any crystalline compound to be measured. For example, if directions for a reagent specify 8.0 g of Na_2SO_4 (anhydrous), but the available salt is $Na_2SO_4 \cdot 10\ H_2O$, a simple ratio using the molecular weights of the 2 salt forms adjusts the instructions.

$$\frac{X\text{ g}}{8\text{ g}} = \frac{322\text{ g/mol}}{142\text{ g/mol}}$$

$$X = 18.1\text{ g}$$

Dilution

In the daily routine of the laboratory, a technologist frequently makes dilutions of samples and solutions. This may be done as part of a test procedure, as when a protein-free filtrate is made from serum or when a concentrated stock standard is diluted to make a less stable daily working standard. Dilution of patient specimens (serum, urine) may be required when some constituent is too concentrated to be accurately measured in the routine procedure—a high blood glucose in a diabetic patient, for example.

In either case, the new concentration may be calculated by a simple formula, which is based on the fact that the diluted sample contains the same total amount of the constituent as the original concentrated sample. If 1 mL of the diabetic patient's serum is diluted to 5 mL, the total amount of glucose in the specimen before and after dilution is the same. Assuming an original serum concentration of 500 mg/dL, the glucose concentration in the diluted solution is calculated as follows:

$$V_1C_1 = V_2C_2$$

$$1\text{ mL} \times 500\text{ mg/dL} = 5\text{ mL} \times C_2$$

$$C_2 = 500\text{ mg/dL} \times \frac{1}{5}$$

$$C_2 = 100\text{ mg/dL}$$

Dilutions are generally expressed as a *ratio of the original volume to the total final volume;* this example is described as a 1:5 dilution of the patient's serum. The formula $V_1C_1 = V_2C_2$ is valid as long as the same units of volume and concentration are used for both the original and the final solutions.

Example: What is the resulting concentration of Na^+ if a serum that contains 140 mmol/L is diluted 1:100?

$$1 \times 140 \text{ mmol/L} = 100\ C_2$$

$$C_2 = 1.4 \text{ mmol/L}$$

Example: To what volume should 2 L of a stock HCl solution, 4 mol/L, be diluted to provide a solution of 0.8 mol/L?

$$2 \text{ L} \times 4 \text{ mol/L} = V_2 \times 0.8 \text{ mol/L}$$

$$V_2 = \frac{2 \times 4}{0.8} = 10 \text{ L}$$

It is sometimes necessary to make a large dilution, such as 1:1000, of a solution when some stock standard solutions are diluted to working standards. A 1:1000 dilution may be performed in steps as follows: 1 mL of the original solution is diluted to 100 mL (1:100), and 1 mL of the product is further diluted to 10 mL (1:10). The concentration of the final solution would be 1/100 × 1/10, or 1/1000 of the original concentration.

When a patient's specimen is diluted for a laboratory test procedure, the dilution factor must be included in the calculation of the final results. In the example of the diabetic patient whose serum was diluted 1:5, the test result on the diluted serum indicates a glucose concentration of 100 mg/dL. Before the result is reported, the concentration of the *original* serum must be obtained by multiplying by the dilution factor:

$$100 \text{ mg/dL} \times 5 = 500 \text{ mg/dL}$$

Example: 3 mL of urine were diluted with 2 mL of water, and an aliquot was assayed for creatinine. To calculate the creatinine concentration of the original urine:

$$\text{The dilution factor} = \text{final volume/original volume} = 5/3$$

$$\text{Creatinine concentration} = \text{concentration in diluted aliquot} \times 5/3$$

Conversion of Units

Concentrations of solutions may be expressed in many ways, and a technologist must be able to convert units of concentration from one form to another. The concentration of some blood constituents, such as glucose, has been traditionally reported as milligrams per deciliter (mg/dL), and others, such as chloride, in milliequivalents or mmol per liter. For some blood constituents, however, no one traditional method of reporting is used, so that the units may vary from laboratory to laboratory. Magnesium and calcium concentrations are still reported in meq/L or mg/dL even though the SI recommends mmol/L. When

using data from another laboratory, the technologist may have to convert the data from one set of units to another.

Such a conversion is a straightforward task of multiplying by a series of factors, with each factor designed to convert one dimension from the given unit to the desired unit. For example, one factor may change the volume measurement from deciliters to liters, whereas another converts the weight or mass from milligrams to milliequivalents.

Example: A serum calcium concentration is 10 mg/dL. Express this concentration in meq/L. (Calcium: atomic weight 40, equivalent weight 20.)

$$10 \text{ mg/dL} \times \frac{10 \text{ dL}}{1 \text{ L}} \times \frac{1 \text{ meq}}{20 \text{ mg}} = 5 \text{ meq/L}$$

Example: The same principle applies in converting a magnesium concentration of 2 meq/L to mg/dL. (Mg: atomic weight 24, equivalent weight 12.)

$$2 \text{ meq/L} \times \frac{1 \text{ L}}{10 \text{ dL}} \times \frac{12 \text{ mg}}{1 \text{ meq}} = 2.4 \text{ mg/dL}$$

When the conversion factors are all set up correctly, the original units of dimension are canceled, leaving only the units desired for the final result.

As SI units become more widely used, technologists must be able to convert concentrations from other units to mol or mmol per liter and vice versa.

Example: The normal range for serum glucose levels in a laboratory is 70 to 105 mg/dL. What is the same normal range expressed in SI units? (The molecular weight of glucose, $C_6H_{12}O_6$, is 180.)
For the lower limit:

$$\text{mol/L} = 70 \text{ mg/dL} \times \frac{10 \text{ dL}}{1 \text{ L}} \times \frac{1 \text{ g}}{1000 \text{ mg}} \times \frac{1 \text{ mol}}{180 \text{ g}}$$

$$= 70 \times \frac{1}{100} \times \frac{1}{180} \text{ mol/L}$$

$$= 0.00388 \text{ mol/L} = 3.88 \text{ mmol/L}$$

For the higher limit:

$$\text{mmol/L} = 105 \text{ mg/dL} \times \frac{10 \text{ dL}}{1 \text{ L}} \times \frac{1 \text{ mmol}}{180 \text{ mg}} = 5.83 \text{ mmol/L}$$

Thus, the normal range for glucose is 3.88 to 5.83 mmol/L.

Example: In SI units, a serum uric acid concentration is 0.75 mmol/L. Express this concentration in mg/dL.

The uric acid molecule has the formula $C_5H_4O_3N_4$, Mol wt 168.

$$0.75 \text{ mmol/L} \times \frac{1 \text{ L}}{10 \text{ dL}} \times \frac{168 \text{ mg}}{1 \text{ mmol}} = 12.6 \text{ mg/dL}$$

For some serum constituents, molecular weights cannot be unequivocably known. Serum proteins are quantitated as a family of compounds consisting of molecules that vary greatly in size. No single molecular weight can represent all protein molecules; an "average" molecular weight becomes meaningless as the proportions of different classes shift with individuals and with disease states. Protein concentration is reported in g/dL in conventional systems and g/L in the SI.

Molarity of Concentrated Reagents

The common acids and ammonium hydroxide are supplied as concentrated solutions. Each lot of a reagent is assayed by the manufacturer, and data are provided concerning the specific gravity and percentage by weight of the reagent; from these data the concentration in mol/L of the reagent is readily calculated.

Example: A concentrated H_2SO_4 solution has the following composition:

Specific gravity	1.84
H_2SO_4	95% by weight

To express the concentration in mol/L

1. Determine density in g/L:
   1.84 g/mL × 1000 ml/L = 1840 g/L
2. Because 95% of the total weight is H_2SO_4:
   1840 g/L × 0.95 = 1748 g H_2SO_4/L
3. Convert to molar concentration:
   (H_2SO_4 has a molecular weight of 98.1 g)
   $$1748 \text{ g/L} \times \frac{1 \text{ mol}}{98.1 \text{ g}} = 17.8 \text{ mol/L}$$

Although each lot of the reagent may differ slightly in composition from the others, the range of concentrations for each individual acid or base is quite narrow; a concentrated solution of sulfuric acid, for example, always contains about 18 moles of H_2SO_4 per liter. Approximate concentrations of the common concentrated solutions are shown in Appendix 2.

Hydrogen Ion Concentration

The following definitions are used in the review of acids and buffers:

1. *Acid:* Any substance that can dissociate to form protons (hydrogen ions, H^+). For example,
   $$H_2CO_3 \rightleftharpoons H^+ + HCO_3^-$$
   An acid is termed a *strong acid* if its aqueous solutions are highly dissociated (for example, HCl, H_2SO_4, HNO_3). It is termed a *weak acid* if the degree of dissociation in aqueous solution is low. Stated another way, weak acids have small dissociation constants (less than 10^{-4}). Carbonic acid, H_2CO_3, is a weak acid, with a dissociation constant of 4.3×10^{-7} for the first H^+.
2. *Base:* Any substance that can accept H^+. For example, NH_3:
   $$NH_3 + H^+ \rightleftharpoons NH_4^+$$

3. *pH:* Hydrogen ion concentration of plasma is about 0.00000004 mol/L (4×10^{-8}). People working with biologic fluids needed a simpler method for expressing the H^+ concentration and adopted the method of expressing it in terms of pH, where $pH = \log 1/[H^+] = -\log [H^+]$. Because the hydrogen ion concentration $[H^+]$ of pure water is 10^{-7} mol/L, the pH = 7.0.

   Each increase of 1 pH unit represents a tenfold decrease in the concentration of hydrogen ion. If adding a base to a solution causes a pH change from 4 to 7, there is a shift of 3 pH units and the $[H^+]$ is decreased from 10^{-4} to 10^{-7} mol/L, a 1000-fold dilution.

   The calculation of $[H^+]$ when the pH is not a whole number requires the use of a logarithm table or an electronic calculator. The process requires that the mantissa, or decimal part of the logarithm, always be converted to a positive number before it can be found in a table of logarithms. Also, multiplication is carried out by the process of adding logarithms. For example, to find the concentration of hydrogen ion in a solution of pH 5.2:

$$pH = -\log [H^+] = 5.2, \text{ or } \log [H^+] = -5.2$$

$$[H^+] = \text{antilog} (-5.2) = \text{antilog} (-6 + 0.8)$$

   From log tables, the antilog of 0.800 is 6.31, and the antilog of −6 is 10^{-6}, so

$$[H^+] = 6.3 \times 10^{-6}$$

   If the hydrogen ion concentration of a solution is known, the pH may be calculated. For a solution in which

$$[H^+] = 0.0005 \text{ mol/L} = 5 \times 10^{-4}$$

$$pH = -\log (5 \times 10^{-4}) = -\log 5 - (-4) = 4 - \log 5$$

   From log tables, log 5 = 0.699, so

$$pH = 4 - 0.699 = 3.3$$

   As previously mentioned, the hydrogen ion concentration of plasma is 4 $\times 10^{-8}$. Calculate the pH and check your result against the reference value for pH, which is given in Appendix 1.

Buffer Solutions

The rates of many chemical reactions, particularly those involving enzymes, depend on the hydrogen ion concentration. Most enzymes have a narrow pH range for optimal action, and activity falls off rapidly on either side of the optimum. Buffer salts stabilize the pH of solutions by accepting or donating protons to prevent a fall or rise in the pH.

A buffer is a mixture of a weak acid with its salt of a strong base or of a weak base with its salt of a strong acid. Buffers resist a change in pH when acid or alkali is added by forming weak acids or bases that consist mostly of undissociated molecules; hence, they are only slightly ionized. Adding HCl to an acetic acid-sodium acetate buffer results in the formation of more of the weak acetic acid and neutral NaCl. Adding NaOH to the same buffer generates sodium acetate and water. The regulation of the pH of body fluids to within narrow limits is made possible only by the presence of various buffer systems.

The equation for the reversible dissociation of a weak acid, HA, may be written:

$$\text{Equation 1: } HA \rightleftharpoons H^+ + A^-$$

and expressed mathematically as:

$$\text{Equation 2: } K_a = \frac{[H^+][A^-]}{[HA]}$$

where K_a is the symbol for the dissociation constant of the acid HA at equilibrium. If $K_a = 10^{-5}$ for the acid HA, and concentration of the acid is 0.1 mol/L,* then

$$10^{-5} = \frac{[H^+][A^-]}{10^{-1}}$$

$$10^{-6} = [H^+][A^-]$$

$$[H^+] = [A^-] \text{ so } [H^+]^2 = 10^{-6}$$

$$[H^+] = 10^{-3} \text{ mol/L}$$

The pH of the 0.1 mol/L solution of the weak acid HA is 3, and the $[H^+] = 10^{-3}$ mol/L. Because the concentration of HA is 0.1 mol/L, it is 1% ionized ($10^{-3} \div 0.1 \times 100 = 1\%$).

According to Equation 2, the dissociation constant K_a determines the product of the concentrations of the hydrogen ion and the anion (A^-). Because K_a is a constant for the acid HA, the hydrogen ion concentration may be varied by changing the concentration of the anion, that is, by adding a salt of the acid to the solution. This principle, known as the "common-ion effect," is the basis for many widely used buffers.

To illustrate, Equation 2 may be rearranged to:

$$\text{Equation 3: } [H^+] = K_a\frac{[HA]}{[A^-]}$$

If 0.1 mol/L of the sodium salt of HA is added to the 0.1 mol/L of HA, $[A^-]$ is increased, resulting in a change in $[H^+]$. The total concentration of A^- consists of the $[A^-]$ provided by the completely ionized salt plus the $[A^-]$ produced by the dissociation of HA. The dissociation of HA is usually so slight that the final concentration of A^- is, for practical purposes, the same as the initial concentration of NaA.

By the same reasoning, [HA] at equilibrium is equal to the initial concentration of acid, minus the amount that dissociates into H^+ and A^-. Because our hypothetic acid is a weak acid ($K_a = 10^{-5}$), the loss of [HA] from dissociation

*[HA] is actually equal to 0.1 mol/L minus the concentration of the dissociated acid or $[H^+]$. In this instance, $[H^+] = 10^{-3}$ and $[HA] = 0.1 - 0.001$ or 0.099 mol/L, a difference of 1%. In the preparation of a buffer, the normality of the acid would not be determined by titration; a 1% variation is within the limits of error and is considered negligible.

is extremely small compared to the initial concentration of HA, and the concentration of HA at equilibrium is essentially equal to the initial concentration of HA.

From Equation 3:

$$[H^+] = (10^{-5})\frac{10^{-1}}{10^{-1}} = 10^{-5}$$

$$\text{and pH} = 5$$

Thus, it was possible to reduce the hydrogen ion concentration of HA one hundredfold from 10^{-3} to 10^{-5} mol/L by the addition of the salt.

Addition of a strong acid (HCl) drives the reaction in Equation 1 to the left as the hydrogen ions from the HCl combine with the buffer salt anions to form a weak acid. Cl ion replaces A ion in solution. Conversely, a strong base, such as NaOH, would shift the equilibrium to the right, releasing more hydrogen ions to neutralize the hydroxyl ions. Again this would result in the formation of the buffer salt, plus water. Thus, adding either a strong acid or a strong base to a buffer solution results in the formation of a weak acid or a weak base with a relatively small change of hydrogen ion concentration.

Henderson-Hasselbalch Equation: From Equation 3, a useful relationship between pH and the ionization constant may be developed:

$$[H^+] = K_a\frac{[HA]}{[A^-]}$$

$$\log[H^+] = \log K_a + \log\frac{[HA]}{[A^-]}$$

$$-\log[H^+] = -\log K_a - \log\frac{[HA]}{[A^-]}$$

but $pH = -\log[H^+]$ and $pK_a = -\log K_a$.

Substituting pH and pK_a in the preceding equation:

$$pH = pK_a - \log\frac{[HA]}{[A^-]}$$

A negative logarithm is equal to the log of the reciprocal of the number, so the preceding expression can be written in the usual form of the Henderson-Hasselbalch equation:

$$\text{Equation 4: } pH = pK_a + \log\frac{[A^-]}{[HA]}$$

In a mixture of a weak acid and its sodium salt, the anion concentration depends on the concentration of the salt, which is completely ionized. The equation may be transformed to:

$$\text{Equation 5: } pH = pK_a + \log\frac{[\text{salt}]}{[\text{acid}]}$$

The following examples illustrate the effects of varying the acid and salt concentrations of HA whose $pK_a = 5.0$.

With concentrations of both HA and NaA at 0.1 mol/L.

$$pH = 5 + \log \frac{0.1}{0.1} = 5 + \log 1 = 5 + 0 = 5$$

Thus, when the salt concentration and the acid concentration are equal, the pH is equal to the pK_a and the buffering capacity of the solution is at its maximum.

When 0.1 mol/L salt and 0.01 mol/L acid concentrations are used:

$$pH = 5 + \log \frac{0.1}{0.01} = 5 + \log 10 = 5 + 1 = 6$$

With 0.01 mol/L salt and 0.1 mol/L acid:

$$pH = 5 + \log \frac{0.01}{0.1} = 5 + \log 10^{-1} = 5 - 1 = 4$$

The Henderson-Hasselbalch equation may also be used to calculate the proportions of salt and acid necessary to make a buffer of the desired pH. In selecting a buffer system, the technologist should choose a buffer with a pK_a value close to the desired pH. If a buffer with a pH of 5.0 is desired, an acetic acid-sodium acetate (HAc-NaAc) buffer may be selected. The pK_a of this system is 4.7. This means that when the salt and the acid concentrations are equal, the solution has a pH of 4.7:

$$pH = pK_a + \log \frac{[Ac^-]}{[HAc]}$$

$$pH = 4.7 + \log \frac{0.1}{0.1} = 4.7 + 0 = 4.7$$

To make a buffer of pH 5.0, the Henderson-Hasselbalch equation is used:

$$5.0 = 4.7 + \log \frac{[Ac^-]}{[HAc]}$$

$$\log \frac{[Ac^-]}{[HAc]} = 0.3$$

$$\frac{[Ac^-]}{[HAc]} = \text{antilog } 0.3 = 2$$

Whenever the sodium acetate is twice as concentrated as the acetic acid in the buffer solution, the pH is 5.0. The concentrations finally chosen depend on other requirements of the system: the buffering capacity needed, the ionic strength, and solubility limitations. Usually, the buffer for an enzyme reaction system is chosen by selecting the one that gives the highest enzyme activity when varying the pH and ionic strength.

QUALITY ASSURANCE

Medical decisions are based, in part, upon the results of laboratory tests. The validity of test results cannot be taken for granted, but must be supported by convincing evidence that the figures are reliable. The assurance of accurate analytical work is only one facet of the problem; quality assurance involves every step of the process, from the initial ordering of a test and collection of a patient's sample, to the analysis, and finally to the distribution of test results to the proper destination. A quality assurance program not only involves every person in the laboratory, from the director to the laboratory helpers, but also includes everyone who has contributed to the enterprise, such as the phlebotomy team (blood collectors) and data processors. The process is intricate, and constant vigilance is required at all levels to ensure that accurate results are delivered to the physicians in a timely manner.

The clinical chemistry laboratory has an important and specific role in the quality assurance program. The director and supervisor have the following responsibilities:

1. To select the most accurate and precise analytical methods for performing the tests, in a time period that is most helpful to the physicians.
2. To adequately train and supervise the activities of laboratory personnel.
3. To make available printed procedures for each method, with explicit directions, an explanation of the chemical principles, a listing of reference values (normal range), and a listing of common conditions in which the test results may be high and those in which they are low. This information packet should also include data on the linearity, precision, and sensitivity of the method, as well as a list of substances that may interfere in the assay.
4. To select good instruments and institute a regular maintenance program.
5. To institute a good quality control program, make available appropriate control sera, regularly inspect the control charts, and instill in the entire staff an awareness of the importance of a good quality control system.
6. To conduct continuing education sessions with the technologists to sustain their professional interests and keep them abreast of changing technology.
7. To document the preceding steps for an appropriate accrediting group (for example, College of American Pathologists).

The technologists must:

1. Follow assay directions explicitly.
2. Use the proper control serum for each run, chart the results, and take appropriate action when the control serum result is beyond the established limits.
3. Always use sound analytical techniques.
4. Be conscientious in instrument maintenance.
5. Notify the supervisor immediately when analytical problems develop, when the run is out of control, and when results indicate the presence of a life-threatening situation.

Quality Control

The integrity of an analytical process is continually assessed by a quality control program that documents the precision of each analytical run and that detects variations in precision that may be caused by random or systematic errors (bias in a method). The goal of quality control is the generation of test results that are reliable and precise.

Precision and Accuracy

The first step in the establishment of a quality control program is to ascertain the limits of uncertainty for each test. Every measurement and every analysis carries with it a degree of uncertainty, a variability in the answer as the test is performed repeatedly. It is essential to determine the *precision* of each test, which reflects the *reproducibility* of the test (the agreement of results among themselves when the specimen is assayed many times). The less the variation, the greater is the precision. Precision must not be confused with *accuracy*, which is the deviation from the *true* result. An analytical method may be precise but inaccurate because of a bias in the test method; an example was the Folin-Wu glucose method, which determined certain nonglucose-reducing substances as glucose.

Standard Deviation

The degree of precision of a measurement is determined from statistical considerations of the distribution of random error; it is best expressed in terms of the *standard deviation*. A normal frequency curve (bell-shaped, Gaussian curve) is obtained by plotting the values from multiple analyses of a sample against the frequency of occurrence, as shown in Figure 2.3. From statistical considerations, the standard deviation (s) is derived from the following formula:

$$\text{Equation 6: } s = \sqrt{\frac{\Sigma(\bar{x} - x)^2}{N - 1}}$$

where s = 1 standard deviation, Σ = sum of, $\bar{x}$ = mean (average value), x = any single observed value, and N = total number of observed values. With a normal

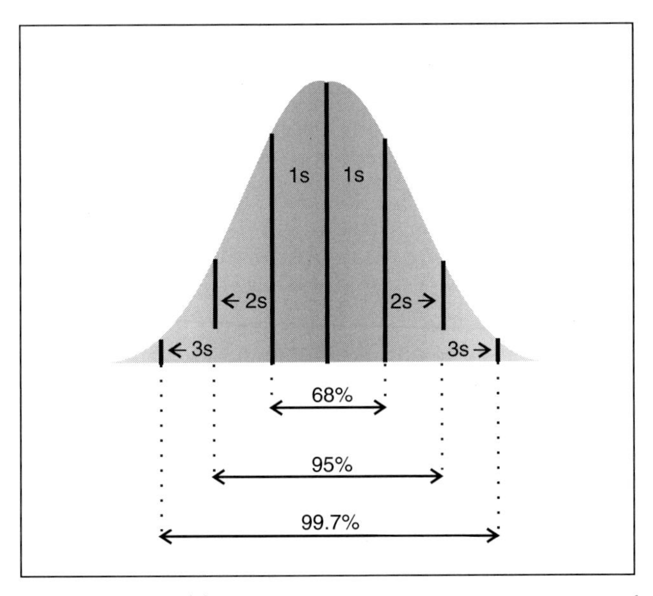

FIG. 2.3. Normal frequency curve, ± 1 *s* encompasses 68% of the values, ± 2 *s* 95%, and ± 3 *s* 99.7%.

TABLE 2.3
CALCULATION OF STANDARD DEVIATION (S) AND COEFFICIENT OF VARIATION (CV)

ASSAY VALUES	$\bar{x} - x$	$(\bar{x} - x)^2$
18	2	4
20	0	0
21	1	1
17	3	9
22	2	4
19	1	1
20	0	0
20	0	0
21	1	1
22	2	4

$$N = 10$$

$$\bar{x} = 20$$

$$\Sigma (\bar{x} - x)^2 = 24$$

$$s = \sqrt{\frac{\Sigma(\bar{x} - x)^2}{N - 1}}$$

$$s = \sqrt{\frac{24}{9}} = 1.6$$

$$CV = \frac{s}{\bar{x}} \times 100 = \frac{1.6}{20} \times 100 = 8\%$$

distribution, 68% of the values are encompassed by $\bar{x} \pm 1$ s, 95% by $\bar{x} \pm 2$ s, and 99.7% by $\bar{x} \pm 3$ s.

The procedure for calculating the s of the example shown in Table 2.3 follows:

1. Calculate the mean of all values.
2. Find the difference of each individual value from the mean (Column 2). Use the absolute value; ignore negative or positive differences.
3. Square the differences (Column 3).
4. Add the entries in Column 3 to obtain the sum of the squares of the differences.
5. Find the standard deviation (s) by using Equation 6.

For the sake of brevity, the example given in the table includes only 10 values. The precision data for a test should be acquired from at least 20 separate assays.

In the example given in Table 2.3, 68% of the future assays are expected to fall within ± 1.6 units of the mean value of 20, and 95% of the values within 20 ± 3.2, when the method is functioning well. Thus, 1 assay in 20 may produce a value farther than 3.2 units from the mean, even when reagents, standards, and instruments are all acceptable, and there is *no* technologist error.

Coefficient of Variation

The standard deviation is greater when a method is less precise; it is generally greater when the mean of the assay values is a larger number. A larger deviation from the mean (and therefore a larger s) should be expected when the mean value is 200 than when it is 20 or 2. The *coefficient of variation (CV)* expresses the standard deviation as a percentage of the mean value and is a more reliable means for comparing the precision at different concentration levels:

$$CV = \frac{s}{\bar{x}} \times 100 \text{ and is expressed as } \%$$

The precision of a method varies inversely with the CV; the lower the CV, the greater is the precision.

Control Specimens

The reliability of a procedure cannot be taken for granted in its daily use because many things can go wrong if vigilance is relaxed. The instrument may be out of calibration, reagents may be deteriorating, or a technologist may have made an error. Therefore, every assay must be checked.*

The simplest, most straightforward way to check the reproducibility of a method is by including control specimens in the run. These are samples for which the correct answer is known. If control sera or urines are included with patients' samples, and the observed results are the expected results, we can feel confident about the assay and can probably safely assume that the results on the patients' samples are also correct. (Of course, one must always watch for aberrant results—those inconsistent with a patient's history, beyond linear range, or suspect for other reasons.)

Control specimens may be purchased from many commercial sources. Often they consist of lyophilized sera or urines; when reconstituted, they contain the constituents of interest at normal levels (within the reference range) or at abnormal concentrations (high or low) to represent concentrations that may be encountered in some disease states. Usually the controls included in a run are selected to represent a normal concentration and concentrations representative of medical decision levels. For example, glucose assays may include control sera at 80 mg/dL (in reference range), 50 mg/dL (suspect hypoglycemia), and 150 mg/dL (hyperglycemia, possible diabetes mellitus).

Lyophilized control sera and urines are available for most assays. The routinely determined constituents are present in most of the general controls. Special control specimens are available that contain trace metals and some of the more esoteric compounds (both therapeutic drugs and hormones) at normal and abnormal concentrations.

Control specimens may be purchased in two forms: assayed (constituents analyzed by a group of reference laboratories, with the range of values on the label) or unassayed. The unassayed form is cheaper, but the laboratory itself bears the responsibility for determining the concentration of the various constituents. Those who purchase control serum should obtain at least 1 year's supply of the same lot number to avoid the extra labor and cost of performing multiple analyses to establish new control values (mean and standard deviations) at more frequent intervals. A laboratory could prepare its own control serum by collecting a large pool sufficient to last 1 year, freezing aliquots, and using it daily. It must be analyzed at least 20 times to obtain the mean value for each constituent and the standard deviations; these values should be recalculated every 2 months as more data are accumulated.

Always treat control serum (and any patient's sample) as a potential source of infectious disease, even though it may have been tested for the antigens or antibodies indicative of a specific disease.

Control sera are usually included with every run of each assay; with long runs they should be interspersed at regular intervals among the patients'

*Some highly automated instruments using prepackaged reagents require the use of controls only once during each shift, according to the FDA.

samples. Inclusion of control sera at both abnormal and normal concentrations provides information about the performance of the assay over a range of values. The selection of a control serum with values conforming to medical decision levels can provide reassurance to both the laboratory and the physician. If a drug is considered toxic above a certain serum concentration, the laboratory should strive to verify its results by inclusion of a control serum containing the critical drug concentration.

The manufacturers of many of the newer instruments state that the number and frequency of control samples may be greatly reduced, requiring only one assay of controls per shift, regardless of the number of analytical runs made during this period. Although imprecision has been greatly reduced in some of these instruments by automation, more stable electronics, and the use of prepackaged reagents, laboratory technologists should be skeptical of such claims. The more conservative approach to quality control (control samples included with every assay run) should be implemented until the laboratory can demonstrate that manufacturers' suggestions for assaying fewer control specimens provide comparable accuracy and precision.

Quality Control Charts

A quality control chart is established for each constituent in the control serum. In the most commonly used form, the Levey-Jennings chart,[8] the concentration is plotted on the ordinate, with a black line drawn across the chart at the mean value, blue lines at $\pm$ 1 s, orange lines at $\pm$ 2 s, and red lines at $\pm$ 3 s. The days of the month are plotted on the abscissa. The chart is hung in a convenient location or kept in a notebook at the workbench, and each value obtained on the control serum is recorded on the chart every time an analysis is made. A permanent record must be kept of each value that is out of control and the corrective action that was taken (Fig. 2.4A).

Sometimes when control values are plotted regularly, one can see that a method is getting out of control even while the values are still within 2 s of the mean (Fig. 2.4B and C). Figure 2.4B shows a shift in the values as more than six consecutive results have suddenly fallen below the mean. Such a shift is an early warning that something has gone wrong, possibly an improperly made or contaminated reagent or standard or an incorrectly calibrated or functioning instrument. Corrective action is necessary. Figure 2.4C illustrates a trend in control values that may indicate a steadily deteriorating component of the system. Thus, the control charts become a source of help in detecting problems and a source of reassurance when all goes well. They have to be kept up daily and inspected regularly.

The mean and standard deviation of a control chart should be updated monthly as more analytical results come in; they become more meaningful when they rest upon a larger number of analyses. In many laboratories, patient results are reported by means of a computer; all control values are also entered and the computer is programmed to calculate the mean, standard deviation, and coefficient of variation. These calculations may include 1 month's data or several months' data, depending on the storage capacity of the computer as well as the needs of the laboratory.

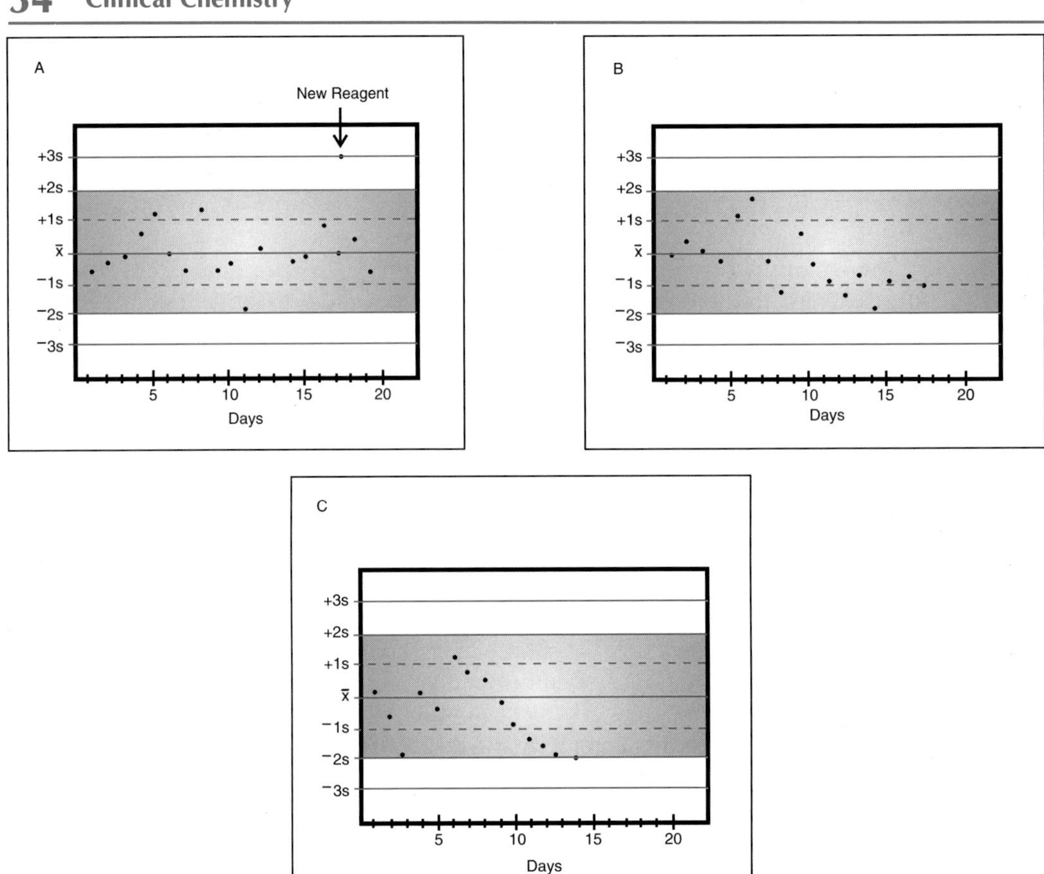

FIG. 2.4. Control charts. A, One unacceptable value, with notation of the correction made. B, A shift in values. C, A downward trend in the control serum values.

Interpretation of Control Sera Results

Expanded quality assurance programs have kept pace with advances in technology and have led to more sophisticated quality control programs. The common practice now is to run two control samples with every batch, either one in a normal and one in an abnormal range, or one high and one low value. A single control sample does not suffice. When this practice is linked with the multi-rule procedure proposed by Westgard and co-workers[9] for interpreting and acting upon the control data, much greater discrimination is achieved. The result is fewer false rejections of an analytical run, as well as indicators that suggest the type of error—whether random or systematic—when such rejections do occur.

Multi-rule Procedure

Analyze one sample each of two different control sera with each analytical run. Plot these results on a Levey-Jennings chart that has lines drawn for the mean and for ± 1 s, ± 2 s, and ± 3 s.

If the value for each control serum is within ± 2 s, accept the run. When one control serum is within the ± 2 s limit and the other is between the 2 s and the 3 s limit, however, hold the run while applying the following rules:

1. 1_{3s} *Rule:* If 1 control observation exceeds $\bar{x} \pm 3$ s, reject the run for probable random error.
2. 2_{2s} *Rule:* If 2 consecutive control values are greater than $\bar{x} + 2$ s or $\bar{x} - 2$ s, reject the run for probable systematic error.
3. R_{4s} *Rule:* If the difference between the 2 controls is equal to or greater than 4 s, reject the run. R stands for the difference between the 2 control sera values. If 1 control value is $\bar{x} + 2$ s and the other is $\bar{x} - 2$ s, reject the run. The error is probably random.
4. 4_{1s} *Rule:* If 4 consecutive control values exceed the same $\bar{x}$ + s or $\bar{x}$ − s limit, reject the run for probable systematic error.
5. *10$\bar{x}$ Rule:* If 10 consecutive control values fall on 1 side of $\bar{x}$ (either + or −), reject the run for probable systematic error.

If any of the five rules advises rejection, the run should be repeated. Accept the run only if all five rules conclude "acceptable." If a batch of test results is held up because of a quality control problem, grossly abnormal results on patients' specimens coming from the emergency service should be telephoned to the attending physician, with the statement that the results are provisional, pending a repetition of the test. The provisional result may confirm a diagnosis and permit the start of treatment.

Although the most widely used system in the clinical laboratories, the Levey-Jennings control chart just described is not the only system. Cumulative sum (Cusum) charting was introduced in the 1960s.[10] Deviation from the target is plotted in a cumulative manner so that each point represents the sum of all the deviations to date from the mean or target value. For example, if the target value for the serum control is 100 mmol/L, and your first result was 97 mmol/L, your first plot point would be at −3, representing 3 units below the established value. If the next day's result was 98 mmol/L, you would consider that −2 and plot your second value at −5 on the chart (2 units below the −3 point plotted previously). A control result greater than the target value of 100 mmol/L would be necessary to reverse the downward trend of the curve. For example, a result of 103 mmol/L would be plotted as 3 units above the last entry (−5), at −2 on the graph. This method of plotting exaggerates trends in the data, and makes shifts of the mean much more obvious than by other plots. The rules for using the Cusum system for quality control are less well defined than those for the Levey-Jennings system, although a proposal has been made for combining the two systems to establish a good quality control system.[11]

Proficiency Testing

Proficiency testing in clinical chemistry has been used in the United States for many years as a voluntary means of improving analytical accuracy. With passage of the Clinical Laboratory Improvement Act of 1988 (CLIA '88, effective in January, 1992), testing became mandatory for all but the smallest clinical

TABLE 2.4
SAMPLE OF ACCEPTABLE PERFORMANCE* FOR EVALUATING PROFICIENCY TESTING

ANALYTE OR TEST	CRITERIA FOR ACCEPTABLE PERFORMANCE (TARGET VALUE)
Albumin	± 10%
Alkaline phosphatase	± 30%
Aspartate aminotransferase (AST)	± 20%
Bilirubin (total)	± 0.4 mg/dL or ± 20% (greater)
Blood gas, pO_2	± 3 s
Blood gas, pCO_2	± 5 mm Hg or ± 8% (greater)
Blood gas, pH	± 0.04
Cholesterol	± 10%
Glucose	± 6 mg/dL or ± 10% (greater)
Potassium	± 0.5 mmol/L
Sodium	± 4 mmol/L
Hormones (T_3, Free T_4, T_3 Uptake, TSH)	± 3 s
Hormones (T_4)	± 20% or ± 1.0 μg/dL (greater)

*Condensed from §493.929 in the Federal Register, Vol. 57, No. 40, pp. 7157–7159, Friday, Feb. 28, 1992, which defines acceptable performance for clinical chemistry laboratories under CLIA '88.

laboratories, and penalties were attached to failure to achieve a passing grade. Most laboratories, including those in physicians' offices, must participate in an approved proficiency testing program administered by a federal, state, or private agency. Each proficiency test presents the challenge of analyzing a set of five different test sera (the testing event) for all the common analytes that each laboratory measures. Three or four such testing events occur during the year, and each laboratory must receive a passing grade in each testing event for each analyte.

The testing program establishes the "*correct answer*" or *target value* for each analyte by selecting the value reached by 90% of 10 or more referee laboratories or by using a value in harmony with 90% of the results reported by all the participating laboratories. The test results of all participating laboratories, irrespective of method used, are compared with the target values (TV) and graded as either acceptable or nonacceptable. The criteria for acceptability vary among the specific analytes, and may be TV ± X%, TV ± X mmol/L or X mg/dL, or TV ± 3 *s*. See Table 2.4 for some specific examples.

Acceptable performance for each analyte in each testing event is defined as achieving test results that are within the limits prescribed in the Federal Register (Table 2.4) for at least 4 of the 5 samples in each testing event, that is, achieving a test score of 80 to 100% acceptable for each set of 5 sera analyzed. CLIA '88 specifies penalties for laboratories that fail to achieve a passing grade of 80% for each analyte.

Proficiency testing provides assurance in the analytical performance when all goes well and calls attention to problem areas when results are out of line. Some of the agencies engaged in proficiency testing are the American Association for Clinical Chemistry (Therapeutic Drug Monitoring), the College of American Pathologists, the Institute for Clinical Science, and the National Institute for Drug Abuse. Some states have their own proficiency testing service for laboratories within their states.

Preventive Maintenance

Another phase of quality control requires a regular maintenance program for the various laboratory instruments to ensure that they are in top working condition.[12] Such maintenance includes regular calibration of spectrophotometer wavelengths, continuous recording of refrigerator and freezer temperatures to ensure that requisite cold temperatures are maintained, testing of water purity with a resistance meter, checking water bath temperatures regularly, and calibration of micropipets. These seemingly small details may greatly affect performance.

The maintenance of a good quality control program is costly in both time and money. Control serum is not inexpensive, and much of it is used in a year. The time invested by technologists in carrying out tasks that bring in no revenue (analyzing control serum, repeat testing of samples in a run not in control, and calibrating glassware) is considerable. A good quality control system adds approximately 20 to 25% to a laboratory's operating cost.

Evaluation of New Methodology

Every laboratory seeks to improve its methodology as technology advances, new techniques are designed, and new assays are introduced. The proposed method may be faster, more accurate, more sensitive, or less expensive than its predecessor; it may measure a component that the laboratory had not previously included among its tests.

The new method may be included with new instrumentation, a kit provided by a commercial company, an adaptation of a method described in a journal, or a test developed within the laboratory itself. In any case, the new method cannot be accepted on faith; it must be rigorously tested and shown to meet the criteria established by the clinical laboratory.

When a method has been tentatively selected, precision studies using replicate measurements of a single sample should be carried out. The measurement of within-run precision is the first step; if this step is satisfactory, day-to-day precision at normal and abnormal levels should be measured over a 20-day period. During this period, other method evaluation procedures can be performed. The linear range is determined, and a protocol for diluting samples, if necessary, is established.

The new method is then compared to current method, a reference test method, or a reference laboratory's assay results. Patients' samples (a minimum of 40) are split and assayed by the 2 procedures. In this correlation study, the paired results are plotted on a graph, with the reference method as the abscissa (x-axis) and the new test method as the ordinate (y-axis). The regression line is calculated (calculators and computers do this readily) and the method results are compared.

Figure 2.5 shows a plot representing excellent agreement between two methods; the plot points deviate little from the regression line. This scatter results from random error and must be expected. The regression line has a slope of 1.0 and an intercept at (0.0), so the results by both methods agree closely within the tested range.

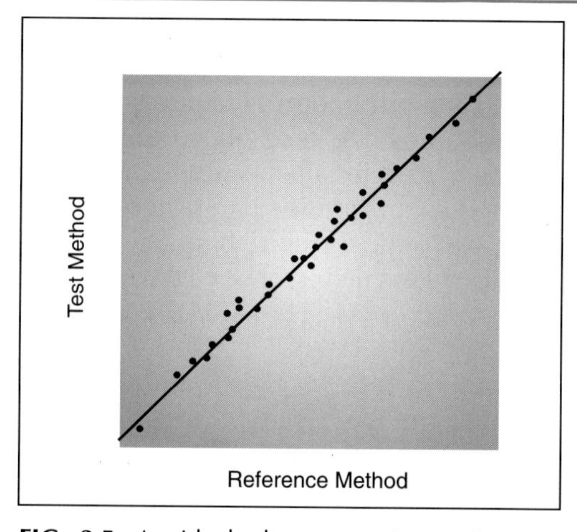

FIG. 2.5. An ideal plot comparing patient test results by a new methodology (test method) and an approved reference method. Slope of the line, which passes through the origin, is 1, thereby showing good agreement between the 2 methods. Units for both coordinates are identical.

Figure 2.6 illustrates the regression lines in two other comparison studies, where the results obtained by the two methods might differ appreciably from each other. Example A reveals a constant error, or bias, between the methods; the new test method consistently gives a result that is about 7 units lower than that of the reference method. This difference suggests some interference in one of the methods: the reference method may be measuring some component in addition to the desired analyte, or some constituent may be inhibiting the reaction in the new test method. Figure 2.6B contains a regression line showing a proportional error; the error is small at low levels, but becomes larger with increasing concentrations of analyte. A proportional error might be caused by an incorrect standard concentration or an incorrect instrument setting. In Figure 2.6, only the regression lines (actual and ideal) are drawn. The individual plotted points and their scatter are not shown.

In method comparisons, a correlation coefficient, r, is usually calculated to state numerically agreement between 2 methods. An r of 1.0 represents perfect correlation between the 2 methods; −1.0 represents a perfect negative correlation (the new test result decreases with increased concentration of the analyte measured by the reference method); and 0.0 correlation means *no* agreement at all (a completely random scatter of points).

In Figure 2.5, a correlation coefficient of 1.0 would have resulted if all the points had fallen exactly on line; the actual scatter lowers the r-value. Excellent correlation could be represented by an r greater than 0.90 or greater than 0.98, depending on the complexity of the tests and their similarity to each other. An inherent danger exists in accepting the correlation coefficient as an indicator of perfect agreement between results obtained by two methods.[13] As previously mentioned, r is affected by random error. It is *not* sensitive to either constant error or proportional error. The regression lines shown in Figure 2.6 would

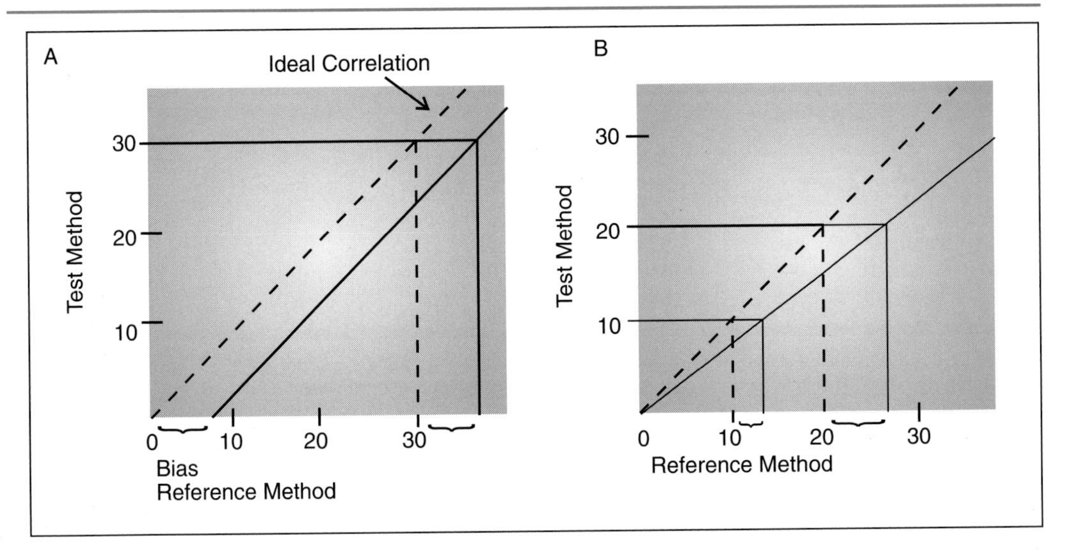

FIG. 2.6. Two plots representing patient sample comparisons. A, Case of constant error, with the new method giving results that are lower by a consistent difference. B, Case of proportional error where the discrepancy increases as the concentration is increased. In each graph, the ideal lines are broken; actual lines are solid.

both give a calculated r-value of 1.0. A correlation coefficient of 1.0 means merely that the points generate a straight line, no matter what the slope or y-intercept is. Inclusion of scattered points would lower r.

The correlation coefficient is also sensitive to the range of values represented. If all values are in a narrow range (for example, serum calcium: 7.00 to 11.00 mg/dL), r is considerably lower than if the range were broadened by adding 1 high value and/or 1 low (for example, urine calciums, 4.0 and 19 mg/dL). Thus, the correlation coefficient may be a valuable tool, but its limitations must always be considered.

The best way to assess the accuracy of a new method is to compare it with a recognized reference method in a manner similar to that just described.

Another approach to accuracy is to assess the percentage of recovery of pure substance added to a serum pool that has been analyzed repeatedly.[13] An example is shown in Table 2.5, in which the glucose concentration of aliquots of a serum pool is increased by 50, 100, and 150 mg/dL; the original glucose concentration of the pool was 45 mg/dL. After analysis of the 3 supplemented aliquots in triplicate, as well as of the original pool, the mean concentration of glucose in the serum pool is subtracted from the glucose concentration found in the supplemented samples to give the total actual recovery in mg/dL of glucose.

$$\text{The \% recovery} = \frac{\text{actual recovery}}{\text{added glucose}} \times 100$$

In the example in Table 2.5, with 50 mg/dL added:

$$\text{\% recovery} = \frac{92 - 45}{50} \times 100 = 94\%$$

A recovery of 95% or better is generally considered acceptable.

TABLE 2.5
RECOVERY OF SERUM GLUCOSE

SAMPLE	ASSAYED VALUE (mg/dL)	GLUCOSE RECOVERED (mg/dL)	GLUCOSE ADDED (mg/dL)	% RECOVERED
Serum Pool	45	—	—	—
Pool + 50 mg/dL glucose	92	47	50	94
Pool + 100 mg/dL glucose	145	100	100	100
Pool + 150 mg/dL glucose	199	154	150	103

The mean glucose value of a serum pool was 45 mg/dL. In three separate experiments 50, 100, and 150 mg/dL, respectively, were added to it, assays were made, and percentages recovered were calculated.

A brief summary of the major steps in evaluating a new method has been presented here. In practice, data must be collected and analyzed, including information on interferences, limits of error at critical medical decision levels, and comparisons with the "state of the art." An excellent series of articles by J.O. Westgard presents a practical plan of action for the selection and evaluation of new methods.[14]

Sometimes the new method being adopted by the laboratory has been studied extensively, with much information published. Still the method must be checked in your laboratory to be sure that the desired precision, accuracy, and linearity can be obtained and that patients' results are comparable. Watts[15] suggests, however, that the testing need not be so rigorous with a well-documented method; fewer samples can be run if your results agree with published data.

SPECIMEN COLLECTION

Proper sample acquisition, storage, and handling are essential, but frequently neglected, steps that precede the chemistry analysis. One must ensure that the sample to be used for a test has been collected from the right patient under the proper conditions and in the appropriate blood collection tube, and has been stored in a manner that maintains sample integrity.

Several good references that review the mechanics of blood drawing are available.[16,17] The following sections review some basic principles of sample acquisition.

Blood Collection

When collecting blood, one must choose the type of blood to draw.

Arterial Blood

An arterial blood sample is drawn from an artery. Most often, arterial blood draws are performed by physicians. Arterial blood is needed to assess pH, carbon dioxide, and oxygen concentrations as measured in a blood gas determination (see Chap. 5), but is used in few other analyses.

Venous Blood

Venous blood is collected from a vein, is obtained more readily than arterial blood, and is the primary source for clinical laboratory specimens. Although some constituents may vary slightly in venous versus arterial blood (ammonia, lactate, pyruvate, glucose), reference ranges are most often established for venous samples. The many ways of collecting venous blood depend on the type of analysis to be performed on the sample.

Blood collected in a plain tube with no anticoagulant yields serum upon clotting and centrifugation. When blood coagulates, the soluble fibrinogen is converted to an insoluble fibrin clot, which enmeshes the red blood cells; clear serum is obtained upon centrifugation. If blood is collected with an anticoagulant (a chemical that prevents the blood from clotting, such as heparin, ethylenediaminetetraacetic acid [EDTA], oxalate, or citrate), plasma is obtained upon centrifugation. Most anticoagulants function by removing calcium ions from the plasma, either bound to an undissociated complex (citrate, EDTA) or as an insoluble salt (oxalate). Calcium is essential to the clotting process. Heparin functions as an antithrombin and prevents the conversion of prothrombin to thrombin.

Plasma is a complex fluid that contains all the dissolved substances in whole blood. Inorganic compounds, small organic molecules, such as glucose and urea, and hundreds of different proteins, including fibrinogen, the main clotting protein, are all dissolved in plasma. Plasma is 92% water by weight.

One of the advantages of collecting blood with an anticoagulant is the ability to spin the blood down immediately, rather than waiting a minimum of 20 minutes for the blood to clot. This time factor becomes important when monitoring critically ill patients whose treatment awaits laboratory values.

Anticoagulants must be carefully chosen for compatibility with test procedures. For example, sodium heparin is a common anticoagulant used for many chemistry tests. Ammonium heparin (another salt of the same anticoagulant), however, would be inappropriate for a blood draw if ammonia values are requested, as it would artificially elevate the results. EDTA, the most common sample used by the hematology laboratory for a complete blood count, would be inappropriate for enzyme analysis because of the binding of metal cofactors necessary for enzyme function. Blood drawing tubes are color coded so that one can recognize the type of anticoagulant by the color of the stopper on the tube.

Many blood collection tubes now contain gels that separate the red cells from the serum or plasma after centrifugation. Because many analytes change when the serum or plasma sits on the red cells too long (glucose, for example), this gel barrier can be a convenient way to maintain sample integrity without having to remove the serum or plasma from the red cells by a manual transfer. The gels are made of an inert material, which theoretically should not cause an interference in analysis. They are not recommended for long-term storage of all tests, however. Again, the sample tube must be verified to be acceptable for the method of analysis to be used.

"Capillary" Samples

This type of specimen is used widely with pediatric patients, in whom analysis must be performed on the smallest amount of blood possible. Blood is obtained from puncture near the lateral and medial portion of the plantar surface of the heel or from a fingertip. Arteriolization of the area by warming to 39 to 40° C for 3 to 5 minutes allows for better blood flow. The sample is actually a mixture of venous and arterial blood and, depending on sample technique, can have varying concentrations of interstitial fluids. Squeezing an infant's heel to obtain blood causes contamination with excess tissue fluids. References on this particular type of blood collection are worth reviewing if one is working with the pediatric population.[18,19]

Certain analyses, such as blood gases, pH, lactate, and ammonia, are influenced by glycolytic processes occurring in red blood cells. These changes can be prevented by immersing the specimen tubes in ice water and keeping them at 0° C until analyzed. Most serum or plasma samples may be kept at room temperature for several hours until analyzed. Short-term storage of most samples can be carried out safely at 4° C, but long-term storage requires preservation at −20° C. Tissue samples should be frozen and stored at −20 or −70° C.

Urine Collection

Although in most cases urine is an easier sample to obtain than is blood, collection requirements should be adhered to strictly. A random urine sample varies significantly in concentration during the day. For this reason, "timed" or 24-hr urine specimens are commonly collected for chemistry tests. All urine samples must be preserved to prevent bacterial contamination. The type of preservative depends on the chemistry test requested. Refrigeration is adequate for some tests; different types of acids are used for others. Laboratories usually provide patients with urine containers with the appropriate preservative added so that a 24-hr urine can be collected at home.

Twenty-four-hour urine collections are usually started in the morning. The patient should be instructed to discard the first morning urine and start the timing at that point. For exactly 24 hr after, all urine should be collected. The bladder is emptied at the end of the 24-hr period, and that urine is added to the collection. The exact start and finish time should be recorded. Shorter times may be requested for a urine collection (4, 6, or 12 hr). The same care and method of timing should be carried out.

Urine collection in infants can be more challenging. For random specimens, urine bags, such as "U-Bag" from Hollister, are taped onto the infant's skin. A 24-hr collection can be made using these bags, if they are changed often. Care must be taken to prevent loss of urine and to prevent contamination of urine with feces, as either event will influence the accuracy of the test.

SAFETY

People working in clinical laboratories are exposed to potential hazards. Risks can be minimized by eliminating dangerous situations when possible, establishing clean and safe work habits, taking proper precautions at all times, and maintaining awareness of good safety practices.

The National Fire Protection Association (NFPA) has developed a series of symbols to warn against fire, biologic, chemical, and radiation hazards. An example is shown in Figure 2.7. These symbols help to provide safety information.

A federal law, the Occupational Safety and Health Act (OSHA), is designed to make places of work, including clinical laboratories, safer. Many states have passed similar legislation to supplement the Federal laws. Although these laws help to force recalcitrant or dilatory employers to take the necessary steps to make the work place a safer environment, no clinical laboratory is completely safe unless laboratory workers themselves become aware of the potential hazards and their remedies and actively participate in an ongoing safety program. Each laboratory also should have a safety manual that lists all hazards and the precautions to take to minimize the dangers.

For further review, see the Safety chapter by Forrey, Delaney, and Ruff in *Standard Methods of Clinical Chemistry*.[20]

The main possible dangers in a clinical chemistry laboratory are (1) fire, (2) infection, (3) contact with corrosive chemicals, (4) exposure to toxic fumes, (5) cuts or punctures from broken glassware or other sharp objects, (6) exposure to

Instant Warning System for

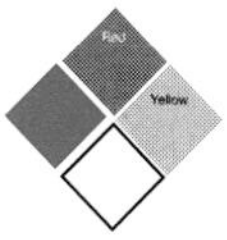

Red
Yellow

DANGEROUS MATERIALS

The National Fire Protection Association (NFPA) has developed a numerical system for the identification of the fire hazards by materials. The numbers given in the health, flammability and reactivity columns have been taken from NFPA Fire Protection Guide on Hazardous Materials, 7th Edition.

Abbreviated definitions are as follows:

Health (Blue)
4 Can cause death or major injury despite medical treatment.
3 Can cause serious injury despite medical treatment.
2 Can cause injury. Requires prompt treatment.
1 Can cause irritation if not treated.
0 No hazard.

Flammability (Red)
4 Very flammable gases or very volatile flammable liquids.
3 Can be ignited at all normal temperatures.
2 Ignites if moderately heated.
1 Ignites after considerable preheating.
0 Will not burn.

Reactivity (Stability) (Yellow)
4 Readily detonates or explodes.
3 Can detonate or explode but requires strong initiating force or heating under confinement.
2 Normally unstable but will not detonate.
1 Normally stable. Unstable at high temperature and pressure. Reacts with water.
0 Normally stable. Not reactive with water.

Special Notice Key:
Pol - Polymerizes under normal conditions.
W - Water reactive
Oxy - Oxidizing Agent
Exp - Explosive-heat or shock sensitive

FIG. 2.7. NFPA symbols for the identification of hazardous materials. In addition to the color code, substances are rated from 0 to 4 to indicate the degree of hazard. (Courtesy of Laboratory Safety Supply, Inc.)

carcinogenic compounds, and (7) possible exposure to low-level radioactivity. These hazards are reviewed more fully in the following sections.

Fire

The danger of fire is always present as long as volatile, flammable solvents are in the laboratory; a static spark or the spark from a refrigerator thermostat can set off an explosion if the proper mixture of ethyl ether or acetone and air is present. The danger of fire (or explosion) with flammable solvents varies inversely with the boiling point; the more volatile the solvent, the greater is its vapor pressure at any given temperature, and hence the more quickly the vapor concentration reaches a combustible level. Of the flammable solvents commonly present in a clinical chemistry laboratory, the following are listed in ascending order of their boiling points: ethyl ether (34° C), petroleum ether (40 to 60° C fraction), acetone (56° C), methanol (65° C), ethyl acetate (77° C), ethanol (78.5° C), isopropanol (96° C), toluene (111° C), and xylene (139° C).

A fire in a chemistry laboratory can cause tremendous damage because of the quantities of flammable liquids that are frequently used and the presence of gas cylinders under high pressure. A raging fire in the laboratory could cause these cylinders to explode.

Precautions

The following precautions are advised:

1. When possible, substitute solvents with higher boiling points for the more dangerous solvents with lower boiling points. Avoid the use of ethyl ether if at all possible.
2. Always work in a fume hood with the blower on when using flammable solvents.
3. Keep the laboratory supply of flammable solvents to a minimum, and store them in a flameproof cabinet.
4. Do not permit smoking in the laboratory.
5. Use fiberglass heating mantles when possible to heat flasks containing volatile solvents. Work in the fume hood and do not use gas flames. Hot plates can also be dangerous if the solution should bump (superheat) and boil over. Flammable solutions usually ignite on touching the hot plate.
6. When flame is necessary, use a mechanical igniter instead of matches to light the flame.
7. Place CO_2 fire extinguishers at strategic locations in the laboratory and have drills to make certain that all personnel know where they are located and how to use them. Seconds are precious in extinguishing an incipient fire or preventing its spread.
8. All connections of flammable gases to instruments, such as the flame photometer, must be leakproof. After making a new connection, check for leaks by painting the threads with soap solution and looking for bubbles.

9. All gas cylinders, whether full or empty, must be chained securely so that they cannot be knocked over.
10. Never store flammable solvents in a refrigerator or freezer in which the thermostat is inside the compartment; a spark may explode the vapor mixture with tremendous force. Laboratory refrigerators and freezers should be converted so that thermostats are outside the cold compartment.
11. The laboratory should cooperate with fire inspectors to reduce all dangers to a minimum, educate the staff to be constantly alert and careful, and hold periodic fire drills.

Infection

Laboratory personnel handle and analyze blood and body fluids that may contain infective agents (viral or bacterial) from sick and seemingly well patients. The greatest danger of disease transmission arises from an accidental puncture from a contaminated hypodermic needle, from exposure of cuts or skin abrasions to infective fluids that are spilled or allowed to contaminate the outer surface of their containers, or from aerosols that escape when opening a vacutainer or after a centrifuge accident. Transmission of disease through accident or mishandling of biohazards can be kept to a minimum by practicing the "*universal precautions*" recommended by the Centers for Disease Control[21,22] that appear in a following section.

Precautions Against Hepatitis and AIDS

Although a rare occurrence, the most likely infections to be transmitted to laboratory personnel by accident or carelessness are hepatitis and the human immunodeficiency virus (HIV) infection. The two types of hepatitis that are transmitted primarily by the parenteral (by injection or through cuts in skin) and sexual routes are hepatitis B (sometimes called "serum hepatitis") and hepatitis C (see Chap. 11, section on hepatitis). At present, about 90% of the hepatitis transmitted by contaminated blood or blood products at blood banks is hepatitis C, but the number of such instances should decrease with the present use of tests for its detection. After a long incubation period (up to 5 years or so), infection with HIV may develop into the acquired immunodeficiency syndrome (AIDS), but such development is not universal. HIV infection is defined as the finding of antibodies against HIV in plasma (seroconversion). The disease AIDS is applied only to those HIV-positive persons who exhibit a compromised immune system, as demonstrated by the presence of opportunistic infections, or who have a low number and ratio of T4 cells (killer lymphocytes).

The routes for accidental transmission of HIV and the viruses of hepatitis B or C (HBV or HCV) are the same. The infective viral particles must enter the body through punctures, cuts, or abrasions or through mucous membranes. They are not transmitted by casual contact. HBV (or HCV) is far more infective than HIV because the average patient with hepatitis may have as many as 1 million more viral particles per milliliter of blood than does the usual patient with HIV infection.[21] Infection with HIV, as measured by seroconversion, occurred in less than 1% of workers accidentally stuck with a needle containing blood from a person with HIV. The risk of HBV infection is much greater, but a vaccine is available and its use is recommended for laboratory personnel.

Universal Precautions

The National Committee for Clinical Laboratory Standards (NCCLS)[23] and the U.S. Department of Labor (OSHA Instructions)[24] have tentatively accepted with modifications the universal precautions of the Center for Disease Control. A summary of the pertinent precautions follows:

1. *Treat all laboratory specimens as infectious.* Make no exceptions, and do not rely on warning labels to designate contaminated specimens.
2. Use a protective barrier, for example, gloves, when handling blood and body fluids. Wear a gown or laboratory coat and change when contaminated. Cover all cuts and abrasions with adhesive tape or bandage.
3. *Avoid handling of hypodermic needles. Dispose of used needles in a rigid container.*
4. Avoid an aerosol or droplets when opening a vacuum tube containing blood. Point tube away from yourself and open slowly, preferably in a hood.
5. Use Pasteur pipets or other devices for transferring fluid samples. Do not pour from one tube to another, because a drop or two may contaminate the outer surface of the tube.
6. Never pipet by mouth.
7. Minimize spills and spatters. If they should occur, absorb the liquid with disposable, absorbent material, clean with detergent, and disinfect with a hypochlorite solution (1 part household bleach plus 9 parts water).
8. Wash hands after contamination, after removing gloves, and always before eating.
9. Dispose of samples properly when no longer needed for possible reanalysis. Samples should be autoclaved before discarding.
10. Place warning signs on all known biohazards.

Corrosive Chemicals

Strong acids and alkalies are the most common corrosive chemicals to which clinical technologists are exposed. Most accidents occur through the splashing of reagents during their preparation. Injury to the eyes is the greatest danger because it takes time to wash away the chemical completely, and the cornea is easily injured. Chemicals are even more difficult to remove quickly if a person is wearing contact lenses. Protective goggles are a necessity whenever one is preparing caustic or irritating solutions of any type. Care must be taken in the pouring of reagents to minimize splashing. Corrosive chemicals should be appropriately labeled.

Toxic Fumes

In some sections of a laboratory, one must occasionally prepare extracts with volatile solvents whose vapors are toxic. This toxicity is particularly true of chlorinated hydrocarbons (CH_2Cl_2, $CHCl_3$, CCl_4) or benzene used as solvents. Continued exposure to these agents may cause liver damage; some may depress bone marrow activity or affect other organ systems. A few simple precautions allow work with safety:

1. Always work in a fume hood with good ventilation whenever evaporating or using organic solvents.
2. Avoid contamination of the skin with solvents because they are slowly absorbed. If solvent gets on the skin, wash it off with soap and water.
3. Make sure that the laboratory is well ventilated. Exposure to a low concentration of vapor for a long period of time is also dangerous.

Broken Glassware

Modern automation and the practice of buying prepared reagents have greatly reduced the use of glassware; thus, broken glassware is much less of a hazard now than formerly. Nevertheless, any flasks or beakers with broken lips are a source of danger; they should be destroyed and replaced.

Carcinogens

Because a few chemicals used in the laboratory, particularly aromatic amines, may be carcinogenic, a few basic precautions are in order. Substitute a noncarcinogenic compound for a known carcinogen when possible. For example, benzidine, a commonly used reagent for the detection of hemoglobin, is a carcinogen that can be replaced by an all orthosubstituted tetramethyl benzidine that poses little danger. Likewise, dimethylaminoazobenzene, another carcinogen that is a component of Töpfer's reagent (for measurement of gastric acidity), can be replaced by the indicator, methyl orange, or by measurement using a pH electrode. When substitution is not possible, the infrequent handling of carcinogenic reagents can still be carried out with minimal danger if care is taken to avoid spilling of reagents or creating dust when weighing or transferring chemicals. Wear rubber gloves and a face mask when necessary. Clean the work area when finished, and wash hands thoroughly with soap and water.

Radioactivity

Federal and state regulations permit the use of radioactive isotopes only in facilities under the direction of a person authorized to use them. Carefully specified procedures for storage, work space, monitoring programs, waste disposal, and operations must be followed. The details may appear onerous, but they are designed for the protection of personnel and containment of radioactivity. A course in radiation safety is a prerequisite for all persons working with radionuclides.

For most clinical chemistry laboratory test procedures in many institutions using radionuclides, for example, in radioimmunoassay (Chap. 4), only tracer doses are used. Thus, the level of radioactivity is low and the exposure danger is small if proper precautions are taken in storing, handling, and disposing of the radioactive material. Gamma rays are penetrating, so gamma emitters should be stored behind a lead shield. Beta particles have much less penetrating power, so they are safe if stored in their containers in an out-of-the-way corner of the laboratory or in the refrigerator.

Disposable gloves should be worn when handling the concentrated material and care taken to avoid contamination or spills of any kind. Periodic monitoring of personnel, clothes, and work areas for contamination should be standard practice. Careful disposal of the waste is a necessity. Danger is minimal when good, careful work habits are practiced and exposure is kept to a minimum.

Hazardous Waste Disposal

The sound management of hazardous chemical wastes is a priority. Recent legislation passed by Congress has directed the Environmental Protection Agency (EPA) to develop and implement a program to protect human health and the environment from improper hazardous waste disposal. Most clinical laboratories are included under the EPA hazardous waste regulatory system and OSHA 40CFR 261.21—261.24.

Sound management of hazardous waste implies maximum reduction of waste generation at the source. Whenever possible, the laboratory should substitute a nonhazardous chemical for a dangerous chemical. Reducing the scale of a laboratory procedure also reduces the volume of chemicals used. Many hazardous chemicals, such as metallic mercury, photographic fixer, and some solvents, should be collected for purification and recycling.

A variety of chemicals can be sufficiently neutralized or inactivated in the laboratory to permit safe disposal in the sewer system. Wastes that cannot be recycled or rendered sewer-disposable, however, should be collected, properly stored and labeled, and responsibly transported at intervals to an official disposal center.

REFERENCES

1. Quantities and Units of Clinical Chemistry, International Union of Pure and Applied Chemistry and International Federation of Clinical Chemistry. Information Bulletin Number 20. Oxford, IUPAC, 1972.
2. Young, D.S.: Implementation of SI units for clinical laboratory data. Ann. Intern. Med., *106*: 114, 1987.
3. Commission on Laboratory Inspection and Accreditation: Reagent Water. College of American Pathologists, Skokie, IL, 1978.

4. National Committee for Clinical Laboratory Standards: Preparation and Testing of Reagent Water in the Clinical Laboratory. 2nd Edition. Proposed Guidelines Code C3-P2. Villanova, PA, 1985.
5. Young, D.S., and Mears, T.W.: Measurement and standard reference materials in clinical chemistry. Clin. Chem., *14:* 929, 1968.
6. Meinke, W.W.: Standard reference materials for clinical measurements. Anal. Chem., *43:* 28A, 1971.
7. Catalog of NBS Standard Reference Materials. NBS Special Publication 260, U.S. Department of Commerce/National Bureau of Standards.
8. Levey, S., and Jennings, E.R.: The use of control charts in clinical laboratories. Am. J. Clin. Pathol., *20:* 1059, 1950.
9. Westgard, J.O., Barry, P.L., Hunt, M.R., and Groth, T.: A multi-rule Shewhart chart for quality control in clinical chemistry. Clin. Chem., *27:* 493, 1981.
10. Griffin, D.F.: Systems control by cumulative sum method. Am. J. Med. Tech., *34:* 644, 1968.
11. Westgard, J.O., Groth, T., Aronsson, T., and de Verdier, C-H.: Combined Shewhart-Cusum control chart for improved quality control in clinical chemistry. Clin. Chem., *23:* 1881, 1977.
12. Dharan, M.: Preventive maintenance of instruments and equipment. *In* Selected Methods of Clinical Chemistry. Vol. 9. Washington, D.C., American Association for Clinical Chemistry, 1982.
13. Grannis, G.F., and Caragher, T.E.: Quality control programs in clinical chemistry. Crit. Rev. Clin. Lab. Sci., *7(4):* 327, 1977.
14. Westgard, J.O., et al.: Concepts and practices in the evaluation of laboratory methods. Am. J. Med. Technol., *44:* 290, 420, 552, 727, 1978.
15. Watts, M.T., and Watts, H.W.: A different view of method performance evaluation: the minimized evaluation. J. Med. Technol., *1:* 138, 1984.
16. National Committee for Clinical Laboratory Standards Publication H3-A2: Procedures for collection of diagnostic blood specimens by venipuncture. 2nd Edition. Villanova, PA, NCCLS, 1984.
17. Clark, B.A.: Getting those blood samples right. RN, *44:* 36, 1981.
18. Meites, S.: Skin puncture and blood collection techniques for infants: update and problems. Clin. Chem., *34:* 1890, 1988.
19. National Committee for Clinical Laboratory Standards Publication H4-A2: Procedures for the collection of diagnostic blood specimen by skin puncture. 2nd Edition. Approved standard. Villanova, PA, NCCLS, 1986.
20. Forrey, A.W., Delaney, C.J., and Ruff, W.L.: Safety. *In* Standard Methods of Clinical Chemistry. Vol. 9. Edited by W.R. Faulkner and S. Meites. Washington, D.C., American Association for Clinical Chemistry, 1982.
21. Centers for Disease Control: Recommendations for prevention of HIV transmission in health care settings. Morbidity and Mortality Weekly Report. *36* (2S):1S, 1987.
22. Centers for Disease Control: Summary statement for human immunodeficiency virus and report on laboratory acquired infection with human immunodeficiency virus. Morbidity and Mortality Weekly Report, *37* (S4):1988.
23. National Committee for Clinical Laboratory Standards (NCCLS) Document M29-T, Tentative guideline, Protection of laboratory workers from infectious disease transmitted by blood, body fluids, and tissue, 9:1, Jan., 1989.
24. OSHA: Enforcement procedures for occupational exposure to hepatitis B virus (HBV) and human immunodeficiency virus (HIV). Instruction CPL 2-2, 44A, Aug. 15, 1988.

QUESTIONS

1. A patient has a blood glucose concentration of 90 mg/dL. What is the concentration in SI? The formula of glucose is $C_6H_{12}O_6$.

A. *5.0 mmol/L* **B.** *4.5 mmol/L* **C.** *0.5 mmol/L* **D.** *2.0 mmol/L*

2. A value for creatinine is reported as 106 μmol/L. What is its concentration in mg/dL? The formula for creatinine is $C_4H_7N_3O$.

A. *1.2 mg/dL* **B.** *0.94 mg/dL* **C.** *2.4 mg/dL* **D.** *9.4 mg/dL*

3. The concentration of a sodium chloride solution is labeled as 6% (w/v). How may its concentration also be stated?

A. *60 g/L* **B.** *600 mg/dL* **C.** *6.0 g/L* **D.** *0.6 g/L*

4. What is the H^+ concentration of a solution that was made with 0.1 mol/L of acetic acid and 1.0 mol/L of sodium acetate? The pK_a of acetic acid is 5.0.

A. 10^{-6} *M* **B.** 10^{-7} *M* **C.** 10^{-5} *M* **D.** 10^{-4} *M*

5. You have analyzed a control serum 25 times and found the mean concentration and standard deviation of substance X to be 80 ± 2.4 mmol/L. What is the coefficient of variation?

A. *3.0%* **B.** *3.3%* **C.** *1.9%* **D.** *3.0 mmol/L*

3 Basic Instrumentation

OUTLINE

VI. **AUTOMATED INSTRUMENTS** *(continued)*

C. Notes on Instrument Selection

VII. **LABORATORY COMPUTERS**

A. Components

B. Types of Laboratory Computers

1. Computerized instruments
2. Personal computers
3. Laboratory Information Systems

OBJECTIVES

After reading this chapter, the student will be able to:

1. Name the regions of the electromagnetic spectrum and the colors of the visible spectrum in order of increasing energy and decreasing wavelength.
2. State Beer's law in terms of absorbance and concentration and apply it in quantitative calculations.
3. Describe the essential components of a UV-visible spectrophotometer and their functions.
4. Describe the principle of flame photometry and the basic components of a flame photometer.
5. Describe the basic principle of atomic absorption spectrophotometry and the basic components of the instrument.
6. Describe the principle of fluorometry and the basic components of a fluorometer. Explain why fluorescence measurements may be more specific and sensitive than spectrophotometric measurements.
7. List the different types of chromatography and the basis for separation used in each.
8. Explain the principle of gas-liquid chromatography and describe several different types of detectors.
9. Describe the principle of electrophoresis and explain how it is commonly used in a clinical laboratory.
10. Describe the principle and components of a pCO_2 electrode and a pO_2 electrode.
11. Discuss the principle of an ion-selective electrode technology and list the ions commonly measured by it in a clinical laboratory.
12. Describe common features found within automated instruments.

The clinical chemistry laboratory could not function without its instrumentation. Instruments permit numerous patients' samples to be processed and reliably assayed in a remarkably short time. Much of the instrumentation is automated. Some instruments are designed to be flexible and to adapt to the changing needs of the clinical laboratory, whereas others are dedicated to performing a specific test or group of tests. The instruments all rely on the basic analytical principles that allow measurement of changes in light intensity, voltage, current, or time. At present, the most powerful automated clinical chemistry analyzers involve several basic measuring devices combined within a single frame.

This chapter provides an overview of the basic principles of instrumental analysis and reviews briefly some adaptations to automated systems.[1-5]

PHOTOMETRY

Photometry means "measurement of light." A camera light meter is an example of a photometer that quantifies the total light intensity striking the photocell. In the laboratory, relative light intensity is usually measured after passage of light through a solution. Because light can undergo absorption, scatter, reflection, emission, or fluorescence, photometry is the most frequent measurement in the laboratory. A review of some of the basic characteristics of light may make this section more understandable.

Light is a form of electromagnetic energy that appears to travel in waves. The wavelength, or distance between the peaks of a light wave, is a function of its energy. High-energy gamma rays, which are associated with nuclear reactions, have short wavelengths, in the order of 0.1 nm. At the other end of the scale are the long-wavelength, low-energy radiowaves, 25 cm or more in length. Visible light, light that can be seen by the human eye, constitutes a small portion of the electromagnetic radiation and is limited to the region of 380 to 750 nm (Table 3.1). A good spectrophotometer may measure light from 180 to 200 nm in the near-ultraviolet (UV) region to about 1000 nm in the near-infrared region.

The color of light is a function of its wavelength (Table 3.2). For example, when a strip of filter paper is placed in a spectrophotometer cuvet so that the paper interrupts the light beam, the spectral colors on the paper change as the wavelength decreases from about 800 nm to less than 400 nm. The first light that appears is red; red gradually becomes orange and then moves through yellow, green, blue, and finally violet before entering the invisible UV range near 380 nm.

TABLE 3.1
WAVELENGTHS OF VARIOUS TYPES OF RADIATION

E	ν	λ	TYPES OF RADIATION	APPROXIMATE WAVELENGTH (nm)
↑	↑		Gamma	<0.1
			X-rays	0.1–10
			Ultraviolet	<380
E	ν	λ	Visible	380–750
			Infrared	>750
		↓	Radiowaves	over 25×10^7

A listing of types of electromagnetic radiation in the order of the energy involved. E = energy; ν = frequency; λ = wavelength. The vertical arrows indicate the direction of increase in magnitude.

TABLE 3.2
THE VISIBLE SPECTRUM

APPROXIMATE WAVELENGTH (nm)	COLOR OF LIGHT ABSORBED	COLOR OF LIGHT REFLECTED
400–435	Violet	Green-yellow
435–500	Blue	Yellow
500–570	Green	Red
570–600	Yellow	Blue
600–630	Orange	Green-blue
630–700	Red	Green

The wavelengths are approximate. The colors of the light absorbed and reflected change gradually from one color to the next with no clear line of demarcation. The sum of the colors of the reflected light forms the apparent color of the object to the viewer.

Light from the sun or from incandescent light bulbs contains the entire visible spectrum: this continuum of light appears "white" or colorless. Objects that appear colored absorb light at particular wavelengths and reflect the other parts of the visible spectrum, resulting in many shades of color. For example, a substance that absorbs violet light at 400 nm reflects all other light and appears as yellow-green, whereas a substance absorbing yellow light at 590 nm is seen as blue, which is the sum of the reflected light (Table 3.2).

Spectrophotometry takes advantage of the property of colored solutions to absorb light of specific wavelengths. To measure the concentration of a blue solution, light is passed through it at about 590 nm. The amount of yellow light absorbed varies directly in proportion to the concentration of the blue substances in the solution. This is an example of Beer's law.

Beer's Law

When light of an appropriate wavelength strikes a cuvet that contains a colored sample, some of the light is absorbed by the solution; the rest is transmitted through the sample to the detector. The proportion of the light that reaches the detector is known as the percent transmittance (%T) and is represented by the equation

$$\frac{I_t}{I_o} \times 100 = \%T$$

in which I_o is the intensity of light striking the sample and I_t is the intensity of the light transmitted through the sample. In actual practice, the light transmitted by a Blank (I_B) is substituted for I_o. The Blank may be water or the entire reagent mixture except for the sample.

As a concentration of the colored solution in the cuvet is increased, I_t, and consequently %T, is decreased. The relationship between concentration and %T is not linear, as shown in Figure 3.1A, but if the logarithm of the %T is plotted against the concentration, a straight line is obtained. The term

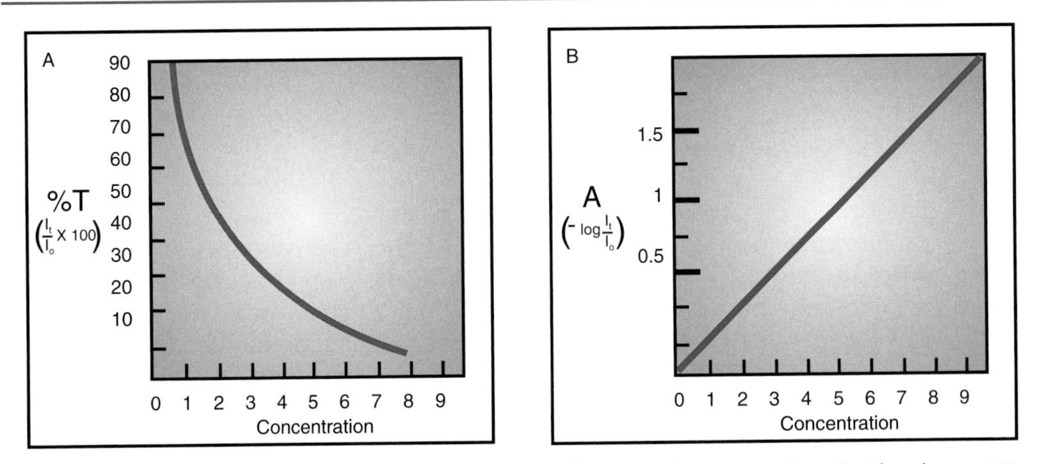

FIG. 3.1. A, Percent transmittance (%T) as a function of concentration. B, Absorbance (A) as a function of concentration.

absorbance is used to represent the logarithm of 1/%T or (−log%T). Absorbance increases linearly with concentration (Fig. 3.1B).

The relationship of absorbance to concentration is expressed in the equation known as the Beer-Lambert law (often referred to as Beer's law):

$$\text{Equation 1: } A = a\, b\, C$$

in which A is absorbance, C is the concentration of the colored compound, *a* is the absorptivity coefficient (a constant) of the colored compound, and *b* is the length of the light path through the solution. Because *a* and *b* are constants in an assay, A is directly proportional to C.

In spectrophotometric procedures, the absorbance of an unknown concentration of a particular constituent is compared with that of a known concentration (a standard), which is reacted in the same way to produce a colored solution. The following relation holds:

$$\text{Equation 2: } \frac{A_u}{A_s} = \frac{C_u}{C_s}$$

in which A_u and A_s are the respective absorbances of unknown and standard, and C_u and C_s are their respective concentrations. To solve the equation for C_u, the equation is rearranged to

$$\text{Equation 3: } C_u = \frac{A_u}{A_s} \times C_s$$

Equation 3 is routinely used for calculations in spectrophotometric assays.

No assay gives a linear response between absorbance and concentration for all concentrations, going from small to large, because sooner or later, some reactant becomes limiting and less color is formed. The range of linearity may be wide for some constituents and narrow for others. All procedures require one to determine whether spectrophotometric response is linear for the usual concentrations to be measured and to determine the limits of linearity.

UV-Visible Spectrophotometry

The simple spectrophotometer or colorimeter has long been the workhorse of the hospital chemistry laboratory. Today, these instruments still serve that purpose, but often as detection systems in more complex automated instruments. *Colorimeter* has been the traditional name for an instrument that isolates specific wavelengths of light with interchangeable filters for the visible portions of the spectrum. In contrast, *spectrophotometers* have a continuously adjustable monochromator (prism or grating) and can often measure the intensity of light from the UV range through the visible.

The components of most spectrophotometers are basically the same. They consist of a power supply that provides current at the proper voltage and the components shown in Figure 3.2: (1) a lamp as a light source; (2) a monochromator to isolate the desired wavelength; (3) a light-beam exit slit; (4) a sample holder or cuvet; (5) a photodetector, which produces a current in response to the light impinging upon it; and (6) a meter, computer, or other readout device. These components are all described in detail in the following sections:

Power Supply

The power supply may be a simple transformer to convert line voltage to a constant low voltage required for the lamp, or it may provide current (both AC and DC) of different voltages for several components, such as the detector and readout devices, as well as the lamps.

Light Sources

The light source is usually a tungsten lamp for wavelengths in the visible range (320 to 700 nm) and a deuterium lamp for UV light (below 350 nm). The lamps are positioned in the instrument in such a way that an intense beam of light is directed through the monochromator and the sample. A *selector switch* is used

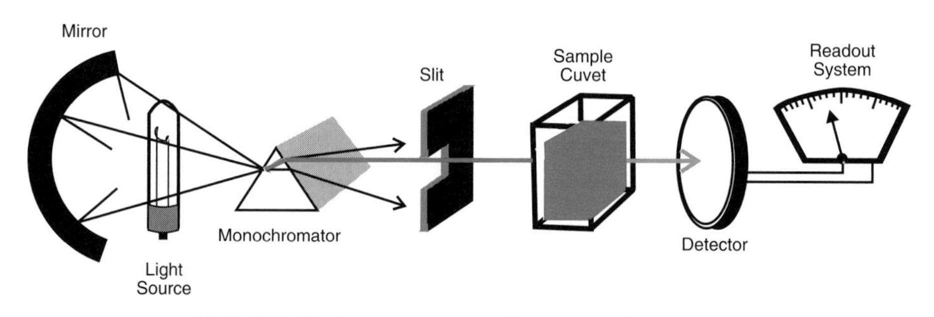

FIG. 3.2. Basic components of a spectrophotometer.

to shift a mirror to reflect the light from either the deuterium or the tungsten lamp into the monochromator as needed for the UV or visible light measurement. Although the lamps are prefocused and their positioning may be predetermined, intensities vary among lamps, and a spectrophotometer should be recalibrated whenever the lamp is changed. The calibration should be checked daily at 1 wavelength with a didymium filter and the spectrophotometers should be checked monthly for linearity at 3 wavelengths, using a commercial set of calibration standards. Wavelength-scanning instruments are checked against a holmium oxide filter for accuracy of both wavelength and absorbance.

Monochromators

The next component in the system is the monochromator. Early colorimeters used glass filters that transmitted a wide segment of the spectrum (50 nm or more). Newer instruments use interference filters that consist of a thin layer of a magnesium fluoride crystal with a semitransparent coating of silver on each side. The crystal transmits only light for which an exact multiple of the wavelength is equal to the thickness of the crystal. All other wavelengths are blocked. Interference filters have a bandpass of 5 to 8 nm.

The term "bandpass" defines the width of the segment of the spectrum that is isolated by a monochromator; it is the range of wavelengths between the points at which the transmittance is equal to one-half the peak of light intensity (Fig. 3.3). The peak of light intensity in Figure 3.3 occurs at 550 nm, the nominal wavelength setting of the monochromator. The bandpass is 20 nm because the intensity of light at 540 nm and 560 nm is one-half that of the 550-nm peak. The inexpensive spectrophotometers commonly used in the

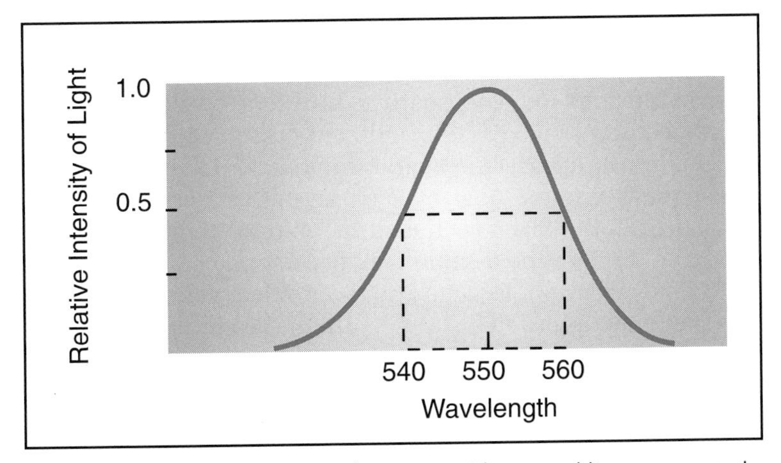

FIG. 3.3. Bandpass of a spectrophotometer. The curved line represents the intensity of the light at various wavelengths transmitted by a monochromator when it is set at 550 nm. The wavelengths at which the light intensity is one-half as great as at peak height (550 nm) are 540 nm and 560 nm. These two wavelengths encompass the bandpass of the instrument (20 nm).

laboratory often have a bandpass of about 20 nm, but the more sophisticated instruments may have a bandpass of 0.5 nm or less.

The monochromator consists of an entrance slit to exclude unwanted or "stray" light and a prism or diffraction grating preceded by a series of light focusing lenses. An exit slit allows only a narrow fraction of the spectrum to reach the sample cuvet.

Polychromatic light that enters a prism is refracted at the air-glass interface. The shorter wavelengths are refracted more than the longer wavelengths; consequently, light is dispersed into a continuous spectrum. This dispersion permits the isolation of a particular wavelength band. Glass prisms and lenses are suitable for work in the visible range, but quartz or fused silica is essential for the UV range because glass does not transmit light efficiently at wavelengths shorter than 340 nm.

Diffraction gratings consist of a series of parallel grooves cut into a surface, all at exactly the same angle; there may be 10,000 to 50,000 grooves per inch. Light striking the grooves is diffracted and dispersed according to wavelength; a continuous spectrum is produced. If a grating is made from polished metal, the diffracted light is reflected and may be used for UV, visible, or infrared measurements. Diffraction gratings with the grooves cut into glass or quartz transmit the light through the grating.

Good-quality gratings are difficult to cut and therefore are expensive. A carefully made grating, however, can serve as a template for many replicas, thereby lowering the cost. A coating of epoxy is layered over the master grating and assumes the form of the original surface. The replicate surface is coated with a silvered layer to produce a reflectance grating. These replicas are relatively simple to make, so that even inexpensive instruments may contain diffraction gratings of good quality.

Exit Slit

At the end of the monochromator chamber, an exit slit controls the width of the light beam and, hence, the bandpass of the spectrophotometer. As the grating of prism is turned in response to the wavelength selector, different wavelengths of the spectrum strike the open slit and pass through it. A narrower slit produces a smaller bandpass. One would think that all instruments could have the ideal "narrow bandpass" just by making the exit slit small. In actual practice, a practical limit exists. As the slit is narrowed, the total amount of light energy passing through the sample may become too small for accurate measurement.

Some spectrophotometers have a nonmovable slit; others have slits that can be manually adjusted. As the width of the slit is increased, more light reaches the detector, and the bandpass is also increased. Because many compounds have broad absorption peaks, this width poses no problem for routine tests.

In some instruments, the width of the slit varies automatically as the wavelength is changed. This compensates for variations in the intensity of the different wavelengths of light produced by the lamp, and also for variations in the response of the phototube to different wavelengths. In such instruments, the sensitivity remains relatively constant over the whole wavelength range, although the bandpass varies.

Sample Cuvets

For accurate and precise readings, the cuvet must be clean; fingerprints, scratches, or any spills must not be allowed on the optical surfaces; and bubbles should not adhere to the inner surface of the filled cuvet. The solution in the cuvets must be thoroughly premixed and, when possible, at room temperature so that moisture does not condense on the exterior optical surface.

Square or rectangular cuvets with flat optical surfaces are most desirable. They generally have an inside dimension (pathlength) of 1 cm and a capacity of 3 to 4 mL. For smaller sample volumes, cuvets are available with thicker sidewalls that make them long, narrow, and rectangular. The pathlength is still 1 cm, but the required volume of solution is reduced to 1 mL or less.

Glass cuvets are used for readings in the visible light range. For measurements in the UV range, the cuvet must be made of quartz or fused silica.

Many of the less expensive spectrophotometers use round cuvets and have adaptors to accommodate cuvets of differing diameters. The round cuvets can be matched to about 1% tolerance in light transmission, which is adequate for routine testing.

A convenient accessory for the spectrophotometer is a flow-through or flush-out cuvet. (Each sample can be flushed out of the cuvet after it has been read.) Care must be taken to ensure that the cuvet is clean. Frequent checking of blanks for zero absorbance, thorough rinsing after use, and immediate cleanup after overfilling are important. The cuvet should be checked frequently for visible signs of contamination or cloudiness.

For some tests requiring the periodic reading of absorbance, such as kinetic enzyme assays, cuvets are retained in the sample compartment for a prolonged time period, during which the sample must be maintained at a constant temperature. A constant temperature is provided in either of two ways: (1) the sample compartment is surrounded by a water jacket through which water is pumped from a constant-temperature water circulator or (2) the compartment is warmed by thermostatically controlled heating coils.

Detectors

A wide range of photodetectors varying in sensitivity, amplification, and cost are available. All contain a light-sensitive surface that releases electrons in numbers proportional to the intensity of the light impinging upon it.

A *photocell* (barrier layer cell, selenide cell) is the simplest of the detectors. It consists of three layers sealed in a protective casing: (1) the bottom support layer, consisting of a conductive metal, such as iron; (2) a photosensitive layer of selenium or cadmium on top of the metal support; and (3) a transparent conductive layer covering the light-sensitive material. Light passing through the transparent layer to the selenium causes the release of electrons from the selenium. The emitted electrons move to the clear conductive layer and create a weak electric current as they complete a circuit by flowing to the metal support. The measured current thus generated is a direct function of the light intensity striking the photocell. Photocells are simple and sturdy and seldom need replacement. They are generally not sensitive, and their output is not readily amplified, so they cannot be used for measuring low light levels or small changes in intensity. They are sensitive to temperature change (such as the heat of the lamp) and are slow to respond to changes in light intensity. The current produced by a photocell at any one wavelength is quite linear with light intensity, although the response to light in the shorter and middle wavelengths of the visible light range is much greater than that for the longer wavelengths. The instrument and photocell should be warmed up for several minutes before use.

More sensitive than the photocell is the *phototube*, which consists of a curved cathode of metal coated with a photosensitive material. When light strikes its surface, the cathode emits electrons that are attracted to a positively charged anode, thereby causing a current flow that is proportional to the light intensity. The response, or intensity of current, can be increased by increasing the voltage across the terminals. The phototube may be a vacuum tube or gas-filled. In the gas-filled phototube, electrons emitted from the cathode strike gas molecules that release more electrons and thus amplify the current response to the light radiation. Although the increased sensitivity is counterbalanced by a shorter tube life, the advantages outweigh this limitation.

When low levels of light or quick bursts of light must be measured, a *photomultiplier tube* is required. The photomultiplier tube is sensitive and fast in its response. It consists of a photosensitive cathode and an anode with several intermediate faces called dynodes. Each dynode is slightly more positive than the preceding dynode, and it attracts the electrons from the preceding dynode because of this increasing charge. The principle is simple: light striking the cathode causes emission of electrons that are attracted to the first dynode. Each impinging electron causes the release of secondary electrons; these electrons are attracted by the more positive charge on the second dynode, and so on. The process is repeated at several successive dynodes, with each step increasing the electron amplification. The total amplification factor may exceed 10^6. Photomultiplier tubes are used in spectrophotometers with a narrow bandpass (low light level) and in instruments that must record fast changes in light emission or absorbance—scanning spectrophotometers, for example.

Phototransistors and *photodiodes* are the newest entries among the light detectors. These are constructed of two types of semiconductors joined together that resist current flow between them. As light strikes the junction, the resistance is overcome, and current flows across the junction. Detectors of this type are small, durable, and capable of high amplification. They are relatively inexpensive and have wide application in spectrophotometry. One adaptation is the "diode array detector," as used in a Hewlett Packard spectrophotometer. It consists of several hundred diodes, each of which is associated with a capacitor. As current flows through a diode in response to light energy, the accompanying capacitor is discharged. The capacitors are recharged several times each second; the current required for the recharging is directly proportional to the quantity of light striking the diode detectors.

Readout Devices

The magnitude of the current generated by a detector may be measured by any of several types of readout devices: a galvanometer, an ammeter with a meter needle, a recorder, or a digital readout. Alternatively, the signal may be transmitted to a computer or printout device. The information from the readout may be presented in absorbance units, presented in percent transmittance, or transformed by calculation directly into the concentration of the constituent by a computerized instrument.

Turbidimetry and Nephelometry

Light that strikes a molecule can be absorbed, transmitted unchanged, or reflected and scattered. This review thus far has focused on photometric quantification of soluble samples, which absorb light. Suspensions of insoluble particles, however, reflect light. They appear cloudy or "turbid" and can be quantified using a turbidometer. A turbidometer has the same design as a spectrophotometer (Fig. 3.2), and also measures the fraction of light that passes through the sample. The use of turbidimetry is usually reserved for the measurement of abundant large particles, such as bacteria in solution. Alternatively, the photodetector can be placed to the side of the light beam to measure the amount of light scattered by the sample when the insoluble particles are present in lower concentrations. The measurement of the scattered light is performed during nephelometry. This method is often used in the clinical laboratory to quantitate the rate of insoluble antibody:antigen complex formation during the assay of specific serum proteins.

Double-Beam Spectrophotometer

Double-beam spectrophotometers operate like single-beam spectrophotometers except that they are designed to compensate for possible variations in intensity of the light source. This compensation is accomplished by "splitting" the light beam from the lamp and directing one portion to a reference cuvet and the other to the sample cuvet. Any change in light intensity affects both cuvets simultaneously and thus is canceled out.

There are two types of dual-beam systems. In the dual beam in space, illustrated in Figure 3.4A, the light beam is split so that half of the monochromatic light is directed through the reference cuvet while the other half is directed through the sample cuvet. Separate detectors monitor the respective light intensities, and the ratio of the light intensity of the sample to that of the reference cuvet is measured. The two detectors must be matched.

The configuration of a double beam in time is illustrated in Figure 3.4B. The light from the monochromator is directed to a rotating semicircular mirror. Half of the time, light strikes the mirror and is reflected through the reference cuvet; otherwise, the light passes directly through to the sample cuvet. The rotating motion of the semicircular mirror directs the light alternately through either the sample cell or the reference cell to a single detector, where the ratio of the sample light intensity to the reference light intensity is measured.

Double-beam spectrophotometers are ideally suited for making spectral scans because the instrument automatically corrects for the change in light transmission through the reference cuvet as the wavelength is changed. If a single-beam instrument were used, each change in wavelength would require

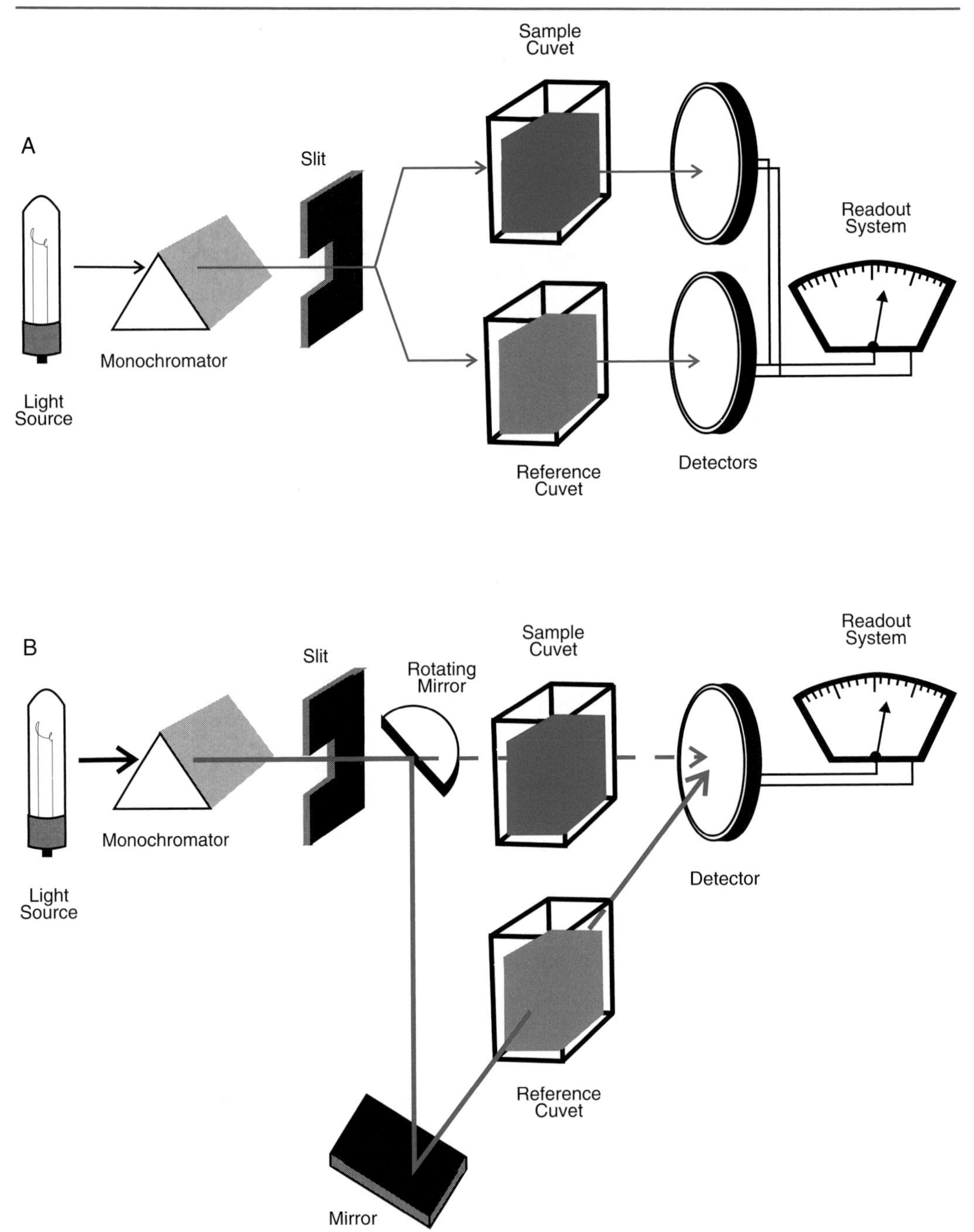

FIG. 3.4. Two configurations for double-beam spectrophotometers. A, Double beam in space. B, Double beam in time.

an adjustment of the instrument to zero absorbance for a reagent blank before reading the absorbance of the sample.

The principles and components described pertain directly to absorption spectrophotometers, although they also apply to the other types of spectrophotometric instruments, such as the reflectance photometer, the flame photometer, and the atomic absorption spectrophotometer.

Reflectance Photometry and Reflectance Densitometry

Reflectance photometry is the measurement of light reflected from solid surfaces. For example, the absorption of yellow light can cause a dye to appear as "blue ink" on white paper (Table 3.2). Several reflectance photometric reactions in the clinical chemistry laboratory do not require instrumentation, for example, an unaided human eye can judge the intensity of colored light reflected from the reagent pads of urine dipsticks (Chap. 6). Reflectance densitometers usually perform this task faster and more consistently, however, and have become popular components of whole blood glucose monitoring devices designed for use outside the laboratory.

Principle: A beam of light is directed onto the unpolished solid surface of the sample to be analyzed. A portion of the light is absorbed by the sample, and the remainder is either reflected or scattered (Fig. 3.5). A lens is used to focus a fraction of the reflected light onto a detector where it can be measured. A filter is used to select the most appropriate wavelength of light for analysis, either before or after the light reflection. As the concentration of analyte in the sample increases, more of the light is absorbed, less light is reflected, and the sample appears "darker." Although the components and principles of reflectance photometers and spectrophotometers are similar, unlike absorbance, the

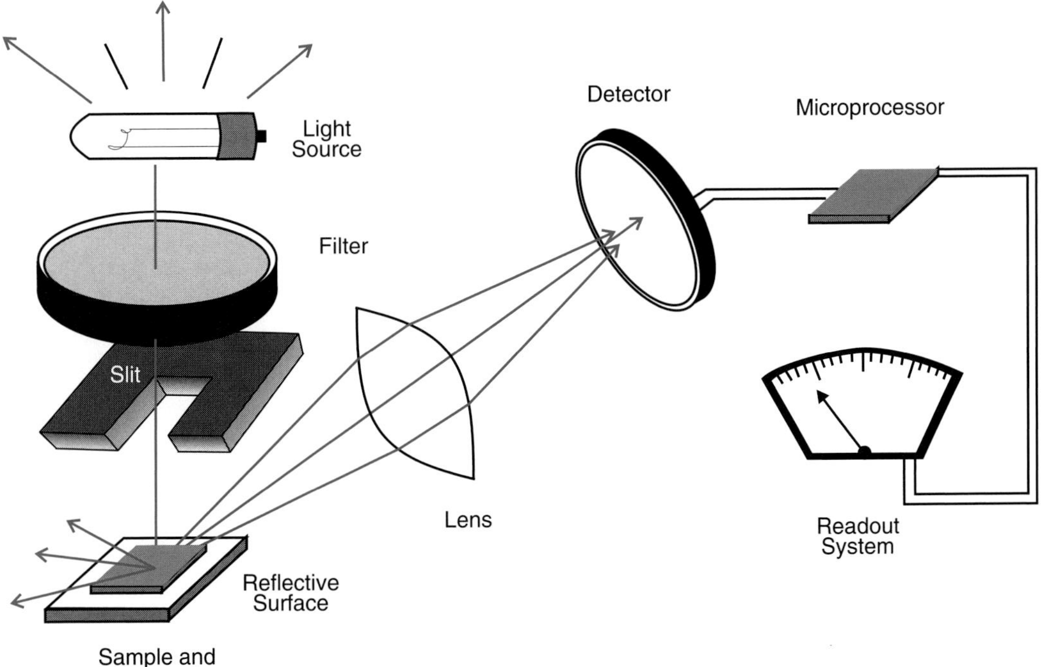

FIG. 3.5. Basic components of a reflectance photometer.

reflective density* has a nonlinear relationship with the concentration of the absorbing analyte. To assist in the nonlinear calculations, a microprocessor is often incorporated into instruments to transform the data into linear responses for both calibration and analyses.

Atomic Emission Spectrophotometry (Flame Photometry)

The perfection of flame photometry as a clinical chemistry technique revolutionized the determination of the serum Na and K concentrations during the 1960s by providing results within a few minutes. Although ion-selective electrode methods are most frequently used today, atomic emission spectrophotometry has remained the reference method for Na and K analysis.

Components

The components of a flame photometer, illustrated schematically in Figure 3.6, consist of an aspirator, premix burner, flame, monochromator, detector, and readout system. The monochromator, detector, and readout system are similar to those of an ordinary spectrophotometer, but the flame serves as a light source and sample compartment. Also, entrance and exit slits in the monochromator narrow the beam of light.

The monochromatic light is obtained by means of an interference filter. The detector is a phototube, the signal is amplified, and results are usually displayed by direct readout.

Principle

The aspiration into a flame of a salt solution initiates the process shown in Figure 3.7. A small percentage of the atoms is transformed to a temporary excited state; the atoms immediately return to the ground state and, in the process, release light. The wavelength of the emitted light is specific for each

*Reflective light density is defined as $R_D = -\log\left(\frac{I_R}{I_O}\right)$, where I_R is the intensity of the reflected light and I_O is the intensity of the incident light.

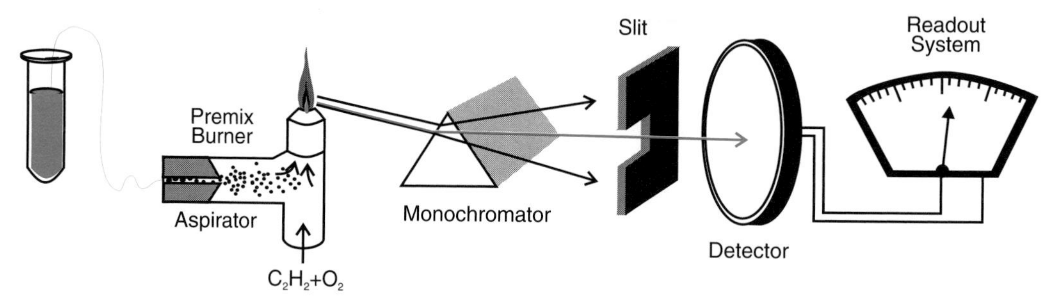

FIG. 3.6. Components of a flame photometer (atomic emission spectrophotometer).

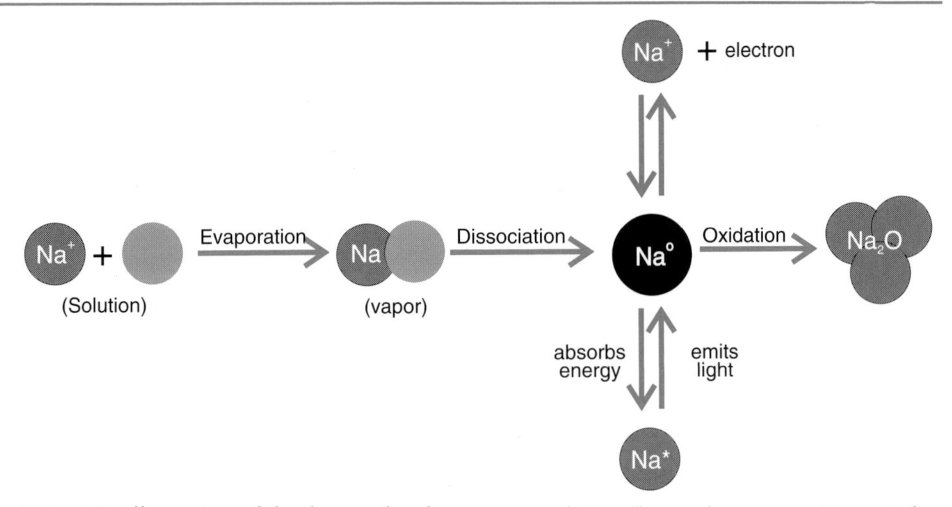

FIG. 3.7. Illustration of the forms of sodium present during flame photometry. See text for explanation of the production of atoms in the excited state and their emission of light as they return to ground state. Na* = an excited atom; E = energy; and Na^0 = atomic sodium.

excited state of each element and can be quantified under carefully controlled conditions.

Procedure

An analysis for Na and K is begun by diluting the sample (serum, urine, or other fluid) with a dilute solution of a nonionic detergent (wetting agent) containing a specified concentration of a cesium or lithium salt. The detergent reduces the viscosity of the solution and improves aspiration of the sample. The cesium serves as an internal reference, described later.

The diluted specimen is aspirated into a premix chamber by passing a stream of air at high velocity (compressed air) over the open end of a capillary tube that dips into the specimen solution. The sample is aspirated into the airstream by the partial vacuum produced at the capillary jet, broken into fine droplets, and carried as a mist to the premix chamber, where the fuel (natural gas or propane) is mixed with the air. Larger droplets settle out and are drained off while the fine mist (about 2 to 5% of the aspirated specimen) is drawn into the flame with the fuel mixture.

The heat of the flame evaporates the water and vaporizes the salts. For NaCl, the Na^+ ion is converted to its elemental state, the Na^o atom. If the flame is too hot, some of the Na^o becomes ionized as electrons are lost. In a small percentage of the Na^o atoms, a valence electron absorbs some of the heat energy and temporarily moves into a higher orbit (the excited state). Light is emitted as excited atoms return to the ground state (Fig. 3.7).

The light emitted by the excited atoms is measured at a wavelength specific for that element. The characteristic emission spectral line for each element chosen for measurement is distinct from that of other elements that may be present in the solution. The special lines chosen for the Na and K analyses are 589 and 766 nm, respectively.

The concentrations of Na and K are usually measured simultaneously in current instruments. Cesium* or lithium is incorporated as an internal standard in the diluting fluid to compensate for variations in sample feed, gas pressure, or fuel:air ratio that may change the intensity of the emitted light, the background light, or both. A separate filter and phototube are required for each of the three elements as the photometer measures changes in the light intensity ratios of Na:Cs and K:Cs.

There is no way of calculating from theory the light intensity that a certain concentration of Na or K should emit when aspirated into a flame. Standard solutions of the salts to be analyzed, in concentrations close to the unknown solutions, must be used in the same run to calibrate the instrument. The flame photometers are adjusted to read zero while aspirating the diluent and to give the expected values when aspirating the standards. The unknowns then are analyzed.

Lithium salts are commonly administered as medication for manic-depressive states; in such instances, the need to monitor the blood lithium concentration is imperative (Chap. 14). Cesium has essentially replaced lithium as the internal standard in flame photometers, thus allowing measurement of Na, K, and Li against a common internal standard.

Atomic Absorption Spectrophotometry

Atomic absorption spectrophotometry (AAS) is the measurement of the absorption of light by free metallic atoms. AAS is similar to atomic emission spectrophotometry in that it uses the heat of a flame to dissociate molecules to free atoms, primarily at their lowest energy level (ground state), but it differs drastically in what is measured. Flame photometry measures the intensity of *emitted* light when an activated atom returns to the ground state, whereas AAS measures the *absorption* of light of a unique wavelength by atoms in the ground state. The unique wavelength absorbed corresponds to the particular line spectrum for that element. With the proper light source, a particular cation can be analyzed in a mixture of many cations. Because at least 99.998% of the atoms are in the ground state at flame temperatures, AAS is far more sensitive than flame photometry.

Principle

Monochromatic light for a particular element is produced by a hollow cathode lamp using that element as the cathode. The monochromatic light is beamed through a long flame into which is aspirated the solution to be analyzed. The

*In addition to serving as an internal standard, the cesium salt serves as a *radiation buffer*. If cesium is not present, the emission of light from K atoms is increased by the Na* atoms present. A solution of K in the presence of Na emits more light at 766 nm because a transfer of energy occurs from excited Na atoms directly to K atoms, thereby increasing the number of excited K atoms and increasing light emission from K*. When a high concentration of cesium atoms is present, the excited Na atoms interact with the cesium rather than with K, and the Na:K interaction is minimized.

heat energy dissociates the molecules and converts the components to atoms. Although some atoms are activated, most atoms remain in the ground state at the temperatures commonly used. The ground state atoms of the same element as that in the hollow cathode cup absorb their own resonance lines; the amount of light absorbed varies directly with their concentration in the flame. The transmitted light that is not absorbed reaches the monochromator, which passes only the wavelengths close to the selected resonance lines of the particular element to be assayed. The transmitted light strikes a detector, and the decrease in transmitted light is measured.

Components

The components of an atomic absorption spectrophotometer parallel those of a good-quality UV-visible light spectrophotometer (Fig. 3.8). The monochromator, photomultiplier tube, and readout devices are identical.

The light source, the hollow cathode lamp, is unique to the atomic absorption spectrophotometer (Fig. 3.9). The lamp is filled with an inert gas, such as argon or neon, at a low pressure. When the lamp is illuminated, gas atoms between the electrodes are ionized and strike the cathode with high velocity, thereby causing the metal atoms to be "sputtered" from the cathode surface. Further collisions with gas ions produce excited metal atoms, which then emit light at the characteristic wavelengths (resonance lines) for that element. The emitted resonance lines are much narrower in wavelength than those that could be achieved with the finest monochromator.

The second unique component in the system is the burner through which the sample is introduced (Fig. 3.10). The burner has three parts: nebulizer, premix chamber, and burner head. A sample in the solution is aspirated and nebulized (reduced to a fine spray) by a stream of oxidant (air or oxygen) flowing across a sample capillary tube, as in flame photometry. The mist is mixed with oxidant and fuel (commonly acetylene) in the premix chamber, where large droplets are trapped and drained off. The mixture of gases and sample is

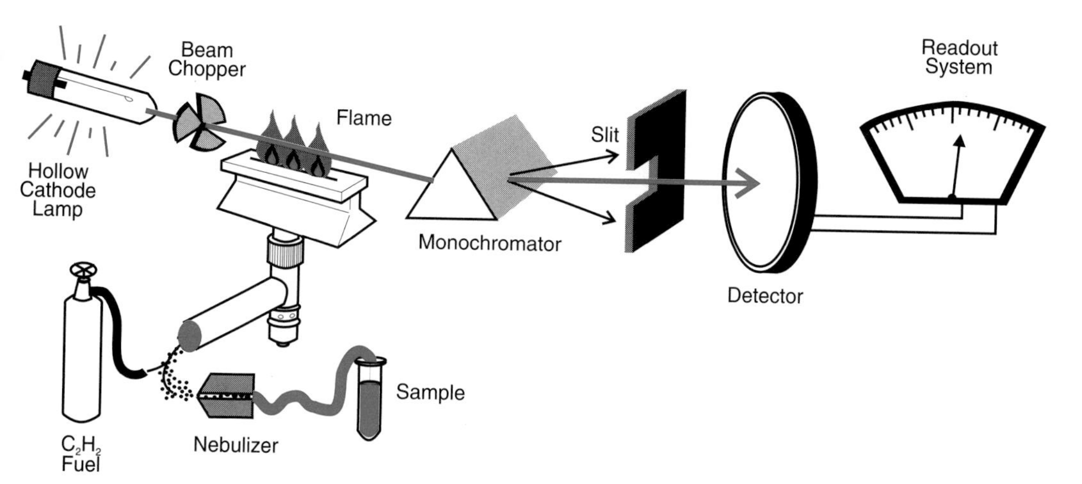

FIG. 3.8. Components of an atomic absorption spectrophotometer.

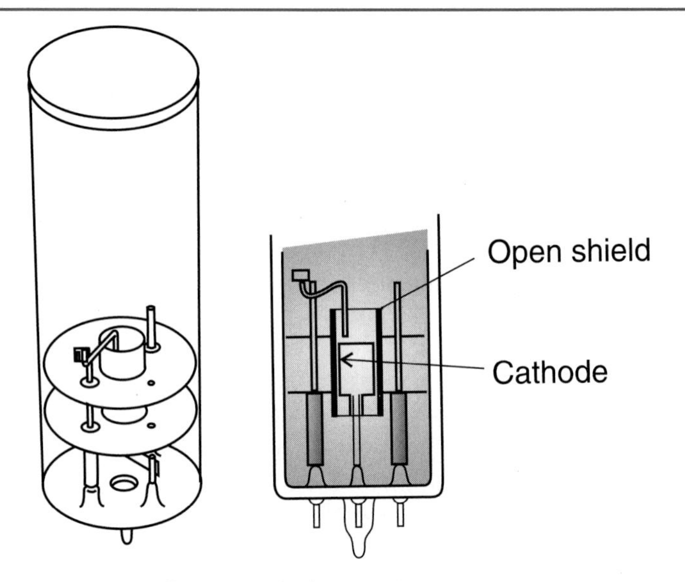

FIG. 3.9. Hollow cathode lamp. (Courtesy of Perkin-Elmer Corp.)

directed into the burner head and the flame. The burner, specially designed for AAS, has a long, flat-topped head positioned directly below and parallel with the beam of light from the lamp. The gases flow through a 10-cm-long slot in the top of the burner head so that a long, thin curtain of flame is produced. The light from the hollow cathode lamp passes through the full 10-cm length of the flame, thereby greatly enhancing the absorption of light by the ground state atoms in the flame. The narrowness of the slot also concentrates the atoms and results in greater efficiency of light absorption. The aspirator is more efficient than that in flame photometry; 10 to 20% of the aspirated sample is nebulized and enters the flame.

The function of the monochromator is to block light of wavelengths different from the desired resonance line from reaching the detector. The monochromatic light strikes the light-sensitive cathode of a photomultiplier tube, and the resulting current is converted to absorbance or concentration units. Like the flame photometer, the instrument must be calibrated with known standards.

One problem is unique to atomic absorption. While many of the analyte atoms in the flame are absorbing light from the hollow cathode lamp, others are being excited by the heat of the flame and emit light of the same wavelengths as the lamp. Light reaching the detector, then, is the sum of the light transmitted through the sample plus the light emitted by the sample in the flame. The two sources emit light at the same wavelengths, so they cannot be separated by a monochromator. In practice, the light from the lamp is pulsed either by a chopper that cuts off the light signal at brief intervals or by a modulator that periodically interrupts the electric signal. The amplifier recognizes only the pulsed signal originating from the source lamp and rejects the steady light emitted from the flame. This effectively separates the essential light signal from the extraneous light signal.

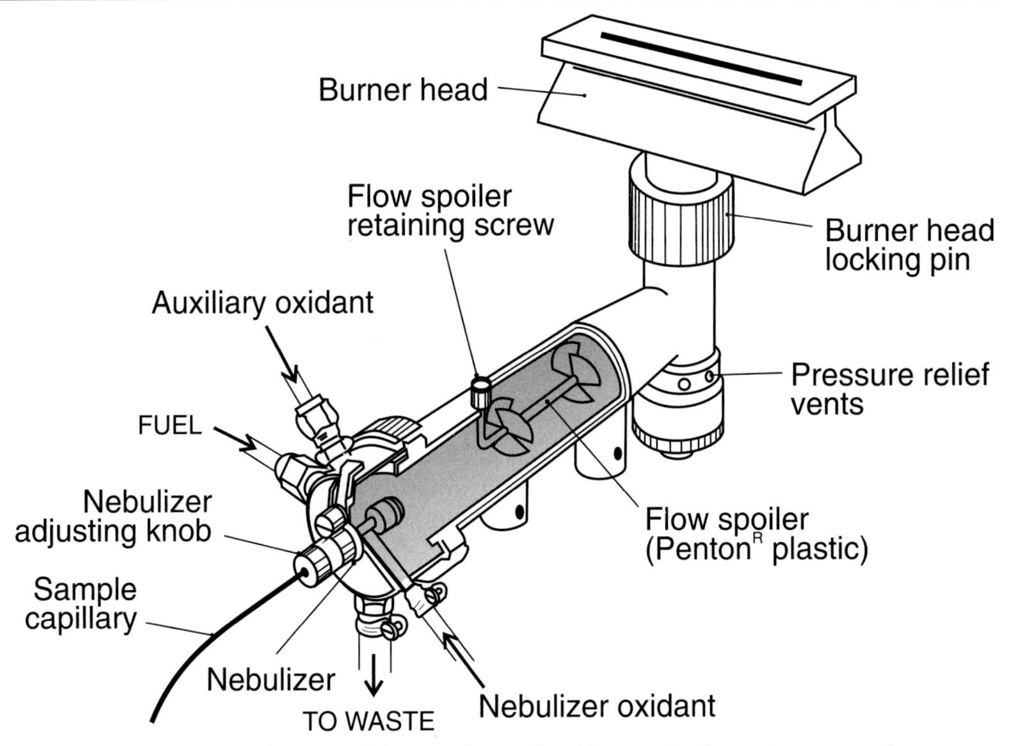

FIG. 3.10. Premix burner with laminar burner head for atomic absorption spectrophotometry. (Courtesy of Perkin-Elmer Corp.)

Special Applications

AAS is frequently used for detecting small amounts of an element when concentrations are too small to be accurately measured by standard chemical means. Consequently, several means of enhancing the signal have been introduced. For each assay procedure, the fuel:air ratio, gas flow rate, sample aspiration rate, and burner height are optimized so that the greatest possible number of free atoms is present in the optical path. The diluent with which the sample is mixed is also an important factor. For example, Ca as it occurs in serum may be ionized, bound to protein, or complexed with citrate, phosphate, or other anions. Some of the complexes (the phosphates) are only slightly dissociated by the heat of the flame; therefore, the Ca atoms in these complexes are not measured proportionally with other Ca forms. To eliminate this problem, the diluent for Ca determinations contains lanthanum chloride. Lanthanum, which is closely related to Ca in the periodic table, displaces Ca from the complexes in the solution and converts the Ca to a more readily dissociated and, therefore, measurable form.

A specific lamp is necessary for every element to be assayed by AAS. The alternative is a multielement lamp in which several different metals are incorporated to the same extent in the cathode. A single lamp is common for both Ca and Mg. Double-beam instruments are also used for AAS; the reference beam is directed around the flame.

Flameless Techniques

Increased analytical sensitivity in AAS has been achieved by incorporating flameless techniques into the system. A graphite furnace (a hollow carbon tube) replaces the burner head and is positioned so that the light beam is directed through the hollow cylinder to the detector. The sample is inserted into the cylinder through a port in the cylinder wall. An electric current is applied in three stages: small at first to evaporate the liquid, increased to ash the sample, and then large to instantaneously vaporize the entire sample and release the whole cloud of atoms in one burst. The absorption of light from the hollow cathode lamp occurs for one brief instant, too quickly for digital displays or meter needles to respond. A recorder is used to record the signal, a sharp peak whose height or area is proportional to the concentration of atoms in the sample. The release of the atoms in an instantaneous cloud enhances the sensitivity of the method by as much as a thousandfold over the flame method, with its steady flow of sample through the flame. Thus, metals that occur at low concentrations in biologic fluids can be measured, and much smaller samples are required for analysis. This small sample size can be of major importance when testing pediatric populations for metal poisoning or nutritional deficiencies. A series of standards is always included in the run.

In flameless systems, the surge of current used to vaporize the sample may also release a cloud of smoke or nonspecific vapor that blocks some light from reaching the detector. A background correction capability is advisable. For this correction, a deuterium continuum light may be also directed through the flame, and its light is monitored at a wavelength other than that of the assay. Because extraneous vapor blocks light of all wavelengths, the instrument compensates for the amount of light deflected at both wavelengths and measures changes that occur only at the wavelength of the resonance line of interest.*

Another more effective method of background correction is based on the Zeeman effect: when atoms are placed in a strong magnetic field, a profound change occurs in their ability to absorb radiant energy. Only vertically polarized light can be absorbed at the analytical wavelength, and horizontally polarized light can be absorbed only at wavelengths above and below the analytical wavelength (outside the window of the monochromator). One of the systems (Perkin-Elmer) uses this principle in the following manner: the magnet is rapidly turned on and off, and the light beam entering the furnace passes through a dichroic filter, which transmits only horizontally polarized light. Thus, while the magnet is turned on, the atoms cannot absorb the energy from the lamp, and only the extraneous "background" blockage of light is measured. Alternatively, when the magnet is off, the detector measures the total drop in transmitted light caused by sample absorption plus background. The difference between the two readings represents the light absorbed by the sample.

*Some instruments correct for background absorption at the wavelength of interest by passing light from the hollow cathode and deuterium lamps through the same monochromator. The lamps are pulsed at different frequencies, a procedure that permits electronic isolation of the two signals.

Fluorometry

Principles

Some molecules fluoresce, or emit light, after exposure to light of a certain wavelength. Light is emitted within a brief time (10^{-9} to 10^{-6} s) and is of lower energy (longer wavelength) than the light absorbed. The intensity of the fluorescence varies directly as the concentration of the solute; the sensitivity of the fluorometric assay may be 10^3 times that of absorption spectrophotometry. Because light of a particular wavelength is required for excitation of a given molecule and the emitted, fluorescent light is also of a wavelength characteristic for that molecule, a fluorescent analysis is more specific than the usual spectrophotometric method.

The process of fluorescence is illustrated in Figure 3.11. A molecule at the ground state energy level is excited by light absorption to a higher excited energy level (E_1). Vibrational energy losses (collisions, heat loss) drop the molecule to a lower, yet still excited, energy level (E_2); no light emission accompanies this drop. As the molecule quickly returns to the more stable ground state level, light is emitted. This process, when completed within 10^{-6} s, is known as *fluorescence.*

Some molecules undergo a slower transition involving changes in electron spin before emitting light. The time required for these changes is at least 10^{-4} s; the emitted light is called *phosphorescence*. Several Lanthanum series metal ions fluoresce in the time interval 10^{-6} to 10^{-4} s after reaching the excited E_2 level. This delay allows the emission of light at the fluorescent wavelength to occur when the exciting light is no longer present and is known as *time-resolved fluorescence*. At present, phosphorescence measurements are

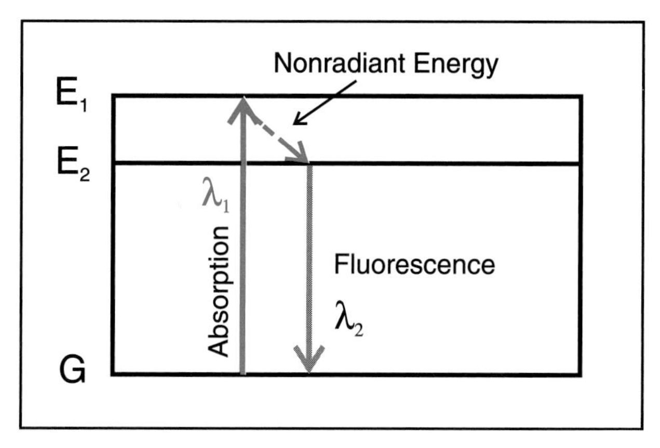

FIG. 3.11. Illustration of the principle of fluorescence. Light of the requisite wavelength (λ_1) striking a molecule at the ground state energy level (G) immediately raises it to an excited energy level (E_1). Some energy is lost by collision with other molecules or as heat, thus dropping the molecule to a lower energy level (E_2). The molecule falls back to the ground state level, thereby releasing energy as fluorescent light with a longer wavelength (λ_2).

not widely used in clinical chemistry laboratories, but time-resolved fluorescence has been applied in several automated immunoassay systems.

The number of organic molecules that can fluoresce is limited; they are usually cyclic molecules with conjugated double bonds (—C═C—C═C—). Some molecules that cannot fluoresce by themselves may be chemically converted to derivatives that are fluorescent. This conversion is usually accomplished by adding side chains that increase the freedom or mobility of the double-bond electrons. As an example, addition of an $-NH_2$ group enhances fluorescence, but adding an $-NO_2$ group depresses the ability to fluoresce by attracting the electrons and reducing their mobility.

Components

An instrument for measuring fluorescence differs from the other spectrophotometers reviewed earlier by having two monochromators, one to regulate the wavelength of the light striking the sample and one to isolate the wavelength emitted from the sample. The basic components of a fluorometer are shown in Figure 3.12. The lamp is usually either a mercury arc discharge lamp or a xenon arc tube so that light in the UV range is provided. If no excitation at wavelengths shorter than 350 nm is required, a tungsten lamp is adequate. The monochromators may both be diffraction gratings (a fluorescence spectrophotometer) or filters in a simple filter fluorometer. A fluorescence spectrophotometer has the advantage of providing narrow bandpass isolation at any wavelength of the spectrum for both exciting and emitted light, but a simple filter fluorometer is a far less expensive instrument that can meet most of the needs of the clinical chemistry laboratory. The primary filter (or grating) allows the passage of light of the proper wavelength for absorption by the molecule; the secondary filter transmits light of the specific wavelengths emitted by the sample. Because the light is emitted equally in all directions, the detector is placed at right angles to the beam of light from the lamp to the sample; this position prevents transmitted light originating in the lamp from reaching the

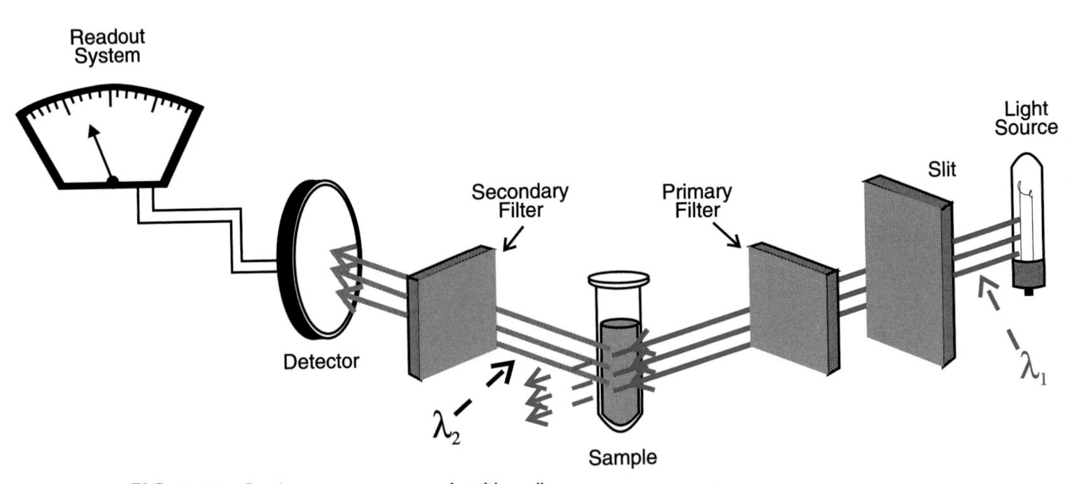

FIG. 3.12. Basic components of a filter fluorometer.

detector. As the whole sample is illuminated with light, the cuvet glows like a lamp, thus minimizing the effects of any surface scratches or imperfections. Glass cuvets can be used for exciting wavelengths greater than 350 nm in length; otherwise, quartz or fused silica is required. A phototube or photomultiplier tube is used as the detector, depending on the sensitivity required.

Factors Affecting Fluorescence

The intensity of fluorescence is theoretically linear with the concentration of the fluorescing molecule, but the following factors also affect the fluorescence:

1. *pH:* Changes in pH may induce changes in the ionic state of a molecule, with accompanying changes in fluorescing properties.
2. *Temperature:* Increased temperature enhances molecular motion, thereby increasing molecular collisions and decreasing fluorescence.
3. *Length of time of light exposure:* Because molecules are excited more quickly than they fluoresce, an increase in the number of excited molecules, and consequently an increase in fluorescence, occurs with time.
4. *Concentration:* Fluorescence increases with concentration only in dilute solution. In more concentrated solutions, the light emitted may be absorbed by other molecules of the same compound as it passes through the solution in the cuvet. If many absorbing molecules are present, a significant decrease in fluorescence (quenching) results.

Quenching of fluorescence may also rise from the action of other molecular species present in solution; this may involve absorption of either the exciting radiation or the fluorescent emission. Some common solvents, such as acetone and benzene, absorb light in the UV region, where excitation occurs. Dichromate, used for cleaning glassware, absorbs light at 275 and 350 nm and may easily interfere with the excitation or emitted light of fluorescing compounds. Extraction and elution solvents and glassware cleaning solutions must therefore be chosen with care.

Clinical Applications

Fluorometry is used in the clinical laboratory for certain classes of compounds, particularly when great sensitivity is required. Porphyrins, drugs, and several enzymatic reaction products are commonly measured by fluorometry. *Time-resolved fluorescence* has been most frequently applied to the immunologic measurement of proteins and peptide hormones.

Fluorescence Polarization

This method is based on the measurement of the intensity of polarized light emitted by fluorescent compounds. Most light sources are unpolarized, and the light is composed of waves that vibrate at all angles. Light waves vibrate up and down in only a single plane when polarized (like the waves on a pond as water moves up and down within the single plane of the surface). When a polarized light beam excites a rigid fluorescent molecule, the excited molecule cannot move, and the fluorescent light emitted is also polarized. Polarized light can excite large fluorescent molecules, which rotate slowly in solution and emit polarized fluorescent light as if they were rigid molecules. When excited by polarized light, the small excited molecules can rotate quickly and randomly in solution and emit the fluorescent light in an unpolarized manner. In brief, the size of a molecule is related to its ability to emit polarized fluorescent light. This characteristic is the basis of fluorescence polarization immunoassay methods used in the measurement of drugs and hormones.

An antibody directed against a drug is incubated with the patient's serum plus a small amount of the drug, which has been tagged with fluorescein. The drug molecules in the patient's serum compete with the tagged drug for binding sites on the limited number of antibody molecules. The greater the concentration of the drug in the patient's serum, the smaller is the amount of fluorescein-tagged drug bound to the large antibody molecules. After the incubation period, the reaction mixture is irradiated with polarized light. This irradiation excites the fluorescein, and light is emitted. When little drug is present in the patient's serum, the tagged drug molecules are bound by the large antibody molecules, which rotate slowly, and the emitted light is polarized. When a large amount of drug is in the patient's serum, the small, tagged drug molecules remain unbound and can rotate rapidly; consequently, the emitted light is not polarized. The amount of polarization present is a measure of the antibody-bound tagged drug, and this measurement is inversely related to the concentration of the drug in the patient's serum.

Chemiluminescence

Chemiluminescence is the term for the process of exciting molecules by chemical means and measuring the light emitted as the molecules return to their stable, unexcited state. When chemiluminescent reactions are catalyzed by enzymes, the reaction is termed *bioluminescence*. Light can be emitted either directly from the reacting molecule or indirectly as energy is transferred to a molecule capable of light emission. Because the excitation is carried out by chemical means, the instruments consist only of a photodetector (a luminometer) to measure the intensity of the light emitted from the reaction tube. Although the determination of total chemically bound nitrogen has been made for many years by NO_2 chemiluminescence, novel immunoassay designs are the most common clinical chemistry application of chemiluminescence today.

CHROMATOGRAPHY

Chromatography is the separation of soluble components in a solution by differences in migration rate as the solution (mobile phase) carries the solutes over or through a stationary phase. The separation may use one or more of the following physicochemical properties, depending on the particular chromatographic system: differences in adsorption to the medium (the sorbent*), differences in the relative solubilities between a liquid (stationary phase) coated on inert particles and the liquid (mobile phase) percolating through the column of coated particles, differences in ionic attraction to the sorbent, and differences in penetration into a sorbent gel as a result of different molecular sizes (molecular sieving). Various types of chromatography are described in the following sections.

Chromatography is a highly efficient separation technique and is widely used in the clinical chemistry laboratory for the identification in serum or urine of drugs (Chap. 14), sugars (Chap. 7), and amino acids (Chap. 16). The process may be used for the purification of materials, for identification of compounds, and for quantification.

Thin-Layer Chromatography

Thin-layer chromatography (TLC) is a simple technique that has been used in laboratories for separation and identification of urine sugars, amino acids, drugs, and other groups of compounds. In TLC, the stationary phase consists of a thin layer of a finely divided substance applied to a sheet of glass or plastic backing. Sorbents commonly used, and commercially available as finished plates, include alumina, silica gel, and cellulose. Samples are applied as small streaks or spots near one edge of the plate. The application area must be kept small; if necessary, repeated applications may be superimposed upon each other, with drying between applications. The plates are then placed in a closed chamber containing an appropriate solvent mixture (mobile phase) (Fig. 3.13). The TLC plate is positioned so that the applied samples are along the lower edge, slightly above the level of the mobile phase. As the solvent moves up the plate, the samples move upward too, separated by differing attractions between the sorbent and the components of the solvent mixture. The mobile phase is usually an aqueous solution, which may be modified with NH_4OH or acetic acid (to adjust pH) or with water-miscible organic solvents (for example, methanol, ethanol) to increase the solubility of the components of the sample spotted on the TLC plate. When the solvent front has almost reached the top, the plate is removed from the chamber. The solvent front is marked and the plate is dried. The spots sometimes are colored and therefore visible, but more likely, they must be visualized by viewing under UV light or by spraying with color-producing reagents. The spots may be identified by their color and by their relative distance of migration from the point of application (*R_f value*), which is a characteristic for each substance when chromatographed in a certain solvent system. The R_f value is calculated by the following formula:

$$R_f = \frac{\text{distance traveled by compound from the origin}}{\text{distance traveled by solvent from the origin}}$$

Known compounds should be applied to the same plate as the unknowns for

**Sorbent* is the term applied to the stationary phase, which may be liquid or solid. Solid sorbents are usually chosen for *adsorption, ion exchange*, or *molecular sieving* properties; a liquid sorbent is usually chosen for the *differential solubilities* of substances between it and the mobile phase.

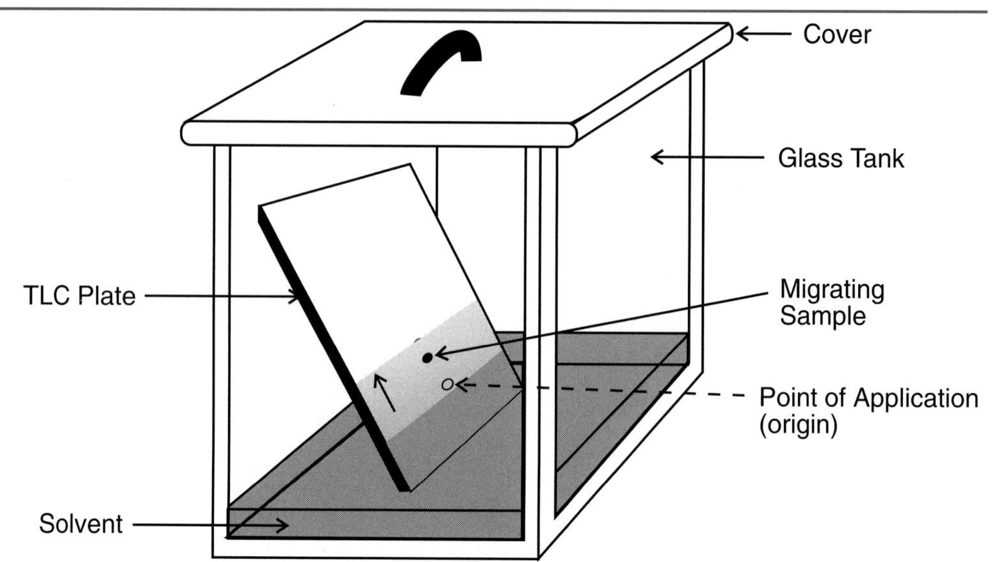

FIG. 3.13. Apparatus for thin-layer chromatography (TLC). Note that the TLC plate is the stationary phase and the solvent is the mobile phase. Components within a sample are carried up the plate with the solvent. The components are separated during chromatography when they have different solubilities in the mobile phase or different strengths of adhesion to the stationary phase.

comparison and confirmation of identification. A typical TLC plate is shown in Figure 3.14.

If the identity of the substance is known, a series of standards may be applied to the plate to estimate the concentration of the unknown specimen by comparing the size and density of the spot with the standards. A more quantitative result may be obtained by scraping the spots off the plates, eluting the colored compound, and reading the absorbance in a spectrophotometer.

Liquid Chromatography

The original work on chromatography used an inert material (sorbent), similar to powdered sugar, packed in a column. Plant pigments dissolved in heptane were placed on the column and strongly adsorbed to the sorbent. A more polar solvent (a mixture of heptane and ethanol, the mobile phase) was passed through the column, and the pigment bands began to separate on the basis of their relative solubilities in the heptane-ethanol mixture. The more soluble pigments traveled farther down the column. This separation of components by differential solubility is also known as *partition* chromatography. The final separation product is the *chromatogram*. The chromatogram may be extruded from the column and the different bands cut from it, or the materials may be eluted with an appropriate solvent and the liquid collected in a series of tubes. The resolution obtained in liquid chromatography depends on the pH and ionic strength of the mobile phase and on the relative solubilities of the constituents in the two phases.

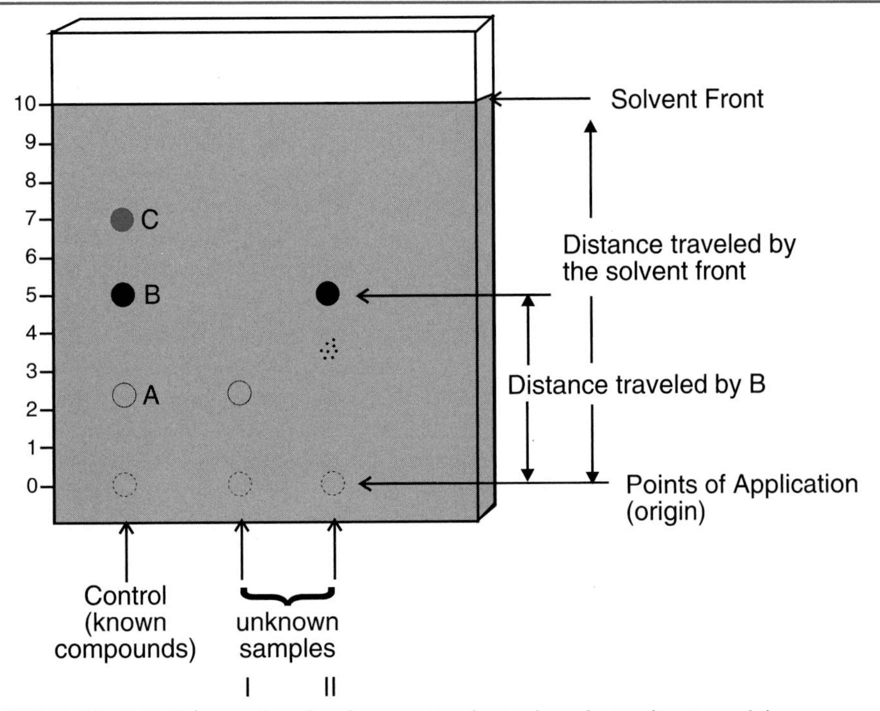

FIG. 3.14. A TLC plate after development in the tank and visualization of the spots. Sample I contains compound A; Sample II contains compound B and an unidentified compound. R_f for the compound B = 0.5 in this solvent system.

In liquid chromatography, the mobile phase may be passed through a vertical glass column by gravity alone or assisted by a low-pressure pump (column chromatography); it may also be forced through a narrow-bore stainless steel column by a high-pressure pump (high-performance liquid chromatography, HPLC). Because the principles used in separating components are identical in column chromatography and HPLC, these techniques are reviewed together after a brief description of the apparatus used for each.

Column Chromatography

As the name implies, the system usually consists of a vertical glass column that has been carefully filled with an evenly distributed packing material that has no channels coursing through any part of it. The column is equilibrated with solvent before the sample is gently layered on top. Eluting solution is introduced above the sample, flows through the column by gravity, and carries along the components of the sample according to relative solubilities. The reservoir of eluting solvent is raised above the top of the column to create a "head" of pressure for faster flow. The eluting fluid may be unchanged throughout the separation, or a gradient flow may be used by progressively increasing the concentration of one component of the eluting solution during the separation.

A low-pressure pump may be introduced into the system before the column to increase the rate of eluent flow.

High-Performance Liquid Chromatography

In HPLC, a high-pressure pump is used to force the solvent and sample through a relatively short (often about 30 cm) narrow-bore column (2 to 4 mm). Pressures of about 1000 to 3000 psi are frequently used in routine separations.

A diagram showing the major components of a typical HPLC system appears in Figure 3.15. The elution solvent must be filtered and degassed before use and is constantly stirred as it is forced into the system by a reciprocating pump at a constant flow rate (for example, 2 ml/min). A high-precision reciprocating pump maintains the smooth, constant flow. Various packing types and particle sizes are available commercially in prepacked columns.

The column contains the packing appropriate for the particular separation. The packing is fine (3 to 10 μm in diameter) to greatly increase the surface area per unit volume. These columns may be packed by the user but are usually purchased as prepacked units because of the expertise required for packing.

The most common detector is a UV photometer. Most models are set to read absorbance at 254 nm. A high-intensity emission wavelength of a mercury lamp is used as a light source. Other detectors include UV-visible photometers (which use both deuterium and tungsten lamps as light sources and allow the user to select specific wavelengths between 200 and 700 nm), refractometers, and fluorometers. Specialized detectors that measure radioactivity or oxidation-reduction (electrochemical) characteristics of the sample are available.

Normally, the mobile phase is pumped through a longer coil (A), as shown in Figure 3.15. When a sample is to be injected, the solvent flow is bypassed

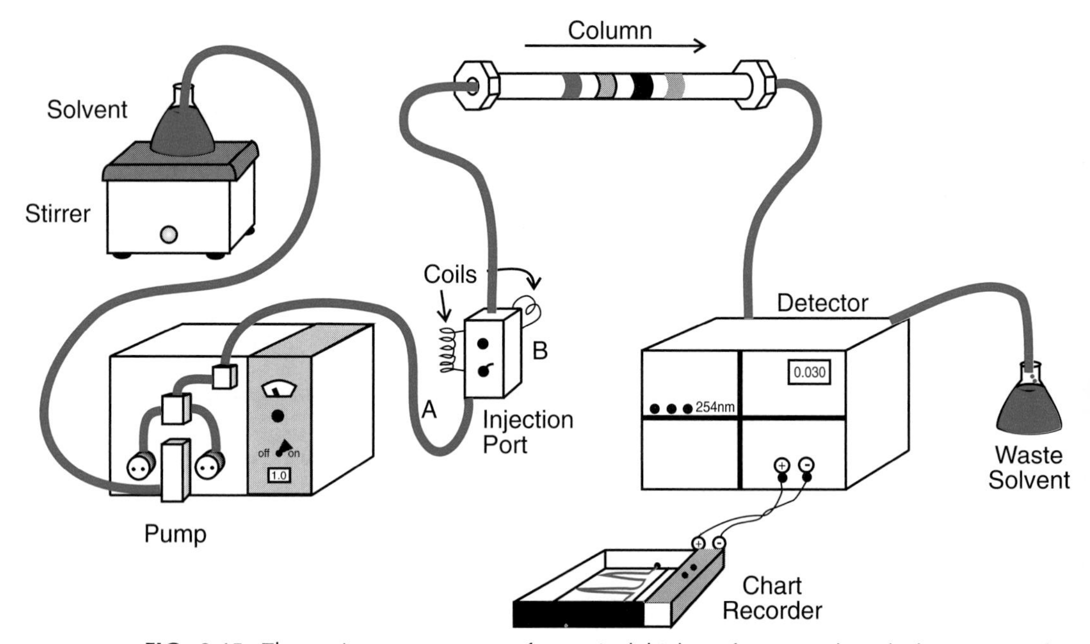

FIG. 3.15. The major components of a typical high-performance liquid chromatography system.

through the shorter coil (B). Then the sample (a small volume, frequently less than 10 μl) is injected into the longer coil. When the flow is returned to the regular route through the longer coil, the sample is swept by the solvent into the column. As the separated components elute from the column, they pass through the detector, such as a UV detector, where they produce an absorbance change. This signal is recorded as a chromatogram, with absorbance of the eluting mobile phase plotted against time (see Fig. 14.3).

The main advantages of HPLC are good resolution, which may be obtained by the proper selection of packing, solvent, and flow rate, and fast separation requiring minutes rather than hours for each sample.

An internal standard (similar in composition to the analyte) is generally added to each sample to compensate for variations in the size of sample injected, solvent evaporation, and other possible irregularities. The ratio of peak height between the analyte and internal standard on the chromatogram is plotted against the concentration of the analyte-standard to construct the standard curve and to analyze unknown patient samples (Fig. 3.16).

$$\text{Peak height ratio} = \frac{\text{Height of analyte peak}}{\text{Height of internal standard peak}}$$

Gel Filtration (Size-Exclusion Chromatography)

Gel filtration (size-exclusion chromatography) separates molecules according to their size (molecular weight), although the shape of the molecule affects the filtration to some extent. The gels are in the form of beads containing a network of openings or pores through which small molecules may pass. The beads are available with different pore sizes; all molecules larger than the pores are excluded from the beads. Molecules much smaller than the openings enter the pores readily. The intermediate sizes may also flow into the pores, although less frequently. The molecules too large to enter the pores flow with the solvent between the beads and are eluted quickly. The small molecules take a long, devious route by moving in and out of many beads. Intermediate sizes enter fewer pores; their route is intermediate in length, and the time required for their passage through the column is a function of their molecular size. Three types of gel are used for molecular filtration: cross-linked dextran (Sephadex, Pharmacia), agarose (Sepharose, Pharmacia), and polyacrylamide (Bio-Gel P, Bio-Rad).* Each comes in a variety of pore sizes for separating compounds within different ranges of molecular weight.

The gels listed cannot withstand the high pressures involved in HPLC; the most common size-exclusion packing for use under pressure is a silica-based product (Waters), which is more resistant to pressure. Pressures greater than 800 psi tend to compact the gels, so they are seldom used in size-exclusion chromatography.

*Complete names and addresses of sources of materials and equipment are given in Appendix 4.

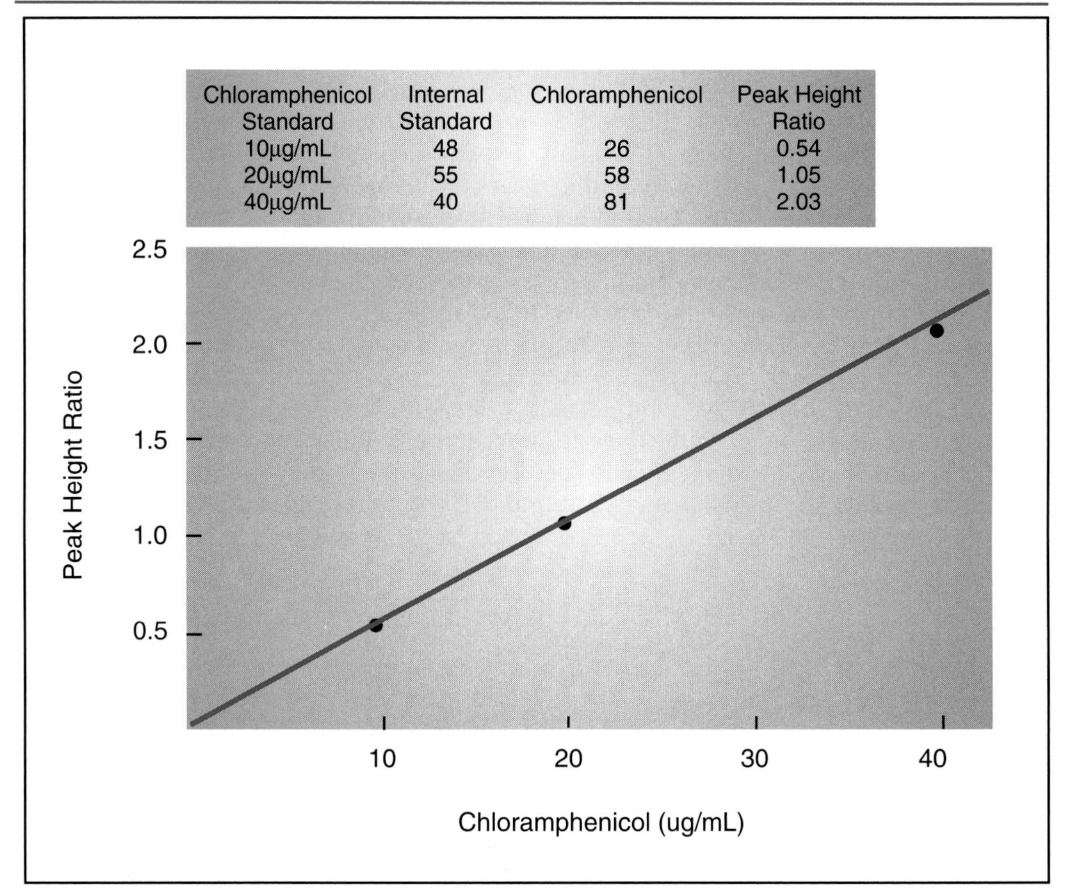

Chloramphenicol Standard	Internal Standard	Chloramphenicol	Peak Height Ratio
10µg/mL	48	26	0.54
20µg/mL	55	58	1.05
40µg/mL	40	81	2.03

FIG. 3.16. A standard curve for an HPLC analysis of the antibiotic chloramphenicol. The values of peak heights listed under chloramphenicol and internal standard are in arbitrary units. A sample chromatogram for the 20 µg/mL standard of chloramphenicol is shown in Figure 14.3.

Ion Exchange Chromatography

Molecules can be separated by their ionic charge in a process known as ion exchange chromatography. The sorbent or stationary phase consists of polymers with covalently bound ions. In cation exchange resins, these tightly bound ions are negatively charged and are associated with positive ions that are loosely attached by electrostatic charges. The positively charged substances to be separated from a mixture are first adsorbed to the sorbent, thereby displacing the cations present in the resin. The mixture is buffered at a pH that facilitates binding, and the column is washed with the same buffer to remove the nonbinding fractions. A slight change in the pH or ionic strength of the eluting buffer causes the separation from the column of the adsorbed cationic substance as a result of competition from the buffer ions for the anionic sites. The buffers must be carefully chosen for good resolution. An anion exchanger operates in exactly the same way, except that its covalently bound ions are positively charged to attract the anions from the solution.

Partition Chromatography

In partition chromatography, the solute components are distributed between the mobile and stationary phases according to their relative solubility in each. Molecules that are readily soluble in the mobile phase are eluted quickly, whereas those with greater affinity for the stationary phase are retained on the column for a longer time.

Historically, the early separations used a polar stationary phase with a less polar organic mobile phase, and this pattern has become known as *normal phase* chromatography. For the separation of drugs and other nonpolar substances, *reverse phase* chromatography has proved more useful. Reverse phase systems use a nonpolar stationary phase with a relatively polar mobile phase. In the clinical chemistry laboratory, a common reverse phase chromatography system would involve octadecyl-silane bonded to a silica stationary phase (C18) and a methanolic aqueous buffer as the polar mobile phase.

Gas Liquid Chromatography

Gas liquid chromatography (GLC) or gas chromatography is designed to separate and quantitate volatile materials in a heated column consisting of an inert support material, such as diatomaceous earth, coated with the stationary phase. The stationary phase must be a liquid at the elevated temperatures (up to 400°C) of the column, nonvolatile, and nonreactive with the samples or solvents moving through the column. The mobile phase is an inert carrier gas (for example, helium or nitrogen) that sweeps the volatile compounds through the column to a detector. The more volatile molecules, those with the lowest boiling points, tend to be swept most rapidly along by the carrier gas. Those that are soluble in the liquid phase are retained for periods in the liquid, and their forward movement is slowed. Therefore, both volatility and solubility differences affect the rate of flow of sample molecules through the GC column.

A diagram of the main components of a gas chromatograph is shown in Figure 3.17. The carrier gas, which must be pure and dry, flows through the system to the detector. The sample enters the system through the injection port, a heated chamber at the beginning of the column. The temperature at the injection chamber is high enough to flash volatilize the sample so it can be swept into the column by the carrier gas. The column may be 1 to several meters in length and about 0.5 cm in diameter, although both dimensions may vary widely. A separation may be run isothermally, with a constant oven temperature maintained, or the temperature may be steadily increased to speed the forward movement of the less volatile components.

The search for greater resolution led to the development of quartz capillary tubing. These capillary columns have a layer of the stationary phase coated on the inside surface of an exceedingly thin quartz tubing. The high tensile strength of the columns allows lengths of as many as several hundred meters to be wound into a bundle less than 15 cm in diameter (the same size as a conventional packed column). Capillary columns are used frequently for the analysis of pesticide residues and for drug screening (Chap. 14).

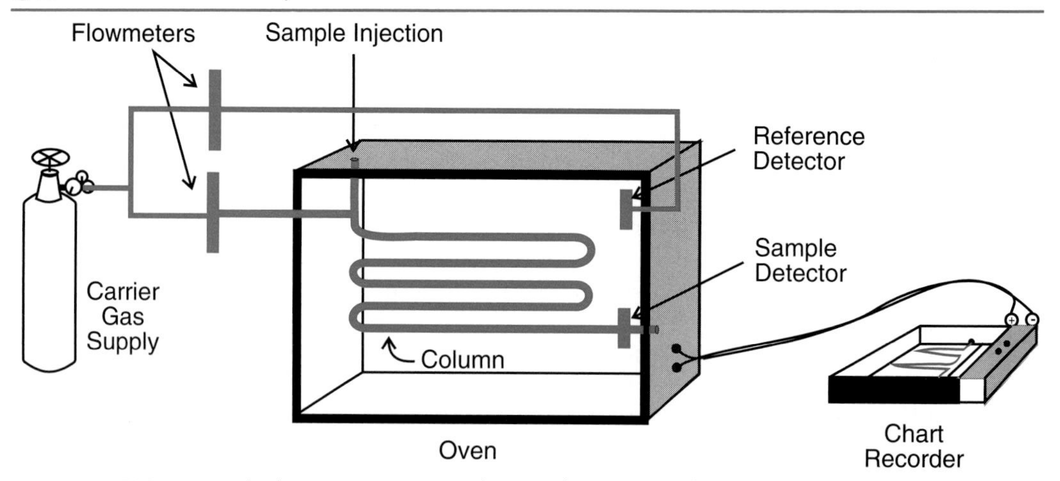

FIG. 3.17. The basic components of a gas chromatograph.

After separation on the column, the sample components enter the detector. There are several types of detectors, but the most commonly used are the thermal conductivity and flame ionization detectors. The thermal conductivity detector consists of a metal block with two separate filaments similar to an incandescent lamp filament heated by an electric current. As the carrier gas flows over the reference filament, the gas conducts heat away from the filament, maintaining it at a cooler temperature. When a sample constituent eluted from the column is mixed with the carrier gas, the conductivity of the gas is reduced, the cooling effect on the sample filament is lessened, and its temperature rises. The difference in temperature changes the resistance and causes a current flow that is amplified and recorded as a "peak" by the recorder. The peak is proportional in size to the quantity of the eluted compound.

The flame ionization detector contains a small burner in which hydrogen gas and air are mixed with the gas from the GC column and burned. As a carbon-containing compound from the sample is burned, negative ions are formed that are attracted to a positively charged wire. The current produced is measured and recorded.

A more sophisticated detector, the mass spectrometer, can be combined with GLC. Because of recent advances in this technology, the reliability has greatly increased and the cost has decreased: consequently, gas chromatography/mass spectrometry (GC/MS) systems are being purchased by larger clinical laboratories. In GC/MS, as the components of a mixture elute from the column into a vacuum chamber of the mass spectrometer, they are bombarded by a stream of electrons. The resulting collisions knock electrons from the sample molecules to produce positively charged ions, which rapidly fragment into a collection of lower-mass, positively charged daughter ions. This group of charged particles is passed through an electrical and/or magnetic field, where it is sorted by molecular weight. The particles then impinge upon a photomultiplier tube, where they are counted. These "counts" are presented as a plot (mass spectrogram) of the number of particles (y-axis) versus molecular weight (x-axis). Each type of molecule gives a unique and reproducible fragmentation pattern, thereby allowing unambiguous identification of the components of a mixture. GC/MS frequently uses capillary columns to achieve high resolution of the sample components, before they elute into the mass spectral detector. GC/MS systems offer high sensitivity and specificity and are used for screening urine specimens for drugs of abuse (Chap. 14) and for identifying low-molecular-weight branched-chain ketoacids, as are characteristic in the metabolic defect called maple syrup urine disease (Chap. 16).

Each of the detectors senses sample constituents as they emerge from the column. The signals are recorded as peaks: the distance from the solvent peak (eluted first) to sample peak designates the *retention time* for the compound, the time required for that constituent to pass through the column. For a given column with its packing, retention time for a substance is a constant, as long as all conditions (temperature, flow rate) are unchanged. The area of a peak is proportional to the amount of the substance eluted, and thus to its concentration in the original sample. Thus, the position of a peak identifies a substance; the area of the peak is used for quantitation. Because most peaks are symmetric, peak height is often used for convenience.

Sample injections of 1 μL in size are not sufficiently reproducible, especially when they are in a volatile solvent. To compensate for this variability in sample size, an internal standard similar to the constituent to be quantitated is added to each sample-solvent mixture at a fixed concentration. If the sample size is slightly increased, all peaks from its constituents, including the internal standard, will be proportionally increased. Standard solutions containing known amounts of the compound being assayed are also chromatographed with an internal standard. A standard curve is drawn, using the peak-height ratio of sample to internal standard as the ordinate and concentration as the abscissa. The concentration of the unknown samples may be read directly from the curve.

ELECTROPHORESIS

Principle and Components

Electrophoresis is the process of separating the charged constituents of a sample by means of an electrical current. Charged particles placed in an electrical field migrate toward either the anode or cathode, depending on their net charge. The rate of migration varies with the net charge, the strength of the electrical field, and the weight or size of each particle. In electrophoretic analysis, the apparatus is designed so that an electrical current passes between the two electrodes and through the sample and its support medium. The current flow is partially carried by the migration of the charged components of the sample. In a constant electrical field, a separation of particles with different net charges or weights occurs because of different migration rates.

As shown in Figure 3.18, an electrophoresis chamber consists of two buffer compartments separated by a dividing wall: one side contains the anode and the other the cathode (platinum wire or carbon electrodes). Each compartment is filled to the same height with a buffer. An electrical "bridge" across the top of the dividing wall is created by the support material for the separation. The support material may be composed of an agarose gel, an acrylamide gel, or a

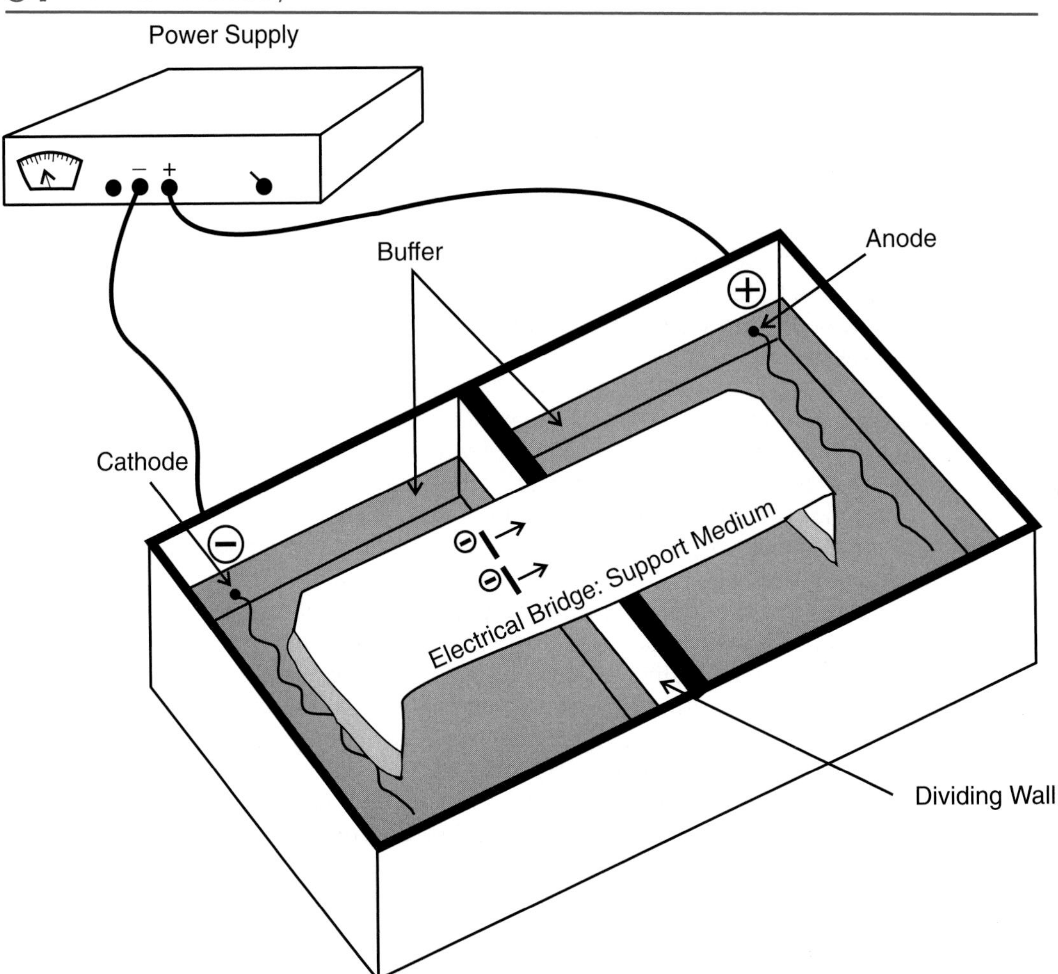

FIG. 3.18. Diagram of an electrophoresis chamber. The two buffer compartments are completely separated from each other except for an electrical bridge created by the moist electrophoretic support material.

moist cellulose acetate membrane that is in contact with the buffers in each compartment. The only electrical connection between the two compartments is by way of the support material bridge. The sample is gently applied to the surface of the bridge, and a voltage is applied to the electrodes. The current is carried across the bridge by both the buffer ions and the migrating charged particles in the sample.

When a buffer of pH 8.6 is used to electrophoretically separate a serum sample, all the serum proteins carry a net negative charge and migrate toward the anode. Albumin is relatively small and carries the largest charge; it therefore moves the fastest. The gamma globulins are large proteins and have the smallest net charge; they move the least distance (Chap. 9). Electrophoresis is often used to separate and analyze DNA and RNA, which are highly negatively charged molecules that migrate toward the anode and separate on the basis of length and size (Chap. 16).

The actual distance of migration and the resolution of the separate bands are affected by many variables in the system. On large-pore support materials, the separation is primarily determined by the charge on the molecule (for example, cellulose acetate, agar, and paper). On a fine-pore, highly cross-linked gel, the separation is affected additionally by the size of the molecules because the gel acts as a molecular sieve (for example, acrylamide, agarose). Serum proteins separate into 5 bands on cellulose acetate, 10 to 15 bands on agarose gel, and more than 20 bands in acrylamide.

Selection of the proper buffer is important; it must provide the desired pH and must not react with the sample. As the buffer concentration is increased, the migration rate is decreased, but resolution of the components is improved. The reduction in sample migration distance is caused by competition from the increased number of buffer ions available to carry the current. As the *ionic strength* of the buffer (the number of available ions) is increased, more current can be carried. The increased current results in elevated temperatures, which may denature some proteins and affect the separation.

The power supplies used for electrophoresis can be adjusted to provide either constant current or constant voltage. When the voltage is kept constant during the separation, the current gradually increases, accompanied by an increase in temperature. For short-term electrophoresis (one-half hour or less), constant voltage is often used. For longer periods of time, constant current is preferable because the temperature does not increase appreciably during the run. Increased temperature, as previously mentioned, may result in poorer resolution of the separated components.

The support medium, buffer, and electrical field affect the final separation of components during electrophoresis. Other influences, not as readily apparent, also modify the results of electrophoretic migration. *Electroendosmosis* is a process that takes place concurrently with sample migration. The support medium (cellulose acetate, for example) develops a negative charge itself when exposed to an alkaline buffer. Associated with the negatively charged groups on the membrane are positive ions from the buffer. As these positive buffer ions move toward the cathode, they exert a counterflow to the protein molecules that are migrating toward the anode. All the protein molecules thus are slowed in their progress, and the most weakly charged proteins, the gamma globulins, may be moved with the buffer, thereby ending their migration at a point closer to the cathode than the point of application. The net effect of protein migration and electroendosmosis is similar to that of rowboats headed upstream against a river current: the strongest oarsmen make progress against the current if their forward motion is greater than the river flow, and the weakest oarsmen are moved downstream in spite of their rowing in an upstream direction. Between these two extremes are the boats that make little progress or can only hold their own against the current. The force of electroendosmosis does not prevent separation of the charged molecules, but it does affect their final position on the membrane.

Another force affecting the migration of the protein molecules is *wick flow*. The evaporation of water from the membrane is constant at all points above the liquid level in the cell. As the water evaporates, buffer flows upward from the cell through both ends of the membrane to replace the lost moisture. This replenishment results in a buffer flow toward the center of the membrane from the ends; the flow moves faster near the ends and more slowly as it approaches the center. Wick flow thus adds distance to the migrating molecules approaching the center, and retards those moving outward from the center. The net effect is a compression of the final pattern, with a reduction in the separation between the components. A good electrophoretic procedure, then, is a satisfactory compromise among all the competing forces.

Electrophoresis is versatile analytical tool because the factors influencing the migration and resolution of the charged particles can be altered and optimized easily. Such agents as sodium dodecyl sulfate (SDS) can mask the natural protein charges and allow separation on the basis of protein molecular size alone. Alternatively, in the method of isoelectric focusing (IEF), proteins migrate within a pH gradient in the support medium until no net charge remains. IEF electrophoresis is ideal for separating proteins of identical size, with different net charges. When serum samples are sequentially analyzed by IEF in one dimension and by SDS electrophoresis in a second dimension, they can be resolved into hundreds of bands.

Densitometry

A densitometer is a specialized colorimeter designed to scan and quantitate electrophoresis patterns. Instruments read various gels, as well as cellulose acetate membranes, which are used here as an illustration of the principle.

After the serum proteins have been separated by electrophoresis, the proteins are precipitated and visualized by immersion in a fixative dye, such as Ponceau-S. After removal of excess dye and dehydration of the membrane, the membrane is cleared (from an opaque white) by carefully timed immersion in a dehydrating agent and a mixture of organic solvents. Heating the membrane to evaporate the solvents leaves a thin, transparent film with the protein bands stained by the red dye (see Fig. 9.9).

The cleared membrane is then scanned in a densitometer that contains a light beam passing through a filter to a photodetector. As the clear portion of the membrane passes through the light beam, the instrument records a baseline level. As the bands pass through the light beam, less light reaches the detector, and the instrument records the absorbance of light of the protein bands as a peak on the graph. Serum protein separations on cellulose acetate membranes generally have five bands that appear as five peaks on the densitometer graph (see Fig. 9.8). The densitometer also integrates the area under each peak so that the relative concentration of each fraction (% of total protein) is recorded; if the concentration of the total protein is known, the actual concentration of each protein fraction (g/dL) is calculated.

Clinical Applications

Electrophoresis is used in the clinical laboratory for separation of serum proteins, isoenzymes, hemoglobins, DNA, RNA, and other classes of macromolecules. The separation is similar in each case, although buffer, pH, voltage, and support media may differ. The method of visualizing the bands is determined by the constituent to be measured.

ELECTRODES

A major advance in clinical chemistry was the development of the pH electrode and, subsequently, other electrodes that selectively measure the concentration of various analytes. Many analytes can undergo electrochemical oxidative-reductive reactions, and when they do, the electrical characteristics of the chemical reaction can be measured with an electrode-containing instrument. Electrical current, voltage (also called the potential), and resistance or conductivity have been measured by electrodes in clinical chemistry laboratories; the most common electrode is designed to measure the potential or voltage.

A brief overview of electrodes is presented in the next section. For more detail, the reader can refer to textbooks on laboratory instrumentation.[1-3]

Principle

Analytical electrochemical methods used in a laboratory involve a minimum of two linked electrochemical reactions: an analytical electrode and a reference electrode. A common electrochemical reaction is as follows. When a silver wire is immersed in a concentrated solution of potassium chloride, an electrochemical reaction immediately begins. Silver atoms give up electrons to the wire and become silver ions, which rapidly react to form silver chloride. When this reaction occurs, the silver wire develops a negative charge because of the excess of electrons. The formation of silver ions and silver chloride is soon halted, because the negative charge of the wire attracts, binds, and neutralizes the positively charged silver ions. The negative charge prevents additional silver atoms from giving up electrons. To measure the potential (or voltage) associated with the silver atoms giving up electrons, one must perform this measurement when the electrons have not accumulated within the silver wire. The potential of the silver wire to give up or "push away" its electrons can only be measured by electrically connecting this wire to a second electrochemical reaction that will accept the electrons. Although the second reaction may not impartially accept the electrons, it may "pull" the electrons of the silver wire down a potential gradient and consume the electrons in an electrochemical reaction. As a consequence, when the voltage of two linked electrochemical reactions is measured, the sum of the "push" and "pull" influences on the electrons from both reactions is recorded; the contribution of the silver wire alone cannot be directly measured. Thus, a fundamental concept in electrode design follows: *each analytical electrode must be electrically connected to a second electrode, called the reference electrode*. Two electrochemical reactions are completely connected when the metals are connected by a wire to allow for the flow of electrons and the solutions are connected by a salt bridge to allow for the flow of ions.

A second fundamental concept is that *the voltage or potential contributed by a single electrode (half-cell) depends on the concentration of the ions participating in its electrochemical reaction.** The potential for silver atoms to give up electrons depends on the concentration of silver ions in the solution. In this special case, the concentration of silver ions is tightly controlled by the concentration of chloride available to form silver chloride, thereby allowing the voltage from this electrode to depend on the chloride concentration (Fig. 3.19).

A third fundamental concept is that *a voltage or potential applied to a membrane can be added to the potential contributed by the electrochemical reactions.* Membrane voltages can be added to half-cell voltages in the same

*A more accurate statement is that the *ion activity* actually modifies the half-cell potential. Ion activity is closely approximated by ion concentration in dilute solutions. The influence of ion activity on half-cell potential is described by the Nernst equation.[5]

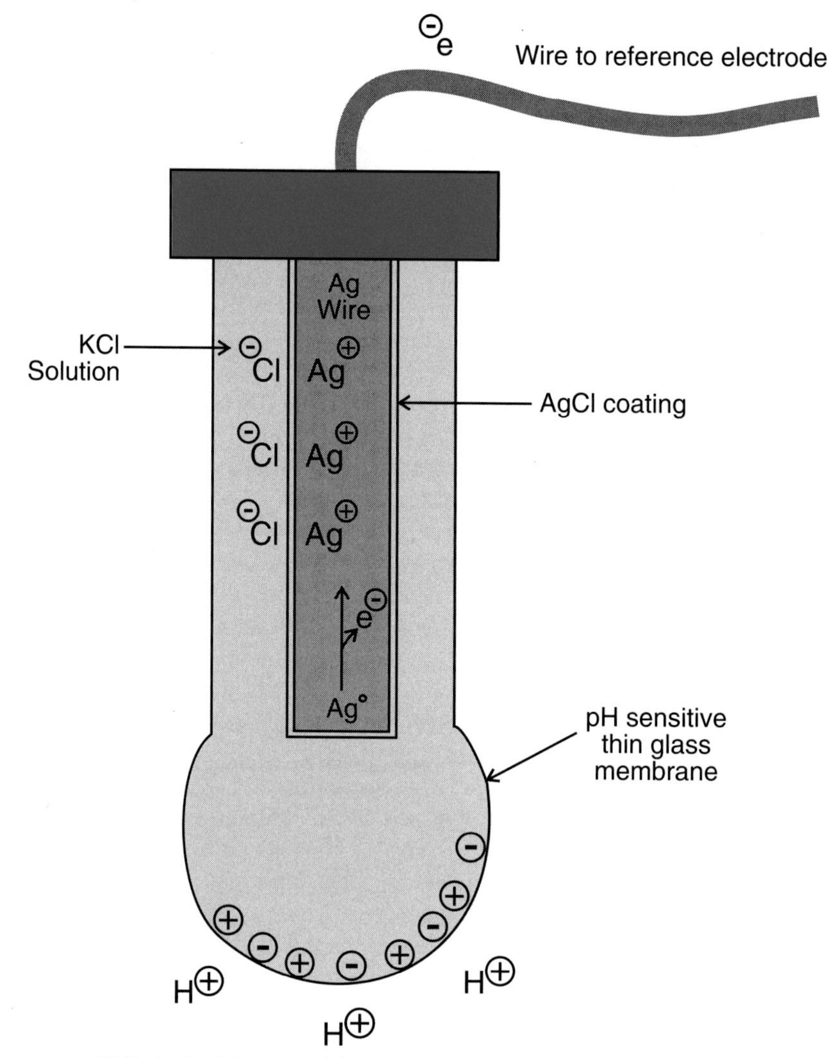

FIG. 3.19. Diagram of the components within a pH electrode.

manner that two batteries can be aligned to obtain a greater voltage in a flashlight. If the silver-silver chloride reaction of the example is contained within a pH-sensitive glass membrane, as the pH of the solution surrounding the membrane changes, the membrane will impart an extra voltage or potential to the half-cell reaction. This arrangement is commonly used in a pH electrode and pH meter design.

This explanation of the principles of electrode analysis is basic and greatly simplified. These three fundamental concepts can be seen in the design and application of all forms of electrodes.

Electrodes within instruments can be categorized as either *direct* or *indirect*. During direct analysis, an undiluted sample is in contact with the electrode, whereas predilution of the sample occurs during indirect analysis. Each method is subject to different biases under specific conditions.

Reference Electrodes

Some half-cell potentials are extremely stable and easy to reproduce. Because the potential of these half-cells can be established accurately, they serve as reference potentials against which unknown voltages are measured. The two most widely used reference electrodes (reference half-cells) are the silver-silver chloride electrode previously mentioned and the calomel (Hg_2Cl_2) electrode. The calomel electrode contains mercury in contact with mercurous chloride (calomel), which in turn is in contact with a KCl solution of known concentration. The silver-silver chloride electrode consists of a silver wire coated with a deposit of silver chloride, which is immersed into a solution of known KCl concentration. Both reference electrodes possess a small ceramic plug or fiber saturated with the KCl at the tip of the glass envelope to carry the ionic current into the surrounding solution and complete the circuit.

When a reference electrode and the appropriate analytical electrode are immersed in a test solution and the circuit is completed, the voltage difference between them can be measured. The potentiometer is calibrated against a solution of known ion concentration, and the unknown solution is then introduced. The measured voltage changes are related to the change in ion concentration. For H^+, a change of 1 pH unit causes a voltage change of 59.15 millivolts at 25° C.

Frequently, a miniature reference electrode can be included in the glass cylinder with the measuring electrode. This "combination" electrode is convenient and requires a smaller volume of solution than that required when two separate electrodes are used.

The pH Electrode

The first ion-selective electrode to be widely used was the pH electrode designed to measure hydrogen ion activity (Fig. 3.19). This electrode was made possible by the development of a special pH-sensitive glass. When a thin membrane of this glass separates two solutions of differing hydrogen ion concentrations, a hydrogen ion exchange takes place in the outer hydrated layers of the glass, thereby causing a potential to develop across the glass membrane. The size of the membrane potential or voltage varies with the difference in hydrogen ion activity between the two solutions, and this voltage is added to the voltage between the indicating and reference electrodes. As mentioned in the review of electrode principles, the indicating electrode for pH analysis commonly consists of a silver wire in a solution of KCl surrounded by pH-sensitive glass. When this electrode is combined with a reference electrode to a pH meter, the instrument can measure the potential difference (or voltage) between the electrodes and convert this measurement to pH units.

Whole blood pH meters are based on exactly the same principles, but use electrodes adapted to analyze small volumes of blood. An analysis of blood pH is usually performed with oxygen and carbon dioxide analysis. Because these measurements must be made anaerobically, the blood sample is drawn directly into a fine capillary tube. The tube itself is constructed of pH-sensitive glass and is surrounded by a standard KCl solution and silver wire. The system is usually maintained at 37° C to determine the pH of the blood as it exists in the patient's body. The temperature control is important: pH is temperature-dependent and decreases about 0.015 units for each degree rise in temperature. Blood gas instruments in the laboratory are designed to measure the partial pressures of carbon dioxide (pCO_2) and of oxygen (pO_2), as well as blood pH. Specialized electrodes for pH and for each gas determination are placed within the same instrument along a common capillary tube so that one small blood sample suffices for measurements of all three parameters (pH, pCO_2, pO_2).

The pCO_2 Electrode

The pCO_2 electrode consists of a pH electrode with a CO_2-permeable membrane covering the glass membrane surface. Between the two is a thin layer of dilute bicarbonate buffer. The aspirated blood sample is in contact with the CO_2-permeable membrane, and as CO_2 diffuses from the blood into the buffer, the pH of the buffer is lowered. The change of pH is proportional to the concentration of dissolved CO_2 in the blood. The glass electrode responds to the buffer pH change, and the meter is calibrated to read the pCO_2 in mm of mercury. This type of pCO_2 electrode is known as the Severinghaus electrode (Fig. 3.20).

In a patient's blood, the values pH, pCO_2, and bicarbonate are all interrelated according to the Henderson-Hasselbalch equation (Equation 2.4). If any two of the values are known, the third can be calculated. Because blood gas instruments designed today measure the pH and pCO_2, the bicarbonate value may be calculated. Often a "total CO_2," which includes dissolved CO_2, carbonic acid, and bicarbonate is ordered. The bicarbonate concentration can again be calculated from the data: it is slightly lower than total CO_2. A quick substitution of the value into the Henderson-Hasselbalch equation can then serve as an extra cross-check on the laboratory results.

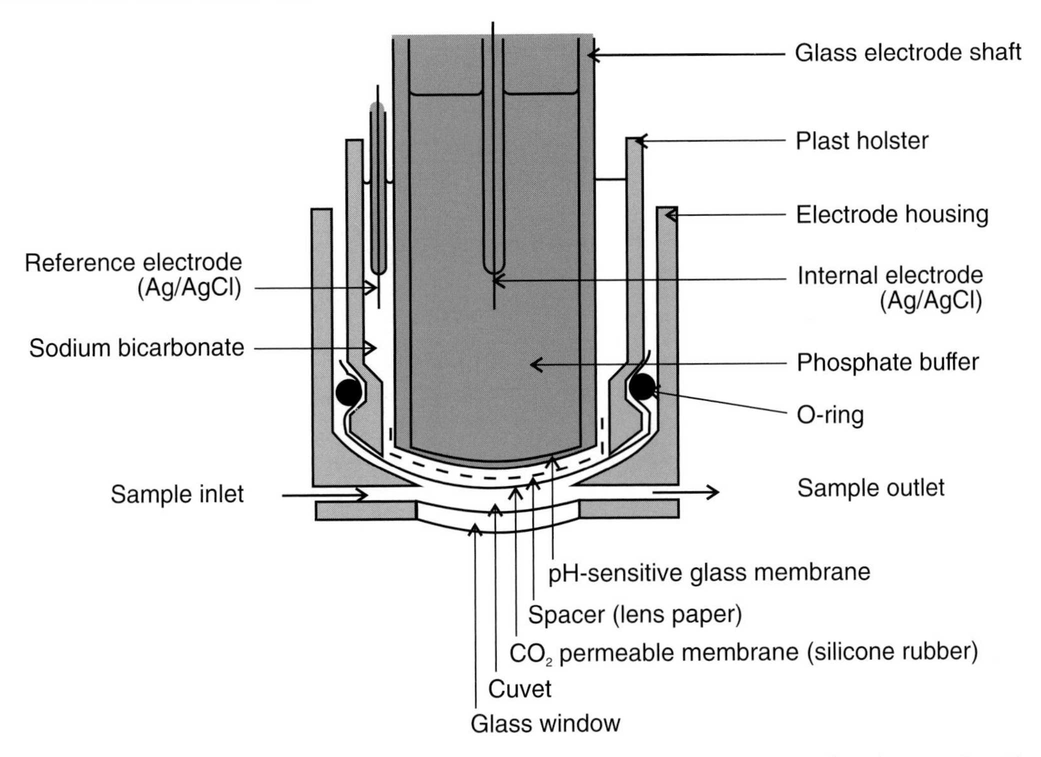

FIG. 3.20. Diagram of the pCO_2 electrode. (Adapted from Siggaard-Andersen, O.: The Acid-Base Status of the Blood. 4th Edition. Copenhagen, Munksgaar, 1974.

The pO_2 Electrode

The Clark electrode for measuring the partial pressure of oxygen in the blood is based on a different principle from that of pH measurement. The latter measures a voltage difference when no current is flowing; the pO_2 electrode measures the current that flows when a constant voltage is applied to the system. The current is the stream of electrons that flows as the oxygen molecules are reduced at the cathode:

$$\tfrac{1}{2}\,O_2 + H_2O + 2e^- \rightarrow 2OH^-$$

The source of the electrons is the silver-silver chloride anode where the silver molecules are oxidized:

$$Ag \rightarrow Ag^+ + e^-$$

The amount of current that flows through the system is a direct measure of the number of electrons released to the oxygen and is consequently a measure of the number of oxygen molecules available for reduction. The current is directly linear with O_2 concentration as long as the constant voltage is maintained. The blood gas instrument measures the current flow produced by the loss or gain of electrons when the system is subjected to a polarizing current.

The electrode may be constructed as shown in Figure 3.21. A platinum wire forms the cathode; the anode is a silver wire in AgCl. The contact between the poles is an electrolyte solution that is separated from the test sample (blood) by a membrane permeable to O_2 molecules. Dissolved O_2 diffuses from the blood through the membrane and is reduced at the cathode. The rate-limiting factor in the system is the diffusion of oxygen molecules through the membrane. The diffusion rate depends directly on the pO_2 of the sample, so the change in current flow offers a direct measurement of the pO_2.

Blood Gas Instruments

Blood gas instruments must be monitored constantly and calibrated frequently. All three electrodes (pH, pO_2, and pCO_2) are calibrated by setting with two standard concentrations. Two buffers in the physiologic range are used for pH calibration, and two gases (high and low concentrations of O_2 and CO_2) are used for the gas electrodes. The gases are bubbled through water in the instrument to saturate them with water vapor; gases dissolved in the blood would be comparably saturated. Corrections must be made for water vapor pressure and for the barometric pressure, which must be checked regularly throughout the day.

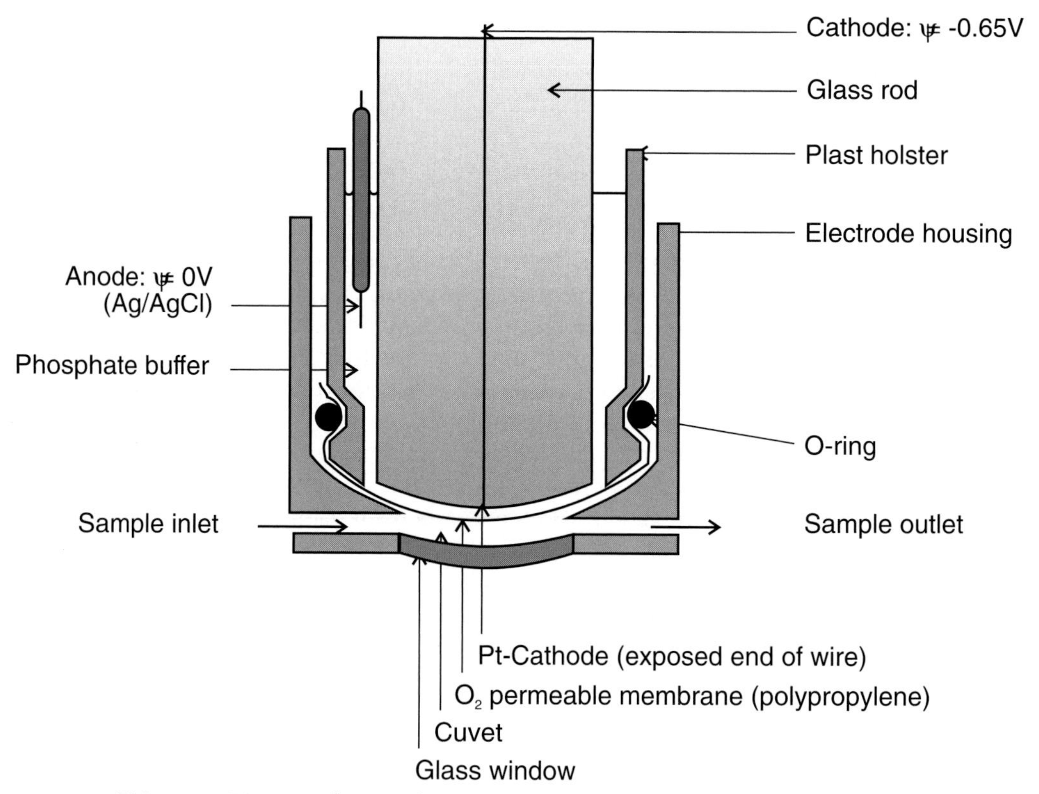

FIG. 3.21. Diagram of a pO_2 electrode. (Adapted from Siggaard-Andersen, O.: The Acid-Base Status of the Blood. 4th Edition. Copenhagen, Munksgaar, 1974.

Because the pressure of gases and pH depends on temperature, the temperature of the bath surrounding the electrodes must be carefully monitored and closely controlled. Usually the bath is maintained at 37° ± 0.1° C.

Both the pO_2 and pCO_2 electrodes require regular maintenance to keep the membranes intact, taut, and clean. Obstruction to diffusion, such as protein buildup on the membrane, slows down the response and may give low results.

Ion-Selective Electrodes

An ion-selective electrode (ISE) is composed of an electrochemical half-cell and an ion-specific membrane. In fact, the pH electrode previously described is an example of an ISE composed of a silver-silver chloride half-cell and a hydrogen-ion-specific glass membrane. The development of other ion-selective membranes has resulted in many other types of potentiometric electrodes. In each case, when an ion-specific membrane separates two solutions that differ in the concentration of that ion, a potential is developed across the membrane. The size of the potential depends on the difference in the ion concentrations. Ion-selective membranes may be composed of a crystal, an immobilized precipitate, or a liquid layer.

The activity of any ion can be determined potentiometrically if an electrode can be developed that responds selectively to the ion of interest. A sodium-sensitive glass has been introduced that is insensitive to hydrogen ions and shows a selectivity for sodium over potassium of about 300 to 1. The potential is developed as the sodium ions undergo an ion exchange in the hydrated layer of the glass membrane. A potassium electrode incorporating a valinomycin membrane shows a selectivity for potassium over sodium of 1000 to 1. This specificity for potassium is developed because the ion exchange cavities in the valinomycin membrane are nearly equal in size to potassium ions. In these and similar systems, the potential developed is a function of the concentration of the ion being measured.

The development of an electrode selective for calcium ions has made possible the direct measurement of ionized calcium, the physiologically active form. Although total calcium measurements in the laboratory still far outnumber assays for ionized calcium, the latter determinations are becoming more important in the clinical picture.

Ion-selective electrodes have also been developed for many other ions measured in the clinical laboratory, including lithium, lead, copper, ammonia, and chloride. Chloride electrodes have long been used in the laboratory for the measurement of chloride in sweat.

The use of selective electrodes in the clinical chemistry laboratory has grown by the skillful combination of enzymatic action with selective electrodes. Instruments (Beckman) have been devised for the measurement of serum glucose, urate, and cholesterol by using specific oxidases to oxidize the substrate (glucose, urate, and cholesterol, respectively). Each reaction requires oxygen. The instrument has a pO_2 electrode to measure the rate of O_2 utilization, which is proportional to the specific substrate concentration. In a similar manner, another instrument (Beckman; Fisher) measures serum urea concentration by a combination of an enzyme reaction and a conductivity meter. The enzyme urease splits off NH_3 from urea: the NH_3 diffuses through the membrane and is converted to NH_4^+ in the buffer. The change in conductivity of the solution is measured.

MEASUREMENT OF RADIOACTIVITY

Radionuclides and Radioactive Particles

Radioactivity is a general term used to describe the by-products released during the decay of unstable atomic nuclei, or *radionuclides*. Although the nuclei of most atoms are stable, certain ratios of protons and neutrons in a nucleus are unstable and spontaneously change to a more stable configuration. This nuclear alteration in the neutron:proton ratio is called *radioactive decay*, and results in the release of the excess neutrons, protons, and electrons, as well as of energy from the new stable nuclear configuration. Measurement of radioactivity involves quantifying the atomic particles and energy released during radioactive decay. In the clinical laboratory, biologic molecules modified by the addition of radionuclides become versatile tools that can be used in various biologic assays.

Radioactive decay of a radionuclide is accompanied by at least one of the three possible types of nuclear emissions: alpha (α) particles, beta (β) particles, and gamma (γ) rays. *Alpha particles* are the largest, consisting of two protons and two neutrons (equivalent to a helium nucleus). Their large size limits their penetration into matter, including skin, so danger from external exposure to alpha particles is slight. Internal exposure is very serious, because their considerable energy would be dissipated in a small area. The alpha particles are not measured in the clinical laboratories. The *beta particles* are negatively charged and equivalent to electrons. Emission of an electron transforms a neutron to a proton. Beta particles, because they are smaller and have greater velocity than alpha particles, can penetrate matter for short distances. This property is used in their measurement. Beta particles are an internal radiation hazard; external exposure at high levels may also present a hazard. *Gamma rays* have no measurable mass or charge. They are units of energy emitted to stabilize an atomic nucleus without changing atomic numbers or mass, although they may accompany such changes as beta emission or electron capture. Electron capture is the transfer of an inner orbital electron to a proton in the nucleus, thus converting the proton to a neutron. As previously stated, these nuclear disruptions may be accompanied by emission of high-energy electromagnetic radiation, the gamma rays. Gamma rays, like x-rays, can readily penetrate matter. Because of this property, they can be used in extremely small amounts for *in vivo* scans of internal organs, as well as for *in vitro* laboratory assays. External exposure to high-energy gamma rays can be extremely dangerous.

Half-Life of Radionuclides

The radionuclides used most commonly in the clinical chemistry laboratory for beta particle emission are carbon-14 (^{14}C) and tritium (^{3}H); the usual source of gamma radiation is iodine-125 (^{125}I). The rate of the radioactive decay of a radionuclide is measured in terms of its half-life ($t_{1/2}$). The half-life of a radionuclide is a measure of the time required for half of its radioactive atoms to decay to a stable state. The half-life of ^{125}I is 60 days: thus one half of the radioactivity (measured as disintegrations per min, dpm) in a vial of ^{125}I remains after 60 days and one fourth after 120 days. For assay purposes, one tries to select a radionuclide with a reasonable shelf-life so that its activity will remain sufficiently high for practical measurement. ^{14}C has a half-life of 5000 years; ^{3}H, 12 years; and ^{125}I, 60 days. The use of ^{131}I in radioassays has been virtually eliminated because of its short half-life of 8 days.

Instrumental Analysis of Radioactivity

The decay of a radionuclide is a disintegration "event" that releases at least an alpha or beta particle or a gamma ray. When radioactivity is measured, the frequency of the events is counted and reported as either disintegrations per minute (dpm) or simply as counts per minute (cpm). The relationship between the dpm and cpm is a function of the efficiency of the counting instrument. Not all the particles released during the decay can contact the instrument's sensor.

Beta radiation is measured in *liquid scintillation counters*. For counting, the sample is dissolved or suspended in a scintillation fluid in a vial. The fluid contains chemicals called fluors, which emit flashes of light when struck by beta particles. The beta particles must be closely surrounded by these "scintillants" because their own travel path within the vial is so short. The scintillation vial is positioned between two photomultiplier tubes that count and register the flashes of light from the fluors as cpm.

Gamma radiation is detected by a scintillation crystal, a sodium iodide crystal activated with thallium. The sample tube is lowered into a well within the crystal. A gamma ray striking the crystal results in a flash of light that is registered by an adjacent photomultiplier tube. Gamma rays travel from the sample tube into the crystal for detection.

Both types of scintillation counters contain discriminators to isolate the voltage range typical of the energy emitted by the radionuclide of interest. These discriminators serve a purpose analogous to the monochromator in a spectrophotometer.

Radioactive elements are not indigenous to the biologic samples assayed in the laboratory. "Tracers," molecules into which radioactive atoms have been inserted, are added to the sample for assay. These molecules compete with, and react like, the analogous nonradioactive molecules that are present in the sample. Measurement of the radioactive tracer provides information about the quantity of the nonradioactive species present. A detailed description of the radioimmunoassay incorporating this principle is found in Chapter 4.

AUTOMATED INSTRUMENTS

A customary goal of clinical laboratories is to rapidly produce accurate and precise test results. In general, automated instruments consist of a sophisticated arrangement of tools designed to assist the laboratory staff in achieving this goal *consistently*. The term *automated* implies mechanical or electronic control over an analytical process the instrument performs. When an instrument performs automatically, an analysis requires much less human operator time. Laboratories seldom have an excess of staff and usually depend on a combination of mechanical devices and computer programming within automated instruments to cope with the volume and variety of tests demanded by physicians.

The term *automated instrument* encompasses both large main laboratory analyzers capable of analyzing more than 30 different tests and small handheld instruments capable of analyzing 6 tests. A few components or designs are shared among all instruments. A sample followed through a complicated measurement process in a generic automated analyzer can present most of the terminology and concepts.

Common Features

1. *The sample tube:* Before entering an analyzer, a sample must be properly identified by both the human operator and the analyzer. Serum may be transferred into a special disposable cup and placed in numbered position on a tray by the operator so the serum can be identified by the instrument. Alternatively, when labeled with a *bar code,* the *primary tube* of centrifuged blood can be identified and sampled by the instrument.
2. *Sampling:* The two general categories of automated instruments differ in the method of sampling. *Continuous-flow instruments* withdraw a large aliquot of each serum sample into a piece of tubing. As each sample is propelled through the tubing, every chemical test available is performed on a fraction of each sample. Alternatively, during *discrete analysis,* instruments repetitively pipet small aliquots of serum into reaction chambers, where only a single test is usually performed.
3. *The analyses:* When instruments perform more than one test and produce a set of results for each specimen, they are termed *multi-channel analyzers,* whereas *single-channel analyzers* only perform a single type of reaction. The order of sample analysis can also be varied. During a *batch analysis,* a group of specimens is processed within the same analytical session, with either *sequential analysis* of samples within the batch or *parallel analysis* of groups of samples undergoing the same test at the same time. *Random access analysis* implies that any test can be performed on any sample at any time without concern for the initial organization. This feature is particularly useful if the instrument analyzes both stat and routine tests.
4. *Reagents:* The categories of reagents include *wet* or *dry,* and *open* or *closed.* In general, wet reagents are subject to faster deterioration than are dry reagents, and consequently, those instruments may require more frequent recalibration. Dry reagents may be combined with the sera directly (for example, Ektachem 700, Eastman Kodak) or sonicated into solution prior to the addition of sera (for example, Paramax, Baxter). An instrument is considered a *closed system* if only one supplier is available for the reagents, which are typically packaged in special containers.

Reagent packages may also be bar coded to allow an instrument to identify reagents. Automated systems often monitor the consumption of reagents and warn the operator before a solution runs out.

5. *Pipets:* The contamination of subsequent analysis, known as *carry-over,* is avoided by aliquoting devices designed with internal and external rinsing stations or with disposable pipet tips.
6. *Mixing:* Reagents and samples can be mixed by many devices, including blenders, magnetic stir bars, vibrating rods, and forced liquid displacement. In *centrifugal analyzers,* mixing occurs when the reagents and samples within a rotor cavity are forced together by centrifugal force as the rotor accelerates and decelerates.
7. *Cuvets and reaction vessels:* The cuvets used for photometric analyses may be either re-usable or disposable. When re-usable cuvets are used, they must be thoroughly rinsed between tests and routinely checked for flaws, stains, or other signs of deterioration.
8. *The computer:* A small internal computer can provide many functions for an analyzer. Not only can it coordinate the mechanical components, it can also perform calibration calculations, quality control monitoring, and data transmission directly to and from a laboratory information system (LIS). When linked to a LIS through a *bidirectional interface,* the instrument is directly informed of the tests that have been ordered for each specimen, and the instrument can report a patient's test results back to the LIS without requiring transcription by human hands. A computer can also be programmed to compare the most recent result to previous analyses and to look for sudden changes in concentration. This type of monitoring is known as a *delta check* and can be used to alert instrument operators to a possible need for repeat analysis.
9. *Maintenance:* Although most analyzers are designed to have a long and prosperous lifespan, they all need to be carefully nurtured. A daily and weekly maintenance schedule should be established for each instrument. In addition to poor maintenance, instrument performance can be affected by variation in temperature, humidity, static electricity, quality of the water supply, electrical power fluctuations, and lamp deterioration. Automated instruments can often run unattended for short intervals, but usually require knowledgeable operators to at least maintain the sample and reagent supplies, remove fibrin clots from tubing, scrutinize lipemic or hemolyzed samples, and act on the quality control data.

Instruments

Technologic advances have allowed many tests to be performed with only a handful of automated instruments. Perhaps some day, every test within the laboratory will be performed by a single instrument! Until that day arrives, automated instruments will have more specific tasks: stat or routine chemistry analyses, immunoassays for drugs and hormones, or urinalysis. Several instruments are described in the following sections to illustrate how the concepts of instrument automation have been applied.

Stat and Routine Chemistry Analyzers

These instruments are the main clinical chemistry laboratory analyzers. This form of instrument was described in the previous "Common Features" section. Typically they incorporate spectrophotometers and ion-selective electrodes as the basic analytical instruments, although several alternatives exist.

The DuPont Automatic Clinical Analyzer (aca) system uses a transparent plastic pouch (test pack: Fig. 3.22) as the discrete container for each test. The pouches (about 8 × 13 cm in size) are prepackaged with the reagents for each assay, and the header is coded with specific instructions, which are read by a bar-code scanner, for the test procedure. These instructions include how much sample is added, which buffers are added, on which wavelength the final absorbance is measured, and what stored calibration curve is used to calculate the result. An assay is initiated by placing a serum sample cup in a rack together with patient identification and assay information, which is photographed to accompany the results. A test pack for each procedure is placed on the rack directly behind its specimen. The required volumes of serum are aliquoted into each pack before the pack is moved along the track, where the assay is to take place. Within each pack are small dimples carrying the reagents; the seals to these compartments are broken automatically to initiate addition of the reagent at the proper time. Reagents are added and mixed and reaction mixtures are incubated, all at the direction of the coded information. When the reaction time has been completed, a cuvet with flat optical surfaces is molded from the pack itself. Absorbance of the reaction fluid is read in a colorimeter.

An alternative approach has been introduced by Eastman Kodak Company in its Ektachem System. Small plastic slide mounts holding multilayered dry films of reagents (Fig. 3.23) form the discrete container for this system. A 10-μl drop of serum is automatically applied to the top, or "spreading," layer, thereby causing the sample to spread evenly across the slide before penetrating to the lower layers. The subsequent layers vary, each selected to suit the

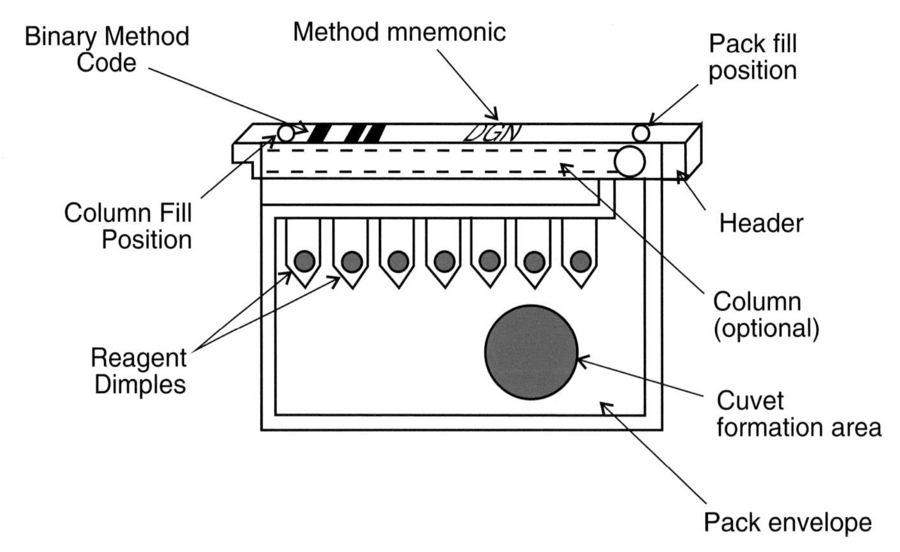

FIG. 3.22. A du Pont aca test pack. (Courtesy of DuPont Co.)

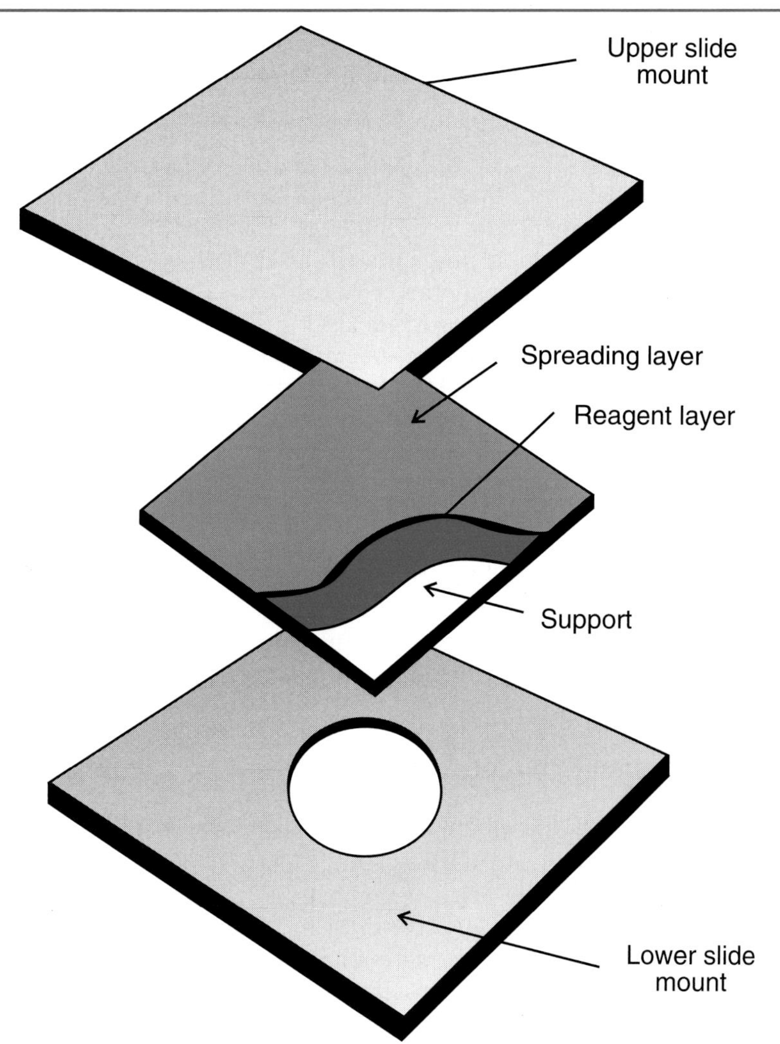

FIG. 3.23. An Ektachem colorimetric slide. (Courtesy of Eastman Kodak Co.)

chemistry required for the test. A barrier layer impervious to large molecules may be included if removal of proteins from the reaction mixture is desired. In all cases, a reagent layer containing the reagents, often enzymes, forms the medium for the assay of the test analyte. As needed, a buffer layer or a semipermeable membrane may allow only the desired products to penetrate to the next layer. After an incubation period, the final colored reaction product descends to a support layer on which the reflectance density is read. Ektachem has also set up another system with disposable electrodes embedded in slides for the measurement of electrolytes. In the automated Ektachem instrument, the slides are stored in cartridges in dispensers, from which they are selected as specific tests are programmed into the system. Kodak also makes a smaller version of the Ektachem, about the size of a typewriter. This instrument, the

DT-60, is designed for use in a physician's office; the slide for each test, as well as the serum sample, is introduced manually by the operator.

Automated Immunoanalyzers

Although relatively new, this group of instruments has already had a noticeable impact on laboratory and medical practices. The analysis of drugs and hormones by manual methods required at least several hours and relied on extensive technical experience of the personnel. In comparison, most current immunoanalyzers can provide assay results with greater precision within an hour and do not depend on special technical skills. Unlike the group of main chemistry analyzers, which often use similar chemistries and spectrophotometry, the automated immunoanalyzers use a diverse array of immunochemical detection methods, including fluorescence, fluorescence polarization, time-resolved fluorescence, chemiluminescence, turbidity, nephelometry, and spectrophotometry. A detailed review of immunochemical methods is presented in the following chapter.

Various automated or semiautomated instruments have been developed for urinalysis, HPLC, column chromatography, and electrophoresis. Because almost any method that is performed in a repetitive fashion can be automated, the automation trend is likely to continue.

Notes on Instrument Selection

The approaches to automation are many and varied; selection of an appropriate system depends on the needs of a particular laboratory. Cost is a primary consideration—the price of the instrument, as well as its cost of operation. A high capital cost may be offset by the large volume of tests generated if the workload is heavy. The operational cost (cost per test) is an important factor, for the instrument may use many disposables, such as cuvets, sample cups, tips, rotors, pouches, columns, and reagents. Reagent costs vary widely and are always expensive when bought already prepared. Some instruments pump reagent through the system continuously, whether or not a sample is being run. In discrete systems, reagent volumes per test can vary from 50 μL to 3 mL. For some instruments, the user may prepare reagents; for others, the reagents are supplied by the manufacturer. Large laboratories may find real savings in making their own reagents; smaller laboratories may find the technologists' time more expensive.

Selection of an instrument that requires only a small volume of serum per test is important. This characteristic is essential for hospitals with a neonatal center and pediatric population and is important for patients undergoing extensive workups. Insufficient serum for a series of tests causes aggravation for everyone.

Other considerations in selecting an automated instrument may be (1) the number of tests that can be run per batch; (2) the number of assays that may be run on each sample; (3) the time required per test; (4) the space (electrical, plumbing connections) required by the instrument; and (5) the ease of including stat samples as they come in.

In summary, then, an instrument must be selected with the goals of the laboratory in mind. Does the laboratory perform large volumes of single tests, so that an instrument capable of running many samples per batch is important? Is a standard profile of tests run on every admission or on many people? Are batches small, or must individual profiles of tests be selected for each patient? Are other instruments available for stat tests, or must the automated cycle be interrupted to insert individual tests? Is the instrument rugged and dependable, or does it break down frequently? Does the company provide good service in a hurry? Does the instrument come in components that can be replaced individually on short notice, or must the whole instrument be shipped to the manufacturer for major repair? Does the company provide instruction classes for your laboratory when a new instrument is purchased? Are good instruction and trouble-shooting manuals available? There are instruments appropriate for many situations and for many laboratories. The task is to select the right tool for the job and then to verify that it is performing assays with precision and accuracy.

LABORATORY COMPUTERS[6]

The modern clinical laboratory contains several different types of computers that support many different laboratory functions. Because technologists in most laboratories must use the computers to some extent, a basic review of the computers and their operation is helpful.

Components

All computer systems are composed of three parts: hardware, software, and personnel. Hardware refers to the physical components of the computer: the screen, keyboard, central processing unit (CPU) (the "brain"), memory, and input and output devices. Software refers to the programs or commands that operate within the hardware. As functions are carried out by a computer, the software sends and retrieves information from the hardware, including the screen, keyboard, printer, and memory. The successful operation of the hardware and software depends on the knowledgeable individuals who use, operate, and maintain it. Although computer operators may be required to perform special functions, such as minor program modifications, medical technologists often serve as laboratory data managers and supervise the system data files and security codes.

Types of Laboratory Computers

Computerized Instruments

Many instruments rely on internal computer programs to coordinate mechanical parts, record measurements, and calculate and print results. These computers are usually composed of small sets of circuits and chips that are dedicated to the functions they perform. A recent trend allows instrument computers to communicate with a central laboratory computer and transfer assay results electronically to minimize human transcription errors.

Personal Computers

Relatively small "personal computers" often serve in administrative functions in laboratory offices, and can be interconnected into local area networks to acquire and distribute information, as well as software. Word processing, statistical analysis, and graphics software are commonly used.

Laboratory Information Systems

A central laboratory information system (LIS) is often the most frequently consulted computer within a laboratory. A LIS computer may be responsible for assigning an analyzer and the tests to be performed on each specimen, following the location of specimens during analysis, acquiring the results directly from the "on-line" analytical instruments, and printing the reports. Many LIS computers are designed to exchange patient information with the hospital information system computer. The LIS often stores and calculates quality control data, assay turnaround times, and workload statistics. The computer system may notify the laboratory staff of tests not performed, overdue results, and lists of pending analyses. In addition, the LIS may organize billing information and other administrative data. A thorough understanding of the structure and limitations of the LIS is a valuable asset for each member of the laboratory staff.

REFERENCES

1. Bender, G.T.: Principles of Chemical Instrumentation. Philadelphia, W.B. Saunders, 1987.
2. Hicks, M.R., Haven, M.C., Schenken, J.R., and McWhorter, C.A.: Laboratory Instrumentation. 4th Edition. New York, Van Nostrand Reinhold, 1992.
3. Lee, L.W., and Schmidt, L.M.: Elementary Principles of Laboratory Instruments. 5th Edition. St. Louis, C.V. Mosby, 1983.
4. Willard, H.H., Merritt, L.L., and Dean, J.A.: Instrumental Methods of Analysis. 7th Edition. New York, D. Van Nostrand, 1988.
5. Tietz, N.W.: Textbook of Clinical Chemistry. Philadelphia, W.B. Saunders, 1986.
6. Elevitch, F.R., and Allen, R.D.: The ABCs of LIS. Chicago, ASCP Press, 1989.

REVIEW QUESTIONS

1. What are the regions of the spectrum and the colors of the visible spectrum in order of increasing wavelength?
2. What are the three variables that influence the light absorbance in Beer's law? State Beer's law.
3. Describe the essential components of a UV-visible spectrophotometer, flame photometer, atomic absorption spectrophotometer, and fluorometer.

4. Why might fluorescence measurements be more specific and sensitive than spectrophotometric measurements?
5. What is the principle for molecular separation involved in gel filtration, ion exchange, partition, and gas liquid chromatography?
6. Describe the different types of detectors used for gas liquid chromatography and high-performance liquid chromatography.
7. What are the three principal factors that influence the migration rate of a protein during electrophoresis?
8. Describe the principle and components of a pCO_2 electrode, a pO_2 electrode, and a pH electrode.

Immunochemical Techniques 4

OUTLINE

OBJECTIVES

After reading this chapter, the student will be able to:

1. Describe how antibodies to a particular protein are formed in animals after several injections of that protein.
2. Define hapten and describe how haptens are used in developing immunoassays for some therapeutic drugs, hormones, and other small molecules.
3. Describe the principle and three components of a radioimmunoassay system.

4. Understand the advantages and disadvantages of isotopic and nonisotopic immunoassays.
5. List three different types of labels employed in immunoassays that are sufficiently sensitive to measure the concentrations of drugs or hormones in small volumes of serum, and describe the kind of signal emitted by each label.
6. Discuss the principle and the advantages and disadvantages of homogeneous immunoassays versus heterogeneous immunoassays.
7. Describe the principle and label used in each of the following homogeneous immunoassay systems: enzyme multiplied immunoassay technique and fluorescence polarization immunoassay.
8. Explain the principles of radial immunodiffusion and the Ouchterlony technique and how they may be utilized for identifying proteins in body fluids.

Various methods using physical-chemical principles (for example, spectrophotometry, fluorometry, and potentiometry) have been adapted to measure compounds of clinical interest. Figure 4.1 shows the relative circulating serum concentrations of selected clinical analytes; on the right side of the figure are bars indicating the approximate measurement ranges for methods that are currently employed to assay for these compounds. Note that as endogenous concentrations of these analytes become low (<1 μmol/L), a relatively new class of method, immunochemical assays,[1,2] is required to achieve the necessary sensitivity for accurate measurement.

Immunochemical assays are based on the highly specific and tight, noncovalent binding of antibodies to target molecules. Antibodies are exquisitely specific in selecting and binding to a complementary target molecule in a complex mixture, such as plasma or serum. Immunochemical techniques have made possible the determination in serum of many compounds of clinical interest, for example, drugs, steroid hormones, peptide hormones, and proteins, because specific antibodies can be raised against them.

PRODUCTION OF ANTIBODIES

The immune system reacts to repeated injections of antigens (foreign macromolecules, such as proteins or carbohydrates, that initiate an immune response from the host). The plasma cells (B-lymphocytes) begin to synthesize antibodies that are complementary to some reactant groups on the surface of the antigen. There is a good "fit" between the reactant groups on the surface of the antigen and the antibodies (Fig. 4.2), similar to that of a key fitting a lock. Antibodies usually do not bind to other antigens, unless they are closely related in structure. Antibodies produced in response to antigen challenge include the IgG class (Fig. 4.2; see Chap. 9). Most of the IgG structure is constant (that is, the

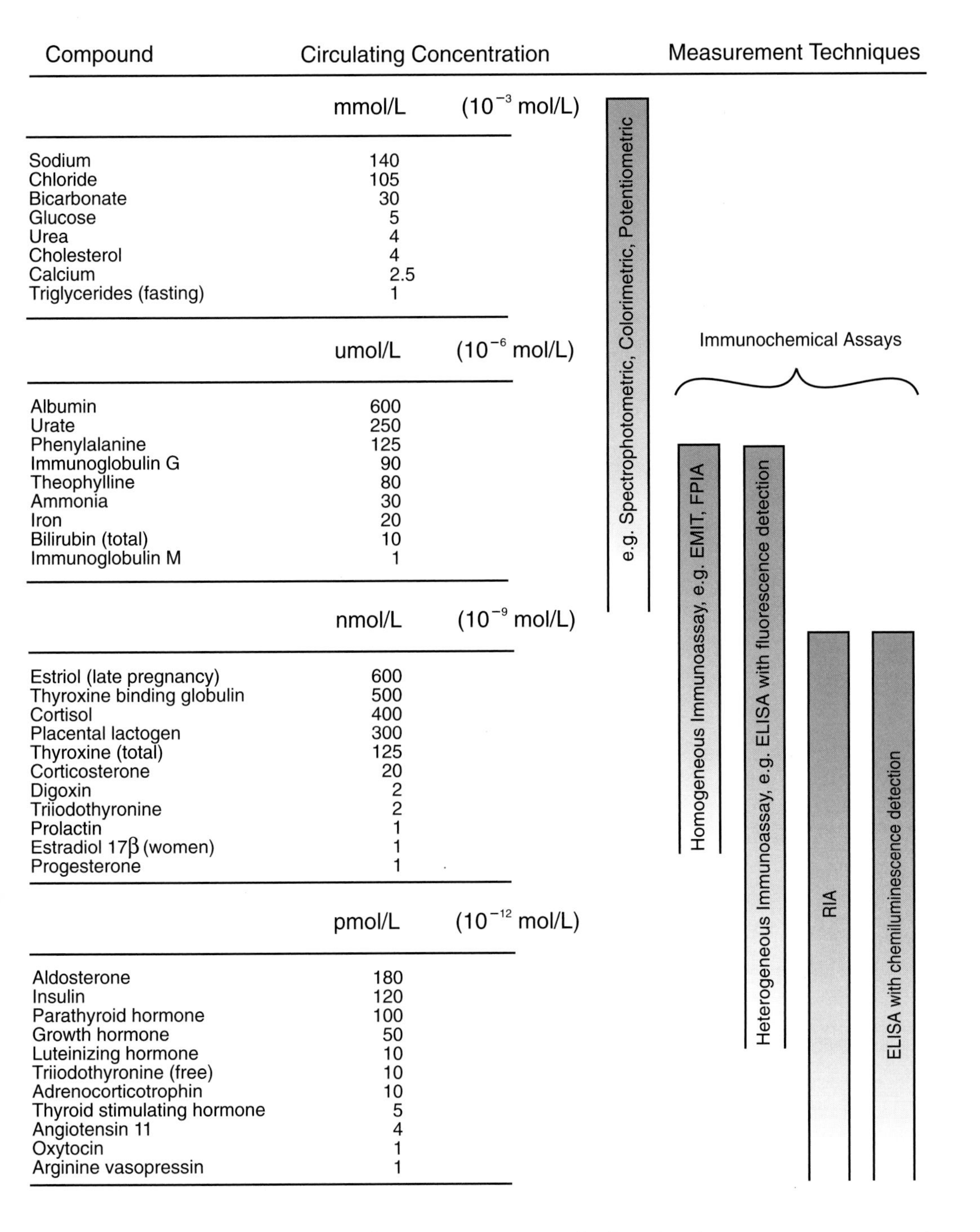

FIG. 4.1. Approximate circulating concentrations of selected endogenous compounds and therapeutic drugs, with the corresponding assay ranges and methods to measure these substances. (Adapted from Langan, J., and Clapp, J.J. (Eds.): Ligand Assay. NY, Masson Publishing Co., 1981.)

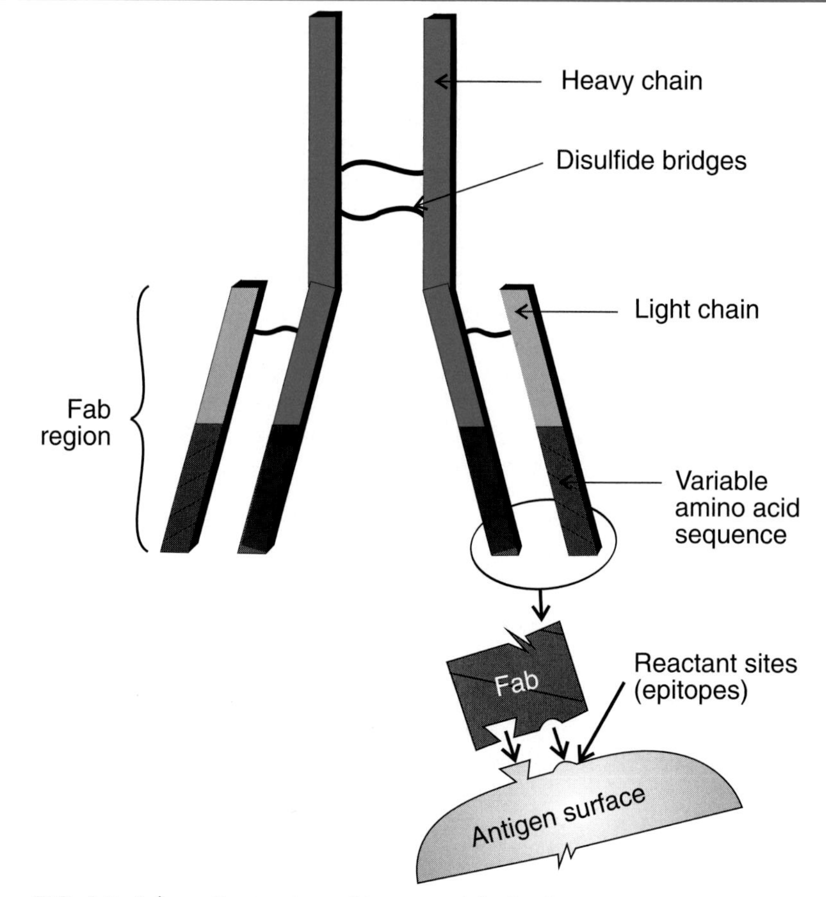

FIG. 4.2. Schematic structure of immunoglobulin G.

amino acid sequence is the same for all IgG molecules), with the exception of a portion of the Fab region. Variability in the amino acid sequence in this region of the molecule allows the production of antibodies that can bind specifically to virtually any molecule. As shown in Figure 4.2, this binding is accomplished by matching the Fab variable portion of the antibody with the complementary structure on the surface of the antigen. Each IgG contains two identical Fab binding sites.

For a foreign molecule to elicit an antigenic response, a minimum molecular weight of about 4000 daltons is required. Lower-molecular-weight molecules, such as drugs or steroids, that by themselves cannot elicit an antigenic response to produce antibodies (such molecules are called haptens) can be made to do so by covalently binding them to an antigenic macromolecule (for example, the protein albumin). With modern techniques, antibodies can be produced against virtually any molecule of medical interest, from macromolecules (proteins and polypeptides) to such small molecules as those of drugs or amino acid derivatives.

A single B-lymphocyte cell produces only one type of antibody (that is, the Fab portion is the same for all IgG molecules made in that cell) in response to the antigen; however, each antibody-producing cell may synthesize an antibody whose variable Fab region is different from that of other B-lymphocytes. As a result, animals produce a mixture of antibodies specific for different regions on the antigen. These mixtures of different antibodies are called polyclonal antibodies. Large quantities of antibody for use in immunochemical assays are produced by injecting antigen into such animals as rabbits or sheep. When the antigenic response has occurred after several weeks, blood is withdrawn from the animal, and the IgG protein fraction is separated from the blood. An antibody suitable for immunoassay should bind only the desired analyte (specificity), and its noncovalent binding should be tight (avidity or high binding constant).

Techniques have been developed to remove a single antibody-producing cell from an animal and to fuse it with a type of cancer cell to create a rapidly multiplying hybrid cell (a hybridoma) that still produces the antibody of the original cell. From such hybrid cell groups (clones), a large amount of a single type of antibody (all Fab portions are identical) can be isolated; these antibodies are called monoclonal antibodies. Monoclonal antibodies are useful in immunoassays for which high specificity is necessary.

RADIOIMMUNOASSAY (RIA)

In 1959, Yalow and Berson[3] devised a quantitative immunoassay for the measurement of the insulin concentration in plasma. This assay was far more sensitive and specific than any method in existence at that time. This technique has since been given the name radioimmunoassay (RIA) because it used the radioisotope ^{125}I. RIA (as well as nonisotopic modifications) now can determine the concentration of nearly all hormones, as well as of many drugs, proteins, and other compounds. The specificity of RIA tests depends on the ability of an antibody to recognize its unique antigen in a heterogeneous mixture. The sensitivity of the assay depends on the specific activity (emission of alpha or beta particles or gamma rays per unit weight of radioisotopically labeled antigen or hapten) of the radiolabel used.

Principle of RIA[4]

Two items are indispensable for an RIA test: a high-potency antibody against the purified antigen or hapten and a radiolabeled antigen or hapten of high specific activity. The most commonly used radioisotope for labeling is ^{125}I, with ^{3}H and ^{14}C used less frequently. The steps are as follows (Fig. 4.3):

1. A limited amount of specific antibody (sufficient to bind 40 to 60% of the radiolabeled antigen) is incubated with a fixed amount of radiolabeled antigen plus the antigen (usually in the biologic matrix, such as serum or urine).
2. The radiolabeled antigen and the unlabeled antigen (in the patient's sample) compete for the limited number of binding sites on the antibodies.

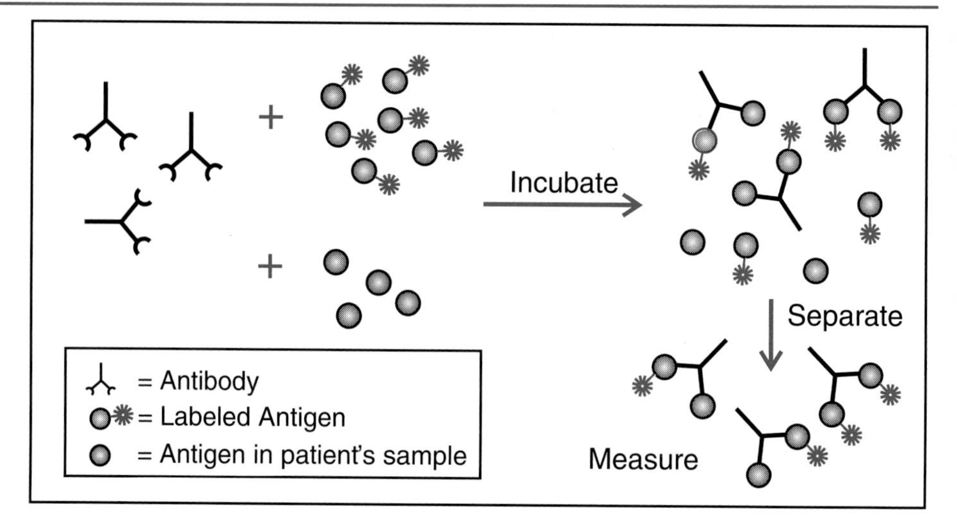

FIG. 4.3. Radioimmunoassay (a heterogeneous immunoassay).

3. When equilibrium is reached, the bound antigen-antibody complex (with both labeled and unlabeled antigens) is separated from the unbound (free) antigens, and the radioactivity is counted. The ratio of antibody-bound labeled antigen to total labeled antigen (B/T) decreases as the concentration of unlabeled antigen (in the patient's sample) increases. The number of radioactive counts in an aliquot of labeled antigen is used to determine the total counts (T).
4. A series of known antigen concentrations is run in the same batch as the unknown samples, and the ratio, B/T, is calculated. A standard curve is prepared by plotting B/T versus the antigen concentration; the concentrations of the unknown samples are then obtained by reading from the standard curve (Fig. 4.4). Alternatively, some laboratories plot the function %B (rather than B/T) versus concentration. The percent bound is calculated:

$$\%B = B/T \times 100$$

Equilibrium in the antigen-antibody binding need not be complete because the conditions are the same for both the standards and the unknown samples. A standard curve must be run with every batch of unknown samples because slight changes in assay conditions from run to run (temperature, pH, ionic strength, amount of antibody or labeled antigen added) may alter the B/T (or %B) and, hence, the standard curve.

Disadvantages of Radioisotopic Immunoassays

The use of radioisotopes as labels is accompanied by the following problems: (1) waste disposal continues to be costly and inconvenient; (2) shelf life of labeled reagents (particularly ^{125}I) is short because of radioactive decay; (3) relatively expensive instruments for counting radioactivity are required; (4) the

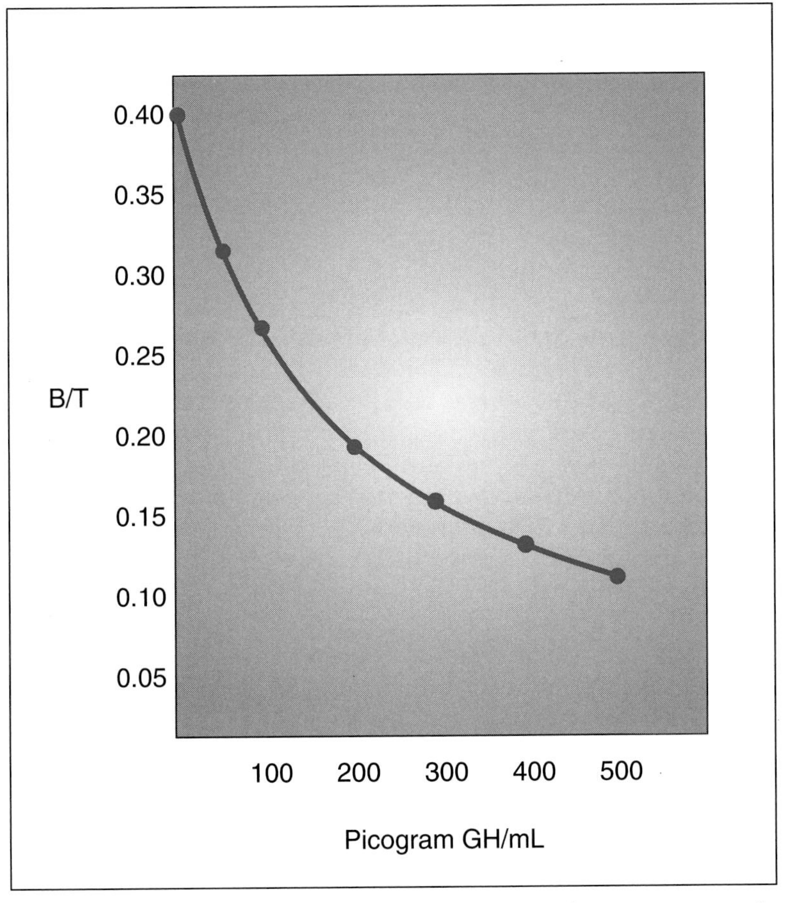

FIG. 4.4 Standard curve for growth hormone (GH) radioimmunoassay. The ratio, B/T, is plotted against picograms per mL of GH.

procedures are relatively labor intensive; (5) bothersome bureaucratic procedures are involved, such as obtaining an appropriate government license to work with radioactivity and meeting all prescribed regulations for safety and disposal; (6) a minimal potential health hazard exists (exposure to radiation). These problems prompted a search for nonisotopic labels that would approach the sensitivity of radioactive tags.

NONISOTOPIC IMMUNOASSAYS[5,6]

Alternate labels to radioisotopes are now available. These labels use enzymatic reactions, fluorescence, chemiluminescence, or other technologies. They are grouped together as nonisotopic immunoassays, some examples of which are described in the following sections. When enzymes are used as labels for haptens or antigens (enzyme immunoassay, EIA), their reaction with substrate may produce compounds that can be detected by UV-visible spectrophotometry or by fluorescence, depending on the particular substrate chosen. When fluorescent molecules are used to tag antigens or haptens, the fluorescent emission signal can be measured directly (fluorescence immunoassay, FIA) or by a change in polarization of the exciting light (fluorescence polarization immunoassay, FPIA). More recently, enzymes have been used in combination with chemiluminescent substrates, which release a photon of light when cleaved.

Enzyme Immunoassay (EIA)[1,2]

The technique of EIA is similar in principle to that of RIA (Fig. 4.3) except that the antigen or hapten is labeled with a stable enzyme instead of a radionuclide. EIA is just as selective as RIA because it uses the specificity of antibodies for recognizing their antigens or haptens; it depends on the competition between enzyme-labeled and unlabeled antigen or hapten for binding to a limited amount of antibody. EIA results may be obtained in a much shorter time than that required by RIA, often within 5 to 10 min compared to 1 to 5 days for RIA. The amplification power of enzymes and the sensitivity of fluorescence measurement enable a combination of the two techniques to reach a degree of sensitivity that is adequate for most of the assays hitherto performed by RIA.

Many different enzymes have been used as the label. The characteristics that govern the choice are stability, cost, and generation of a suitable chromophore or fluorochrome that can be quantitated easily. A few of the enzymes commonly used are glucose-6-phosphate dehydrogenase (generates NADH), alkaline phosphatase (splits p-nitrophenylphosphate into a colored compound and phosphate), and β-galactosidase (splits off a fluorescent product from the substrate).

Fluorescence Immunoassay (FIA)[1,2]

The principles of FIA are similar to those of RIA and EIA; the only difference is in the label. In FIA, the label has the capacity to fluoresce, or an enzyme-labeled hapten liberates a fluorochrome on incubation with its substrate. In all cases, the labeled hapten competes with the serum hapten for binding to a limited amount of antibody. The concentration of hapten is obtained from a standard curve. A further description of FIA appears in Chapter 14.

Enzyme-linked Immunosorbent Assay (ELISA)[7]

In ELISA (Fig. 4.5), the antibody is adsorbed on a solid surface (wall of a plastic test tube, microtiter well, or plastic bead). The antigen in the patient's sample and a fixed amount of enzyme-labeled antigen are added and allowed to compete for the limited number of antibody binding sites. The unbound enzyme-labeled antigen is removed simply by decanting the supernate. Enzymatic activity of the bound labeled antigen is determined by incubation with substrate. The concentration of antigen in the patient's sample is obtained from a standard curve.

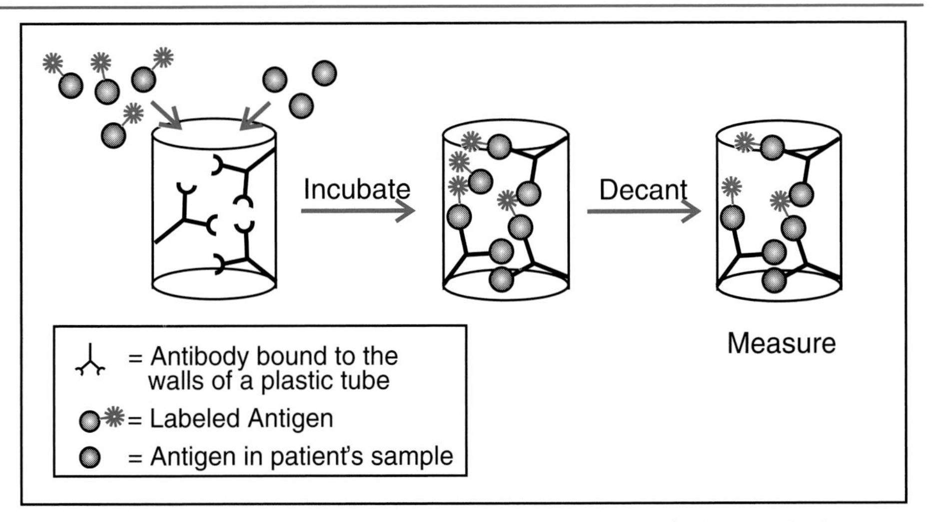

FIG. 4.5. Enzyme-linked immunosorbent assay (ELISA) using antibody-coated tube.

COMPETITIVE BINDING IMMUNOASSAYS

The most frequently used immunoassays in the clinical laboratory can be classified as either competitive immunoassays or immunometric assays. The RIA (Fig. 4.3), as well as the EIA, ELISA, and FIA described previously, are examples of competitive binding immunoassays. In these assays, the amounts of antibody and of labeled antigen or hapten added in each reaction is constant and limiting; thus, the relative amounts of labeled antigen bound to antibody and the amount left free in solution are determined by the amount of unlabeled antigen or hapten present in the patient's sample, standard, or control.

Competitive binding assays can be further subdivided into two classes, heterogeneous and homogeneous. In heterogeneous immunoassays (see Fig. 4.3), the labeled antigen or hapten bound to antibody must be physically separated from the labeled antigen or hapten that remains free in solution. The separation procedure may involve precipitating the antibodies with polyethylene glycol (PEG) or with the addition of a second antibody that binds and precipitates the original antibody. Newer techniques, for example, ELISA (Fig. 4.5), immobilize the antibody to the sides of a plastic test tube or to plastic beads, so that the supernate solution containing the unbound labeled antigen or hapten may simply be poured off. Homogeneous immunoassays (Fig. 4.6), on the other hand, do not require separation of the bound and free labeled antigen and, as illustrated in the figure, can be easily automated.

Homogeneous Immunoassays

Enzyme Multiplied Immunoassay Technique (EMIT)[8]

The distinguishing characteristic of a homogeneous immunoassay system is the elimination of the need to separate the bound from the free hapten; the enzymatic activity is severely reduced when it becomes bound to antibody

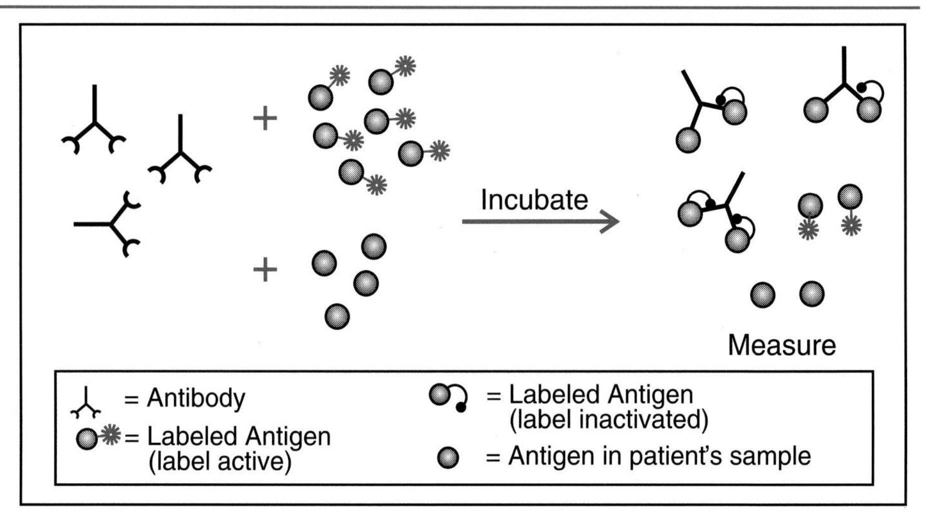

FIG. 4.6. Enzyme multiplied immunoassay technique (EMIT) (a homogeneous immunoassay).

(Fig. 4.6), thus making separation unnecessary. The loss of activity may be explained by steric hindrance (the antibody blocks the binding of substrate to enzyme) or by a change in conformation of the enzyme so that the substrate no longer "fits."

The EMIT technique, using a homogeneous system, was first devised for the quantitation of drugs and hormones in biologic fluids by the Syva Co. A sample of serum or other body fluid is mixed with a solution containing antibody, enzyme-hapten complex, and buffered substrate. The mixture is incubated at 37° C in a cuvet for a short time (frequently 2 min), and the rate of absorbance change (kinetic or rate reaction) is measured. Many of the enzyme labels generate NADH, so the absorbance change at 340 nm is recorded. The free enzyme-antigen molecule is enzymatically active, whereas that bound to antibody has greatly reduced activity (Fig. 4.6). The concentration of analyte (drug or hormone) is obtained from a standard curve derived from analyzing various concentrations of pure antigen in the same manner as patient's samples and plotting the concentration of antigen against absorbance. Although the standard curves are usually stored in the instrument for reuse, control specimens should be run with each batch of samples to validate the standard curve. EMIT assays are best suited for the analysis of small molecules, such as drugs, steroid hormones, or thyroxine (T_4). This technique requires that the antibody, when bound to the antigen, must be sufficiently close to the enzyme to inactivate it; for large antigens, the antibody may bind at a site too far from the enzyme to hinder its activity. A standard curve should be run with each batch of samples. EMIT assays are easily adapted to run on automated spectrophotometers, available in almost every clinical laboratory.

Fluorescence Polarization Immunoassay (FPIA)[9]

Certain organic molecules emit a plane-polarized wave for a short time after exposure to polarized light. Polarization of fluorescence depends on the fluorescence lifetime (time between excitation and emission) and the rotational relaxation time of the molecule (the time required for an oriented molecule to return to a random orientation in solution). Small molecules, such as haptens, rotate rapidly, have short relaxation times, and do not exhibit fluorescence polarization. When conditions are kept constant, rotational relaxation time depends on molecular size. Hence, a fluorescein-labeled hapten (thyroxine or most drugs) has practically zero fluorescence polarization, but when labeled hapten is bound by a large antibody molecule, fluorescence polarization is considerable because the relaxation time is prolonged. These principles are used by Abbott Laboratories in their automated TD_X system to determine the serum concentration of some hormones (see Chap. 13), many therapeutic drugs (see Chap. 14), and as a measure of surfactant in amniotic fluid (see Chap. 15).

Competitive Protein Binding (CPB) Assays

Competitive protein binding (CPB) is not an immunoassay, but is reviewed in this chapter because the principles are nearly the same. In CPB, a specific binding or carrier protein derived from tissues (cell receptors) or plasma replaces the antibody. These binding proteins have essentially the same specificities as antibodies and bind the analyte tightly. The analyte of interest is labeled with either a radionuclide or a nonisotopic label that uses an enzyme, fluorescence, or luminescence. The principle is the same as that described for RIA, EIA, or FIA. The labeled analyte competes with unlabeled analyte for binding to a limited amount of specific binding protein. After separation of the bound from the unbound label, the signal from the label is recorded as described for the other techniques. The concentration of analyte is obtained from a standard curve, where B/T (or %B) is plotted against concentration. The CPB technique is particularly suitable for the analysis of some hormones because of their strong and highly selective binding to plasma carriers or cell receptors.

IMMUNOMETRIC ASSAYS[10]

Immunometric assays determine the presence of antigens in biologic fluids using antibodies that are labeled, instead of the antigen. Further, this type of assay has an excess of antibody, rather than a limited amount as in competitive binding immunoassays. Thus, all antigen in the sample becomes bound to antibody, resulting in a more sensitive assay compared to competitive binding assays. The labels used in immunometric assays to tag the antibody can be the same type as those used in competitive binding assays. In practice, the labels used most frequently are radioisotopes (immunoradiometric assays [IRMA] and enzymes (immunoenzymatic assays [IEMA]). The two types of immunometric assays are one-site and two-site.

In the one-site assay (Fig. 4.7), the antigen complexed with the labeled antibody is selectively precipitated. Labeled antibody is mixed with patients'

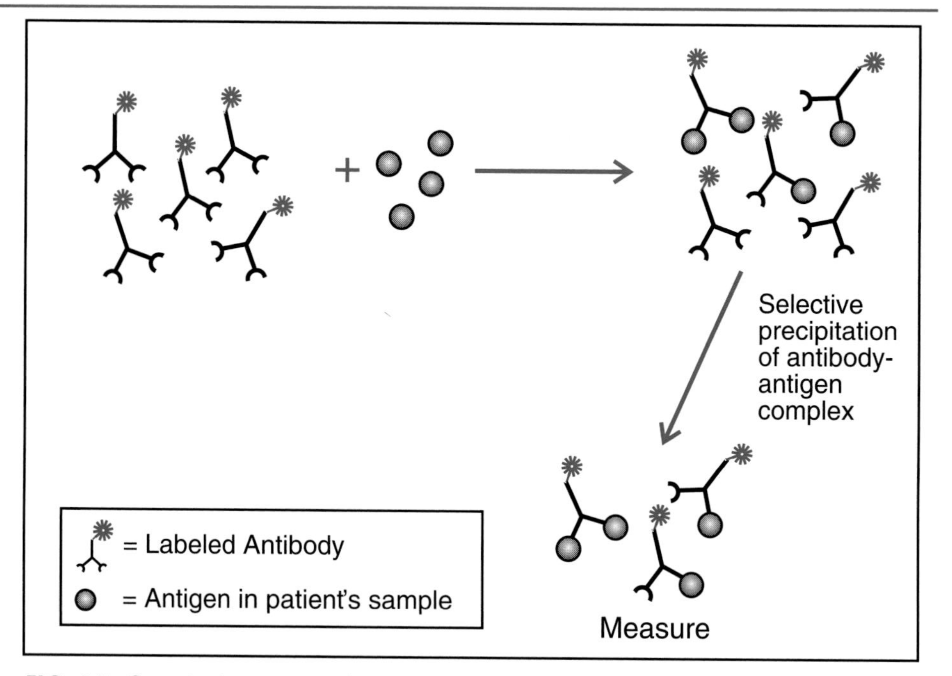

FIG. 4.7. One-site immunometric assay.

samples, standards, and controls, and the resulting antigen-antibody complex is precipitated (for example, with ammonium sulfate). The amount of label in the precipitate is then determined (by counting radioactivity or measuring enzyme activity of the label). The concentration of antigen in a patient's sample can then be determined from a standard curve.

In the two-site method (sometimes called a "sandwich" technique; Fig. 4.8), two antibodies made against different antigenic portions of the antigen are used. The first antibody is immobilized onto a solid support (for example, the walls of a plastic test tube or in a microtiter-plate well). After the antigen has bound to the immobilized antibody, the second (labeled) antibody is added to form a "sandwich" with the antigen between the two antibodies. The excess second (labeled) antibody can be removed by decanting or washing the solid support, and the amount of label remaining (bound) is determined. This method is appropriate for measuring large antigens, but cannot be used to measure haptens because they are too small to allow the binding of two antibodies.

A variant of the two-site assay makes possible the quantitation of antibodies. In this situation, antigen is bound to the solid phase, and the sample with antibody is added. The antibody binds to the antigen. After washing, a labeled antibody directed against the first antibody (from a different species) is added, and after incubation, the unbound second antibody is removed by decanting. Enzymatic activity of the bound enzyme-labeled antibody is determined by measuring its reaction with added substrate. This type of assay is currently used for screening individuals and blood bank specimens for the presence of various antibodies to HIV.

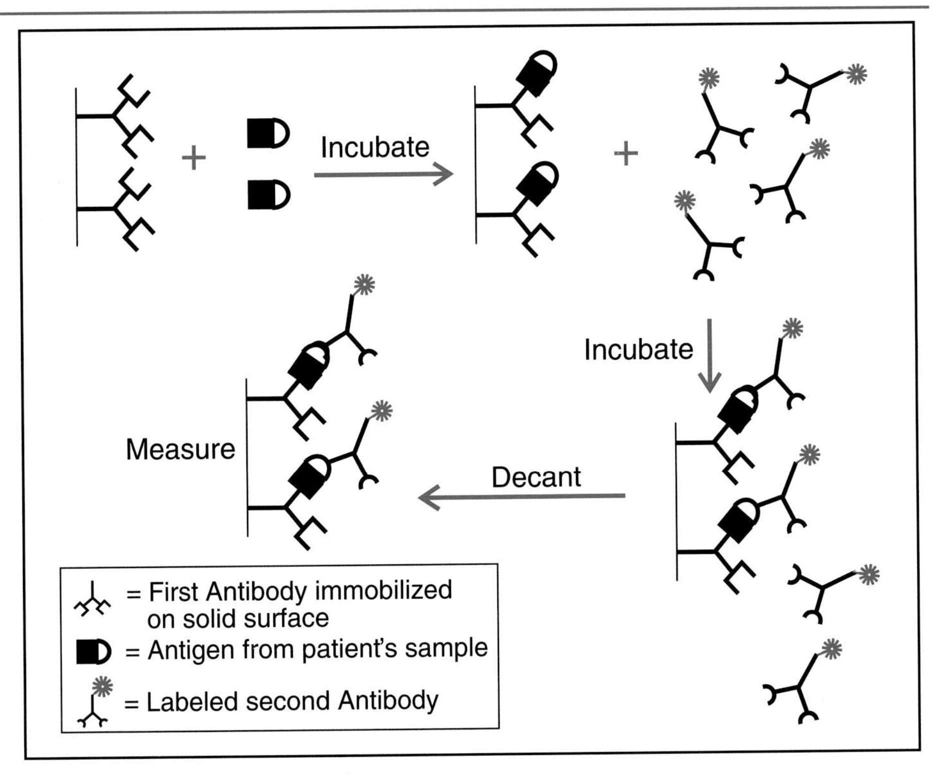

FIG. 4.8. Two-site immunometric assay.

IMMUNODIFFUSION

Double Diffusion in Agar or Agarose[11]

Antibody in solution interacts with its antigen rapidly, but the reaction complex is soluble at first. With the passage of time, the antigen-antibody complex continues to aggregate until visible particles (precipitin) are formed. Most antibodies are divalent (have two separate combining sites for antigen). Aggregation occurs as the divalent antibodies bridge separate antigens and form a lattice. The reaction is complex and is affected by the pH, temperature, ionic strength of the solution, time, and the ratio of antibody to antigen. A critical ratio is necessary for the formation of a precipitin. When either antigen or antibody is present in excess, no precipitation occurs.

The Ouchterlony technique of double diffusion is useful for the identification of particular proteins in body fluids. Agar gels, 3-mm thick, are prepared in Petri dishes with well patterns of various types and designs. Antibody is usually placed in a central well, and antigens or antigen and serum are placed in the other wells. Because both antigens and antibodies diffuse through the gel until they finally meet, the technique is called "double diffusion."

Typical diffusion patterns and precipitation lines are illustrated in Figure 4.9. When antibody to an antigen is placed in the lower central well and its

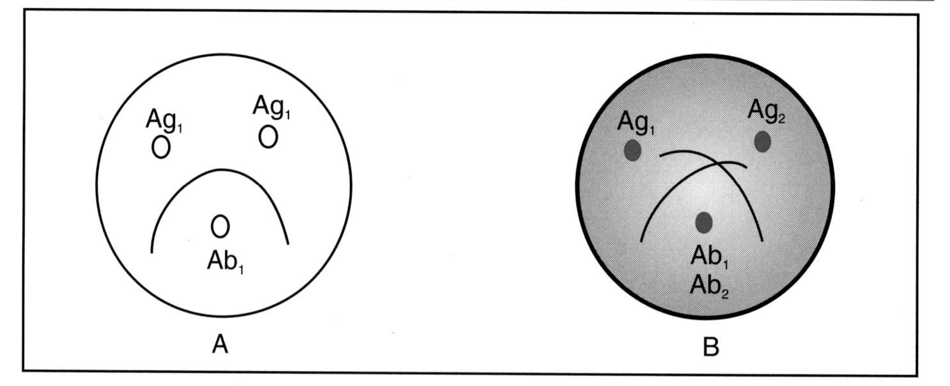

FIG. 4.9. Illustration of double diffusion in agar. A, The same antigen, Ag_1, is placed in each of the upper wells, and the antibody to Ag_1 is placed in the central well. The precipitation pattern is a smooth arc, a line of identity. B, The pattern of nonidentity develops when nonrelated antigens, Ag_1 and Ag_2, are placed in their respective wells and antibodies to each are placed in the central well.

antigen is placed in each of the two upper wells, antigen and antibody diffuse out of their respective wells in concentric circles. In 1 or 2 days, the antibody meets the antigen and combines with it. Diffusion continues, and the process of aggregation of the antigen-antibody complex begins. A line of precipitation of the complex (precipitin) forms an arc as the favorable ratio of antibody to antigen builds up. A line of precipitation can be seen when reading the plates against a black background with side lights. A continuous arc of precipitation is formed, as illustrated in Figure 4.9A, because the antigen in the left well is identical to the antigen in the right well. This smooth arc is called "a line of identity" and signifies that the antigens in the upper two wells are identical. Figure 4.9B illustrates the type of precipitation formed when one antigen (Ag_1) is in one well, another antigen (Ag_2) is in the well adjacent to it, and antibodies to each are in the central well. The pattern of crossed precipitation lines is one of "nonidentity" and signifies nonrelated antigens.

The technique of double diffusion has been modified to reduce serum volumes by using thinner gels. Double-diffusion techniques are widely used to identify particular proteins in serum or other body fluids. Many companies have kits available that contain everything needed for testing (gels, antibodies, antigen standards). A kit is particularly useful for identifying the abnormal immunoglobulin in monoclonal gammopathies or for detecting Bence Jones protein in urine and classifying the type (whether κ or λ chain) (see Chap. 9).

Radial Immunodiffusion[12,13]

Radial immunodiffusion is a single-diffusion technique whereby the antigen diffuses in a gel containing antibodies. Individual proteins or protein classes in serum may be quantitated by incorporating specific antibodies or class-specific antibodies in an agar or agarose gel containing a series of small wells at spaced intervals. Some wells are filled with measured amounts (5 to 10 μL) of solutions of the purified protein at known concentrations; others are filled with patient's serum. The antigen diffuses out of the wells, reacts with the antibody, and forms a diffuse precipitation zone around the well. This process continues until an equilibrium is reached in 2 or 3 days. The logarithm of the antigen concentration varies with the diameter of the precipitation zone in equilibrium techniques. In shorter, timed methods, the concentration varies directly with the diameter squared (Fig. 4.10). Radial immunodiffusion kits are available commercially for quantitating several individual proteins or classes of proteins present in serum, cerebrospinal fluid, or other fluids. A calibrated eyepiece should be used for measuring diameters to increase precision.

IMMUNOELECTROPHORESIS (IEP)[14]

Many proteins in such body fluids as serum or cerebrospinal fluid can be separated and identified by immunoelectrophoresis (IEP). The proteins are separated electrophoretically in the usual manner (see Chap. 9) on an agarose film. Antibody solution is then placed in a long, narrow trough parallel to the direction of electrophoresis, and incubation proceeds as in the double-diffusion technique. The proteins diffuse outward from their position after electrophoresis, and the antibodies diffuse outward from the trough. The antibodies meet their specific antigens and precipitate them in an arc, as shown in Figure 4.11. More than 20 identifiable precipitation lines can be identified in

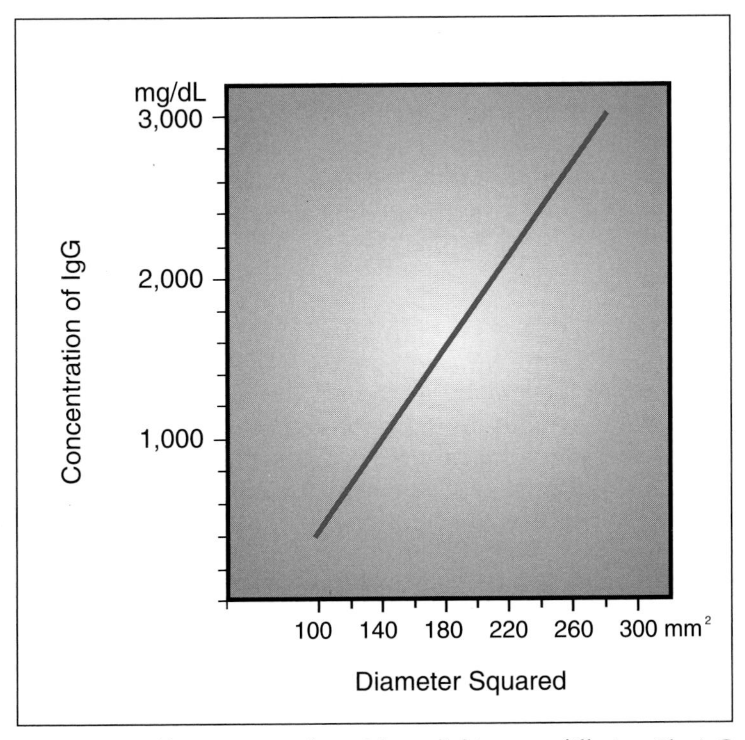

FIG. 4.10. Calibration curve for IgG by radial immunodiffusion. The IgG concentration is plotted against the diffusion ring diameter.

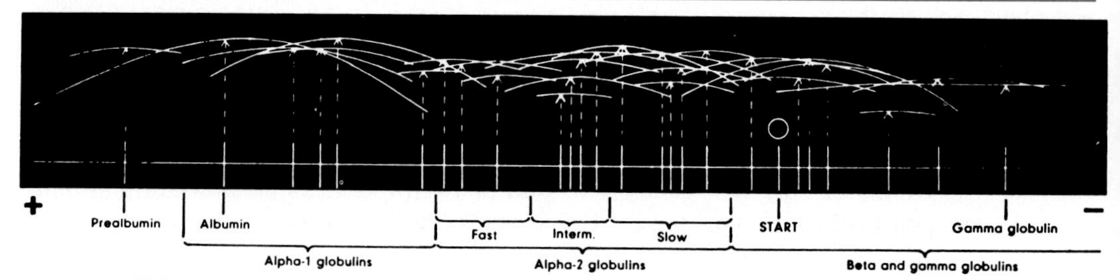

FIG. 4.11. Diagrammatic representation of an immunoelectrophoretic pattern of a normal human serum. (Courtesy of Hirschfeld, J.: Immunoelectrophoresis—procedure and application to the study of group-specific variations in sera. Science Tools, *7*:18, 1960.)

human serum when using an antihuman serum preparation in the trough. If a single antibody, such as an anti-IgG solution, is used, the only precipitation arc is that of the IgG-antibody complex.

Laurell[14] modified the IEP technique by performing the electrophoresis of the serum proteins in agarose gels containing specific antibodies. As the electrophoresis continues, the proteins migrate in the electrical field. The antigen reacts with the antibody and the complex precipitates farther and farther from the starting point as the electrophoretic migration continues until the supply of the particular antigen (protein) is exhausted (completely precipitated). The precipitation pattern is in the form of a rocket, the length of which is proportional to the protein concentration. By running a series of standards containing pure proteins, quantitation is simple. This technique is known as the rocket technique to differentiate it from IEP, which is qualitative in nature.

Immunofixation is a variant of IEP. Electrophoresis of serum or other fluids is carried out in agarose gel as for IEP. Particular proteins are precipitated in place as an antibody complex by flooding the film with specific antibody solution. After incubation, washing removes the unbound proteins and allows visualization of the bound proteins by an appropriate dye or stain.

IMMUNOASSAY BY NEPHELOMETRY OR TURBIDIMETRY[15,16]

In solution, the precipitin formed by an antigen-antibody complex scatters light. When conditions are optimized and standardized, the intensity of the light scattered (or transmitted) is proportional to the antigen concentration and is quantitated by a nephelometer. A nephelometer is similar to a spectrophotometer except that the detector is placed at an angle to the incident light and measures the intensity of scattered light instead of that transmitted. A spectrophotometer or colorimeter is used for turbidimetric assays in which the intensity of the transmitted light is compared to that of the incident light. Both techniques are compatible with automation by continuous flow or by various discrete instruments, including centrifugal analyzers. Some nephelometers use a laser beam as a light source to increase sensitivity and reduce stray light; they are not sufficiently sensitive, however, to measure the serum concentration of polypeptide hormones. Many serum proteins, such as albumin, immunoglobulins, transferrin, haptoglobin, complement components, and others, are successfully assayed by nephelometry or turbidimetry.

IMMUNOCHROMATOGRAPHY

This recently introduced technique (see Chap. 14) uses antibodies, immobilized on a chromatographic support, as the stationary phase of a thin-layer chromatography system. The technique has been applied, in a noninstrumented format, to measure certain therapeutic drugs and is simple enough to be performed in a physician's office.

REFERENCES

1. Langan, J., and Clapp, J.J. (Eds.): Ligand Assay. New York, Masson Publishing, 1981.
2. Chan, D.W.: Immunoassays: A Practical Guide. 2nd Edition. New York, Academic Press, 1992.
3. Yalow, R.S., and Berson, S.A.: Assay of plasma insulin in human subjects by immunological methods. Nature, *184*:1648, 1959.
4. Oxley, D.K.: Radioimmunoassay. *In* Interpretations in Therapeutic Drug Monitoring. Edited by D.M. Baer and W.R. Dito. Chicago, American Society of Clinical Pathologists, 1981.
5. Schall, R.F., Jr., and Tenoso, H.J.: Alternatives to radioimmunoassay: labels and methods. Clin. Chem., *27*:1157, 1981.
6. Miyai, K.: Advances in nonisotopic immunoassay. *In* Advances in Clinical Chemistry. Vol. 24. Edited by H.E. Spiegel. New York, Academic Press, 1985.
7. Voller, A., Barlett, A., and Bidwell, D.E.: Enzyme immunoassays with special reference to ELISA techniques. J. Clin. Pathol., *31*:507, 1978.
8. Schneider, R.S.: Recent advances in homogeneous enzyme immunoassay. *In* Ligand Assay. Edited by J. Langan and J.J. Clapp. New York, Masson Publishing, 1981.
9. Jolley, M.E., Stroupe, S.D., Wang, C.J., et al.: Fluorescence polarization immunoassay. I. Monitoring aminoglycoside antibiotics in serum and plasma. Clin. Chem., *27*:1190, 1981.
10. Miles, L.E.M., and Hales, C.N.: Labeled antibodies and immunological assay system. Nature, *219*:186, 1968.
11. Ouchterlony, O.: Handbook of Immunodiffusion and Immunoelectrophoresis. Ann Arbor, MI, Ann Arbor Science Publishers, 1968.
12. Mancini, G., Carbonara, A.O., and Heremans, J.F.: Immunochemical quantitation of antigens by single radial immunodiffusion. Immunochem., *2*:235, 1965.
13. Becker, W.: Determination of antisera titres using the single, radial immunodiffusion method. Immunochem., *6*:539, 1969.
14. Laurell, C.B.: Quantitative estimation of proteins by electrophoresis in agarose gel containing antibodies. Anal. Biochem., *15*:45, 1966.
15. Savory, J.: Nephelometric immunoassay techniques. *In* Ligand Assay. Edited by J. Langan and J.J. Clapp. New York, Masson Publishing, 1981.
16. Killingsworth, L.M.: Plasma protein patterns in health and disease. CRC Critical Rev. Clin. Lab. Sci., *10*:1, 1979.

QUESTIONS

1. What features in the structures of immunoglobulins make possible the potential for forming antibodies to so many different antigens?

2. How are antibodies to such small molecules (haptens) as triiodothyronine or digoxin produced for use in some form of immunoassay?

3. What are the major disadvantages of radioimmunoassay? Why is radioimmunoassay still widely used?
4. What are some of the labels used in nonisotopic immunoassays?
5. What are the differences between homogeneous and heterogeneous systems of immunoassay?
6. What are the principles involved in the fluorescence polarization immunoassay and the enzyme multiplied immunoassay technique?
7. What is a two-site or "sandwich" immunometric assay?

5 Water Balance, Osmolality, Blood Gases, pH, and Electrolytes

OUTLINE

I. BODY WATER
- A. Water compartments
- B. Water balance
- C. Electrolyte distribution
- D. Plasma volume
- E. Correction of disturbed osmolality

II. SODIUM
- A. Reference values
- B. Increased concentration (hypernatremia)
- C. Decreased concentration (hyponatremia)

III. POTASSIUM
- A. Reference values
- B. Increased concentration (hyperkalemia)
- C. Decreased concentration (hypokalemia)

IV. SERUM SODIUM AND POTASSIUM DETERMINATION
- A. Ion-selective electrodes
  - 1. Principle
- B. Flame photometry
  - 1. Principle
  - 2. Precautions
- C. Sodium and potassium in body fluids

V. CHLORIDE
- A. Serum chloride
  - 1. Determination by ISE
    - a. Reference values
    - b. Increased concentration
    - c. Decreased concentration
  - 2. Determination by coulometric-amperometric methods
    - a. Principle
- B. Cerebrospinal fluid (CSF) chloride
- C. Urinary chloride
  - 1. Procedure
- D. Sweat chloride
  - 1. Coulometric-amperometric method
  - 2. Ion-selective electrode method
  - 3. Reference values
  - 4. Elevated values

VI. ACID-BASE HOMEOSTASIS AND BLOOD BUFFER SYSTEMS
- A. Blood buffer systems
  - 1. Bicarbonate/carbonic acid buffer system
  - 2. Hemoglobin buffer system
  - 3. Phosphate buffer system
- B. Effects of CO_2, HCO_3^-, and H_2CO_3 on pH
- C. Compensation of acid-base disturbances

OBJECTIVES

After reading this chapter, the student will be able to:

1. Identify the different fluid compartments, discuss the distribution of body water between the different compartments, and determine which substances are mainly responsible for this distribution.
2. Identify the mechanisms involved in the regulation of water balance.
3. Discuss the distribution of electrolytes between the different compartments.
4. Calculate the anion gap and use the value in the differential diagnosis of acid-base disorders.
5. Calculate the approximate osmolality of serum when given the concentrations of Na^+, glucose, and blood urea nitrogen.
6. Describe the three main blood buffer systems and understand how they function.
7. Describe how CO_2 is eliminated from the body, starting with its generation in muscle.
8. Explain the different types of acidosis and alkalosis.
9. Define pH and calculate the pH when given the hydrogen ion concentration.
10. Understand the principles used for determining each constituent by means of a blood gas instrument, and discuss compensatory mechanisms that tend to counteract blood gas disorders.
11. Define oxygen content and oxygen saturation and describe methods for their measurement.
12. Understand the principles involved in the noninvasive determination of oxygen saturation.

Hospital laboratories often place top priority on providing accurate analyses of blood gases, pH, electrolytes, and osmolality with rapid turnaround time, available 24 hours each day. This goal has top priority because the vital physiologic processes of acute care patients are unstable, and important clinical decisions are based on blood gas, pH, and electrolyte test results, often on a minute-to-minute basis. Laboratory workers that provide acute care "stat" analyses not only must have a thorough understanding of the stat instruments and methods, but also must comprehend the significance of grossly abnormal test results to cope with the pressures and demands made by other members of the health care team. This chapter reviews the principal concepts in the regulation of water balance, osmolality, electrolytes, pH, and blood gases, as well as their laboratory analysis.

BODY WATER

In a clinical chemistry laboratory, water is seldom considered by itself; rather the constituents of aqueous fluids are analyzed: whole blood, plasma, serum, urine, and cerebrospinal fluid. When interpreting test results, however, the patient's body water and the forces that have acted on it must be considered. These forces move body water into and out of blood vessels, capillaries, cells, and other compartments and cause either dilution or concentration of all analytes. The ability of patients to recover from many acute illnesses or trauma is often determined by the ability of an unhealthy body to balance the forces acting on body water and electrolytes.

Water Compartments

In an average human, about 60% of body weight can be attributed to water (Table 5.1). In general, water is found both inside and outside cells and can be divided into two main spaces or compartments: *intracellular water* within cells and *extracellular water* outside cells. As shown in Table 5.1, the intracellular component accounts for 40% of body weight and the extracellular portion for

TABLE 5.1
DISTRIBUTION OF BODY WATER IN THE ADULT*

COMPARTMENT	PERCENT OF BODY WEIGHT	VOLUME (L)	PERCENT OF TOTAL BODY WATER
Extracellular			
Plasma	5.0	3.5	8
Interstitial	15.0	10.5	25
Intracellular	40.0	28	67
Total body water	60.0	42	100

*For the average 70-kg person. Body water as percent of body weight varies inversely with the fat content; an obese person has the same *amount* of water but at a lower *percentage of body weight.*

20%. The extracellular water is further subdivided into a vascular compartment (blood in the arteries, veins, and capillaries), small compartments of joint and cerebrospinal fluids, and the interstitial fluid between cells. The blood volume is approximately 5% of body weight (3.5 L in an adult), and the interstitial fluid constitutes about 15% of body weight (10.5 L in an adult). Although the pools of body water are constantly undergoing exchange, the distribution of water among compartments is remarkably constant in healthy individuals as a result of the coordination of water and electrolyte balance.

Water Balance

Each compartment of body water remains at constant volume, which implies that the rate of water loss must equal the rate of gain. This concept illustrates *water balance*. An adult drinks and takes in with food about 2500 mL of water daily. The same amount is lost daily in urine, feces, sweat, and expired air (Table 5.2). Individuals remain in water balance by regulating the intake of fluids to compensate for the daily losses. An exquisitely sensitive thirst mechanism normally regulates the desire for fluids so that the balance is maintained within relatively close limits.

The kidneys function either to excrete excess water or to reabsorb water depending on the physiologic needs (see Chap. 6). A state of fluid imbalance occurs when patients are unable to swallow, have excessive fluid losses, or are unable or unwilling to drink or retain fluids. Fluid loss may occur by bleeding, vomiting, diarrhea, excessive urination, excessive sweating, or exuding through severely damaged skin (burns or other injuries). Fluid gain or redistribution is associated with systemic edema, pulmonary edema, and ascites.

Electrolyte Distribution

The concentration of electrolytes in extracellular and intracellular water is strikingly different (Table 5.3). Na^+ is the principal cation of the extracellular fluid and comprises more than 90% of the total cations. Na^+ has a low concentration in the intracellular water, however, and constitutes only 8% of the total intracellular cations. K^+, by contrast, is the principal cation within

TABLE 5.2
DAILY WATER LOSS

SITE	VOL/DAY mL
Skin	500
Expired air	350
Urine	1,500
Feces	150
TOTAL	2,500

TABLE 5.3
CONCENTRATION OF CATIONS AND ANIONS IN EXTRACELLULAR AND INTRACELLULAR WATER

	CONCENTRATION IN WATER			CONCENTRATION IN WATER	
CATION	EXTRACELLULAR mmol/L	INTRACELLULAR mmol/L	ANION	EXTRACELLULAR mmol/L	INTRACELLULAR mmol/L
Na^+	154	15	HCO_3^-	29	10
K^+	5	150	Cl^-	111	1
Ca^{2+}	2.7	1	HPO_4^{2-}	1	50
Mg^{2+}	1.3	13.5	SO_4^{2-}	0.5	10
			Organic acid	7	
			Protein*	17	63

*Anion equivalents per liter.

cells and has a low concentration in extracellular fluid. Similar differences exist with the anions: Cl^- and HCO_3^- predominate in the fluid bathing the cells, whereas phosphate is the principal anion within cells. These differences in ionic concentration are maintained within cells by the expenditure of energy in the form of active transport mechanisms. For example, the sodium pump removes much of the Na^+ that diffuses into cells and replaces it with K^+. Energy is required to move these ions against concentration gradients.

The osmotic pressure of extracellular fluids is primarily determined by the concentration of Na^+ and its associated anions, which together account for more than 90% of the plasma particle concentration (osmolality). Na^+ concentration likewise determines the extracellular fluid volume, because water flows from or into other compartments to restore osmotic homeostasis if disturbed. Water that shifts in and out of cells to equalize osmolality affects cell size and, if extensive, can affect cell function. K^+ similarly determines intracellular osmolality to a large extent, and cells shrink (lose water) when large K^+ losses are incurred (for example, by prolonged vomiting, diarrhea, or hypersecretion of aldosterone).

Plasma Volume

The blood plasma is a subdivision of the extracellular fluid compartment; as such, its volume is determined primarily, but not entirely, by the concentration of Na^+. The capillary walls are freely permeable to ions and water, and the Na^+ concentration in plasma is essentially the same as that in the rest of the extracellular water. The hydrostatic blood pressure is the force that pushes blood through arterioles and capillaries, and it tends also to force plasma water out of the capillary bed. The hydrostatic pressure is counterbalanced by an osmotic pressure generated by plasma molecules that cannot diffuse out of the capillaries. The osmotic pressure produced by plasma proteins (usually called oncotic pressure) attracts water to the capillary lumen and is sufficient to retain fluid within capillaries. Albumin is the plasma protein that contributes the most to oncotic pressure because of its high plasma concentration and relatively low molecular weight (most mol/L of proteins in plasma). The oncotic pressure contribution of the plasma proteins amounts to only 0.5% of the total plasma osmotic pressure, however; the body content of Na^+ is the main determinant of extracellular fluid volume and plasma volume. The concentration of plasma albumin must fall by about 50% before the change in plasma oncotic pressure becomes evident as edema (passage of sufficient plasma water into the interstitial spaces to become noticeable as swelling).

Correction of Disturbed Osmolality

The body responds to disturbances in osmolality by instituting changes in water intake and excretion and not by changing body electrolytes. Dehydration is associated with elevated blood osmolality and promotes the feeling of thirst so that the subject will drink fluids if they are available. The high osmolality also stimulates the posterior pituitary secretion of antidiuretic hormone (ADH) (see Chap. 6), which promotes renal tubular water reabsorption (less water excreted as urine). Overhydration (low osmolality) inhibits the thirst mechanism to impede fluid intake and depress the secretion of ADH, thereby resulting in the excretion of a large volume of dilute urine. The following is a summary of osmolality changes and water shifts:

1. Movement of water from one compartment to another is rapid and follows osmotic pressure gradients from low to high. No energy is required.
2. The concentration of ions in intracellular water is markedly different from that in the extracellular fluid. Movement of ions against a concentration gradient is made possible by active transport. Some of the ions within cells are bound by protein and cannot move freely.
3. Cell volume varies directly with its water content; large changes can handicap cell function or destroy cells.
4. The plasma Na^+ concentration is the principal determinant of extracellular fluid osmolality and, hence, body water osmolality; a steady state exists, and the osmolality of all compartments equalizes by water shifts.
5. The body content of Na^+ determines the extracellular fluid volume, including the plasma volume.
6. The oncotic pressure of the plasma proteins, although small, counterbalances the blood pressure in the capillaries and helps to maintain the plasma volume.
7. The body responds to disturbances in osmolality by regulating water intake and excretion.

A further review of serum and urine osmolality is presented in Chapter 6.

SODIUM

The sodium ion has few physiologic properties that do not focus on its important influence on fluid distribution between the compartments. Apart from fluid balance, the absorption and excretion of the sodium ion should be recognized in context with the counter ions (Cl^- and HCO_3^-) and competing cations (K^+) within the body.

The body of the average-sized adult contains about 80 g of sodium, 35 g of which are present in the extracellular fluids. The amount of sodium in the body is relatively constant, despite variation in intake. Although the average person ingests about 3 g daily of sodium as the chloride, sulfate, or other salt, this amount is also excreted daily. Because the sodium in plasma is in equilibrium with that in the interstitial fluid, the determination of serum sodium concentration is representative of its extracellular fluid concentration.

When an ingested sodium salt is absorbed, a temporary increase in extracellular fluid volume occurs as the absorbed sodium ions (and the water that follows) equilibrate between plasma and interstitial fluid. A small, temporary exchange of sodium for potassium occurs inside the cell. Table 5.3 shows that the concentration of sodium in the fluid outside the cells is about 10 times that of sodium inside the cell. The cell, however, is permeable to sodium ion, and this differential concentration is maintained by a "sodium pump" by which Na^+ is pumped out of the cell and K^+ is pumped into the cell against concentration gradients. Adenosine triphosphate (ATP) supplies the necessary energy.

Plasma sodium is filtered in the renal glomerulus, and approximately 70% is reabsorbed in the proximal tubule; most of the remainder is absorbed in the distal tubule under the influence of aldosterone, a hormone secreted by the adrenal cortex in response to lowered blood volume (see Fig. 6.2). Aldosterone accelerates the exchange of Na^+ for K^+ across all cell walls, including those of the distal tubule. This exchange promotes the retention of Na^+ and the excretion of K^+. The reverse situation occurs with a deficiency of aldosterone. Some gonadal steroids may cause a temporary retention of salt and water; such retention sometimes occurs premenstrually.

The heart is seldom considered an endocrine organ, but the atrial myocardium releases a peptide hormone that causes a natriuresis (urinary excretion of sodium) and relaxation of vascular smooth muscle when injected into animals.[1] The hormone is referred to as atrial natriuretic factor (ANF). The precise physiologic role of ANF, the stimuli that cause its release, and its interaction with aldosterone and ADH are currently under investigation.

Reference Values. The concentration of sodium in normal serum is 136 to 145 mmol/L.

Increased Concentration (Hypernatremia). Elevated levels of serum sodium occur in (1) severe dehydration owing to inadequate intake of water, irrespective of cause, or to excessive water loss; (2) hyperadrenalism (Cushing's syndrome), in which excessive reabsorption of sodium in renal tubules occurs as a result of overproduction of adrenal steroids; (3) comatose diabetic patients after treatment with insulin as some Na^+ in cells is replaced by K^+; (4) hypothalamic injury interfering with thirst mechanisms; (5) patients fed nasogastrically with solutions containing a high concentration of protein, without sufficient fluid intake; and (6) diabetes insipidus (deficiency of ADH) without sufficient intake of water to cover the fluid loss.

Decreased Concentration (Hyponatremia). Most low serum sodium values occur in (1) a large loss of gastrointestinal secretions resulting from diarrhea, intestinal fistulas, or severe gastrointestinal disturbances of any sort; (2) the acidosis of diabetes mellitus before the coma stage, when large amounts of Na^+ and K^+ are excreted into the urine as salts of the ketoacids, with replacement of water because of thirst; (3) renal disease with malfunction of the tubular ion-exchange system of Na^+ for H^+ and K^+ (salt-losing nephritis); (4) Addison's disease, with depressed secretion of aldosterone and corticosteroids; and (5) diabetes insipidus (posterior pituitary deficiency) with compensatory intake of water.

Na^+ and K^+ are usually determined simultaneously with ion-selective electrodes (ISE) or an emission flame photometer. Details of the determination method are given at the end of the next section.

POTASSIUM

Potassium is the cation with the highest concentration within cells because it is constantly transported into the cell by the sodium pump (see the preceding) and its outward diffusion through the cell membrane is slow. Approximately 2 to 3 g of potassium are ingested and excreted daily in the form of salts. Potassium salts in the diet are absorbed rapidly from the intestinal lumen but have little effect on the plasma concentration; the rise is slight and transitory. After tissue needs are met, the remainder is excreted by the kidney. The excretion process consists of glomerular filtration, absorption in the proximal tubule, and, finally, excretion primarily by exchange for sodium ion in the distal tubules (see Chap. 6). The kidney cannot reduce the potassium excretion to nearly zero, as it can for sodium.

The close control of the concentration of potassium in extracellular fluids is essential because elevated concentrations of K^+ ($>$ 7.5 mmol/L) may seriously inhibit the irritability of muscle, including the heart, to the point of paralysis or cessation of heartbeat. Low serum potassium values ($<$ 3.0 mmol/L) are also dangerous because they increase muscle irritability and can cause cessation of the heartbeat in systole (contraction); low serum potassium concentration can be rectified by intravenous injection of appropriate solutions. The laboratory must notify the attending physician immediately whenever a seriously high or low potassium value occurs so that appropriate action can be taken in time. These changes in cardiac muscle irritability caused by either high or low potassium concentration may be reflected in altered electrocardiographic patterns.

Reference Values. The normal serum concentration of potassium varies from about 3.5 to 5.5 mmol/L. It may be a little higher in the newborn, but it soon adjusts to adult values.

Increased Concentration (Hyperkalemia). Because the concentration of potassium within cells is so great, its concentration in plasma rises when it leaves the cells at a greater rate than the kidney can excrete it. This overload occurs in conditions of anoxia and in both metabolic and renal tubular acidosis. It also occurs with normal intake of potassium when output of urine is decreased. Conditions of shock or circulatory failure usually produce hyperkalemia. Adrenal cortical insufficiency, particularly a decreased production of aldosterone, is accompanied by an elevation of serum potassium. Elevated serum potassium values commonly accompany chronic renal insufficiency because a tubular malfunction interferes with the exchange of sodium for hydrogen or potassium ion and promotes potassium ion retention.

An elevation of blood potassium may be an artifact, for example, a consequence of red blood cell lysis during phlebotomy, specimen transport, or centrifugation. Inspect the color of the serum or plasma and report the presence or absence of hemolysis.

Decreased Concentration (Hypokalemia). A decreased concentration of potassium in serum occurs as a result of either a low intake over a period of time or an increased loss of potassium through vomiting, diarrhea, gastrointestinal fistulas, or long-term therapy with diuretics. The fluids of the gastrointestinal tract contain relatively high concentrations of potassium, and their removal or loss can produce serious deficits. Increased secretion of adrenal steroids, primarily aldosterone, results in excessive potassium loss through the kidneys and a low serum potassium concentration. Certain carcinomas that secrete ACTH (adrenocorticotropic hormone) cause a lowering of serum K^+ concentration through stimulation of the adrenal cortex to produce excessive amounts of steroids (see Chap. 9). Some diuretics promote K^+ excretion.

SERUM SODIUM AND POTASSIUM DETERMINATION

Ion-Selective Electrodes

The serum or blood concentrations of Na^+, K^+, and Cl^- are readily and rapidly measured by ion-selective electrodes (ISE) available from many manufacturers. An ISE measurement corresponds to the activity of the ion (its dissociated or completely ionized fraction per unit volume of water) and not to its mass concentration (total amount, both ionized and un-ionized per unit volume of fluid).[2,3] Thus, measurements may be made directly (undiluted sample) on whole blood or serum; this process requires that the standards used must be similar in ionic strength to that of the blood samples.

With some instruments, serum or plasma samples are diluted before assay (indirect method). The values obtained agree closely with those analyzed by flame photometry, during which sample dilution is always carried out. In a dilute solution, the activity of an ion is approximately the same as its concentration.

In most clinical situations, the ISE results for Na^+ and K^+ obtained with undiluted specimens agree within 2% of those obtained by ISE on diluted samples or analyzed by flame photometry. A serious discrepancy occurs, however, in clinical situations of hyperlipidemia or in hyperproteinemia (for example, in advanced multiple myeloma). The direct ISE values (undiluted) are higher than the indirect values because the ions in these abnormal blood samples are distributed in a smaller volume of water. The volume occupied by the increased concentration of lipids or proteins can be considerable. Ion concentrations obtained by flame photometry and by indirect ISE in these situations are in error (too low) unless some correction is made for the volume displacement.[2,3]

Principle. Whole blood, plasma, or serum is analyzed undiluted in some instruments. In others, serum or plasma is diluted in a high-ionic-strength buffer solution. The Na^+ electrode has a glass membrane that selectively exchanges Na^+ 300 times as rapidly as K^+ and that is insensitive to H^+. By contrast, the K^+ electrode is coated with a valinomycin membrane that exchanges K^+ about 1000 times as fast as Na^+ and that is insensitive to H^+. As Na^+ and K^+ associate with their respective ISE membranes, a potential (voltage) change occurs that is directly proportional to the selective ion concentration. The electrodes are calibrated with solutions of known concentrations.

Flame Photometry

The sodium concentration in plasma or serum is also determined by means of a flame emission spectrophotometer. Most photometers measure the sodium and potassium concentrations simultaneously, using lithium or cesium as an internal standard. The directions for carrying out the tests vary somewhat with the make of the instrument, so the manufacturer's directions should be followed. The principles are the same, however, for all makes and types of flame photometers.

Principle. A dilute solution of plasma or serum is aspirated into a hot flame. Some of the atoms of the alkali metal group (sodium, potassium, lithium) are temporarily activated in the hot flame as the electrons move into a higher orbit. Upon return to the ground state, they emit the light characteristic for the particular element. With a sodium filter (590 nm), the intensity of the emitted light is proportional to the concentration of the sodium ion. Most of the instruments have three separate phototubes with interference filters for Na^+, K^+, and Li^+ or Cs^+, respectively, positioned around the emitted light beam. The Na^+ and K^+ concentrations are measured simultaneously, using the intensity of the lithium or cesium light as an internal standard.

Serum is usually diluted 1:100 or 1:200 with a dilute solution of nonionic detergent, depending on the particular instrument used. When lithium is used as an internal standard, a specified concentration of $LiNO_3$ or LiCl is incorporated into the diluting solution. The detergent facilitates the aspiration of sample and the formation of an aerosol before introducing the dilute sample into the flame, and the internal standard serves to minimize the effects of fluctuating gas pressure upon emission light intensity.

The main areas in which differences may be found among commercially available flame photometers are:

1. Use of lithium or cesium as an internal standard.
2. Manual or automatic dilution of serum.
3. Type of fuel mixture (natural gas or propane with air or oxygen).
4. Manual or automated introduction of sample.

Precautions. No matter which instrument is used, the following precautions must be observed:

1. All glassware used in preparing solutions and standards must be scrupulously clean and rinsed with deionized water. Bottles of polyethylene or borosilicate glass should be used for storage of solutions. Soft glass containers must not be used because sodium ion can be leached out of the glass.
2. Disposable plastic cups are recommended for sample handling to reduce the possibility of contamination by washing solution.
3. The aspirator line and burner should be flushed and cleaned periodically.
4. Proper pressures of gas and air (or oxygen) must be maintained.
5. As with all quantitative analyses, the pipetting of sample and diluent must be accurate and reproducible.
6. Standards and control sera must be run frequently to check on the various aspects of the system.

Sodium and Potassium in Body Fluids

The concentration of Na^+ and K^+ in other body fluids is determined by flame photometry or ISE in a manner similar to that in serum; the dilution may have to be modified according to the concentration of these ions. Usually, the Na^+ and K^+ concentrations in cerebrospinal fluid, exudates, transudates, and juices collected from various types of fistulas (pancreatic, duodenal, bile) are within the range of the instrument if treated as serum. In urine, however, the amount excreted depends more directly on the intake, particularly for Na^+.

In an average diet, the 24-hr excretion of K^+ usually varies between 30 and 90 mmol, but can go higher. For Na^+, the 24-hr excretion on a usual diet may vary from 40 to 220 mmol, but it can fall to low levels for the patient with a severely restricted salt intake. With some instruments, the urinalysis for Na^+ and K^+ can be performed exactly as for serum except that different standards, close to the urine concentration, are used.

CHLORIDE

Chloride is the extracellular anion in the highest concentration in serum; it plays an important role in maintaining electrolyte balance, hydration, and osmotic pressure. Because Cl^- cannot accept H^+ at physiologic pH (HCl is a strong acid), it cannot act as a buffer. Its concentration does vary inversely with HCO_3^- at times because electrochemical neutrality must be maintained always. In metabolic acidosis, the $[Cl^-]$ rises as the $[HCO_3^-]$ decreases, whereas the reverse is true in metabolic alkalosis. With the exception of the red cell, chloride ion does not enter cells and is confined to the extracellular space.

About 2.5 g of chloride are ingested daily in the normal diet as a salt of Na^+, K^+, Ca^{2+}, or Mg^{2+}. Chloride is readily absorbed in the intestine and is removed from the body by excretion in the urine and in sweat. Excessive amounts of both Na^+ and Cl^- may be lost during periods of intense perspiration, thus requiring a supplementary intake of NaCl to prevent deficit.

Serum chloride concentration is usually measured when an electrolyte determination is requested, but it provides the least clinical information of the four constituents. The best estimate of the plasma osmolality is represented by the $[Na^+]$, which usually constitutes about 90% of the total cations. The Cl^- determination, however, makes possible the calculation of the anion gap, an estimate of unmeasured anions, such as sulfate, phosphate, or organic acids.

$$\text{Anion gap} = Na^+ - (Cl^- + HCO_3^-).$$

Under normal circumstances, the anion gap is about 8 to 12 mmol/L, but can exceed 25 mmol/L in severe ketoacidosis or lactic acidosis, in which ketoacids (β-hydroxybutyric and acetoacetic) or lactic, respectively, are greatly increased. A large anion gap should be expected in severe renal disease, diabetic ketoacidosis, lactic acidosis, or acute poisoning with aspirin, ethylene glycol, and methanol, in which acid metabolites are formed. Calculation of the anion gap also aids in quality control of electrolyte determinations.[4] If the anion gap is low or high in a series of patients, a consistent analytical error in at least one of the constituents is highly probable. The $[Cl^-]$ in hypokalemic alkalosis should be measured because the condition cannot be corrected without the addition of both potassium and chloride ions. When the plasma $[K^+]$ is low, K^+ diffuses from tissue cells, including the renal tubular cells, and is replaced by H^+. Thus, more H^+ and less K^+ are available for excretion into the urine. An increase of $[H^+]$ in the urine leads to an effective increase in the reabsorption of HCO_3^-. The loss of H^+ and the increase of HCO_3^- cause the alkalosis, which cannot be reversed until K^+ becomes available for entry into cells and Cl^- is increased to displace HCO_3^-.

Serum Chloride

According to the 1992 Comprehensive Survey of the College of American Pathologists,[5] more participating laboratories determined serum chloride concentration by ISE than by any other method, by a large margin. The predominant technologies were ISE (88%), coulometric-amperometric method (5%), and spectrophotometric methods with mercuric thiocyanate (5%) or ferric chloride (2%).

Determination by ISE

Many instruments have ISEs that determine the concentration of serum (or whole blood) Cl^- at the same time as Na^+ and K^+ are measured by their respective electrodes. The same problem of volume displacement (in hyperlipidemia and hyperproteinemia) exists for the assay of serum Cl^- by the indirect ISE technique as is described for Na^+ in the previous section.

The membrane of the Cl^- ISE is a composite of Ag_2S and AgCl. It selectively admits all halogen ions (F^-, Cl^-, Br^-, and I^-), but the only possible interference comes from Br^- in individuals who have been taking medications containing bromide for an extended period of time. At the upper levels of the therapeutic range, the serum Br^- concentration approaches 15 mmol/L (1200 μg/mL). All serum Cl^- methods,[6] however, fail to distinguish Br^- from Cl^-.

Reference Values. The normal Cl^- concentration is 98 to 108 mmol/L.

Increased Concentration. High concentrations of Cl^- are usually found in dehydration, certain types of renal tubular acidosis, and in patients who lose CO_2 by hyperventilation (respiratory alkalosis) after stimulation of the respiratory center by drugs, hysteria, anxiety, or fever.

Decreased Concentration. Decreased concentrations of serum Cl^- occur in metabolic acidosis of various types. Uncontrolled diabetes is characterized by an overproduction of ketoacids whose anions replace Cl^-; in renal disease, phosphate ion retention accompanies impaired glomerular filtration, with a concomitant decrease in plasma Cl^- concentration. A deficit of body Cl^- and a decreased serum $[Cl^-]$ accompany prolonged vomiting caused by pyloric stenosis or high intestinal obstruction. Gastric secretions contain a high concentration of H^+ and Cl^-. Low serum values are also found in salt-losing nephritis and in metabolic alkalosis, in which the $[HCO_3^-]$ is increased and $[Cl^-]$ falls reciprocally. Low values are usually encountered during a crisis in Addison's disease (adrenal cortical deficiency); both Na^+ and Cl^- are low.

Determination by Coulometric-Amperometric Methods

Principle. A coulometric-amperometric method uses an instrument that generates silver ions at a constant rate from a silver wire anode immersed in a solution containing the sample to be measured (serum, urine, cerebrospinal fluid).[7] The silver ions combine with the chloride ions in the sample to form insoluble AgCl salt. When Cl^- is completely precipitated, the first excess of Ag^+ greatly increases the conductivity of the solution, and the subsequent current surge triggers a relay circuit to shut off the current and stop an electrical timer. This provides an accurate measure of the time of the current flow or, when properly calibrated, can give a direct readout of the chloride concentration. The amount of silver ion generated is proportional to the time of current flow when the current strength is kept constant; the $[Cl^-]$ is equal to the $[Ag^+]$ generated when corrected for a blank determination.

Cerebrospinal Fluid Chloride

The chloride concentration in normal cerebrospinal fluid is higher than that in serum because the protein concentration in cerebrospinal fluid is low; hence, there are practically no proteinate anions. The normal concentration of Cl^- in cerebrospinal fluid ranges from 115 to 132 mmol/L. The $[Cl^-]$ in cerebrospinal fluid falls to approximately that in serum in cases of bacterial meningitis when the protein concentration in cerebrospinal fluid becomes greatly elevated. The cerebrospinal fluid Cl^- concentration is determined as in serum.

Urinary Chloride

The amount of urinary Cl^- varies greatly with the intake. An adult eating an average diet may excrete from 110 to 250 mmol/d. Patients on low-salt diets, however, excrete little Cl^-.

Procedure

The test should only be performed upon an accurately timed collection. Test the pH of the urine and adjust to approximately pH 3 with dilute nitric acid. Then measure the chloride concentration as for serum. For a sample containing a high concentration of chloride, dilute as necessary to bring into the range at which an accurate analysis can be carried out.

Sweat Chloride

Significant amounts of Na^+ and Cl^- appear in sweat, although the concentration is much lower than that in serum. In the congenital disease cystic fibrosis, the concentration of these ions in sweat is elevated because of a defect in the gene and corresponding protein known as the cystic fibrosis transmembranous conductance regulator. A sweat Cl^- test is usually requested to screen for this disease when suspected, primarily in children.

An adequate sweat sample (> 50 mg) is difficult to collect from a small child, and an attempt to induce total body sweating is dangerous. An instrument is available, however, to introduce a sweat-inducing drug, pilocarpine, into a limited area of skin by means of an electric current flowing between two electrodes attached to a limb (child) or the back (infant). This technique, called iontophoresis, moves the pilocarpine from a pad under the positive electrode into the skin toward the negative electrode. Local sweating is induced in the skin area where the pilocarpine penetrates. After 5 min of iontophoresis, the current is turned off, and the electrodes are removed. The pilocarpine area is quickly sponged off and treated as follows, depending on the method of measuring the sweat chloride.

Coulometric-Amperometric Method (Recommended by the Cystic Fibrosis Foundation for Cl^- in sweat)[8]

Instruments are commercially available that carry pilocarpine into a small area of skin by means of an electric current (iontophoresis); the pilocarpine induces local sweating. Carry out the following steps:

1. On removing the electrode after the iontophoresis, sponge the skin and wipe dry.

2. With forceps, place an accurately weighed (to 0.1 mg) 25-mm filter paper disk on the pilocarpine area, cover with the plastic cap, and tape down. Before doing so, the filter paper is weighed in a closed vial or weighing bottle.
3. After 15 min, quickly remove the plastic cap, and with forceps, place the damp filter paper in the weighing bottle, close the cover, and weigh. Calculate the weight of the sweat. The Cystic Fibrosis Foundation requires a minimum of 50 mg of sweat before the test is considered valid. Add 2 mL of acid-gelatin or acid-polyvinyl alcohol to the weighing bottle containing the filter paper. After 15 min with agitation, the filter paper is squeezed and removed, and the chloride concentration is determined as for serum on the chloridometer.

Ion-Selective Electrode Method[9]

Orion has an iontophoresis instrument and an ISE for Cl^-. Perform the following steps:

1. Introduce pilocarpine into the skin (forearm, leg, or back) by means of iontophoresis.
2. Place a small plastic cap over the pilocarpine area; tape down so that the sweat can accumulate and not evaporate.
3. Standardize the electrode with known Cl^- standards of 20 and 100 mmol/L. Then test a series containing 10, 40, and 60 mmol/L to confirm whether it is reading properly.
4. After 10 min, remove the cap from the skin, quickly place the electrode on the moist skin, and take a reading.

Reference Values. Normal sweat chloride concentration varies from about 5 to 40 mmol/L.

Elevated Values. The sweat Cl^- concentration of most of the patients with cystic fibrosis is above 60 mmol/L and may reach as high as 100 to 140 mmol/L. Those with a concentration in the range of 35 to 50 or 60 mmol/L should be rechecked on several occasions to verify the result; cystic fibrosis is too serious a diagnosis to make without adequate laboratory verification.

Molecular biology methods have been applied to confirm positive sweat Cl^- screening tests (see Chap. 16).

ACID-BASE HOMEOSTASIS AND BLOOD BUFFER SYSTEMS

Acid-base disorders are common life-threatening states in patients with a wide variety of diseases or injuries. Why is acid-base imbalance such a common consequence of illness? Probably because so many different target tissues and homeostatic mechanisms can be influenced by different diseases or trauma. Every metabolic imbalance is a consequence of *a formation rate NOT EQUALING a removal rate*, which leads to either an accumulation or a depletion of metabolites. *To maintain acid-base homeostasis, the rate of acid (or base) production and absorption must equal the rate of acid (or base) excretion, expiration, and metabolism.* The circulation of buffers in the blood and pulmonary and renal mechanisms act in coordination to prevent the accumulation of acid or base metabolites.

When fats, carbohydrates, and proteins are catabolized for energy purposes, the carbon atoms in the molecules are converted to CO_2 if the oxidation is complete. Although CO_2 forms a weak acid, H_2CO_3, when dissolved in water, the reaction is reversed in the lung alveoli, and the CO_2 is rapidly eliminated by the lungs during respiration. This fortunate reversal entails the elimination of approximately 20,000 mmoles of CO_2 per day. Incomplete oxidation of metabolites, however, causes the formation of nonvolatile acids (those that cannot be exhaled). This formation requires the kidney to eliminate about 50 mmoles of acid daily when all systems are functioning normally. The amount of acid may be greatly increased in certain diseases. The kidney is the organ involved in the excretion of nonvolatile acids, and the lung is responsible for the elimination of H_2CO_3 as it decomposes into CO_2 and H_2O.

When fats, proteins, and carbohydrates are catabolized, the hydrogen atoms of the carbon chains are converted to H^+. The H^+ is transported with an electron along a chain of coenzymes in mitochondria (the respiratory chain) until finally the hydrogen ions with electrons are transferred to atmospheric oxygen to form water (Fig. 5.1). This vital process is known as electron transport. Under normal circumstances, the supply of oxygen is adequate and no appreciable increase in H^+ occurs. Strenuous exercise, however, may cause a temporary shortage of O_2, with local buildup in the concentration of H^+.

In conditions of oxygen deficiency (anoxia), acidosis develops owing to the body's inability to transfer all H^+ to O_2 to form water. Anoxia may be caused by poor gaseous exchange in the lungs, an obstruction in the airways, or a poor delivery system (low blood pressure, heart failure, severe anemia). Hence, the development of a system for reducing the concentration of H^+ is a biologic necessity to protect tissue cells and vital functions. Several buffer systems preserve the internal environment by reducing the concentration of H^+. This reduction is accomplished by the reaction of buffer salts with H^+ to form weak (relatively undissociated) acids. The principal body buffer systems follow.

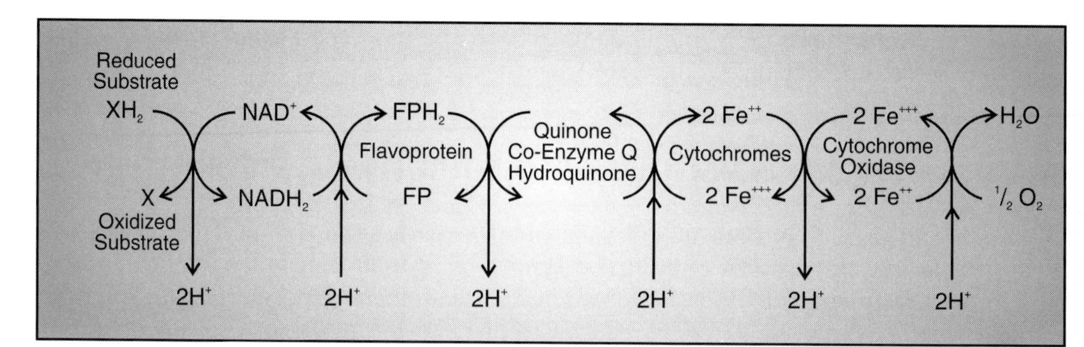

FIG. 5.1. Transport of electrons and H^+ in the respiratory chain in mitochondria. Reduced substrates, such as lactate and malate, are oxidized to pyruvate and oxaloacetate, respectively.

Blood Buffer Systems

Bicarbonate/Carbonic Acid Buffer System

$H_2CO_3 \longleftrightarrow H^+ + HCO_3^-$ is the most important buffer system in plasma because of the high concentration of HCO_3^- and the readiness with which H_2CO_3 may be increased through diminished lung activity or decreased by blowing off CO_2 through increased pulmonary ventilation. The equilibria are

$$\text{Equation 1: } CO_2 + H_2O \overset{(a)}{\longleftrightarrow} H_2CO_3 \overset{(b)}{\longleftrightarrow} H^+ + HCO_3^-$$

Reaction (a) proceeds slowly to the right in plasma, but carbonic anhydrase, an enzyme in erythrocytes and renal tubular cells, greatly speeds up the formation of H_2CO_3. The equilibrium point of reaction (b) is far to the right, so the net effect in renal and red cells is the formation of $H^+ + HCO_3^-$ from CO_2 and water as both reactions proceed to the right. The increased $[HCO_3^-]$ in red blood cells is followed by the diffusion of HCO_3^- into plasma, accompanied by the passage of Cl^- into the erythrocyte to maintain ionic balance. This process, the chloride shift, is reversed in the lungs when the reactions in Equation 1 proceed to the left as CO_2 is exhaled. CO_2 is always being produced through the catabolism of fat, carbohydrate, or protein. Its rate of elimination by the lungs depends on the rate and/or depth of respiration.

Hemoglobin Buffer System

Hemoglobin, the protein present in high concentration in red blood cells (erythrocytes), binds oxygen in the lungs and releases it in tissues. The oxygenated form of hemoglobin is a much stronger acid than the deoxygenated form. Consequently, when the oxygenated hemoglobin ($KHbO_2$) reaches the tissues where CO_2 and, therefore, H_2CO_3 are generated, the following reaction takes place:

$$KHbO_2 + H_2CO_3 \rightarrow HHb + K^+HCO_3^- + O_2$$

The weak acid, deoxygenated hemoglobin (HHb), is formed as O_2 is released and made available to the tissues. As previously described, the HCO_3^- diffuses into plasma and is replaced in the red blood cell by Cl^-. The reaction is reversed in the lungs. This buffer system accounts for about 30% of the buffering capacity of whole blood, and the bicarbonate system accounts for about 65%; in plasma, however, the bicarbonate system supplies about 95% of the buffering capacity.

Phosphate Buffer System

The phosphate buffer system is a minor component of the total blood buffer system, but does play an important role in the elimination of H^+ in the urine.

$$HPO_4^{2-} + H^+ \longleftrightarrow H_2PO_4^-$$

In the plasma at pH 7.4, 80% of the phosphate is in the form of HPO_4^{2-}, but in acid urine, the bulk of it exists as $H_2PO_4^-$ as the preceding reaction shifts to the right.

Plasma proteins can also accept H^+ to a minor extent and, therefore, serve as a buffer, but the total buffering capacity from this source is negligible compared to the contributions of bicarbonate and hemoglobin.

Effects of CO_2, HCO_3^-, and H_2CO_3 on pH

As related before, CO_2 is the ever-present product of oxidative metabolism. CO_2 is also the source of formation of plasma HCO_3^- as H_2CO_3 dissociates into H^+ and HCO_3^-. The H^+ may react with a buffer salt to form an undissociated acid, as happens in erythrocytes and body fluids, or may be excreted by the renal tubular cells into the urine, either in exchange for Na^+ or as an undissociated ketoacid. The H^+ in the lumen usually converts HPO_4^{2-} to the less dissociated form, $H_2PO_4^-$; in acidosis when NH_3 is formed and excreted by tubular cells, H^+ combines with NH_3 to form NH_4^+. Under normal circumstances, no ketoacids are available for excretion, and little NH_3 is formed by the kidney, but both processes assume great importance when acidosis and ketosis are present.

The bicarbonate buffer system is the dominant buffer system in plasma. The pH of the plasma depends on the ratio $[HCO_3^-]/[H_2CO_3]$ as indicated by the Henderson-Hasselbalch equation (see Chap. 2):

$$pH = 6.1 + \log [HCO_3^-]/[H_2CO_3]$$

Because the normal pH of plasma is 7.4, the log of $[HCO_3^-]/[H_2CO_3]$ must be 1.3. The antilog of 1.3 is 20, which means that the pH of plasma is 7.4 when the ratio of $[HCO_3^-]/[H_2CO_3]$ is 20:1. In normal adults, the concentration of H_2CO_3 is usually 1.35 mmol/L; the concentration of HCO_3^- is 20 times as great, or 27 mmol/L.

Deviations from the normal blood pH occur when the ratio of $[HCO_3^-]/[H_2CO_3]$ departs from the 20:1 ratio. In metabolic acidosis, the concentration of HCO_3^- falls more than 20 times as much as the H_2CO_3 concentration, thereby lowering the blood pH. In metabolic alkalosis, the $[HCO_3^-]$ increase greatly exceeds that of $[H_2CO_3]$, with a consequent rise in pH. Thus, metabolic acidosis is characterized by a decrease in $[HCO_3^-]$, and metabolic alkalosis is characterized by an increase of $[HCO_3^-]$.

The concentration of H_2CO_3 responds rapidly to the rate and depth of respiration, which is another way of saying that it is responsive to the rate at which CO_2 is eliminated by the lungs. Changes in pH attributed primarily to increases in $[H_2CO_3]$ are referred to as respiratory acidosis, and as respiratory alkalosis when $[H_2CO_3]$ is decreased. Respiratory acidosis may occur as a result of an impediment to gaseous exchange in the lungs, as in chronic obstructive lung disease (thickening of the alveolar membranes), in blockage of air passages, or in central nervous system depression of respiration, which could result from overdosage with opiate drugs or other depressants. In these instances, the concentration of H_2CO_3 in blood rises proportionately more than that of HCO_3^-, so the ratio $[HCO_3^-]/[H_2CO_3]$ falls, even though the $[HCO_3^-]$ may be higher than normal. The reverse effect, respiratory alkalosis, occurs in hyperventilation caused by anxiety, drugs (as in aspirin overdosage), or central nervous system stimulation. Even though the concentration of HCO_3^- may be lower than normal, the concentration of H_2CO_3 falls even more, relative to the $[HCO_3^-]$, thus causing an increase in the ratio of $[HCO_3^-]/[H_2CO_3]$ and hence in the pH.

Primary changes in $[HCO_3^-]$ arising from alterations in nonvolatile acids or bases cause a disturbance in the $[HCO_3^-]/[H_2CO_3]$ ratio, thereby producing a *metabolic* (nonrespiratory) acidosis or alkalosis.

Compensation of Acid-Base Disturbances

Disturbances in blood pH are usually compensated to a greater or lesser extent by appropriate responses of the respiratory and renal systems, insofar as they are able. For example, an accumulation of nonvolatile acids, as in renal failure or diabetic ketosis, results in a metabolic acidosis manifested by a decreased $[HCO_3^-]$. To minimize the resultant fall in pH, pulmonary ventilation is increased, thereby reducing the $[H_2CO_3]$, increasing the $[HCO_3^-]/[H_2CO_3]$ ratio, and raising the pH. The kidneys also try to compensate by increasing the rate of formation of NH_3 from glutamine and excreting NH_4^+ salts in the urine. Conversely, metabolic alkalosis may be compensated in whole or in part by renal excretion of HCO_3^- and depression of the respiration rate, both of which decrease the $[HCO_3^-]/[H_2CO_3]$ ratio and lower the pH. The compensatory changes produced by alterations in respiratory rate are rapid in contrast to the relatively slow modifications induced by renal changes.

CO_2 is transported in the blood by both erythrocytes and plasma. It exists in the form of HCO_3^-, H_2CO_3, and carbamino-bound CO_2. This latter fraction consists of CO_2 reacting with the free amino group of a protein:

$$CO_2 + R\text{-}NH_2 \longleftrightarrow R\text{-}NHCOOH$$

The amount of CO_2 bound as a carbamino complex in serum proteins is small, but the amount carried as a carbamino group in hemoglobin may amount to 10 to 30% of the total. The carbamino CO_2 cannot be measured as a separate entity in a routine clinical chemistry laboratory, but usually appears lumped together with the concentration of HCO_3^-.

The following relations exist:

$$\text{Equation 2: Total } CO_2 \text{ or } CO_2 \text{ content} = [HCO_3^-] + [H_2CO_3]$$

where H_2CO_3 represents the sum of the concentrations of undissociated H_2CO_3 and CO_2, which is physically dissolved in the plasma. Because both of these concentrations are proportional to the partial pressure of CO_2, designated as pCO_2, then $[H_2CO_3] = k \times pCO_2$. The value for k has been determined and appears in Equation 3.

Equation 3: $[H_2CO_3] = 0.0301 \times pCO_2$

Equation 4: When Equations 2 and 3 are combined, the following is obtained:

$$[HCO_3^-] = CO_2 \text{ content} - [H_2CO_3] = CO_2 \text{ content} - 0.0301 \times pCO_2$$

Substitution of the above in the Henderson-Hasselbalch equation gives:

$$\text{Equation 5: pH} = 6.1 + \log \frac{CO_2 \text{ content} - 0.0301 \times pCO_2}{0.0301 \times pCO_2}$$

All three components, pH, [HCO_3^-], and pCO_2, can be measured in the laboratory. If only two are measured, the third can be obtained by calculation.

The analysis of either HCO_3^- concentration or total CO_2 content is included when electrolytes are ordered as a laboratory test, but the specific form of CO_2 which is measured depends on the particular instrumentation.

BLOOD pH

Blood pH, measured directly on heparinized blood collected anaerobically, is seldom measured alone, and is requested most frequently as part of a blood gas analysis that includes pH, pCO_2, and pO_2. Single glass electrodes are available for the determination of blood pH, however. All the instruments for pH analysis have a probe or capillary tip that leads directly to a micro glass electrode enclosed in a water jacket maintained at 37°C. The principles of operation are the same whether for a single-test or a multi-test instrument.

Principle. In a pH electrode, a thin glass membrane separates the blood from a silver-silver chloride indicator electrode. The glass electrode is connected by a salt bridge to a reference electrode (usually a calomel electrode, $Hg\text{-}Hg_2Cl_2$). A difference in H^+ activity across a glass membrane causes a change in the potential difference between the indicator and reference electrodes. The potential difference is registered by a voltmeter and converted to direct readout of pH (see Chap. 3).[10,11]

The pH meter is calibrated with two buffers, the pH of which is prepared with exactitude (usually pH 6.841 and 7.386). These are available commercially. *Notes:* In all systems, the accuracy of the measurement depends on the following:

1. Accurate standardization of the pH meter with two different standard solutions at pH levels within those compatible with life.
2. Collection of blood anaerobically and without stasis so that no loss of CO_2 from the blood and no buildup of CO_2 in the vein occurs as the result of stagnant flow.
3. Adequate cleaning of the probe and glass electrode to avoid the buildup of a protein film or the growth of bacteria.
4. Accurate and constant maintenance of the temperature of the glass electrode, because the pH is temperature-dependent; the temperature of the water jacket should be maintained at 37 ± 0.05° C. The blood pH also must be corrected to that of the patient's body temperature, because significant deviations occur in patients with high fever or low body temperature. The pH of the blood varies inversely with the temperature; thus, for patients with a fever, the correction is *negative*, and for those in hypothermia (as in open heart surgery), the correction is *positive*. The correction factor for whole blood is 0.0146 × (37.0-t°) and 0.012 (37.0-t°) for plasma, where t° is the body temperature of the patient.

Blood pH Reference Values. The normal range of blood pH is 7.35 to 7.45, with a mean in arterial blood of 7.40. Venous blood is usually 0.03 pH units lower than arterial blood.

Increased pH. The blood pH is increased when a primary increase in [HCO_3^-] occurs, as in metabolic alkalosis (usually following prolonged ingestion of antacids, overdosage with some steroid hormones, or in patients with tumors producing ACTH). Increased blood pH is also present following a primary decrease in [H_2CO_3] or pCO_2 caused by hyperventilation. Some degree of compensation may take place in an attempt to restore a normal ratio of [HCO_3^-]/[H_2CO_3].

Decreased pH. A decrease in blood pH always occurs when the ratio of [HCO_3^-]/[H_2CO_3] falls below 20:1. This fall takes place in conditions of metabolic acidosis typified by a fall in [HCO_3^-] (commonly found in chronic renal disease and uncontrolled diabetes mellitus). The primary defect can also be respiratory, with the greatest change an increase in [H_2CO_3] or pCO_2. Hypoventilation and pulmonary disease are the principal causes.

SERUM CO_2 CONTENT (TOTAL CO_2)

The CO_2 content consists of the sum of the following concentrations: HCO_3^-, undissociated H_2CO_3, dissolved CO_2, and carbamino-bound CO_2; the HCO_3^- fraction is by far the largest (95% of the total) and accounts for all but approximately 2 mmol/L of the CO_2 content. The carbamino fraction is negligible in serum, but is appreciable in whole blood because of the presence of hemoglobin. Plasma or serum is usually taken for the determination of CO_2 content because extraneous factors, such as erythrocyte count or degree of oxygen saturation, do not affect it. The determination of CO_2 content in whole blood, however, is influenced greatly by these factors. In general, the CO_2 content of serum or plasma is obtained by automated methods or pCO_2 electrode.

Principles of Analysis

Automated Enzymatic Method

All forms of CO_2 present in serum are converted to HCO_3^- by the addition of base. The HCO_3^- is then converted to oxaloacetic acid by phosphoenolpyruvate carboxylase. The extent of oxaloacetic acid formation can be monitored spectrophotometrically by measuring its conversion to malate by malate dehydrogenase with a corresponding consumption of NADH*, a UV light chromogen.

*NADH = nicotinamide adenine dinucleotide in reduced form.

Automated Colorimetric Method[12]

The CO_2 appearing in serum as HCO_3^-, H_2CO_3, or carbamino-bound CO_2 is released by the addition of acid. The gaseous CO_2 is dialyzed through a silicone-rubber gas-dialysis membrane into a buffer solution of cresol red at pH 9.2. The CO_2 diffuses through the membrane and lowers the pH of the buffered cresol red solution. The decrease in color intensity is proportional to the CO_2 content and is measured in a spectrophotometer at 430 nm. Other continuous-flow methods have used phenolphthalein as the indicator.

Determination by pCO_2 Electrode

In some instruments, a pCO_2 (Severinghaus) electrode measures the CO_2 released by addition of acid to the serum sample. The electrode consists of a glass electrode surrounded by a weak bicarbonate solution enclosed in a silicone membrane; the membrane is permeable to CO_2 gas and nonpolar compounds but is impermeable to water and ions. The concentration of serum HCO_3^- is obtained by calculation (program in instrument) or by nomogram (Fig. 5.2) after measurement of the pCO_2 and pH. Serum CO_2 diffuses through the membrane and lowers the pH of a bicarbonate solution between the membrane and the pH electrode. The rate of pH change is proportional to the CO_2 content in the sample. The rate signal is compared to an identical electrode used as a reference, and the difference is converted to serum CO_2 concentration.

Reference Values. The CO_2 content of serum varies from about 22 to 30 mmol/L for healthy adults. The concentration in premature newborns varies from 14 to 26 mmol/L; for infants, it ranges from 20 to 26 mmol/L.

Increased Concentration. The CO_2 content of serum is increased in metabolic alkalosis, compensated respiratory acidosis, and frequently the alkalosis accompanying a large potassium deficiency (Table 5.4).

Decreased Concentration. The CO_2 content of serum is decreased in metabolic acidosis and compensated respiratory alkalosis (Table 5.4).

BLOOD GASES: pH, pCO_2, and pO_2

The previous section described how intimately involved the plasma CO_2 content is with acid-base balance and explained the great role that the lung plays in the elimination of CO_2. Pulmonary ventilation is adjusted by reflex action to the body needs for elimination of H_2CO_3 as CO_2 gas; the respiratory control centers are located in the brain (medulla) and in chemoreceptors in the aortic and carotid bodies. The respiratory center and chemoreceptors are sensitive to changes in $[H^+]$ and arterial pCO_2; the pCO_2 receptor is also sensitive to changes in arterial pO_2 and alters pulmonary ventilation according to the need as reflected by the O_2 tension. An oxygen deficiency (anoxia or hypoxia) results in a localized or even general acidosis because of the inability

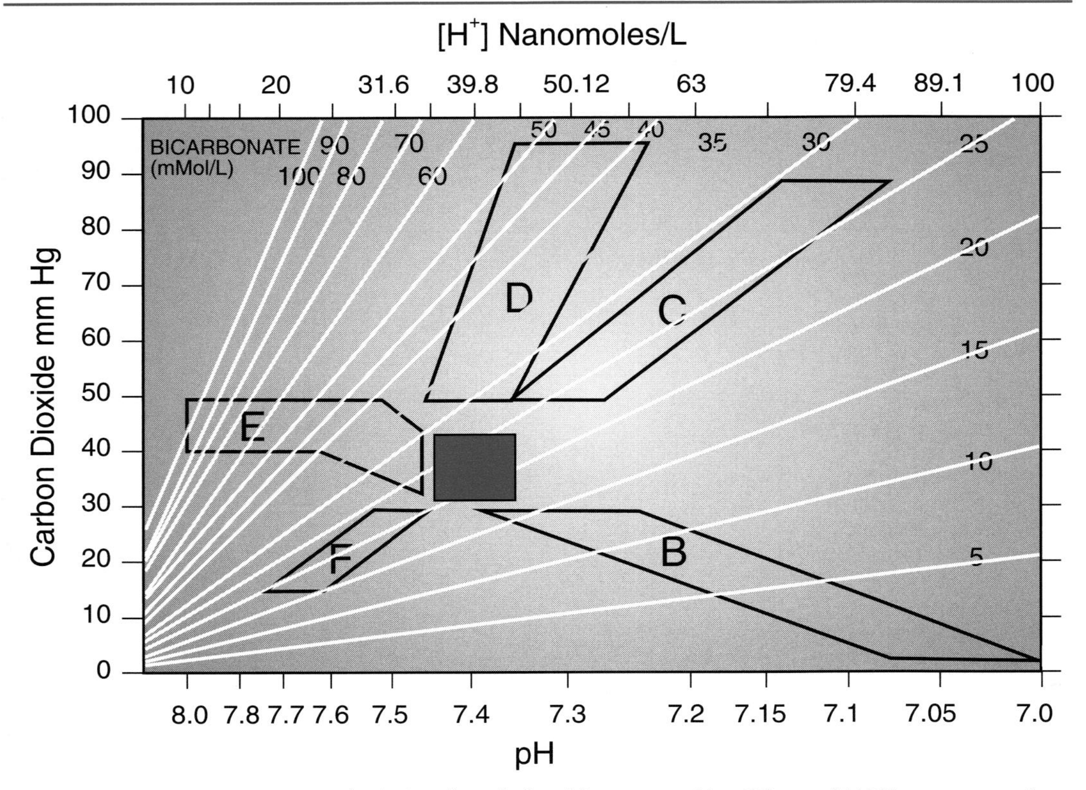

FIG. 5.2. Nomogram depicting the relationship among pH, pCO_2, and HCO_3^- concentrations when values are within the reference range (region A) or during conditions of acid-base imbalance (regions B, C, D, E, and F). The pH values are plotted on the abscissa (x-axis), the pCO_2 values are plotted on the ordinate (y-axis), and the HCO_3^- concentrations are represented by the diagonal white lines. This figure schematically depicts the range of test results associated with metabolic acidosis (B), acute respiratory acidosis (C), chronic respiratory acidosis (renal compensation (D), metabolic alkalosis (E), and respiratory alkalosis (F).

of the electron transport system to transfer all the accumulated hydrogen ions plus electrons to molecular oxygen. Thus, a partial deficiency in the tissue oxygen supply leads to acidosis, and a more severe deprivation may result in loss of consciousness or death.

Three physiologic conditions must be fulfilled to supply the tissues with adequate oxygen:

1. *Introduction into the lung alveoli of an adequate oxygen-containing mixture*. The altitude is an important factor in oxygen intake, because the pO_2 of the atmosphere decreases with the elevation. To bring atmospheric air into the alveoli, the diaphragm and rib muscles of the subject must function properly, and the airway must be free from obstructions.
2. *Properly functioning lung alveoli*. Proper gas exchange must occur between the alveolar blood capillaries and the gas contained in the alveoli. Pathologic changes, such as fluid accumulation in the lungs or

TABLE 5.4
CHANGES IN ACID-BASE PARAMETERS OF WHOLE ARTERIAL BLOOD IN ACID-BASE IMBALANCES

	pH		pCO_2		$[HCO_3^-]$		$[H_2CO_3]$	
PATHOLOGIC STATE	UN-COMP.	COMP.*	UN-COMP.	COMP.	UN-COMP.	COMP.	UN-COMP.	CC
Metabolic Alkalosis								
Vomiting; potassium depletion; primary hyperaldosteronism; excessive alkali ingestion	↑	→	→	↑	↑	↑	→	↑
Respiratory Alkalosis								
Hyperventilation	↑	→	↓	↓	→	↓	↓	↓
Metabolic Acidosis								
Uncontrolled diabetes mellitus; severe renal disease; renal tubular failure; severe diarrhea; acute MI*; starvation; NH_4Cl ingestion	↓	→	→	↓	↓	↓	→	↓
Respiratory Acidosis								
Pneumonia; emphysema; congestive heart failure; lung lesions; drug overdosage (CNS* depressants); anesthesia	↓	→	↑	↑	→	↑	↑	↑

Increase denoted by ↑, decrease by ↓, and no change by →. Arrows enclosed in rectangle show the primary change.
*MI = myocardial infarction; CNS = central nervous system; COMP. = fully compensated

thickened alveolar membranes, greatly interfere with the transfer of oxygen into the blood capillaries and the removal of CO_2 from the blood.

3. *A properly functioning loading and distribution system for the gases.* CO_2 is quite soluble in water, but it does require the presence of the enzyme, *carbonic anhydrase,* in the red blood cells to facilitate the conversion to H_2CO_3. O_2 is rather poorly soluble in plasma and is transported almost entirely by binding to hemoglobin in the erythrocyte. A severe anemia greatly restricts oxygen transport because of the lowered red cell count and the hemoglobin deficiency. Even with a normal erythrocyte count, the cardiovascular system must function properly to deliver sufficient oxygen-laden blood to the tissues. The blood pressure and blood flow must be adequate to fulfill the body's needs; problems arise in the presence of coronary insufficiency (heart failure) or inadequate pulmonary circulation.

Thus, many conditions require a knowledge of the state of the gas exchange of O_2 and CO_2. These conditions include all the respiratory disorders and the situations in which the patient is in a respirator or an oxygen tent, has a tracheotomy, is recuperating from cardiac surgery, or is in a nursery for premature infants, in whom the incidence of respiratory problems is high.

Knowledge of the pCO_2 when measured with the blood pH is useful in the treatment of acidosis.

In a gas mixture, such as air, each constituent gas exerts its partial pressure independently of the others; the total pressure of the mixture is the sum of the component partial pressures. On an average day at sea level, the air pressure is close to 760 mm Hg. The bulk is contributed by nitrogen, an inert gas; the partial pressure of oxygen is about 20% of the total, or 150 mm Hg. Water vapor contributes about 2% of the total pressure, whereas that from atmospheric CO_2 is only a trace. The gases in the alveolar sacs are not completely exchanged or blown out with each respiration; thus, a certain amount of intermixture of newly inspired air occurs with air in the dead space left over from the previous exhalation. This intermixture reduces the pO_2 of the alveolar air to approximately 100 mm Hg. Because the saturation of hemoglobin with oxygen is nearly complete in normal circumstances, arterial pO_2 is approximately 96 mm Hg.

Collection of Blood

Arterial blood is required whenever pO_2 measurement is requested. Venous blood may be taken for just pH and pCO_2 if it is drawn without stasis (no tourniquet) and without the patient clenching the fist. "Arterialized" venous blood may be obtained by heating the hand and forearm in water at 45° C for 5 min and then drawing the blood from the dilated veins on the back of the hand. Capillary blood is arterialized by warming the ear, finger, or heel at 45° C before taking the sample. When blood is drawn with a needle and syringe, the dead space in the needle should be filled with sterile anticoagulant to prevent the formation of an air bubble. To keep the blood anaerobic, one should occlude the tip of the syringe with a plug or rubber stopper immediately after drawing the sample. The syringe with the blood sample should be placed in a tray of ice and sent for analysis to the laboratory, where it should be kept in the ice bath until analyzed. The blood sample should be well mixed before aliquots are taken for measurement.

Principle. Measurements of blood pH, pCO_2, and pO_2 are made simultaneously in a blood gas instrument (Ciba—Corning, Instrumentation Laboratory, Radiometer, Nova).

1. *pH.* The pH is measured by a micro glass electrode as described previously. All electrodes are enclosed and kept at a constant temperature (37° C) by circulating warm water.
2. *pCO_2.* The partial pressure of blood CO_2 is measured by a Severinghaus pCO_2 electrode as described in a previous section. When calibrated with CO_2 gas of known composition, the instrument records the measurement directly as pCO_2. The instruments are controlled by microprocessors programmed to carry out all steps in the process, including calibration and washing.
3. *pO_2.* The partial pressure of blood O_2 is measured amperometrically by a Clark electrode. A membrane (Teflon, polyethylene) permeable to O_2 but not to most of the blood constituents separates the electrode from the

blood. The electrode consists of a platinum cathode and a silver-silver chloride anode. A negative electrical charge is impressed upon the platinum electrode, and the oxygen that diffuses there is rapidly reduced, thereby causing a change in ionic current. The change in current flow between anode and cathode is measured after suitable amplification; conversion of change in current to pO_2 is made by calibrating the electrode with known concentrations of pO_2.

4. *HCO_3^- and CO_2 content.* After measurement of the pH and pCO_2, the HCO_3^- and CO_2 content can be obtained by nomogram (Fig. 5.2) or as a direct readout from a calculation programmed in the instrument. A mathematical relationship exists between total CO_2 concentration, HCO_3^- concentration, pH, and pCO_2 that permits the calculation of one parameter when the three others are known. For the best quality control, the instrument should be checked periodically with known gas mixtures and with tonometered blood of known pCO_2 and pO_2 concentrations.

REFERENCE VALUES (on arterial whole blood). pH = 7.35 to 7.45; pCO_2 = 33 to 45 mm Hg; pO_2 = 85 to 105 mm Hg; and total CO_2 = 22 to 30 mmol/L.

The $\%O_2$ saturation is also related to the pO_2 and pH in a complex, mathematical fashion. If the body temperature, pH, and pO_2 are known, the $\%O_2$ saturation can be calculated from nomograms. Most blood gas instruments have programs that enable calculation of the $\%O_2$ saturation from the analytical data and temperature.

Changes in Acid-Base Parameters

The direction of the changes in values of the acid-base parameters in various states of imbalance are summarized in Table 5.4 and in Figure 5.2. The parameters treated are pH, pO_2, HCO_3^-, and H_2CO_3 concentrations, and the four major states of acid-base imbalance listed are metabolic and respiratory alkalosis and metabolic and respiratory acidosis. Two columns appear under each constituent; the first lists the changes that would be found in the completely uncompensated pathologic state; the second lists the changes when compensation is complete. With many patients, compensation is only partial, and the resulting changes are intermediate between those listed in the uncompensated and the fully compensated columns.

Thus, the common clinical situations in which the pCO_2 is elevated and the pO_2 decreased are the various respiratory disorders, such as pneumonia, hyaline membrane disease of premature newborns, paralysis of the respiratory muscles, depressed respiratory center caused by drug intake, emphysema, and congestive heart failure.

The pCO_2 is depressed in hyperventilation (conscious overbreathing, anxiety states, emotional states) and in various types of fully or partially compensated metabolic acidosis (compensated by overbreathing).

The pO_2 is elevated only when the patients are breathing a gas mixture containing a high percentage of oxygen ($>$ 20% by volume or $>$ 150 mm Hg).

Transcutaneous pO_2 and pCO_2 (T_c pO_2 and T_c pCO_2)

Many premature and newborn infants have respiratory problems (see Chap. 15). Continuous monitoring of pO_2 and pCO_2 on a noninvasive basis is now used frequently in acute care neonatal centers.[13] For this purpose, miniaturized pO_2 and pCO_2 electrodes are taped on thin areas of the infant's skin. The principle of electrode action is the same as described for blood gas determination, but an added complication is variable blood flow through the epidermis. This variability is standardized by warming the skin to 42 to 45° C to ensure vasodilation and maximum blood flow. When done correctly, results are comparable to those obtained by blood gas analysis.

BLOOD OXYGEN CONTENT AND OXYGEN SATURATION

Although not part of the "blood gas package," oxygen content and oxygen saturation values are sometimes requested. Most of the requests come from the heart catheterization laboratory, where the information is used for diagnostic purposes. In heart catheterization, a small catheter is inserted into a large blood vessel in the arm and is threaded into various chambers of the heart or other particular positions in that area. The exact location of the catheter tip is ascertained by fluoroscopy. Heparinized blood samples are withdrawn anaerobically from specific locations and are brought to the laboratory for analysis of their oxygen content. Blood may be withdrawn while the patient is breathing room air or oxygen, depending on the circumstances and the information desired.

The oxygen content and saturation of whole blood may be measured by one of several techniques that vary considerably in the sophistication of the instrumentation and in the length of time required to perform the measurement. These may be done gasometrically, spectrophotometrically, or oximetrically. The classic manometric method devised by Van Slyke was the most widely used gasometric method, but has been supplanted by spectrophotometric and oximetric methods because of greater ease in performance without sacrifice of accuracy.

Determination by Spectrophotometry

The $\%O_2$ saturation can be obtained by measuring the absorbance of a hemolyzed blood sample at 2 wavelengths in a short light-path cuvet (Nahas cuvet, 0.1-mm light path, Waters Corp.[14]).

Principle. Oxyhemoglobin and deoxyhemoglobin (reduced hemoglobin) have different absorption spectra. On plotting absorbance of each against wavelength on the same graph, the absorbance curves of oxyhemoglobin and deoxyhemoglobin cross at several points, called isobestic points, where the absorbances of

oxyhemoglobin and deoxyhemoglobin are identical. The percentage of oxyhemoglobin, which is the same as percentage of oxygen saturation, is obtained by measuring the absorbance of a hemolyzed blood sample at two different wavelengths, one of which is at an isobestic point and one where the difference in absorbance between the oxygenated and deoxygenated states is large. In practice, the choice of 805 nm for the isobestic point and 650 nm for the second wavelength is widely accepted, although other pairs of wavelengths have been used. The absorbance ratio, 650 nm/805 nm, is calculated for each sample. The $\%O_2$ saturation is obtained from a calibration curve (Fig. 5.3). A calibration curve is made by preparing a blood sample with 0% oxyhemoglobin (by blowing nitrogen over a thin blood film) and 100% oxyhemoglobin (by blowing air over a thin blood film) and measuring their absorbances at the 2 wavelengths. They are plotted on a graph, $\%O_2$ saturation versus absorbance ratio, and a straight line is drawn between the 2 points (Fig. 5.3).

The IL CO-oximeter measures the absorbance of hemolyzed blood at three different wavelengths. It solves three simultaneous equations and provides as a readout the $\%O_2$ saturation and the percentage of carboxyhemoglobin (hemoglobin binding of carbon monoxide).

Reference Values. The oxygen content of arterial blood in normal persons is approximately 20 mL/dL blood. Arterial blood is 95 to 98% saturated with oxygen.

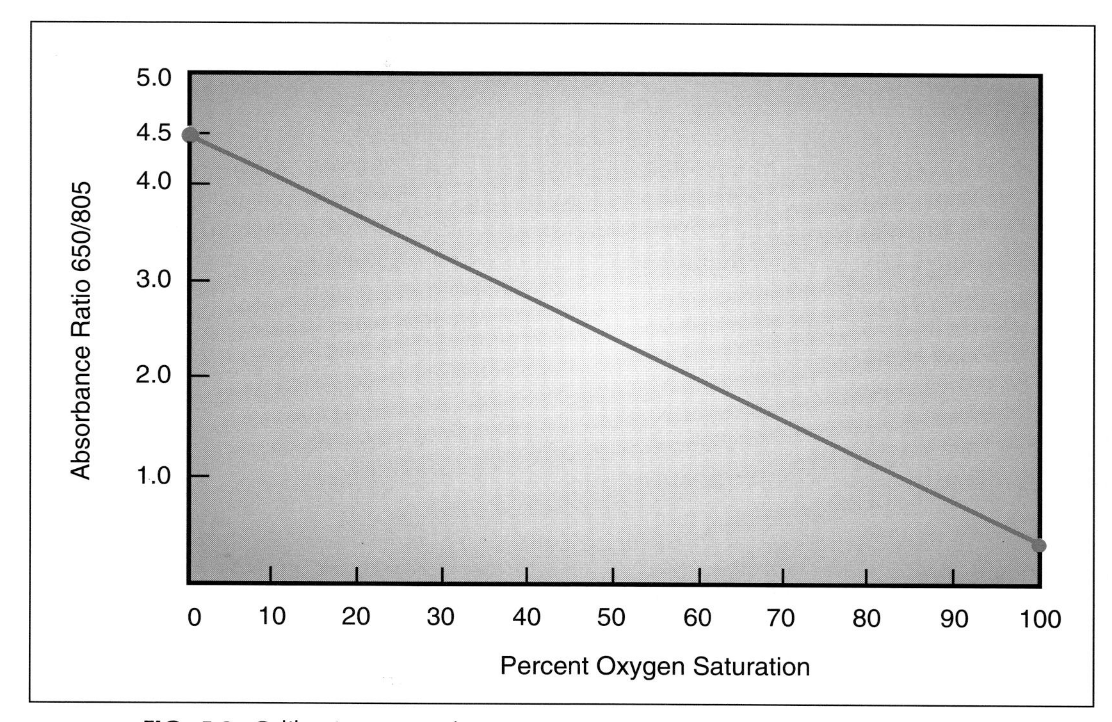

FIG. 5.3. Calibration curve for oxygen saturation. (Adapted from Fleischer, W. R. and Gambino, S. R.: Blood, pH, pO_2 and Oxygen Saturation. Chicago, American Society of Clinical Pathologists, 1972.)

Determination of Oxygen Saturation by Spectrophotometry or Oximetry

The $\%O_2$ saturation may also be measured in a noninvasive manner in infants and adults by pulse oximetry (Nelcor and Novametrixs). A light shines through a finger or the bridge of the nose to a detector. The absorbance of the blood is simultaneously measured at 2 wavelengths as previously described; the $\%O_2$ saturation is obtained from the ratio of the absorbances.

GLOSSARY OF SPECIAL TERMS

Active transport: The transfer of ions or molecules across a cell membrane against a concentration gradient. It involves energy-linked, enzymatic reaction(s) because energy in the form of ATP is usually required.
Anion gap: An estimate of the *unmeasured* anion concentrations, such as sulfate, phosphate, and various organic acids.
Base Excess (BE): The deviation of buffer base from the normal. Normal arterial BE = 0 to ± 2.5 mmol/L.
Buffer: A body buffer is the anion of a weak acid that readily accepts a proton (H^+) to form the weak undissociated acid.
Buffer Base (BB): The sum of the concentrations of buffer anions present in whole blood, principally of bicarbonate, phosphate, hemoglobin, and plasma protein. Normal arterial BB = 46 to 50 mmol/L.
Electrolytes: The term applied in medical usage to the four ions in plasma (Na^+, K^+, Cl^-, and HCO_3^-) that exert the greatest influence on water balance and acid-base relationships. Medical custom is followed in this chapter; other ions that are important in clinical chemistry (Ca^{2+}, Mg^{2+}, Li^+, phosphate, Fe^{3+}, Cu^{2+}, and Zn^{2+}) are reviewed in Chapter 12.
Gastrointestinal (GI) fistula: A tube that drains secretions from the GI tract and exits through an opening in the abdominal wall.
Homeostasis: The maintenance of a steady state in the body, with a relatively constant concentration of ions, pH, and osmolality in the various body fluids.
Internal environment: The concentration of ions and other constituents in all body fluids; includes the H^+ concentration and therefore the pH.
Osmolality: A measure of the number of dissolved particles (ions and undissociated molecules, such as glucose or proteins) per kg of water.
Osmotic pressure: When two compartments of different osmolalities are separated by a semipermeable membrane that allows only small molecules to pass through it, the osmotic pressure represents the hydrostatic pressure that would have to be applied to the compartment of higher osmolality to prevent water passage into it from the compartment of lower osmolality.
Oxygen Content: The volume occupied by the oxygen bound to hemoglobin plus the dissolved oxygen in 100 mL of whole blood, when completely liberated. It is expressed as mL/100 mL blood.
Oxygen Saturation (Capacity): The actual O_2 content of blood expressed as a percentage of the O_2 content of the same blood when fully oxygenated. The fully oxygenated blood is obtained experimentally by exposing the blood in a rotating, thin film to air and then measuring its O_2 content.
Oxygen Tension (pO_2): The partial pressure of oxygen in blood.

REFERENCES

1. Knepper, M.A., Lankford, S.P., and Terada, Y.: Renal tubular actions of ANF. Can. J. Physiol. Pharmacol., *69*:1537, 1991.
2. Maas, A.H.J., Siggaard-Andersen, O., Weisberg, H.F., and Zylstra, W.G.: Ion-selective electrodes for sodium and potassium: A new problem of what is measured and what should be reported. Clin. Chem., *31*:484, 1985.
3. Levy, G.B.: Determination of sodium with ion-selective electrodes. Clin. Chem., *27*:1435, 1981.
4. Cembrowski, G.S., Westgard, J.O., Kurtycz, D.F.I., et al.: Quality control of electrolyte analysers: evaluation of the anion gap average. Am. J. Clin. Pathol., *79*:688, 1983.
5. Comprehensive Chemistry 1992 Survey, College of American Pathologists, Skokie, IL.
6. Elin, R.J., Robertson, E.A., and Johnson, E.: Bromide interferes with determination of chloride by each of four methods. Clin. Chem., *27*:778, 1981.
7. Dietz, A.A., and Bond, E.E.: Chloride, coulometric-amperometric method. *In* Standard Methods of Clinical Chemistry. Vol. 9. Edited by W.R. Faulkner and S. Meites. Washington, D.C., American Association for Clinical Chemistry, 1982.
8. Gibson, L.E., di Sant'Agnese, P.A., and Shwachman, H.: Procedure for the Quantitative Iontophoretic Sweat Test for Cystic Fibrosis. Rockville, MD, Cystic Fibrosis Foundation, 1985.
9. Szabo, L., Kenny, M.A., and Lee, W.: Direct measurement of chloride in sweat with an ion-selective electrode. Clin. Chem., *19*:727, 1973.
10. Maas, A.H.: IFCC reference methods for measurement of pH, gases and electrolytes in blood: reference materials. Eur. J. Clin. Chem. Biochem., *29*:253, 1991.
11. Gambino, S.R.: pH and pCO_2. *In* Standard Methods of Clinical Chemistry. Vol. 5. Edited by S. Meites. New York, Academic Press, 1965.
12. Kenny, M.A., and Cheng, M.H.: Rapid, automated simultaneous determination of serum CO_2 and chloride with the "Auto Analyzer." Clin. Chem., *18*:352, 1972.
13. Mendelson, Y.: Pulse oximetry: theory and applications for noninvasive monitoring. Clin. Chem., *38*:1601, 1992.
14. Fleischer, W.R., and Gambino, S.R.: Blood pH, pO_2 and oxygen saturation. Chicago, American Society of Clinical Pathologists, Commission on Continuing Education, 1972.

QUESTIONS

1. What determines the distribution of body water between cells and the extracellular fluid?
2. What are the principal hormones involved in water and electrolyte balance and how do they act?
3. What substances are mainly responsible for the osmotic pressure of extracellular fluid and the oncotic pressure of blood?
4. What is the mechanism for conserving sodium in the body?
5. What principles are involved in the use of ion-selective electrodes (ISE) for the determination of various ion concentrations?
6. What are the three main blood buffer systems and how do they operate?
7. How does the body try to compensate for acid-base disturbances?
8. If you know the blood pH of a patient, what extra information does a knowledge of his/her pCO_2 provide?
9. What are the different types of acid-base disturbances and how does use of a nomogram help in interpretation?
10. In what situations is the determination of pO_2 clinically useful?
11. What are the advantages of obtaining pO_2 and pCO_2 by transcutaneous measurement?

6 The Kidney and Tests of Renal Function

OUTLINE

(c) Increased concentration
(d) Decreased concentration
3. Uric acid
a. Serum uric acid
(1) Determination by uricase/UV absorbance
(a) Principle
(2) Determination by uricase/H_2O_2-coupled reactions
(a) Principle
(b) Reference values
(c) Increased concentration
(d) Decreased concentration
b. Urine uric acid
E. Urine concentration test
1. Procedure
F. Urinary tract calculi (stones)
1. Principle

OBJECTIVES

After reading this chapter, the student will be able to:

1. Describe the anatomic regions of a kidney and name the functional regions of the nephron.
2. Describe the process of glomerular filtration, the initial composition of the glomerular filtrate, and the factors (pressures) that influence the filtration of blood.
3. Describe the different hormonal processes that promote the concentration of urine and conservation of water and salt.
4. Describe the renal mechanisms for excreting excess acid.
5. Name prerenal, renal, and postrenal causes for abnormally high test results for urea nitrogen and creatinine assays.
6. Explain the principles of the assays of urea nitrogen (with urease), creatinine, and uric acid.
7. State the physiologic principles involved for the creatinine clearance test and for the urine concentration test.
8. Calculate the creatinine clearance.
9. Name three types of urinary casts that signify kidney disease.
10. Name the tests commonly found on the urine dipstick, and name at least one disease state that can be detected with each test.

The kidneys are vital organs responsible for affecting water, pH, and salt balance, as well as eliminating soluble waste products by the formation of urine. The clinical laboratory provides clinicians with valuable information about renal function that is important for evaluating not only patients with kidney disease, but also patients experiencing loss of blood volume, dehydration, head injury, trauma, surgery, and infectious or metabolic diseases. Patients undergoing dialysis treatment are also monitored by laboratory tests to

assess the efficacy of their artificial kidney. The kidneys perform several endocrine functions, such as the synthesis of erythropoietin (a hormone that stimulates red blood cell production) and 1,25-dihydroxy-vitamin D_3 (the biologically active form of vitamin D). This chapter deals only with the renal mechanisms of urine formation and how the clinical chemistry laboratory monitors renal function and renal disease states.

STRUCTURE OF THE KIDNEY

There are two kidneys, one on each side of the abdomen. The central region of the kidney is referred to as the medulla, and the outer layer is the cortex. The kidneys receive approximately one fifth of the blood pumped out with each heartbeat from the renal arteries, and they return approximately 99% of that volume to the renal veins. The small fraction of fluid that ultimately forms urine flows from the collecting ducts into the renal pelvis and through the ureter (one per kidney) to the bladder.

The working unit of the kidney is the *nephron*, shown schematically in Figure 6.1. A nephron is composed of a glomerulus (the filter), a proximal convoluted tubule (primary site of reabsorption), a long loop of Henle (thin-walled descending loop and thick-walled ascending loop), a distal convoluted tubule (secondary site of reabsorption), and a collecting tubule and collecting ducts (sites of water reabsorption and urine concentration). Most regions of each nephron are closely associated with the bloodstream. Whole blood within the afferent arteriole is filtered at a tuft of capillaries within the glomerulus. The valuable components of the filtrate, such as salts, glucose, amino acids, and water, are reabsorbed in the proximal or distal convoluted tubules and returned to the capillaries of the efferent arterioles. The unabsorbed filtrate flows into the ureters and bladder for subsequent excretion as urine.

FORMATION OF URINE

The filtration of plasma by the glomerular capillaries is the first step in the formation of urine. The red and white blood cells, platelets, and molecules of molecular weight $>$ 50,000 daltons (most proteins and constituents tightly bound to proteins) are retained in the capillaries. About 20% of the plasma is filtered through the capillary membrane at the glomerulus and enters the nephron. The diameter of the lumen of each efferent (leaving) arteriole is smaller than the lumen of each afferent (entering) arteriole, thus causing a rise in blood pressure in the glomerular capillaries and aiding the filtration process. For filtration to take place, the effective filtration pressure must exceed the osmotic forces drawing water into the protein-rich plasma (oncotic pressure). Obstruction of urine flow from the tubules decreases the net effective filtration pressure and hinders glomerular filtration.

The filtration process is passive and occurs without energy expenditure. The capillary walls of the glomerular tuft are thin and well adapted to their function

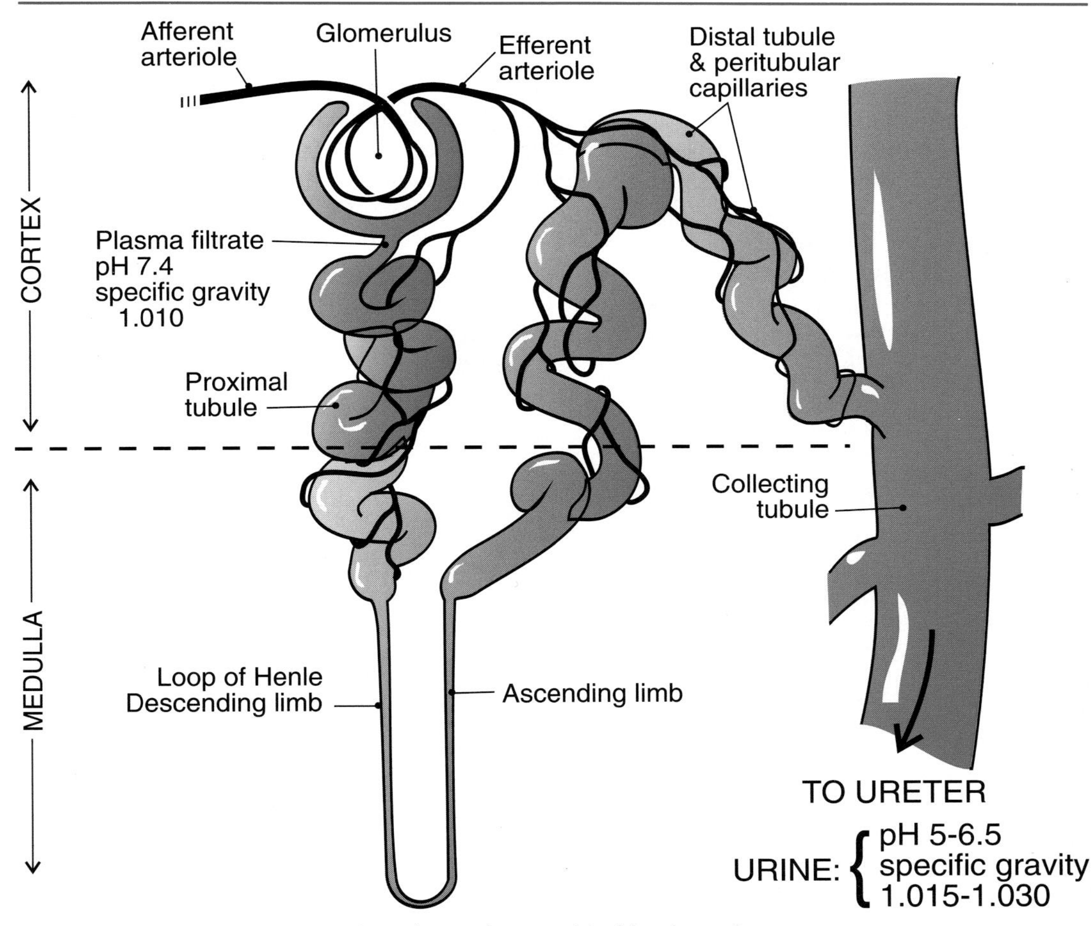

FIG. 6.1. Representation of a nephron and its blood supply.

as a filter; water and small molecules pass easily through the pores to form an ultrafiltrate of plasma. The ultrafiltrate is large in volume (about 180 L/day for the average adult), and its constituents are of the same concentration as those in plasma. Ninety-nine percent of the ultrafiltrate water and a large percentage of its constituents must be reabsorbed before the urine leaves the collecting tubules on its way to the bladder.

The second step in the formation of urine is the passage of the ultrafiltrate down the proximal convoluted tubule, where the various selective processes of absorption begin. The filtrate starts out with the same specific gravity as a protein-free filtrate of plasma, 1.010, and the same pH, 7.4. About 70% of water, Na^+, and Cl^-, and all but negligible amounts of glucose, amino acids, and K^+, are reabsorbed in the proximal tubules. Some substances, such as urea, phosphate, and Ca^{2+}, are incompletely reabsorbed. H^+ is exchanged for Na^+ throughout the tubule, whereas K^+ is exchanged for Na^+ only in the distal tubule, where the K^+-Na^+ exchange is regulated by the hormone aldosterone.

Water in the filtrate is always reabsorbed passively when the osmotic pressure outside a semipermeable tubular membrane is higher than that inside; water follows the osmotic gradient toward restoration of equilibrium. The osmotic gradient is usually produced by active Na^+ transport (the sodium pump), an energy-requiring process. The glomerular filtrate is reduced to about one third of its original volume in the proximal tubule, but because water and salts are absorbed together, no change occurs in the osmotic pressure of the filtrate. Several components of the fluid leaving the proximal tubule are selectively reabsorbed depending on the needs of the body. For example, during the hours of sleep, the physiologic need is to conserve water and to eliminate excess salts and H^+ by producing a hyperosmolar urine that is much more acidic than plasma. This need is met by an interplay of anatomic features, physical forces, and finely regulated hormonal control.

The anatomic features that promote the independent reabsorption of water and Na^+ from the distal and collecting tubules are (1) a thickened wall in the ascending limb of the loop of Henle that is impermeable to water but not to ions; (2) the extension of the loop of Henle and large sections of the distal and collecting tubules from the kidney cortex into the medulla; (3) the passage of the blood capillaries into the interstitial tissue, surrounding in countercurrent fashion the loop of Henle, the distal tubules, and collecting ducts; and (4) the impermeability to water of the walls of the distal and collecting tubules unless acted upon by the antidiuretic hormone.

The interstitial fluid in the kidney increases in osmotic pressure from the cortex to the medulla. This osmotic gradient is an important renal feature that is created and maintained by nephrons and allows them to function. The osmotic gradient is caused by an increasing concentration of Na^+ and Cl^- as the Na^+ is pumped out of the ascending loop of Henle by an active process. The osmotic gradient in the interstitial fluid surrounding the descending loop of Henle is the physical force that greatly accelerates the reabsorption of water from the lumen of the descending loop. As the filtrate moves up the ascending loop, the reabsorption of water stops because the lumen wall is impermeable to water. The reabsorption of Na^+ from the ascending loop is considerable because of pump action; Cl^- moves passively with it. The water and ions that pass into the interstitial fluid are reclaimed by absorption into the blood capillaries surrounding the loops of Henle and the tubules. The net result is the production of a *hyposmolar* urine (greater loss of Na^+ and Cl^- than water) by the time the distal tubule is reached. The interaction of water leaving the descending loop and Na^+ and Cl^- leaving the ascending loop is responsible for maintaining a high osmolality within the kidney medulla while reducing the osmolality of the urine leaving the loop of Henle. This interaction is known as a *countercurrent multiplication* process.

The walls of the distal and collecting tubules are impermeable to water unless acted upon by the *antidiuretic hormone* (ADH). When an excess of water exists within the body, ADH is not secreted, and the dilute fluid in the distal and collecting tubules is passed as urine. When the body needs to retain water, a final reabsorption of water takes place under the influence of ADH, thereby resulting in the production of hyperosmolar, concentrated urine. The hyperosmolar interstitial fluid surrounding the collecting tubules

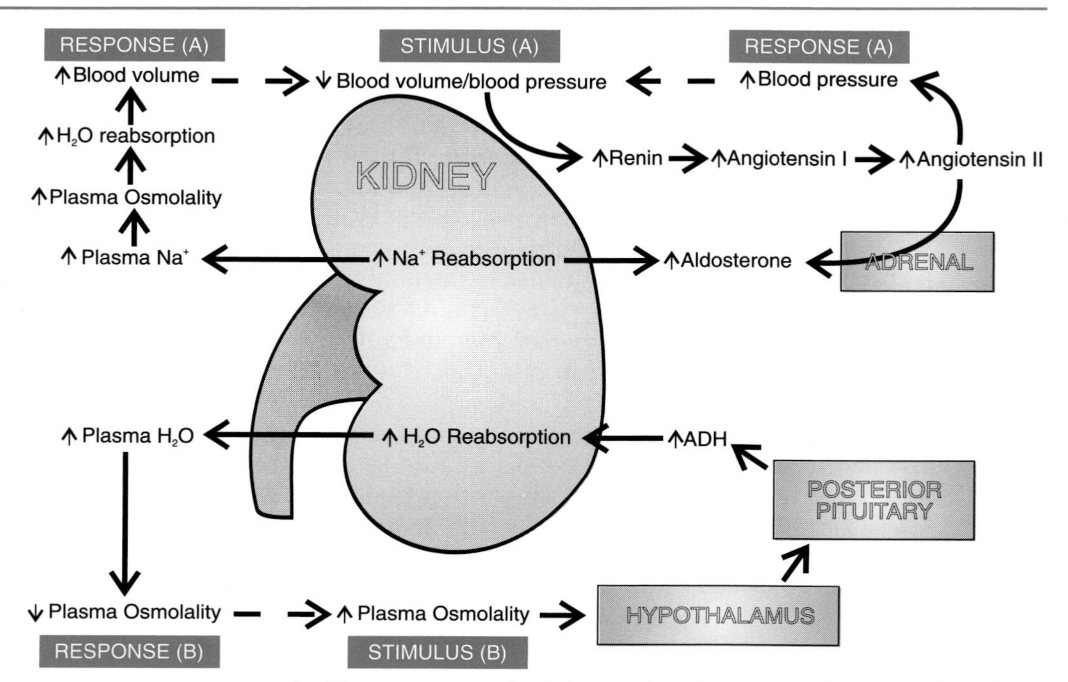

FIG. 6.2. ADH and aldosterone control of the renal reabsorption of water and Na^+. In response to a fall in blood volume or blood pressure (Stimulus A), renin is secreted by the kidney, thereby leading to an angiotensin II-mediated elevation of blood pressure and an aldosterone-mediated increase in Na^+ reabsorption. In response to an increase in plasma osmolality (Stimulus B), ADH is secreted by the posterior pituitary, thereby increasing water reabsorption in the distal and collecting tubules.

accelerates the water uptake. ADH secretion is stimulated by a high plasma osmotic pressure and is inhibited by a decreased osmotic pressure, so that the net result is conservation or excretion of body water according to osmotic need (Fig. 6.2). This process essentially controls the volume of urine finally excreted.

Regulation of ADH Output

ADH is a hormone (vasopressin) produced by the hypothalamus but stored in the posterior pituitary gland. Receptor cells in the hypothalamus are sensitive to changes in osmotic pressure within the bloodstream and transmit nerve impulses to the posterior pituitary. Elevated osmotic pressure stimulates the posterior pituitary gland to secrete ADH and thus promotes water retention by increasing renal reabsorption of water (decreasing urine volume). An elevation of osmotic pressure also stimulates the feeling of thirst that leads to the consumption and absorption of water and a resultant decrease in osmotic pressure. A decrease in osmotic pressure within the bloodstream inhibits the secretion of ADH and promotes water loss by decreasing the renal reabsorption of water (increasing urinary volume).

Reabsorption and Excretion Processes

Many different mechanisms are involved in the various absorption processes. Some are passive and can be explained by concentration gradients or by osmotic pressure differences (for example, water), whereas others require active transport mechanisms involving the expenditure of energy. The active processes are usually coupled with enzyme action to transfer substances from the lumen into cells against a concentration gradient. Water in the filtrate is absorbed passively in the proximal tubule, but glucose and amino acids are almost completely reabsorbed by active transport. About 70% of the sodium is reabsorbed in the proximal tubule by an active process, primarily in conjunction with chloride. In the distal tubule, however, a specialized ion exchange mechanism participates in the conservation of sodium and in the excretion of potassium and hydrogen ions. Bicarbonate is converted to CO_2 and water by the hydrogen ion in the filtrate, and the CO_2 diffuses into the tubular cells, where it is converted by the enzyme carbonic anhydrase back to bicarbonate ion. The effective reabsorption of HCO_3^- occurs throughout the entire length of the tubule. Creatinine, on the other hand, is not absorbed at all by the renal tubule; in fact, small amounts may be secreted by the tubular cells into the urine. Potassium ion is almost completely reabsorbed in the proximal tubule, but a portion is exchanged for Na^+ in the distal tubule, where the potassium ion is secreted into the lumen of the tubule. Phosphate ion is reabsorbed by an active process in the proximal tubule, but the reabsorption is never complete, and fine control is achieved by secretion of the parathyroid hormone, which inhibits phosphate reabsorption. Urea, on the other hand, diffuses passively into the blood as it passes down the tubule. Under ordinary circumstances, about 60% of the urea in the filtrate is excreted in the urine, but this amount varies with the glomerular filtration rate and the rate of filtrate flow in the tubules.

Table 6.1 reveals that the kidneys perform a tremendous amount of work in forming normal urine. Approximately 180 L (180,000 g) of plasma are filtered in a 24-hr period; 178 to 179 L are reabsorbed. Stated another way, about 99.2% of the 180 L of water in the ultrafiltrate returns to body fluids. The amount of solutes in the filtrate varies from 540 g of sodium and 630 g of chloride to 0.06 g of a mixture of proteins (primarily albumin and glycoproteins). Most of the sodium and chloride ions and amino acids are reabsorbed. Glucose is completely reabsorbed in normal circumstances, whereas creatinine is not reabsorbed at all. The reabsorption capacity for glucose has a threshold or limit, however. When its concentration in plasma, and consequently in the glomerular filtrate, exceeds 180 mg/dL, some of the glucose escapes reabsorption and appears in the urine. The high plasma glucose levels in uncontrolled diabetes mellitus explain the customary finding of appreciable amounts of glucose in the urine of diabetic patients.

Table 6.1 also shows in column (3) the amounts of the different substances excreted by a person eating an average diet; in column (4), the amounts reabsorbed or reclaimed for use are given. The degree of reabsorption of the water and salts appearing in the glomerular filtrate is subject to fine control mechanisms responsive to body needs.

TABLE 6.1
AVERAGE FILTRATION, REABSORPTION, AND EXCRETION OF CERTAIN NORMAL CONSTITUENTS OF PLASMA*

(1) Constituent	(2) FILTERED/24 HR (g)	(3) EXCRETED/24 HR† (g)	(4) REABSORBED/24 HR (g)	%
Water	180,000	1,800	178,200	99.0
Chloride	630	5.3	625	99.2
Sodium	540	3.3	537	99.4
Bicarbonate	300	0.3	300	100
Glucose	140	0	140	100
Amino acids	72	1	71	98.6
Urea	53	32	24	39.6
Potassium	28	4	24	85.7
Uric acid	8.5	0.8	7.7	90.6
Phosphate	6.5	1	5.5	84.1
Creatinine	1.4	1.4	0.0	0
Total protein‡	?	0.06	?	?

*Modified after Berliner, R.W., and Giebisch, G.: Best and Taylor's Physiological Basis of Medical Practice. 10th Edition. Baltimore, Williams & Wilkins, 1979.
†Typical normal values, but greatly dependent on dietary intake.
‡Many different proteins appear in the urine in trace amounts. Most of them have a molecular weight lower than 70,000 daltons, but a few have a molecular weight as high as 160,000 daltons.

The control mechanism for the regulation of sodium reabsorption rests with the steroid hormone aldosterone, a mineralocorticoid produced by the adrenal cortex that responds to changes in blood volume. Aldosterone accelerates the reabsorption of sodium in the distal tubule by facilitating its exchange for potassium ions (Fig. 6.2). Following the production and secretion of aldosterone, sodium is retained and potassium is excreted. With a deficiency of this hormone, the reverse takes place; sodium excretion increases as less sodium is retained by the body, and potassium excretion is decreased. The aldosterone-mediated feedback control of sodium excretion makes possible a variation in the excretion of Na^+ from zero to large amounts, according to the needs of the body. The necessity for the renal conservation of Na^+ mediated by aldosterone becomes acute in areas of the world where NaCl is scarce or when the loss of NaCl is great because of profuse sweating with inadequate intake to compensate for the loss. The ability to retain or excrete sodium, and water with it, is an essential life process.

Regulation of Aldosterone Output

Receptor cells in the kidneys are sensitive to changes in renal blood flow and can stimulate the process of fluid retention. When the renal blood volume is reduced by either a reduction in total blood volume or a reduction in blood pressure, the renal receptor cells release *renin*, a protease. Renin acts upon plasma angiotensinogen (an α_2-globulin) and splits off the decapeptide *angiotensin I.* Angiotensin I loses two amino acids from its chain by the action of a lung peptidase and becomes *angiotensin II,* a powerful vasoconstrictor. Angiotensin II not only produces an immediate rise in blood pressure, but also stimulates the adrenal cortex to secrete the steroid hormone *aldosterone.* Aldosterone promotes the retention of Na^+ in the distal tubule, thereby increasing water retention by its osmotic effect.

Thus, the regulation of fluid volume and osmotic pressure is coordinated by the interplay of the two hormones, aldosterone and ADH, that control the degrees of reabsorption of Na^+ and water, respectively, in the distal and collecting tubules. The thirst mechanism is also an important factor in regulation, but sick patients may not have ready access to a supply of fluids and, in some cases, may not be able to retain ingested fluid.

There was no evolutionary pressure to develop a highly refined mechanism to conserve potassium by the kidney. Most foodstuffs contain potassium, thereby making virtually impossible a K^+ deficiency in a mammal eating a natural diet. A K^+ deficiency in humans has appeared only in the last century, with the production of highly refined foods devoid of potassium, the introduction of intravenous saline/glucose therapies, and the use of powerful diuretic drugs for the treatment of high blood pressure. The kidney can excrete large amounts of K^+ when the intake is high, but it cannot reduce the K^+ excretion below a certain level, even if the intake falls to zero. Aldosterone mediates a reabsorption of Na^+ from the filtrate in the distal tubule to the bloodstream by excreting K^+ in exchange. In this manner, elevated levels of aldosterone promote both the retention of sodium and loss of potassium from the body.

Although bicarbonate ion cannot be absorbed directly from the tubular lumen because the wall is impermeable to it, bicarbonate ion is absorbed indirectly. As the glomerular filtrate becomes acidic, HCO_3^- is protonated and converted to carbonic acid (H_2CO_3), which is in equilibrium with CO_2 and H_2O. The CO_2 diffuses into the tubular cell, where it is recombined with H_2O and reaches an equilibrium with H_2CO_3, and H^+ and HCO_3^-. The HCO_3^- is absorbed into the blood capillaries, thereby achieving the same result as direct reabsorption of HCO_3^-. The equilibria are summarized in Figure 6.3.

Under conditions of metabolic alkalosis, however, the urine is alkaline and protonation of bicarbonate does not take place. Appreciable amounts of HCO_3^- are then excreted into the urine, a compensatory mechanism that helps to lower the plasma concentration of HCO_3^- as well as the pH.

Excretion of Acids

The principal renal mechanisms for excreting excess hydrogen ions or protons, summarized in Figure 6.3, follow:

1. Exchange of lumen Na^+ for cellular H^+.
2. Trapping of lumen H^+ by the following reactions:
   a. $HPO_4^{2-} + H^+ \rightarrow H_2PO_4^-$
   In the plasma at pH 7.4, 80% of the phosphate is in the form of

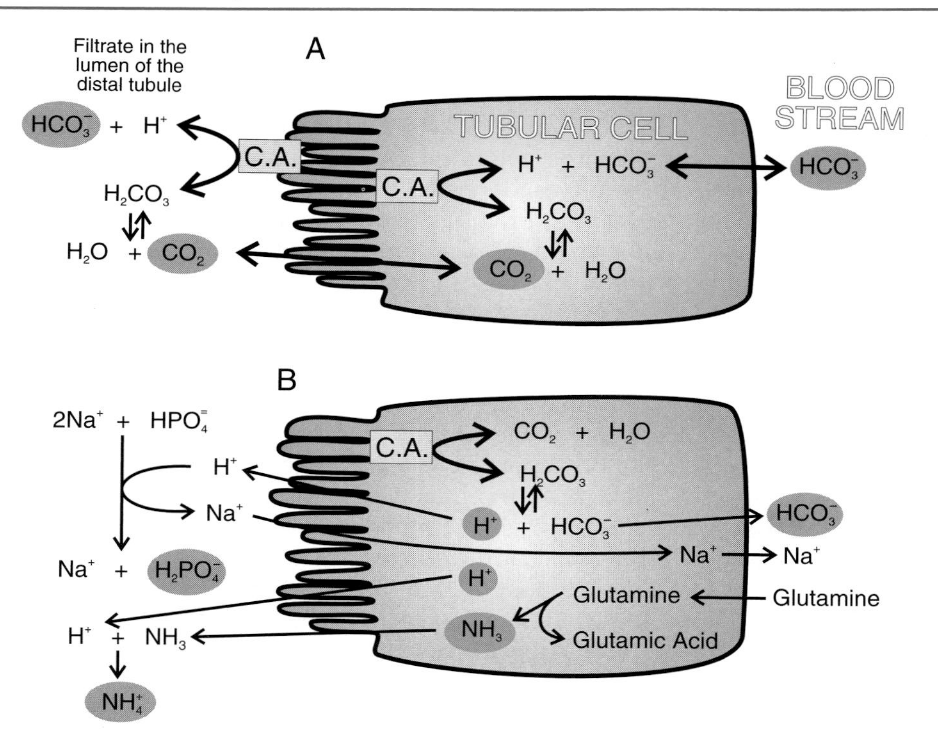

FIG. 6.3. Schematic illustration of the mechanism of bicarbonate reabsorption (A), and of the exchange of Na^+ for H^+ and the trapping of H^+ with ammonia and phosphate in the renal tubule during acidosis (B). C.A. = the enzyme carbonic anhydrase.

HPO_4^{2-}. Protons (H^+) are secreted into the renal tubular lumen, where the preceding reaction takes place. At pH 6.0, practically 100% of the phosphate is in the form of $H_2PO_4^-$.

b. $NH_3 + H^+ \rightarrow NH_4{}^+$

In the tubular cell, glutamine is converted into glutamic acid and ammonia (NH_3) by the enzyme glutaminase. The NH_3 readily diffuses into the lumen, where it reacts with H^+ to form NH_4^+, which is excreted. An equivalent amount of Na^+ is returned to the tubular cell in exchange for the H^+, with the net effect of excreting NH_4^+ instead of Na^+. This mechanism becomes significant for disposing of excess H^+ in chronic acidosis, because the capacity for producing NH_3 from glutamine is greatly increased by synthesis of the glutaminase enzyme, which is stimulated by the state of acidosis.

3. Excretion of undissociated acids, such as ketoacids, in uncontrolled diabetes mellitus or prolonged fasting.

In summary, urine is formed by a continuous process that starts with the ultrafiltration of plasma followed by the reabsorption of the water and other constituents to a greater or lesser extent. Reabsorption may take place by many mechanisms: passive diffusion, special transport mechanisms, energy-linked processes, and ion exchange. The composition and volume of the urine are influenced by the secretion or withholding of essential hormones, notably ADH and aldosterone, that act upon tubular cells. The final composition of the urine is attained by the secretion of some constituents from the blood plasma into the distal tubular lumen. The normal urine contains more dissolved substances than does the plasma ultrafiltrate (higher specific gravity) and is much more acidic. The net effects of this complicated process are the excretion of waste products of metabolism and the preservation of the volume, ionic concentration, balance, and pH of the body fluids.

LABORATORY TESTS OF RENAL FUNCTION

Laboratory tests play an important role in the diagnosis and assessment of renal disease because clinical signs and symptoms may be vague or absent. Some of the renal function tests reveal primarily disturbances in glomerular filtration, and others reflect dysfunction of the tubules, but damage is seldom confined solely to a particular portion of the nephron. Because the anatomic portions of the nephron are closely related and have a common blood supply, damage to one portion gradually involves the nephron as a whole. Eventually both glomerular and tubular portions of the nephron become involved irrespective of the site of the original lesion. All nephrons are not affected by disease at the same time or to the same extent, and a reserve capacity is provided for kidney function. As a consequence, the pathologic condition of the kidney must be considerable before the tests of renal function become abnormal.

The formation of a normal urine at a normal rate requires properly functioning kidneys receiving an adequate blood supply at a sufficiently high blood pressure, with no obstruction to urine outflow. A malfunction in the formation and/or elimination of urine may be attributable to prerenal, renal, or postrenal causes. The prerenal factors affect blood volume, blood flow, or blood pressure, and include such conditions as hemorrhage, shock, dehydration, intestinal obstruction, prolonged diarrhea, and cardiac failure. When these defects are corrected, kidney function usually returns to normal. The renal factors are within the kidneys themselves and may affect the glomerular filtration rate, the various tubular activities, or the renal blood vessels. The postrenal factors that decrease renal function are obstructions to the flow of urine, such as renal calculi (stones); carcinomas or tumors that may compress the ureters, urethra, or the bladder opening; or an enlarged prostate gland that partially occludes the urethra. Chemistry tests alone, however, are incapable of differentiating the three prime causes of renal dysfunction, but a careful history and physical examination of the patient, in combination with a few other tests, can elucidate the problem.

Some of the common renal diseases that produce a diffuse involvement of the kidneys are glomerulonephritis, the nephrotic syndrome, pyelonephritis, and arteriolar nephrosclerosis.

Glomerulonephritis, a diffuse, inflammatory disease, affects the glomeruli first, but rapidly produces degeneration of the tubule. In its *acute* form, the disease is evident suddenly by the appearance of hematuria and proteinuria, with varying degrees of hypertension (high blood pressure), renal insufficiency, and edema. The causative agent for the disease is usually a prior infection with a group A, β-hemolytic streptococcus. The deposition of immune complexes on the renal basement membrane may be responsible for the glomerular damage. Most patients recover completely from the acute phase, but a few undergo progressive loss of renal function and finally die in renal failure.

A *chronic* form of glomerulonephritis varies greatly in its degree of severity. In some patients, the disease is progressive, with continuing loss of renal function; in others, remissions and relapses may go on for many years.

The *nephrotic syndrome* is characterized by heavy proteinuria, hypoalbuminemia, edema, and hyperlipidemia. Lesions in the glomerular membrane allow proteins to escape. Laboratory tests reveal no other impairment of renal function. The syndrome is frequently associated with systemic lupus erythematosus, proliferative glomerulonephritis, amyloidosis, or syphilis, but frequently, the causative factor is not known. Many patients respond to treatment with adrenocortical steroids.

Pyelonephritis is an inflammatory renal disease caused by infectious organisms that have ascended the urinary tract and invaded kidney tissues. Chronic or repeated infections may lead to replacement of renal cells by scar tissue, with some loss in renal function.

Arteriolar nephrosclerosis is characterized by a thickening of the inner lining of the renal arterioles, resulting in a decreased lumen and an increased blood pressure. As the blood vessels become necrotic, progressive loss occurs in both glomerular and tubular function. Scar tissue replaces the damaged cells, and the kidney becomes contracted. Proteinuria is common. In some patients, the high blood pressure cannot be controlled (malignant hypertension), and the impairment in renal function becomes progressive and rapid; the outcome is fatal. In the benign form, the blood pressure can be reduced to reasonable values by appropriate drugs, so the loss in kidney function is relatively mild and slow in its progression.

Renal tubular acidosis (RTA) is a group of disorders in which the renal excretion of acid is reduced far more than the glomerular filtration rate; the defect is primarily confined to the tubules. The several types of RTA are frequently hereditary. The most common type of RTA affects the distal tubules, so that either the H^+ diffuses back from the lumen into the tubule or the distal tubule is unable to transport H^+ from cells to lumen. The net result is an accumulation of H^+ in the body (acidosis), accompanied by decreased HCO_3^- and increased Cl^- concentrations in plasma. The plasma K^+ concentration is usually low. The disorder is seen more frequently in children than in adults. A second type of RTA has a defect in the proximal tubules, resulting in ineffective reabsorption of HCO_3^-. Acidosis is accompanied by K^+ loss and hyperchloremia. The two types of RTA can be differentiated by giving the patient an oral load of NH_4Cl and measuring the urine pH; an individual with distal RTA cannot acidify the urine below pH 5.5, whereas a patient with the proximal defect can do so. Some genetic tubular defects produce, in addition to acidosis, glucosuria, amino aciduria, and low plasma levels of urate and phosphate (Fanconi syndromes).

Because the kidney is the primary organ involved in the excretion of certain nitrogenous wastes (creatinine, urea, uric acid), in the regulation of water and electrolyte balance, and in the excretion of fixed acids, renal dysfunction is characterized by changes or abnormalities in one or more of these parameters. Some of these abnormalities (electrolyte or acid-base imbalance) are nonspecific, because they may also occur in other disease states. Clinical chemistry tests for the diagnosis of renal dysfunction are confined, therefore, to the few that are more indicative of renal disease: serum creatinine, serum urea nitrogen, creatinine clearance, and urinalysis. Some other tests are useful for confirmation or for intelligent management of the patient and may be ordered for these purposes. Urinalysis is described first because it yields useful information and is a routine procedure for all patients.

URINALYSIS[1–5]

Examination of the urine as an aid to diagnosis of many diseases has been carried out for centuries by medical practitioners. Some of the current methods of examination are still traditional, such as noting the appearance and odor of the specimen and making a microscopic examination of the urinary sediment. The main advances have been in providing dipsticks or strips for the semiquantitation of a group of constituents and in measuring urine osmolality as an indication of total solute concentration. Visual and microscopic examination may yield useful clinical information and must not be neglected because it is not "quantitative."

A routine urinalysis usually consists of an examination of a morning specimen (upon arising) for color, odor, specific gravity, or osmolality. Some qualitative or semiquantitative tests are performed for pH, protein, glucose or reducing sugars, ketones, blood and perhaps bilirubin, urobilinogen, leukocyte esterase, and nitrite.

Some hospitals routinely perform a microscopic examination of the urinary sediment, but others do so only when both the nitrite and the leukocyte esterase tests are positive. Tests that are especially useful in evaluating renal function or renal disease are described in detail; those used for the diagnosis of other diseases are treated more fully in the appropriate chapters.

Macroscopic and Physical Examination

Volume

Knowledge of the daily urinary output may be of value in the study of renal disease, but this test requires a timed specimen and a good collection. The daily output of urine depends largely on the fluid intake and many other factors, such as degree of exertion, temperature, salt (NaCl) intake, and hormonal control, but the average excretion of a normal adult is approximately 1 mL/min or about 1400 ± 800 mL/24 hr. A decreased urinary output is called *oliguria;* an increased output is referred to as *polyuria*. Oliguria may be caused by *prerenal* (low blood pressure, shock, hemorrhage, fluid deprivation), *renal* (acute tubular necrosis, certain poisons, renal vascular disease, precipitation of certain compounds in nephrons), or *postrenal* (calculi, tumors compressing ureters, prostatic hypertrophy) factors. Polyuria may be caused by the excretion of a large amount of solutes, with obligatory excretion of water (after excessive salt intake, in diabetes mellitus with glycosuria), by a deficiency or depression of ADH, or by the excessive ingestion of fluids or diuretic substances.

Color

Although infrequent, an abnormally colored urine is important to note. Fresh blood or hemoglobin may impart a reddish color, whereas old blood makes the urine look smoky; both are indications of bleeding in the genitourinary tract. Bile pigments produce a green, brown, or deep yellow color signifying liver or biliary tract disease (see Chap. 11). A dark brown urine may be caused by homogentisic acid excreted in a rare genetic disease, alkaptonuria. Some drugs or dyes may also contribute color to the urine.

Odor

Fresh urine has a characteristic odor that may be affected by foods, such as asparagus. In diabetic acidosis, a fruity odor may be caused by the ketoacids and acetone (see Chap. 7). In maple syrup urine disease, a rare genetic defect, the urine has the odor of caramelized sugar or maple syrup. When urine specimens are old, or when a *Proteus* infection is present, a strong odor of ammonia is usually apparent. A putrid odor usually means that the urine has undergone bacterial decomposition because it has remained too long without refrigeration.

Specific Gravity

The specific gravity of the urine varies directly with the grams of solutes excreted per liter. It provides information on the ability of the kidney to concentrate the glomerular filtrate. The physiologic range of specific gravity varies from 1.003 to 1.032, but the usual range for a 24-hr specimen varies from 1.015 to 1.025. The most concentrated specimen is obtained on arising in the morning. In renal tubular disease, the concentrating ability of the kidney is one of the first functions lost.

The specific gravity of urine may be determined directly with a urinometer or indirectly through measurement of its refractive index.

Measurement By Urinometer (Hydrometer). The urinometer is a hydrometer designed to fit into and float in a narrow cylinder filled with urine. The urinometer has a slender neck, with a specific gravity scale wrapped around it that usually covers the range from 1.000 to 1.040. The urinometer should be calibrated by testing it with a solution of known specific gravity.

Procedure

1. Fill the cylinder about three-fourths full with specimen and place on a level surface.
2. Insert the urinometer into the cylinder and spin slightly so that it floats freely.
3. Read the specific gravity directly from the scale on the stem at the lowest point of the meniscus of the urine surface.

4. *Temperature correction:* If the urine is not at the urinometer calibration temperature, add 0.001 to the specific gravity for every 3° C that the urine is above this temperature and subtract 0.001 for every 3° that the urine is below this temperature.

Measurement By Refractometry. The refractive index of a solution also varies with the amount of dissolved substances, and hence is related to the specific gravity. Because the urine must be clear, this measurement is usually carried out on a centrifuged specimen. Commercial refractometers are available with a scale that gives a direct readout in specific gravity.

Procedure

1. Place a small drop of clear urine on the glass surface of the refractometer and close the lid.
2. Look through the meter directly toward a light source.
3. Record the specific gravity at the point where the line separating the light area from the dark crosses the specific gravity scale.

Measurement By Dipstick. A dipstick that estimates specific gravity is available as part of a multiple test strip (N-Multistix-SG by Ames). The strip uses a polyelectrolyte and an indicator (bromthymol blue) that changes color as H^+ is displaced by Na^+ or K^+ in the patient's urine.[6] The test is good for screening, but measurement by urinometer or by refractometry is more accurate.

Osmolality

Osmolality is a measure of the moles of dissolved particles (undissociated molecules, as well as ions) contained in a kilogram of solvent; it reflects the total concentration of solutes.

When substances are dissolved in a solvent, they affect some of the properties of the solvent and cause some physical changes that can be measured. These changes are a lowering of the freezing point, a decrease in the vapor pressure, and an increase in the boiling point of the pure solvent. Commercial instruments are available that make use of the freezing point and vapor pressure for the measurement of the osmolality of body fluids. The most commonly used instruments use the freezing point depression, whereby one mole of each ionic species and each nonionized solute per kilogram of water lowers the freezing point by 1.86° C.

Procedure

1. Follow the manufacturer's directions for the particular instrument available to you.
2. Calibrate the instrument with known standards.
3. Centrifuge the specimen well to eliminate suspended matter.
4. Measure the lowering of the freezing point or the vapor pressure, as the case may be.
5. Record the osmolality.

Reference Values

Serum:	278 to 305 mosm/kg
Urine—random specimen:	40 to 1350 mosm/kg
On normal fluid intake, 24-hr specimen:	500 to 800 mosm/kg
During maximal urine concentration:	850 to 1350 mosm/kg

Note: The osmolality is a more accurate reflection of the concentration of dissolved substances than is the specific gravity because in various diseases the urine may contain relatively large amounts of glucose or protein. These substances have a molecular weight much higher than that of the salts commonly found in urine, and hence affect the specific gravity much more than they affect the osmolality. The receptors in the body respond to osmolality or changes in solute concentration.

Qualitative or Semiquantitative Tests

A routine urinalysis also includes tests for the measurement of pH and for the detection of protein, glucose or reducing sugars, ketone bodies, and frequently for bilirubin, blood, and other substances. Commercial dipsticks or test strips incorporate chemical reagents in a paper (solid phase) matrix that react with specific urine constituents when dipped into a urine bottle. Semiquantitation is achieved by comparing the color produced with that of a color scale. Some test strips are for a single constituent; others are designed for two or more tests simultaneously. Some dipsticks permit the measurement of nine tests at once (Ames, Boehringer Mannheim). The nine constituents that may be estimated in this manner are pH, protein, glucose, ketones, blood, bilirubin, urobilinogen, nitrite, and leukocyte esterase.

pH

The urine has a physiologic pH range of 4.6 to 8.0, with a mean of approximately 6.0. Starvation and ketosis increase the acidity of the urine. Acid-producing salts are sometimes administered for the treatment of urinary tract infections. The urine is seldom alkaline, but becomes alkaline in alkalosis after the ingestion of alkali over a period of time for the treatment of ulcers or from bacteria in the urine that generate ammonia. The pH is usually measured by means of a paper strip impregnated with an indicator or by using a commercial dipstick that contains a mat of cellulose or paper impregnated with two indicators, such as methyl red and bromthymol blue, to cover the entire physiologic range.

Protein

A small amount of protein (50 to 150 mg/24 hr) appears daily in the normal urine. Some of this protein comes from a small amount of albumin that is filtered in the glomerulus but not reabsorbed in the tubules; the rest is the result of glycoproteins from the linings of the genitourinary tract. Normally, the protein concentration in urine is below 10 mg/dL and is not detectable by the usual urinalysis methods.

Proteinuria (the presence of detectable amounts of protein in the urine) usually indicates injury to the glomerular membrane, which consequently permits the filtration or escape of protein molecules. Proteinuria must be differentiated, however, from a transient proteinuria that may take place during the course of a high fever or from a harmless condition, *orthostatic proteinuria*, which occurs only when a patient is active and on his feet. Because orthostatic proteinuria does not occur when the subject is recumbent, the first urine specimen upon arising in the morning should have no detectable protein. Protein in urine may be measured by dipstick or sulfosalicylic acid tests.

Dipstick. The basis for the protein test is the "protein error" of indicators, a term applied to the change in ionization and color of the indicator, and hence the apparent pH, when an indicator dye is adsorbed to protein. The paper spot in the dipstick is impregnated with citrate buffer (pH 3.0) containing bromphenol blue indicator, which is yellow at pH 3.0 and blue at pH 4.2. At pH 3.0, the indicator is mostly un-ionized. If protein is present in the urine into which it is dipped, the ionized fraction binds to the protein, thereby causing more dye to ionize until equilibrium is reached; hence, the impregnated strip has less yellow and more blue color as the protein concentration increases. This reaction is seen visually as a change from yellow to green (a mixture of yellow dye plus blue dye appears green). The color is compared with that of a color chart, which provides a crude estimation of the protein content from 30 mg/dL to about 1000 mg/dL.

Note: False-positive protein indications by the dipstick method may be obtained with a buffered, alkaline urine. No false-positives occur with x-ray contrast media, sulfonamides, or other drugs or medication that may cause turbidity with other tests for protein (sulfosalicylic acid test or heat precipitation at pH 5.0).

Sulfosalicylic Acid. To perform a chemical test for urine protein, transfer approximately 3 mL of centrifuged urine to a test tube, hold at an angle, and let 3 drops of 250 g/L sulfosalicylic acid run down the side of the tube. The acid forms a layer underneath the urine; do not mix.

Examine the urine-acid interface for turbidity after about 1 min. A barely perceptible turbidity is reported as a "trace" or ±, and heavier amounts are graded from 1+ to 4+. A protein concentration of 5 mg/dL usually registers as a trace, but false-positive sulfosalicylic acid test results may be encountered with urine from patients injected recently with x-ray contrast media or receiving sulfonamides, tolbutamide, or other medications.

Glucose

Although glucose may appear in the urine as a result of renal disease (renal glycosuria), this occurrence is not common. The usual rationale for including a test for glucose in a routine urinalysis is to detect unsuspected diabetes or to check the efficacy of insulin therapy in diabetic patients. This test is described in detail in Chapter 7. Only the dipstick test is described in this section.

The dipstick paper is impregnated with glucose oxidase, peroxidase, buffers, and a chromogen that is colorless in the reduced state and colored when oxidized. Glucose oxidase converts glucose to gluconic acid and produces H_2O_2 in the process. In a coupled reaction, H_2O_2 is decomposed by peroxidase, with the simultaneous production of H_2O and oxidation of the chromogen. One company employs potassium iodide as the chromogen that is oxidized to iodine (brown); another uses *o*-toluidine (colorless) as a hydrogen donor that becomes blue when dehydrogenated (oxidized). The color intensity of the glucose strip is read in 10 sec for the Ames dipsticks and in 30 sec for the Chemistrips of Boehringer Mannheim.

Ketone Bodies

Although ketonuria may accompany situations of carbohydrate deprivation (fasting or high-fat, high-protein diets), it derives its clinical importance primarily from its occurrence in uncontrolled diabetes. Accordingly, ketonuria is reviewed in greater detail in Chapter 7.

The dipsticks contain a strip impregnated with sodium nitroprusside and an alkaline buffer. In the presence of acetoacetate or acetone, a lavender color is produced and compared with that of a color chart.

Blood

The presence of small amounts of occult blood (blood cells or hemoglobin that do not visibly color the urine) may be detected by appropriate dipsticks. The reaction is based on the enzymatic action of hemoglobin in decomposing peroxides, which in the presence of a hydrogen donor, *o*-toluidine, produces a blue color. This end reaction is identical to the color obtained in the glucose oxidase reaction. Hemoglobin in the urine indicates hemolysis in the bloodstream or the lysis of red blood cells in the urinary tract; it is common in various renal disorders or conditions affecting the urinary tract.

The appropriate dipstick (Ames; Boehringer Mannheim) is impregnated with a buffered organic peroxide and *o*-toluidine. A blue color appears within 30 sec if hemoglobin is present and may be graded by the intensity of the color.

Bilirubin, Urobilinogen, Nitrite, and Leukocyte Esterase

Tests for bilirubin, urobilinogen, nitrite, and leukocyte esterase appear on some dipsticks, but not all laboratories test for these substances routinely. Tests for bilirubin and urobilinogen have their greatest application in liver disease and are reviewed in Chapter 11. Nitrite in the urine suggests the possible presence in the urinary tract of bacteria capable of reducing nitrate to nitrite. Leukocyte esterase indicates the presence of leukocytes in the urinary tract, presumably attracted by invading bacteria. A positive test for both nitrite and esterase is presumptive evidence of urinary tract infection.

Microscopic Examination of the Urine Sediment

Some hospitals no longer perform a microscopic examination of urinary sediment on a routine basis. They reserve this test for cases of suspected renal disease or for patients who exhibit positive tests for both nitrite and leukocyte esterase and are thus at risk for urinary tract infection.

Microscopic examination of urinary sediment is important, however, because it yields information that may be helpful in making a diagnosis. For best results, obtain a concentrated urine specimen (upon arising) that has been clean-voided. The specimen should be examined within an hour of voiding because cells deteriorate upon standing; deterioration may be delayed by refrigeration or by the addition of formalin (0.2 mL/100 mL urine). Centrifuge about 12 mL of a well-mixed urine sample for 5 min at 80 × g, and pour off all but 0.2 to 0.3 mL of the urine. Suspend the sediment in the residual fluid by flicking the bottom of the tube on a test tube rack and take a drop for microscopic examination.

This book does not include instructions regarding the use of the microscope and the identification of the various formed elements. Abnormal cells, oval fat bodies, casts, and crystals can be identified if they are present.[2,3]

Normal Findings

1. Squamous and epithelial cells are present in all urine samples, especially those from women; they have no pathologic significance.
2. An occasional or rare red or white cell has no pathologic significance.
3. Some hyaline casts may be found in normal urine, particularly after stress, exercise, or fever, in the absence of renal disease.
4. Bacteria may be present as an external contamination; clean-voided specimens examined when they are fresh help to eliminate possible confusion.

Abnormal Formed Elements[5]

1. *Cells*
   a. *Red blood cells:* More than an occasional red blood cell is abnormal and requires further investigation. Red blood cells may originate from any location in the urinary tract. In women, this finding may also be a contribution from menstruation.
   b. *White blood cells:* A few white cells are not abnormal, but many of these cells in a freshly voided specimen indicate the presence of an infection somewhere in the genitourinary tract.
   c. *Yeasts:* Yeasts are recognized by their budding and oval shape. They are common contaminants, but can cause infections.
2. *Oval fat bodies:* These elements are usually present to some degree in all types of diseases affecting the renal parenchyma but are most numerous in the nephrotic syndrome. They are thought to be degenerated tubular epithelial cells that have become filled with fat droplets.
3. *Casts:* Casts are formed by precipitation of mucoprotein in the lumen of renal tubules and collecting ducts; later, they are passed into the urine. Cellular elements frequently are entrapped in the cast.
   a. *Red blood cell casts:* These casts show presence of red cells in the protein matrix. They are reddish-brown or orange from the hemoglobin leaking out of broken-down red blood cells. The presence of red blood cell casts is always a pathologic condition because it denotes glomer-

ular inflammation and bleeding and is associated with glomerulonephritis, systemic lupus erythematosus with kidney involvement, or other glomerular diseases.

b. *White blood cell casts:* These casts contain imbedded leukocytes and signify an infection (pyelonephritis).

c. *Hyaline casts:* Simple hyaline casts contain protein alone and are clear as glass because they have almost the same refractive index as that of urine. They may be present in the urine of normal individuals, particularly after strenuous exercise, but are also found in the urine in many disease states, usually accompanied by proteinuria. Hyaline casts may contain some cellular debris or fat.

d. *Granular casts:* The presence of epithelial cellular debris in a cast makes it appear granular.

e. *Fatty casts:* These casts appear like oval fat bodies but are cylindric in shape. This abnormal finding is indicative of renal parenchymal disease.

f. *Waxy casts:* These cellular casts have degenerated and look "waxy" or like ground glass. They may be present in any of several kidney diseases.

g. *Broad casts:* These short and wide casts have formed in the broad collecting tubules and are found only in renal failure.

4. *Crystals:* Crystals found in normal urine that have no particular significance are calcium phosphate, triple phosphate, calcium oxalate, amorphous phosphates, sodium or ammonium urate, and sometimes calcium carbonate. Large amounts of urate or uric acid crystals should be noted because they may indicate excessive breakdown of tissue cells (nucleoproteins) or be an accompaniment of gout. Unusual crystals, such as cystine or sulfa drug crystals, must be noted.

Tests for Creatinine, Urea, and Uric Acid

In the early days of clinical chemistry, the nonprotein nitrogen (NPN) constituents of serum were measured as a single group after precipitation of serum proteins. The principal NPN constituents are urea, creatinine, uric acid, and amino acids, all of which are now measured separately in appropriate clinical situations. The two NPN constituents with serum concentrations that yield the most information in kidney disease are creatinine and urea; their serum concentrations are elevated when formation or elimination of urine is impaired, irrespective of the cause. The concentration of blood ammonia, a minor NPN constituent, is sometimes requested when hepatic encephalopathy is suspected (see Chap. 11).

Creatinine[4,5,7,8]

Creatinine is a waste product formed in muscle from a high-energy storage compound, *creatine phosphate (phosphocreatine).* Adenosine triphosphate (ATP) is the immediate source of energy for muscular contraction as it is hydrolyzed to adenosine diphosphate (ADP). ATP cannot be stored in sufficient quantity to meet the energy demand of intense muscular activity; however, creatine phosphate can be stored in muscle at approximately four times the concentration of ATP and is used for this purpose. When needed for energy, creatine phosphate and ADP are converted by enzymatic action to creatine and ATP (Fig. 6.4). A side reaction occurs, however, and a small portion of the creatine phosphate loses its phosphate as phosphate ion, with closure of the ring to form creatinine, as illustrated in Figure 6.4. This reaction is not reversible, and the creatinine is excreted in the urine as a waste product.

The amount of creatinine excreted daily is a function of the muscle mass and is not affected by diet, age, sex, or exercise. It amounts to approximately 2% of the body stores of creatine phosphate and is roughly 1 to 2 g per day for an adult. Women excrete less creatine than do men because of their smaller muscle mass.

A small amount of preformed creatinine is ingested as a constituent of meat, but this amount has little effect on the concentration of creatinine in serum. Elevated concentrations occur only when renal function is impaired.

Creatinine appears in the glomerular filtrate and is not reabsorbed by the tubule; hence, any condition that reduces the glomerular filtration rate results in a lessened excretion from the body, with a consequent rise in the concentration of creatinine in plasma. Because the excretion rate of creatinine is relatively constant (±15% for an individual per day) and its production rate is not influenced by protein catabolism or other external factors, the concentration of creatinine in the serum is a good measure of renal glomerular function.

The serum creatinine concentration is elevated when a reduction in the glomerular filtration rate is significant or when urine elimination is obstructed.

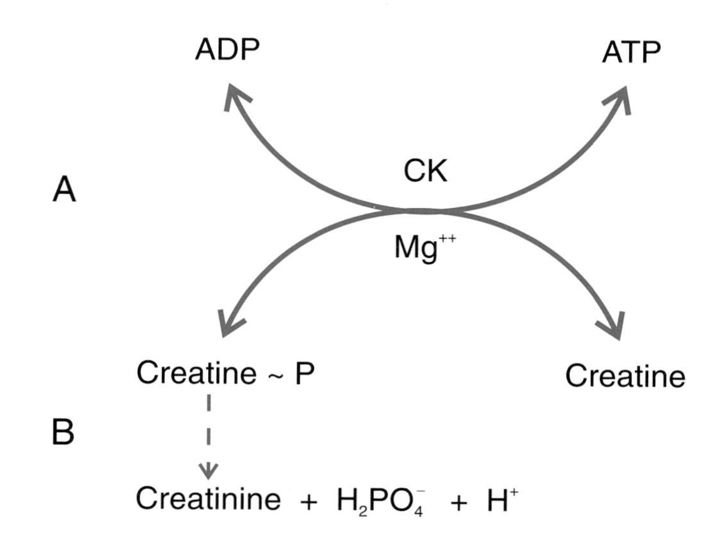

FIG. 6.4. Schema for the formation of creatinine in muscle as a side reaction from the spontaneous breakdown of creatine phosphate. Reaction A illustrates the reversible storage of high-energy phosphate as creatine phosphate; reaction B illustrates the side reaction. CK = creatine kinase; ~ = a high-energy bond.

The kidney reserve is such, however, that about 50% of kidney function must be lost before a rise in the serum concentration of creatinine can be detected. The concentration of serum creatinine is a better indicator of renal function than either that of urea nitrogen or that of uric acid because serum creatinine is not affected by diet, exercise, or hormones, factors that influence the levels of urea nitrogen or uric acid. A small percentage of the creatinine appearing in the urine may be derived from tubular secretion. This amount is negligible at normal serum levels of creatinine but becomes larger as the concentration in serum rises.

Serum Creatinine

The Jaffé reaction was first used for the determination of serum creatinine more than 80 years ago and is still widely used. Although alternative methods based on the enzymatic degradation of creatinine recently have been increasing in popularity, the Comprehensive Chemistry 1992 Survey of the College of American Pathologists (CAP)[9] reveals that 77% of 6400 laboratories estimate serum creatinine concentration by some form of the alkaline picrate reaction (Jaffé reaction). During the Jaffé reaction, creatinine interacts with alkaline picrate to form a red addition product; however, a few other serum constituents, such as ketoacids, glucose, various drugs, and ascorbic acid (to a lesser extent), do the same. The interfering compounds can be removed by adsorbing the creatinine on aluminum silicates or on a cation-exchange resin and washing out the noncreatinine reactants, a time-consuming wash process. Most laboratories use some type of kinetic Jaffé method in which the creatinine complex formation is monitored shortly after mixing the reactants (10 to 60 sec) and is continued for another 20 to 120 sec. This procedure avoids the effect of fast-reacting interfering compounds (primarily acetoacetate) and slow-reacting compounds, such as proteins.[8] The kinetic methods require automation.

Several different enzymatic methods of creatinine determination have been implemented. *Creatinine iminohydrolase* enzymatically degrades creatinine into N-methylhydantoin and releases ammonia. Through the production of ammonia, the pH is altered in proportion to the concentration of creatinine, thereby increasing the absorbance of pH-indicating dyes. The application of *creatinine amidohydrolase* in enzyme-coupled reactions has also been used. In general, the enzymatic methods are more specific than the Jaffé reaction, but they are not free from interferences.[8]

The method described in the following text is a fixed-time Jaffé reaction assay that requires prior removal of proteins by precipitation or dialysis.

Principle. Creatinine in a protein-free filtrate reacts with alkaline picrate to form a red adduct, the absorbance of which is measured at 515 nm.

Reagents

1. Picric acid, 0.036 mol/L. Dissolve 9.16 g reagent-grade picric acid (containing 10 to 12% added water as a safety feature*) in warm water. Cool and make up to 1 L volume.
2. Tungstic acid in polyvinyl alcohol. Place 1 g polyvinyl alcohol (Elvanol 70–05, DuPont) in 100 mL water and heat to dissolve. Do not boil.

*Anhydrous picric acid is a powerful explosive. Picric acid crystals must never by desiccated or heated.

Transfer to a 1-L volumetric flask containing 11.1 g $Na_2WO_4 \cdot 2H_2O$ in 300 mL water. Then add 2.1 mL concentrated H_2SO_4 in 300 mL water. Mix and dilute with water to 1 L volume. This solution is stable at room temperature for 2 years and does not require refrigeration.
3. NaOH, 1.4 mol/L. Dissolve 54 g NaOH in water and dilute to 1 L volume. Store in a polyethylene bottle.
4. Creatinine stock standard, 0.111 mg/mL. Dissolve 111.0 mg of creatinine in 1 L of 0.1 mol/L HCl.
5. Working standard A. Dilute stock standard 1:50 with water to create an equivalent to a serum creatinine of 2.0 mg/dL if 3.0 mL of working standard A is treated as a serum filtrate from 0.5 mL serum as described in the following procedure.
6. Working standard B. Dilute stock standard 1:20 with water, making its concentration equivalent to 5.0 mg/dL if treated as described as follows.

Procedure

1. Precipitate the serum proteins by adding 0.5 mL serum to 4.0 mL tungstic acid solution contained in a 16 × 100 mm test tube. Shake vigorously and centrifuge for 10 min.
2. Set up in test tubes the following: blank—3.0 mL H_2O; standard—3.0 mL standard; unknown—3.0 mL protein-free centrifugate.
3. Add 1.0 mL picric acid to each and mix well.
4. Add 0.5 mL of 1.4 mol/L NaOH to the first tube; mix and set a timer for 15 min. Add NaOH to the remaining tubes at 30 sec intervals.
5. Read absorbances of standards and unknowns against blank at 515 nm exactly 15 min after adding the NaOH. Read tubes at 30-sec intervals.

$$\text{mg creatinine/dL} = \frac{A_u}{A_s} \times C$$

C = concentration of standard (when treated as a filtrate).
A_u and A_s = absorbances of unknown serum and standard, respectively.

$$\text{For SI units: } \mu\text{mol/L} = \text{mg/dL} \times 10 \times \frac{1000}{113} = \text{mg/dL} \times 88.4$$

Reference Values

Men:	0.7 to 1.2 mg/dL or 62 to 106 μmol/L
Women:	0.6 to 1.1 mg/dL or 53 to 97 μmol/L
Children:	0.4 to 1.0 mg/dL or 36 to 88 μmol/L

The values for "true" creatinine are about 0.1 to 0.2 mg/dL lower than those listed.

Increased Concentration. The concentration of serum creatinine rises when formation or excretion of urine is impaired, irrespective of whether the causes are prerenal, renal, or postrenal in origin.

Values that are 2 s above the upper limit of normal suggest possible renal damage, and the test should be repeated; levels of 1.5 mg/dL and greater indicate impairment of renal function. Minor changes in concentration may be significant, and the serum level usually parallels the severity of the disease. Table 6.2 shows the relation of the serum creatinine concentration, the creatinine clearance value, and the patient's status. The prerenal factors causing an increased serum creatinine level are congestive heart failure; shock; salt and water depletion associated with vomiting, diarrhea, or gastrointestinal fistulas; uncontrolled diabetes mellitus; excessive use of diuretics; diabetes insipidus; and excessive sweating with deficient salt intake. Renal factors may involve damage to glomeruli, tubules, renal blood vessels, or interstitial tissue. Postrenal factors may be prostatic hypertrophy, neoplasms compressing the ureters, calculi blocking the ureters, or congenital abnormalities that compress or block the ureters.

The serum creatinine concentration is monitored closely after a renal transplant because a rising concentration, even though small, may be an indication of transplant rejection. Some renal transplant units prefer to monitor a rise in the serum concentration of β_2-microglobulin as the earliest indicator of renal transplant rejection (see Chap. 9, Table 9.4).

Decreased Concentration. Low serum creatinine concentrations have no clinical significance.

Urine Creatinine

Because the concentration of creatinine in urine (approximately 1 mg/mL) is much higher than that in serum, a dilution of urine is in order. For urine of normal protein content, dilute 1:200 with water and treat the same as a serum filtrate, that is, take 3.0 mL for analysis. If proteinuria exists, precipitate the proteins as follows: Add 0.5 mL urine + 0.5 mL water to 4.0 mL tungstic acid solution; mix and centrifuge. Dilute the filtrate 1:20 with water and take 3.0 mL for analysis as for serum. Thus, both types of urine, those with and those without proteinuria, are diluted 1:200. Develop color in standards A and B as for serum. Little noncreatinine chromogen is in urine.

TABLE 6.2
TYPICAL CORRELATION OF SERUM CREATININE CONCENTRATIONS WITH THE CREATININE CLEARANCE AND PATIENT STATUS

SERUM CREATININE (mg/100 mL)	CREATININE CLEARANCE (mL/min)	PATIENT'S STATUS
0.6–1.2	90–140	Normal
		Some patients with proteinuria
1.3–2.4	61–90	Capable of performing usual types of activity
2.5–4.9	24–60	Difficulty in performing strenuous physical activity
5.0–7.9	12–23	Unable to perform all daily physical activities except on part-time basis
		Acidosis, sodium loss, serum Ca ↓
8.0–12	7–12	Severe limitation of physical activity
		Serum K ↑, Ca ↓, Na ↓
>12	6 or less	May be disoriented or in coma

Modification from American Heart Association, Council on Kidney in Cardiovascular Disease.

Calculation

For Standard A: $C = \frac{A_u}{A_s} \times \frac{0.111}{50} \times 3 \times \frac{100}{3/200}$ mg/dL $= \frac{A_u}{A_s} \times 44.4$ mg/dL, where $\frac{0.111}{50} \times 3 =$ mg creatinine in 3 mL dilute standard aliquot, $\frac{3}{200} =$ mL of undiluted urine in 3 mL aliquot, and × 100 converts mL to dL.

Reference Values. The urinary excretion of creatinine depends on the muscle mass of the individual, but the following are rough reference values:

Men:	14–28 mg/kg/d
Women:	11–20 mg/kg/d
Newborn:	7–12 mg/kg/d
0.1–5 years:	8–22 mg/kg/d
10–12 years:	8–30 mg/kg/d

The factor for converting mg/kg/d to mmol/kg/d = 0.00884.

Creatinine Clearance

The most sensitive chemical method of assessing renal function is the creatinine clearance test. This clearance test provides an estimate of the amount of plasma that must have flowed through the kidney glomeruli per minute with complete removal of its content of creatinine to account for the creatinine per minute actually appearing in the urine. The test requires the complete collection of the urine formed in an accurately recorded time period (for calculation of the rate of urine flow) and quantitation of the creatinine concentration in both serum and urine. The creatinine clearance is calculated as

$$\frac{U}{S} \times V$$

where U is the urine concentration of creatinine, S is the serum creatinine concentration, and V is the volume of urine excreted per minute. U and S must be measured in the same concentration units, although it does not matter whether the measurement is in mg/dL or in SI units. The dimension of the clearance thus becomes expressed as mL/min, because the dimensions of U/S cancel each other out. The creatinine clearance is practically the same as the glomerular filtration rate; the tubules excrete a small amount of creatinine, a factor that may become significant when serum creatinine concentrations are elevated. The creatinine clearance value is closer to the glomerular filtration rate when "true" creatinine is measured because noncreatinine chromogens increase the serum creatinine concentration but not the urine creatinine concentration.

The importance of obtaining the total excretion of urine in an accurately timed period cannot be overstressed because any error committed in the collection and measurement of the volume excreted per minute is carried over into the calculation of the clearance. The following points should be emphasized. Those who administer the test must thoroughly understand the procedure and communicate in a simple way with the patient so that he or she understands and cooperates. The test may be carried out over *any accurately timed period* if the urine flow is adequate, even though the usual requests are for 2-, 4-, or 24-hr periods. The elapsed time should be recorded to the minute. One must start timing the test when the bladder is completely empty.

Procedure

1. Promote a good urine flow by having the patient drink about 500 mL of water 10 or 15 min before the test starts.
2. *When the patient is able to void,* have him/her completely empty the bladder, *note down the time exactly,* and *discard this urine.*
3. When the patient seems to have a full bladder again and can void (in approximately 2 hr), have him/her completely empty the bladder into a container, note the time exactly, and send this urine to the laboratory, along with a blood sample, for measurement of the clearance. If the patient's height and weight are provided, the laboratory can correct the clearance for standard body surface area from nomograms in Figure 6.5. Body surface area may also be calculated from the formula:

$$\log A = 0.425 \log W + 0.725 \log H - 2.144$$

where A = body surface in m^2, W = weight in kg, and H = height in cm.
4. *Calculation of Creatinine Clearance*

$$\text{Uncorrected: Clearance} = \frac{U}{S} \times V$$

$$\text{Corrected for surface area: Clearance} = \frac{U}{S} \times V \times \frac{1.73}{A}$$

where U = the urine creatinine concentration, S = the serum creatinine concentration, V = the urine flow in mL/min, A = body surface area in m^2, and 1.73 is the standard body surface area of the average adult.

Notes:

1. The patient must understand what is required, cooperate by completely voiding when necessary, and record the time to the minute. Thus, an order for a 2-hr collection should be recorded as the total elapsed time in min because the time between the 2 voidings is rarely 2 hr exactly.
2. A 24-hr collection is *not* more accurate than a 2-hr collection and is far more inconvenient. Moreover, the chance is good of losing a urine specimen occurring with a bowel movement, or of losing one at night if the patient has not been cautioned or may not be too alert.
3. The total specimen should be sent to the laboratory for measurement. Measurements made on the ward or clinic may not be reliable. If aliquots are to be removed for other tests, such removal should be done by laboratory personnel *after mixing and measuring the complete collection.*
4. One must correct for body surface area in small individuals (infants, children, and small adults) and also in large or obese adults.

Reference Values. The values obtained for the creatinine clearance depend to some extent on the methodology because of a larger percentage of noncreati-

NOMOGRAMS FOR CALCULATING BODY SURFACE AREA*

Nomogram for Calculating the Body Surface Area of Children[1]

Height | Surface area | Weight

Nomogram for Calculating the Body Surface Area of Adults[1]

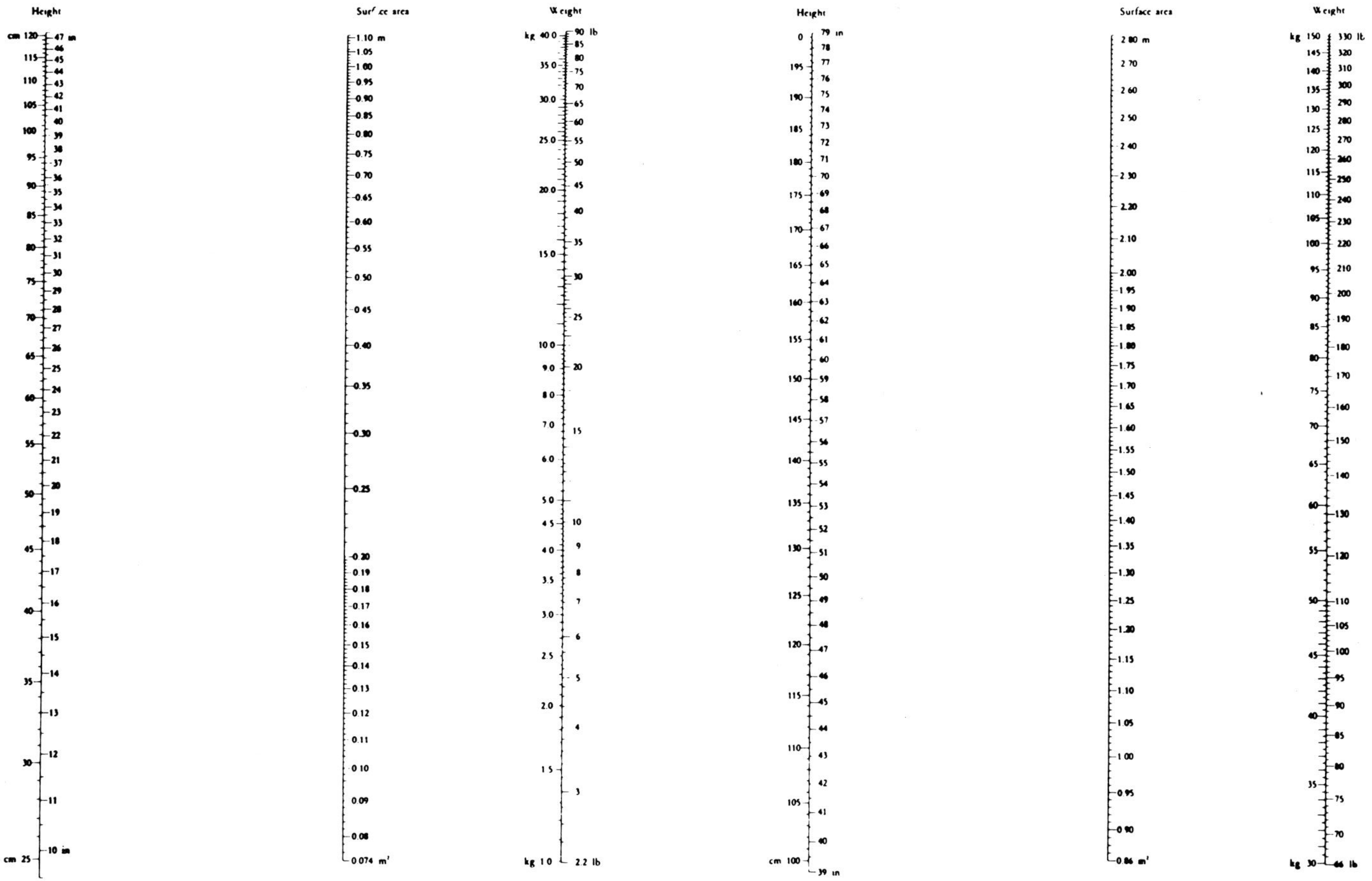

Height | Surface area | Weight

[1] From the formula of Du Bois and Du Bois, *Arch. intern. Med.* **17**, 863 (1916): $S = W^{0.425} \times H^{0.725} \times 71.84$, or $\log S = 0.425 \log W + 0.725 \log H + 1.8564$, where S = body surface area in square centimeters, W = weight in kilograms, H = height in centimeters.

*From *Documenta Geigy Scientific Tables*, 6th ed., pp. 632–633. By permission of J. R. Geigy S.A.

FIG. 6.5. Nomogram for the determination of body surface area. (From Documenta Geigy Scientific Tables, 7th Edition. Basel, Switzerland, Ciba-Geigy Limited, 1970.)

nine chromogen in plasma or serum than in urine. The adult normal creatinine clearance for men is 95 to 140 mL/min and 90 to 130 mL/min for women.

In children over 1.5 years, the creatinine clearance is 55 to 85 mL/min *when corrected for body surface area*. The creatinine clearance of premature and newborn infants is about 35 to 65 mL/min/1.73 m^2.

Increased Creatinine Clearance Values. An increased value has no medical significance. If the value should be greatly increased, suspect some error in the collection and/or timing of the clearance. Perhaps the bladder was not completely empty when the timing was started.

Decreased Creatinine Clearance Values. When the clearance test is carefully executed, a decreased creatinine clearance is a sensitive indicator of a decreased glomerular filtration rate. The reduced glomerular filtration rate may be caused by acute or chronic damage to the glomerulus or to any of its components. Reduced blood flow to the glomeruli may also produce a decreased creatinine clearance. Acute tubular damage may result in a decreased creatinine clearance as blood flow to the glomeruli is drastically reduced in response to osmolar changes. Table 6.2 lists typical values of the creatinine clearance and serum creatinine concentrations in patients with varying degrees of renal damage. In Figure 6.6, creatinine clearance values are plotted against serum creatinine.

Note: If a low creatinine clearance is obtained in a patient with a normal serum creatinine concentration, suspect an error, either in the specimen collection (incomplete urine voiding, incorrect timing, loss of some specimen, incorrect volume measurement) or in the creatinine analyses. An error in specimen collection is more likely.

Iothalamate Clearance

The rate that a molecule present in the bloodstream is cleared into the urine can accurately reflect the glomerular filtration rate if the molecule is solely excreted

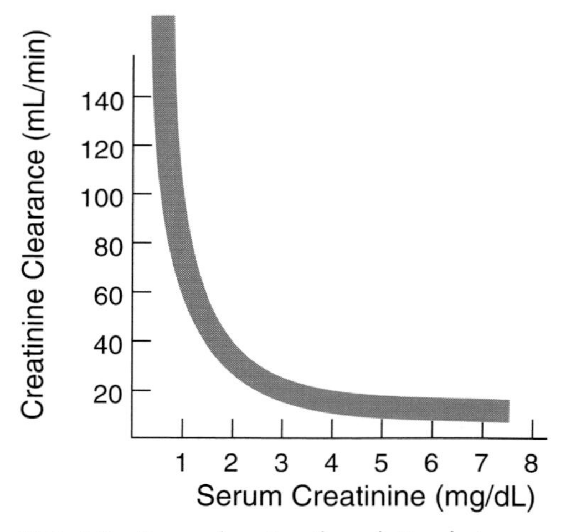

FIG. 6.6. Curve showing the relation between serum creatinine concentration and the creatinine clearance.

by the kidneys and not metabolized, secreted, or reabsorbed. Although the determination of the creatinine clearance is the estimate of glomerular filtration rate most frequently used by clinicians, it is subject to inaccurate timing and incomplete sample collection (as described in the previous section) and is secreted by the distal tubule during renal failure. Clearance of either inulin or iothalamate (Io) is a more accurate indicator of glomerular filtration rate than is creatinine clearance, particularly in children, patients with renal disease, and noncompliant patients. The disadvantages of inulin or Io clearance determinations are that they require intravenous infusions and are far more labor intensive and expensive than a creatinine clearance evaluation. Inulin or Io clearance determinations may be the only way, however, to obtain an accurate clearance value for small children or uncooperative patients. A convenient and sensitive high-performance liquid chromatography (HPLC) assay for Io is available,[10] and a protocol for glomerular filtration rate evaluation has been established.[11]

Principle. The Io level in a patient's blood is maintained at constant concentration (steady state) by subcutaneous Io infusion from a small portable pump, following an intravenous loading dose of Io. While a steady-state condition exists, the rate of Io excretion by the kidney (via glomerular filtration) is equal to the rate of its influx from the pump. A patient's clearance rate of Io can be calculated by dividing the rate of Io infusion by the steady-state plasma Io concentration. The concentration of Io in the patient's blood at steady state and the concentration infused by the pump can be determined by HPLC.

Urea Nitrogen[12]

Ingested proteins are hydrolyzed to amino acids that can be used for anabolic or catabolic purposes. Protein cannot be stored in the body to any appreciable extent, so when the intake is in excess of body requirements for the synthesis of structural and functional components, the surplus amino acids are largely catabolized for energy purposes, as illustrated schematically in Figure 6.7.

The α-amino group of all amino acids that are broken down in the mammalian body, whether derived from the diet or endogenous sources, end up in the compound urea, whose structure is shown in Figure 6.8. Urea is a waste product, soluble in water and excreted solely by the kidney. Urea is the characteristic nitrogenous end product of protein catabolism in mammals, but is not so for many other forms of life. Birds and reptiles excrete uric acid as the end product, and for some bony fish, the chief product is ammonia.

For historic reasons, the serum urea concentration has been customarily expressed in terms of its nitrogen because, in the early days of clinical chemistry, nitrogen-containing substances were analyzed by converting the nitrogen into ammonia, which was trapped and measured. Even though no longer necessary to do so, the urea concentration is still calculated in terms of urea-N when expressed on a mass basis; in SI units, however, the urea concentration is designated as mmol/L of the intact molecule. One mole of urea weighs 60 g and contains 28 g of nitrogen. Thus, a serum concentration of 28 mg/dL of urea-N is equivalent to 60 mg/dL or 10 mmol/L of urea.

Urea appears in the renal glomerular filtrate in the same concentration as in plasma, but unlike creatinine, some urea is absorbed as the filtrate traverses the

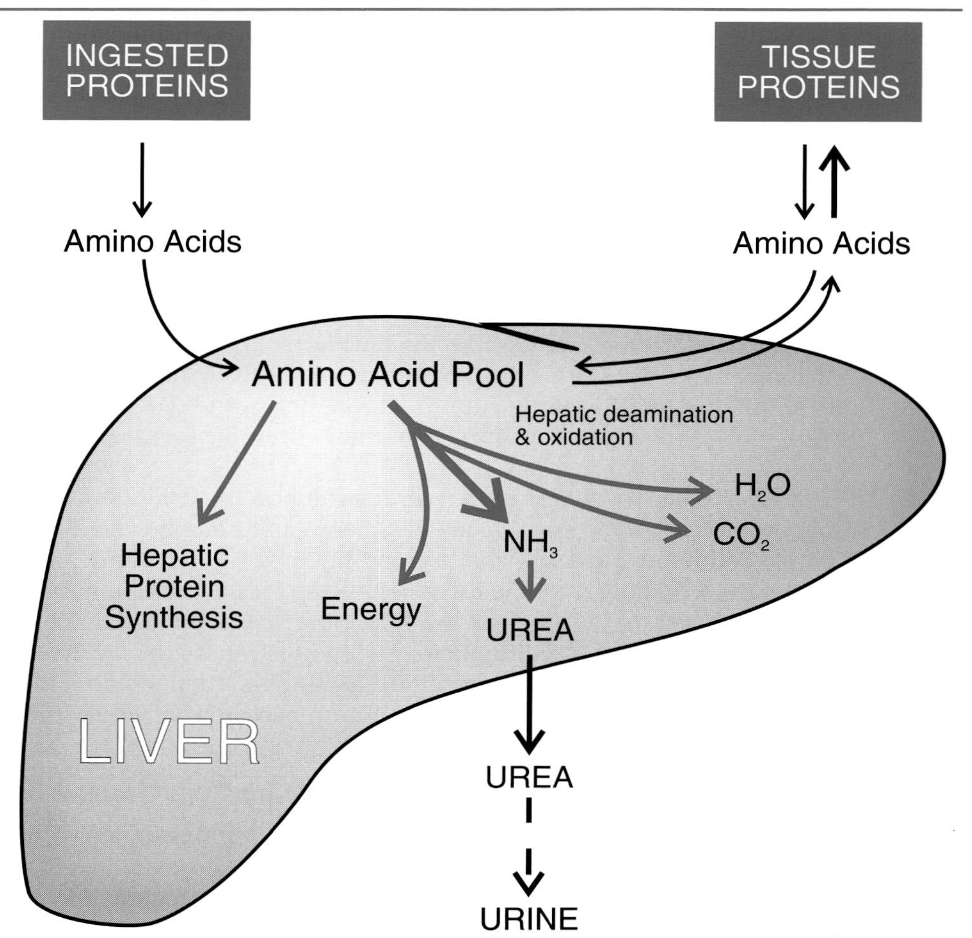

FIG. 6.7. Illustration of the fate of ingested protein. Hydrolysis to amino acids occurs in the intestine. The absorbed amino acids are carried to the liver and other tissues, where they may be used for the synthesis of new proteins or used for energy. When catabolized, the amino group is removed and the ammonia is synthesized into urea, a nontoxic waste product, for excretion by the kidney.

renal tubule. Under conditions of normal flow and normal renal function, about 40% of the filtered urea is reabsorbed; when the flow rate is decreased, the actual and relative amount reabsorbed increases. As with creatinine, the serum concentration of urea (or urea-N) rises with impaired renal function.

The serum concentration of urea-N is influenced by factors not connected with renal function or urine excretion. It is affected strongly by the degree of protein catabolism, whether produced by a high protein diet or by hypersecretion or injection of adrenal steroids that results in the mobilization of protein for energy purposes. In the case of diet, a change to a high protein diet can double the serum urea-N concentration, and a low protein intake can reduce it by half. In like manner, the injection or ingestion of steroids produces a rise in serum urea-N, as do stressful situations (for example, breaking an arm) that

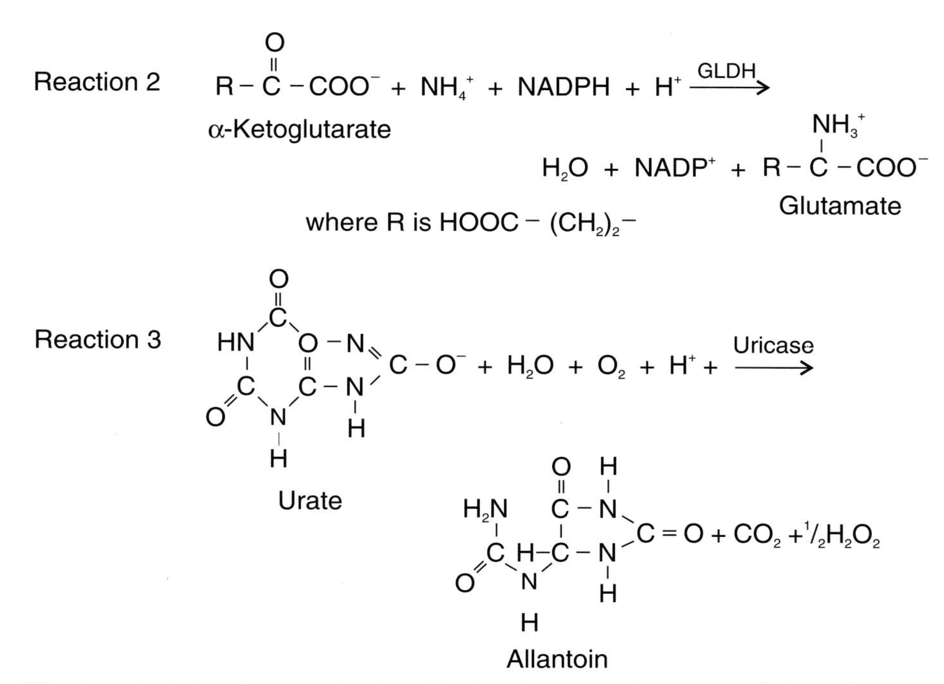

FIG. 6.8. Analysis of urea and uric acid. Reactions 1 and 2 are coupled to allow urea determination by the urease/glutamate dehydrogenase method. Reaction 3 depicts the formation of allantoin and H_2O_2 from uric acid and oxygen.

cause the adrenal gland to secrete additional cortisol. For these reasons, the measurement of serum creatinine is a better indicator of kidney status than is that of urea-N, although in many cases they rise and fall together. The various prerenal, renal, and postrenal factors that affect the concentration of serum creatinine also influence the level of serum urea-N. In the case of serum urea-N, one always must assess the possible influence of dietary or hormonal factors.

Serum Urea-Nitrogen

Most methods for the determination of serum urea-N in the past 75 years have used the enzyme *urease* as the first step in the process. Although still true, according to the CAP Survey,[9] techniques have changed. More than half of the surveyed laboratories used the urease reaction coupled with that of glutamate dehydrogenase; fewer than 1% of the laboratories determined urea directly by heating a serum aliquot with diacetylmonoxime.

Determination by Urease/Glutamate Dehydrogenase.[13] Urea determination by means of the coupled enzymes urease/glutamate dehydrogenase is rapid, precise, and easily automated. Packaged reagents are available from several manufacturers.

Principle. Urease hydrolyzes serum urea, and only urea, into NH_3 and CO_2 as shown in reaction 1 in Figure 6.8. In a coupled reaction, glutamate dehydrogenase (GLDH) in the presence of α-ketoglutarate and NADPH converts the reactants into glutamate and $NADP^+$ respectively. Two moles of NADPH are converted to $NADP^+$ for every mole of urea decomposed. The disappearance of NADPH is measured spectrophotometrically at 340 nm.

Reference Values. 6 to 18 mg/dL with normal protein intake.*
2.1 to 6.4 mmol/L

Increased Concentration. High protein diet, administration of cortisol-like steroids, stressful situations, and prerenal, renal, and postrenal factors as described for creatinine increase the concentration of urea-N.

Decreased Concentration. The urea concentration is low in late pregnancy (when the fetus is growing rapidly and using maternal amino acids), in starvation, and in patients whose diet is grossly deficient in proteins.

Uric Acid

Uric acid is a purine compound that circulates in plasma as sodium urate and is excreted by the kidney. It is derived from the breakdown of nucleic acids that are ingested or come from the destruction of tissue cells; it is also synthesized in the body from simple compounds. The formation of uric acid is illustrated in Figure 6.9. Urate is the end product of purine metabolism in humans and the higher apes, but is oxidized to allantoin by most other mammals. Urate appears in the glomerular filtrate and is partially reabsorbed in the tubules. Urate is of low solubility in plasma, and uric acid is less so; the danger exists of precipitation of uric acid crystals when a local rise in $[H^+]$ occurs in tissues. Urate deposition in the kidney may ultimately lead to renal failure.

*The normal curve is skewed to the right, and some apparently healthy individuals may have a serum urea-N concentration as high as 27 mg/dL.

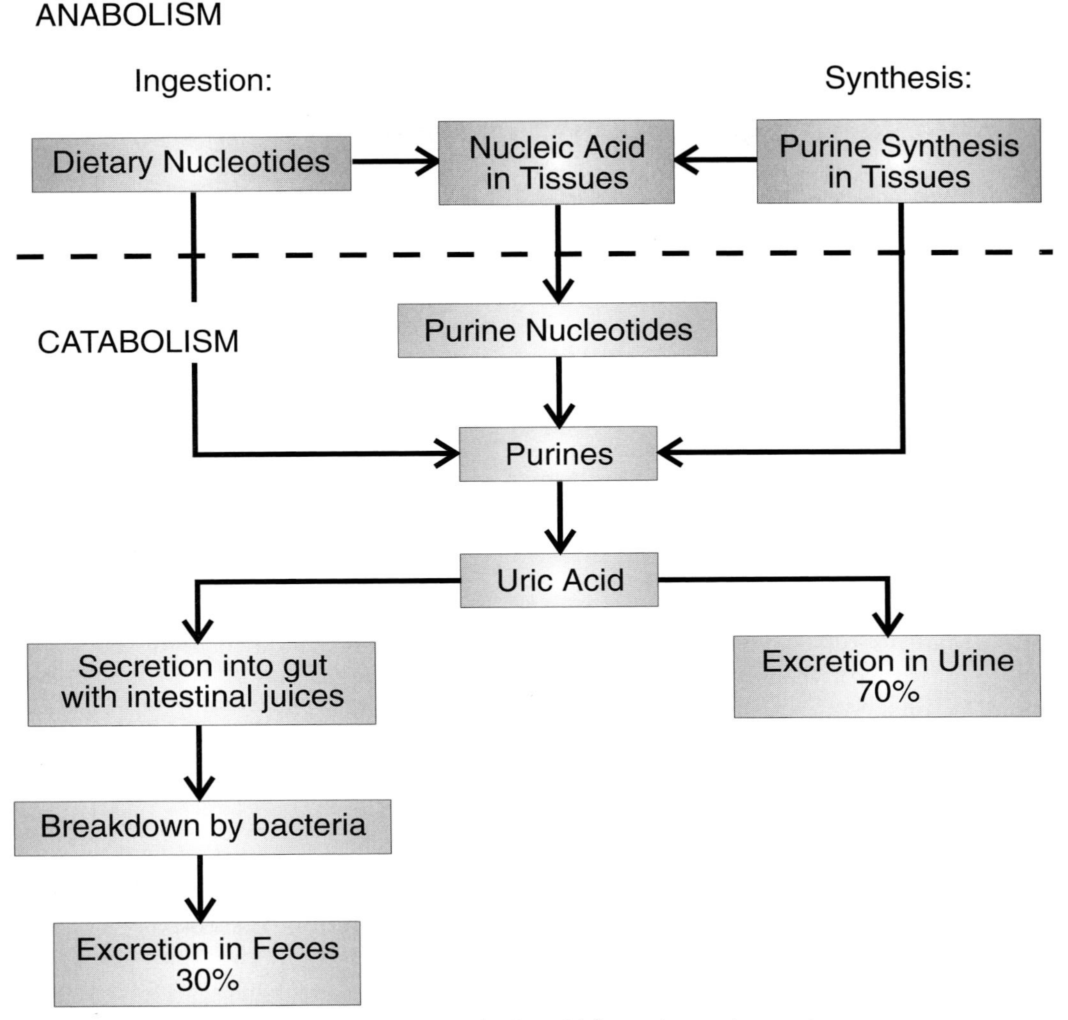

FIG. 6.9. Schematic representation of uric acid formation and excretion.

The measurement of serum uric acid is *not* used as a primary test for the evaluation of kidney function because creatinine and urea serve this purpose much better, but when multichannel analyses are carried out, the concentration of serum urate, in general, should reflect the same changes that occur with creatinine and urea-N. Thus, it may serve as a confirmatory check on the analyses of these constituents; the serum concentration of urate is usually elevated when the formation or excretion of urine is improper, irrespective of the cause.

The main value of the serum uric acid test is in the diagnosis of gout or for following the treatment of patients with this disease. It also is used sometimes as an indicator of a large-scale breakdown of nucleic acid, such as occurs in toxemia of pregnancy, after massive irradiation for tumors, or following the administration of cytotoxic agents in the treatment of malignant disease.

Gout, an ancient and painful disease, is characterized by the precipitation of uric acid crystals in tissues and joints, particularly in joints of the big toe. Although the deposition of crystals in the joints is responsible for the pain and debilitation of gout, the deposition of urate may damage the kidneys and presents the greatest danger to the patient.

Serum Uric Acid

For many years, the principal method for determining the serum urate concentration was based on the reduction of phosphotungstic acid to a blue phosphotungstate complex. The method lacked specificity, however, and other serum constituents also reacted to a small extent to produce the blue complex. Currently, about 99% of the laboratories surveyed by the CAP[9] use the enzyme uricase for catalyzing the oxidation of urate to allantoin and CO_2, with the generation of H_2O_2. The serum urate concentration may be quantitated by (1) measuring the UV absorbance at 292 nm before and after treatment with uricase, (2) measuring the rate of oxygen consumption during the uricase reaction, or (3) measuring the H_2O_2 produced by means of a coupled reaction yielding a colored product.

Determination by Uricase/UV Absorbance

Principle. When oxygen is present, uricase converts urate to allantoin and CO_2, with the production of H_2O_2 as shown in reaction 3 (Fig. 6.8). Urate absorbs light at 292 nm; allantoin and H_2O_2 do not. The serum urate concentration is proportional to the decrease in absorbance.

Determination by Uricase/H_2O_2 Coupled Reactions

Principle. In the presence of peroxidase, hydrogen peroxide reacts oxidatively with 3,5-dichloro-2-hydroxybenzenesulfonic acid and 4-aminophenazone to form a red dye.[14] Other chromogens may also be used to measure the H_2O_2 by a colorimetric reaction. A high serum concentration of ascorbate (vitamin C) usually causes lower values because of competition by ascorbate with the chromogen as a hydrogen donor in the reaction. This competition is eliminated in the method described by including potassium ferricyanide in the reagent to oxidize ascorbate.

Reference Values

Men:	3.5 to 7.5 mg/dL or 0.210 to 0.445 mmol/L
Women:	2.5 to 6.5 mg/dL or 0.150 to 0.390 mmol/L
Children:	2.0 to 5.5 mg/dL or 0.120 to 0.330 mmol/L

For SI units, mmol/L = mg/dL × 0.0595.

Increased Concentration. Serum urate concentration is elevated in most patients with gout, in renal diseases, and after increased breakdown of nucleic acid or nucleoproteins (in leukemia, polycythemia, toxemia of pregnancy, resolving pneumonia, and after irradiation of x-ray-sensitive carcinomas). Some individuals have an elevated concentration of serum urate despite the absence of disease (idiopathic hyperuricemia). Several genetic defects characterized by an enzyme deficiency in the purine nucleotide synthetic pathway are associated with a hyperuricemia. An example of such a defect is the

Lesch-Nyhan syndrome, a rare disorder in which the enzyme that converts hypoxanthine and guanine into their respective monophosphates is lacking.

Decreased Concentration. The concentration of serum urate may be decreased after the administration of adrenocorticotropic hormone (ACTH) or cortisol-like steroids, certain drugs that decrease the reabsorption of urate by renal tubules (aspirin, probenecid, penicillamine), by drugs (allopurinol) that block a step in the formation of uric acid, or by defective renal tubular absorption of uric acid.

The concentration of uric acid in the urine is influenced by the purine content of the diet. The average person on an average diet usually excretes from 250 to 750 mg/24 hr. The method for its determination in urine is identical to that in serum except that (1) the urine must be diluted because its urate concentration is much higher than that in serum and (2) usually no protein precipitate results when the phosphotungstate is added; a 1:10 dilution is usually sufficient.

Urine Concentration Test

The ability of the kidney to concentrate urine is a test of tubular function that can be carried out readily with only minor inconvenience to the patient. A urine concentration test is widely used to test tubular function.

The original concentration tests requiring water deprivation for 24 hr have been replaced by a test with a 14-hr period that gives similar results. The test should *not* be performed on a dehydrated patient.

Procedure

1. The patient eats an early supper and is allowed no food or water after 6 P.M. on the night preceding the test. Discard any urine voided during the night.
2. On the test day, the first specimen is voided at 7 A.M., the bladder is emptied completely, and the specimen is discarded.
3. A second specimen is collected at 8 A.M., 14 hr after the commencement of the test, and the osmolality or specific gravity is measured. If the osmolality exceeds 850 mosm/kg or the specific gravity is > 1.022, the patient has adequate renal concentrating power and the test is over.
4. If the values are < 850 mosm/kg or 1.022 for osmolality and specific gravity, respectively, a urine sample is collected at 9 A.M. and assayed.
5. *Interpretation:* The renal concentrating ability is impaired if neither urine sample reached an osmolality of 850 mosm/kg or a specific gravity of 1.022. Normal values may be as high as 1350 mosm/kg or a specific gravity of 1.032. If the concentrating ability should fail, the specific gravity would be 1.010 and the osmolality close to 300 mosm/kg.

Urinary Tract Calculi (Stones)[1,15]

Calcium and/or magnesium salts frequently form insoluble calculi (stones) in the kidney. Most calculi (75 to 80%) contain calcium, 5 to 8% contain uric acid, and about 1% are composed of cystine. The host is unaware of a stone until it blocks a tubule or a ureter and the dilated vessel produces intense pain. Prolonged deposition causes irreversible renal damage. Stones are formed by the concentric deposition of poorly soluble compounds around some nuclei. The nuclei may be blood clots, fibrin, bacteria, or sloughed epithelial cells. Many times, the precipitation of the relatively insoluble compounds may be initiated or aggravated by an infection, dehydration, excessive intake or production of the compound, urinary obstruction, and other factors. Once formed, the calculi tend to grow by accretion unless they happen to be dislodged and are sufficiently small to travel down the urinary tract to be excreted. The larger calculi may remain in the kidney or become stuck in a ureter, from which they must be removed by surgery. Whether a stone is passed in the urine or removed by surgery, information concerning the type of calculus is of value; changes in diet may be recommended. Treatment varies with the type of stone.

Analysis of calculi is best done by infrared analysis or x-ray diffraction, a test that is usually sent to a specialty center.[1,15] The actual crystal composition of the stone, not just identification of the elements that are present, is obtained in this way.

The chemical analysis of calculi is not difficult if the material with which to work is sufficient, but it is quite time-consuming. Specimens obtained by surgery usually present no problem, but stones passed in the urine are small. When the amount of material to analyze is so small, the analyst should look first for the most likely constituents before running out of specimen (calcium, phosphate, oxalate, magnesium, ammonium, and urate, in decreasing order).

Principle. The calculus is weighed, and size, shape, color, surface appearance, and consistency are noted. The calculus is pulverized in a mortar if it is larger than 25 mg and is crushed in a test tube with a glass rod if it is smaller than 25 mg. The powder is dissolved, if possible, in 1 mol/L HCl, and chemical spot tests are made for different constituents on both the acid solution and the residue or powder.

REFERENCES

1. Freeman, J.A., and Beeler, M.F.: Laboratory Microscopy/Urinalysis. 2nd Edition. Philadelphia, Lea & Febiger, 1983.
2. Hammernyik, P.: Manual on Clinical Urinalysis Procedure. Seattle, Department of Laboratory Medicine, University of Washington, 1984.
3. Ash, K.O., and James, G.P.: Urinalysis. *In* Selected Methods of Clinical Chemistry. Vol. 9. Edited by W.R. Faulkner and S. Meites. Washington, D.C., American Association for Clinical Chemistry, 1982.
4. Santos, F., Orejas, G., Foreman, J.W., and Chan, J.C.M.: Diagnostic Workup of Renal Disorders. Curr. Probl. Pediatr., *21*:48, 1991.
5. Schumann, G.B., and Schweitzer, S.C.: Examination of Urine. *In* Clinical Diagnosis & Management by Laboratory Methods. 18th Edition. Edited by J.B. Henry. Philadelphia, W.B. Saunders, 1991.
6. Burkhardt, A.E., Johnston, K.G., Waszak, C.E., et al.: A reagent strip for measuring the specific gravity of urine. Clin. Chem., *28*:2068, 1982.
7. Perrone, R.D., Madias, N.E., and Levey, A.S.: Serum creatinine as an index of renal function: new insights into old concepts. Clin. Chem., *38*:1933, 1992.
8. Narayanan, S., and Appleton, H.D.: Creatinine: a review. Clin. Chem., *26*:1119, 1980.

9. Comprehensive Chemistry 1992 Survey. Skokie, IL, College of American Pathologists, 1992.
10. Weber, A.F., Lee, D.W., Opheim, K., and Smith, A.L.: Quantitation of iothalamate in serum and urine by high-performance liquid chromatography. J. Chromatogr., *337*:434, 1985.
11. Al-Uzri, A., et al.: An accurate practical method for estimating GFR in clinical studies using a constant subcutaneous infusion. Kidney Int., *41*:1701, 1992.
12. Taylor, A.J., and Vadgama, P.: Analytical reviews in clinical biochemistry: the estimation of urea. Ann. Clin. Biochem., *29*:245, 1992.
13. Sampson, E.J., Baird, M.A., Burtis, C.A., et al.: A coupled-enzyme equilibrium method for measuring urea in serum: optimization and evaluation of the AACC Study Group on urea candidate reference method. Clin. Chem., *26*:816, 1980.
14. Fossati, P., Prencipe, L., and Berti, G.: Use of 3,5-dichloro-2 hydroxybenzenesulfonic acid/4-aminophenazone chromogenic system in direct assay of uric acid in serum and urine. Clin. Chem., *26*:227, 1980.
15. Coe, F.L., and Flavus, M.J.: Nephrolithiasis. *In* Harrison's Principles of Internal Medicine. 12th Edition. Edited by J.D. Wilson, E. Braunwald, et al. New York, McGraw-Hill, 1991.

SUGGESTED READING

Henry, R., Cannon, D.C., and Winkelman, J.W. (Eds.): Clinical Chemistry: Principles and Technics. Hagerstown, MD, Harper & Row, 1974.

Jacobson, M.H., et al.: Urine osmolality. A definitive test of renal function. Arch. Intern. Med., *110*:83, 1962.

QUESTIONS

1. How do the different parts of the nephron function in the production of urine?
2. How do aldosterone and ADH affect urine formation?
3. What are the principal renal mechanisms for excreting protons (hydrogen ions)?
4. What parts of the nephron are attacked in glomerulonephritis and in the nephrotic syndrome? In what manner do the tests of renal function in these diseases differ?
5. How may a careful urinalysis aid in the detection of renal disease? Of early hepatitis? Of unsuspected diabetes?
6. Why is the serum creatinine determination a better indicator of renal status than a serum urea determination?
7. What is the physiologic basis for the creatinine clearance test? What test is beginning to replace it and why?
8. What principles are involved in the determination of serum urea by the urease/glutamate dehydrogenase reactions?
9. Urate is derived from what type of compounds? What is the significance of an elevated serum urate?
10. What are renal calculi composed of?

Carbohydrate Metabolism 7

OUTLINE

I. INTERMEDIARY CARBOHYDRATE METABOLISM

II. HORMONAL REGULATION OF PLASMA GLUCOSE

III. DETERMINATION OF SERUM (PLASMA) OR WHOLE BLOOD GLUCOSE CONCENTRATION

- A. Enzymatic methods
  - 1. Hexokinase method
    - a. Principle
    - b. Reference values
  - 2. Glucose oxidase method
    - a. Principle
- B. Nonenzymatic methods
  - 1. Nelson-Somogyi method
  - 2. Ferricyanide method
  - 3. *o*-toluidine method
- C. Test strips for monitoring of blood glucose
- D. Determination of urine glucose concentration
  - 1. Qualitative or semiquantitative
    - a. Copper reduction
    - b. Glucose-oxidase-impregnated strips
  - 2. Quantitative
- E. Glucose in other body fluids
  - 1. Cerebrospinal fluid
  - 2. Exudates and transudates

IV. KETONE BODIES IN URINE

- A. Impregnated strips or sticks
- B. Reagent tablets

V. IDENTIFICATION OF REDUCING SUGARS IN URINE

- A. Principle
- B. Interpretation

VI. DISORDERS OF CARBOHYDRATE METABOLISM

- A. Diseases associated with hyperglycemia
  - 1. Diabetes mellitus
    - a. Insulin-dependent diabetes mellitus (IDDM or Type 1)
      - (1) Determination of glycohemoglobin
        - (a) Principle
    - b. Noninsulin-dependent diabetes (NIDDM or Type 2)
    - c. Gestational diabetes (GDM)
    - d. Diagnosis of diabetes mellitus
      - (1) Fasting plasma glucose
      - (2) Postprandial plasma glucose
      - (3) Oral glucose tolerance test (OGTT)
        - (a) Procedure
        - (b) Interpretation
  - 2. Other hormonal disorders
- B. Disorders associated with hypoglycemia
  - 1. Hormonal
  - 2. Hepatic
- C. Hereditary disorders (enzymatic defects) with reducing sugars in urine

VII. BLOOD LACTATE AND PYRUVATE

- A. Lactic acidosis
- B. Determination of blood lactate
- C. Determination of blood pyruvate
- D. Reference ranges

OBJECTIVES

After reading this chapter, the student will be able to:

1. Define the terms glycogen, glycogenesis, glycogenolysis, gluconeogenesis, and glycolysis and understand where and how they fit in carbohydrate metabolism.
2. Explain how glucose homeostasis is maintained by the interplay of specific hormones.
3. Describe the principles of the two most widely used methods at present for the determination of serum glucose and be familiar with some of the historical methods.
4. Understand the distinguishing features of both insulin-dependent diabetes mellitus (IDDM) and noninsulin dependent diabetes mellitus (NIDDM).
5. Define glycated hemoglobin, understand how it is formed, and comprehend the significance of finding an elevated concentration in the blood of a diabetic patient.
6. Understand the role of an oral glucose tolerance test in the diagnosis of diabetes mellitus.
7. Describe the principal causes of hypoglycemia.
8. Be aware of the common hereditary disorders associated with a reducing sugar in the urine.
9. Describe some probable causes of lactose intolerance and a laboratory test for same.

Carbohydrates are composed of only three elements: carbon, hydrogen, and oxygen. In nature, carbohydrates exist most abundantly as sugars (mono- and disaccharides) and starches (polymers of pentoses or hexoses). A monosaccharide is a simple sugar containing 3, 4, 5, or 6 carbon atoms, called triose, tetrose, pentose, and hexose, respectively. Each contains an aldehyde or ketone group and is designated an aldose or ketose, respectively.

Carbohydrate, the major component of the human diet in many areas of the world, is an important source of body energy. The capacity of the body to store carbohydrate is limited, however, and is confined to liver (can store up to 10% of its wet weight) and muscle (can store up to 0.5% of its wet weight); the total amount of stored carbohydrate is sufficient to fulfill the energy requirements of a person for only about one-half day.

All tissues can use glucose, the principal and almost exclusive carbohydrate circulating in blood, but under fasting conditions, only a few tissues depend entirely on glucose as a source of energy. These are the brain, by far the most important glucose consumer, followed to a much lesser extent by red blood cells, platelets, leukocytes, and the kidney medulla. Other tissues readily oxidize fatty acids and ketones for energy purposes. As will be shown later in the chapter, plasma glucose levels are maintained during fasting by mobilization of muscle protein; the liver synthesizes glucose (gluconeogenesis) from certain amino acids (primarily alanine) derived from proteolysis. Liver gluconeogenesis is the mechanism during fasting for providing glucose to the tissues that depend completely on glucose for viability.

INTERMEDIARY CARBOHYDRATE METABOLISM[1]

Plasma glucose is derived from the hydrolysis of dietary starch and polysaccharides, from the conversion of other dietary hexoses into glucose by the liver, and from the synthesis of glucose from amino acids or pyruvate.

In times of glucose excess (elevated blood glucose, as after a meal), glucose is enzymatically polymerized in the liver to form glycogen (glycogenesis). When the blood glucose concentration drops, glycogen is converted to glucose (glycogenolysis) by a different set of enzymes. Thus, independent mechanisms exist for regulating the blood glucose level by means of glycogenesis-glycogenolysis reactions. The liver is the main organ for the storage of excess carbohydrate as glycogen, although skeletal and heart muscles can store minor amounts. Excess glucose is converted to fat by adipose cells where it is stored.

The energy stored in the glucose molecules is made available to the organism through several catabolic pathways that mainly generate adenosine triphosphate (ATP). The principal pathway of glucose oxidation consists of two phases, anaerobic and aerobic oxidation (oxidative phosphorylation), both of which are made up of many different enzymatic steps. The anaerobic phase, better known as *glycolysis*, is illustrated in Figure 7.1. The glucose-6-phosphate (G-6-P) is converted through several steps to a triose phosphate and then to pyruvate. All these reactions take place in the cytoplasm. Glycolysis can be reversed, that is, pyruvate can be converted back to G-6-P, but by a partly different pathway. Some of the pyruvate is converted into lactate (Fig. 7.2) by the enzyme lactate dehydrogenase, but the bulk of it enters the tricarboxylic acid (TCA) cycle (Figs. 7.1 and 7.2) when energy is needed and the O_2 supply is sufficient.

When sufficient oxygen is present, the aerobic phase of glucose utilization begins with the decarboxylation of pyruvate (Fig. 7.2). Pyruvate is converted in the mitochondria to acetyl CoA, which is oxidized to CO_2 and H_2O to generate many molecules of ATP. The energy derived from the aerobic oxidation of glucose is about 19 times as great as that obtained by anaerobic glycolysis alone.

Formation of acetyl CoA* is common to the catabolism of glucose, fatty acids, and some amino acids (Fig. 7.3). Acetyl CoA reacts with oxaloacetate in the TCA cycle to form citrate as a first step. With one full turn of the cycle, the citrate is converted back to oxaloacetate by several dehydrogenation and decarboxylation reactions that generate reducing equivalents. The reducing equivalents enter the respiratory chain located on the adjacent mitochondrial

*Acetyl CoA is the compound formed by transfer of an acetyl group to coenzyme A (CoA). CoA is a complex of adenosine with pantothenic acid (a member of the vitamin B complex) and β-mercaptoethylamine. CoA functions as a cofactor in enzymatic acyl transfer reactions.

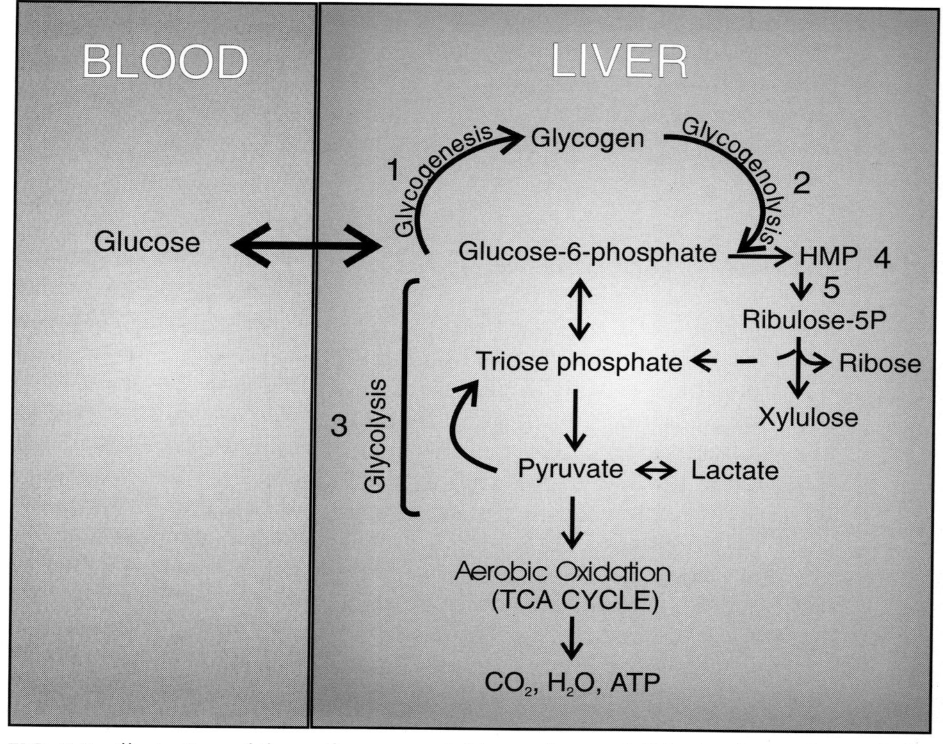

FIG. 7.1. Illustration of the pathways available to glucose-6-P in the liver. The anabolic pathway of glycogenesis is shown as pathway 1, and the regeneration of glucose-6-P by hydrolysis of glycogen is shown as pathway 2. The catabolic pathways are glycolysis (3) and the hexose monophosphate shunt (HMP) (4). The catabolism of glucose-6-P is completed by aerobic oxidation, a minor pathway (5).

membrane (see Chap. 5, Fig. 5.1). The net result of one full turn of the cycle is the generation of many molecules of ATP.

Extrahepatic tissues oxidize the acetyl CoA almost exclusively in the TCA cycle, but other pathways are possible in the liver. Under the combined conditions of depletion of carbohydrate and mobilization of fatty acids, as occurs in starvation and diabetes mellitus, two molecules of acetyl CoA may condense in liver mitochondria to form acetoacetyl CoA (Fig. 7.4). This latter compound then forms ketone bodies or is used for the synthesis of cholesterol (Fig. 7.5). This process is reviewed further in the section on ketone bodies.

G-6-P is also a pivotal point for three possible metabolic pathways illustrated in Figure 7.1. In addition to the glycogenesis and the glycolytic pathways just reviewed, G-6-P may proceed to a small extent ($< 10\%$) by the hexose monophosphate pathway (HMP) to generate NADPH and a pentose phosphate. Further catabolism may proceed to the formation of a triose phosphate that joins the mainstream of glycolysis. The HMP is an important generator of NADPH in erythrocytes that is essential for intracellular lipid synthesis and for preservation of the cell membrane. The pentoses (deoxyribose and ribose),

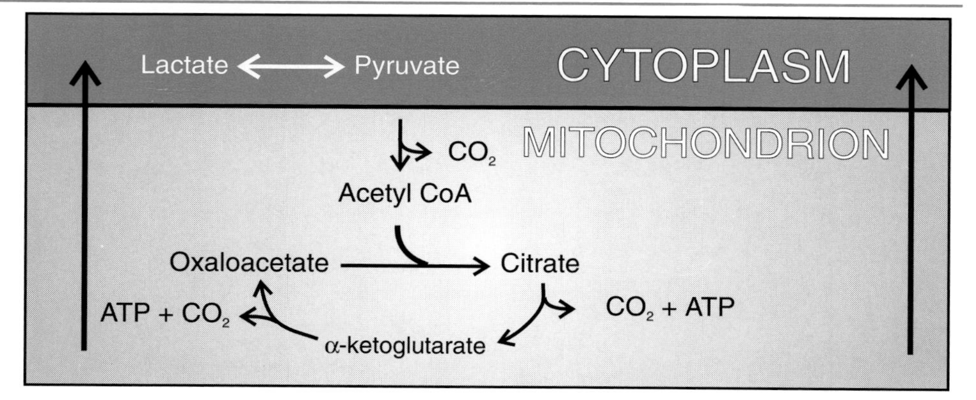

FIG. 7.2. Illustration of the aerobic phase of glucose oxidation. Pyruvate is converted to acetyl CoA in the mitochondrion, where the acetyl CoA enters the tricarboxylic acid cycle (TCA). Only three of the many intermediate compounds of the TCA cycle are shown. Fifteen moles of ATP are produced per mole of pyruvate in the process of oxidative phosphorylation (30 moles ATP per mole glucose) in the respiratory chain.

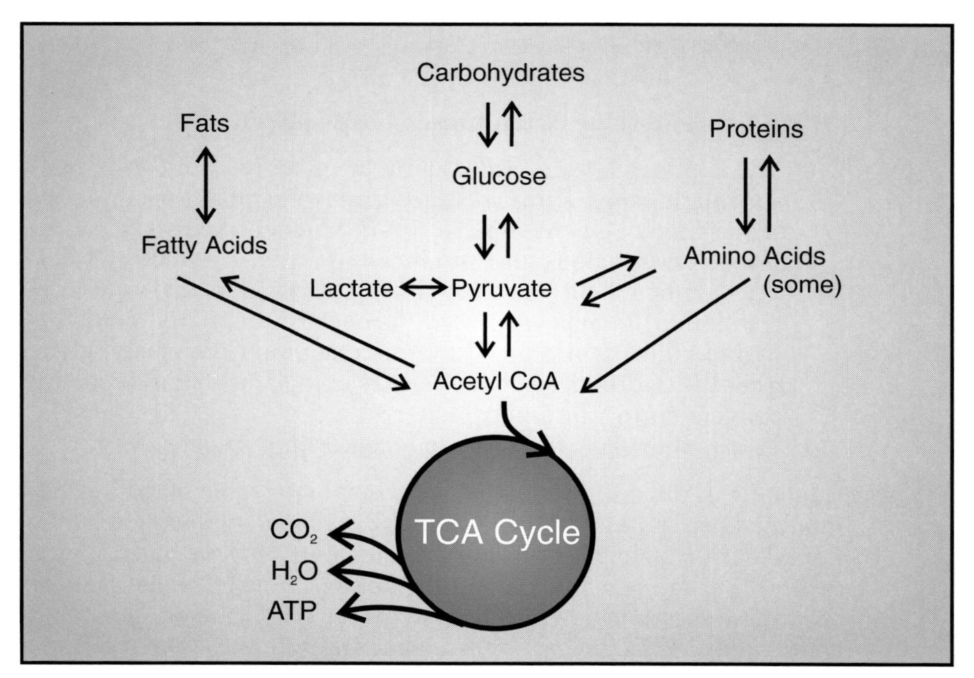

FIG. 7.3. Common meeting point of carbohydrate, protein, and fat metabolism at the level of pyruvate or acetyl CoA. The catabolism of all three foodstuffs produces acetyl CoA as the reduced intermediates are oxidized synchronously in the respiratory chain.

essential for incorporation into nuclear material (deoxyribonucleic acid [DNA] and ribonucleic acid [RNA]), are produced in the HMP. In the liver, the glycolytic pathway (3, Fig. 7.1) accounts for about 90% of the catabolized glucose, whereas the HMP shunt comprises less than 10%.

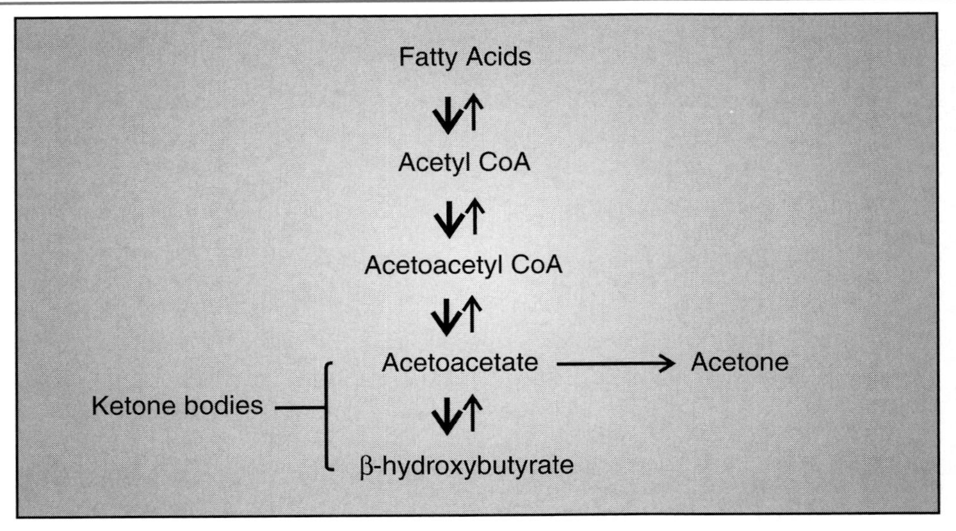

FIG. 7.4. Production of ketone bodies (acetoacetate, β-hydroxybutyrate, and acetone) during fasting states and in diabetes mellitus. Acetoacetate is converted to β-hydroxybutyrate by an enzyme; the conversion of acetoacetate to acetone is a spontaneous reaction that proceeds quite slowly.

Four pathways intersect at triose phosphate (Fig. 7.6):

1. Triose phosphate is produced by the glycolysis of G-6-P, but because the enzymatic steps are reversible, triose phosphate in appropriate circumstances may be converted to G-6-P (gluconeogenesis).
2. Triose phosphate is also produced to a minor extent by the HMP shunt.
3. Triose phosphate may be converted to glycerol in a reversible side reaction. The glycerol used for esterifying fatty acids (synthesis of triglycerides) is produced by this reaction. Conversely, glycerol derived from the hydrolysis of triglycerides enters the glycolytic pathway through conversion to triose phosphate.
4. Triose phosphate may be produced from pyruvate.

In spite of the multiple pathways possible for some of the key compounds in intermediary carbohydrate metabolism, the blood glucose concentration is kept remarkably constant in ordinary circumstances by a multiplicity of delicate control mechanisms for some of the key reactions. The net result is that some enzymatic reactions are inhibited, whereas others are accelerated. Some of the control mechanisms work by feedback inhibition; others are under hormonal control. The operation of the TCA cycle is limited by the available oxygen supply and by the concentration of oxaloacetate, a compound derived from carbohydrate metabolism.

From a clinical viewpoint, most patients with disturbances in carbohydrate metabolism have a defect in hormonal control, although some may be deficient in a particular enzyme.

Ketone bodies are produced when excessive amounts of fatty acids are catabolized and the availability of glucose is limited (Fig. 7.5). These conditions occur most commonly in diabetes mellitus and prolonged fasting.

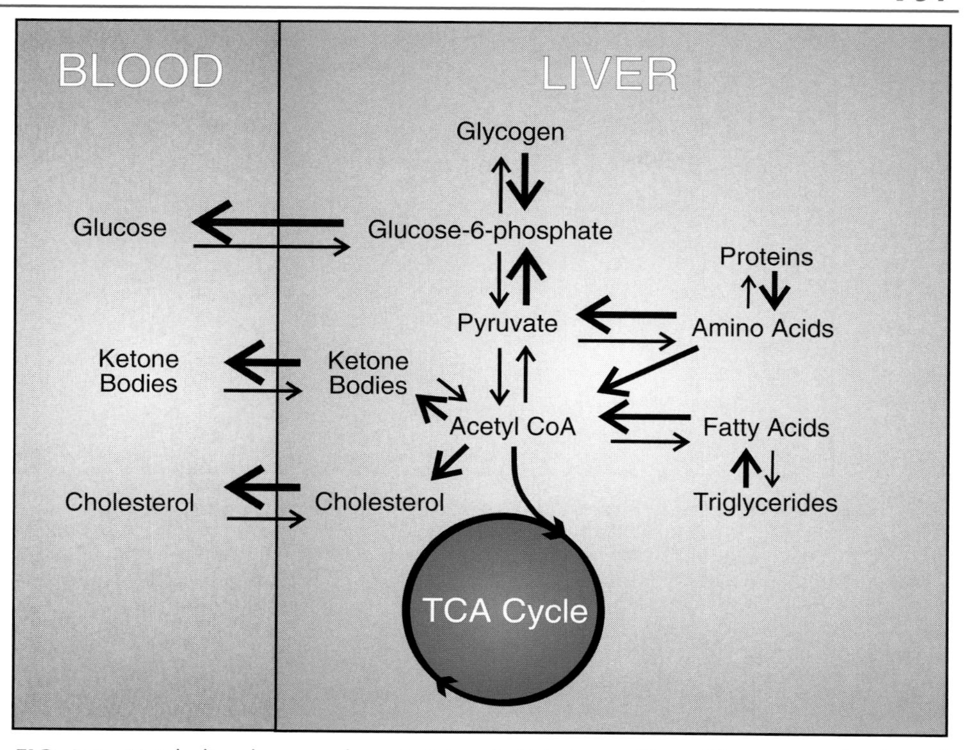

FIG. 7.5. Metabolic changes characteristic of diabetes mellitus. The overproduction of acetyl CoA, which comes from the accelerated catabolism of amino acids and fatty acids, is diverted to the synthesis of ketone bodies and cholesterol instead of entering the TCA cycle. The supply of oxaloacetate in the TCA cycle is limited because of the depressed glucose catabolism.

The three ketone bodies, acetoacetate, β-hydroxybutyrate, and acetone, are shown in Figure 7.4 as arising from the oxidation of fatty acids. When the oxaloacetate is insufficient to drive the TCA cycle, two moles of acetyl CoA join to form one mole of acetoacetyl CoA, which is then converted to acetoacetate. A portion of the acetoacetate is enzymatically reduced to β-hydroxybutyrate, whereas a portion spontaneously decomposes to acetone and CO_2. The proportion of the three ketone bodies in blood and urine is variable.

Heart and skeletal muscle can use acetoacetate as a source of energy to some extent; the brain acquires a limited capacity for oxidizing ketone bodies after about 3 weeks of fasting; the limit is a protective measure to curtail muscle wasting (gluconeogenesis from muscle protein). When the production of ketone bodies exceeds the capacity of the tissues to use them, a condition known as *ketosis*, they are rapidly excreted in the urine. The excretion of ketones into the urine is called *ketonuria*, and an elevated level of ketones in the blood is called *ketonemia*. The overproduction of ketoacids causes acidosis or lowered blood pH.

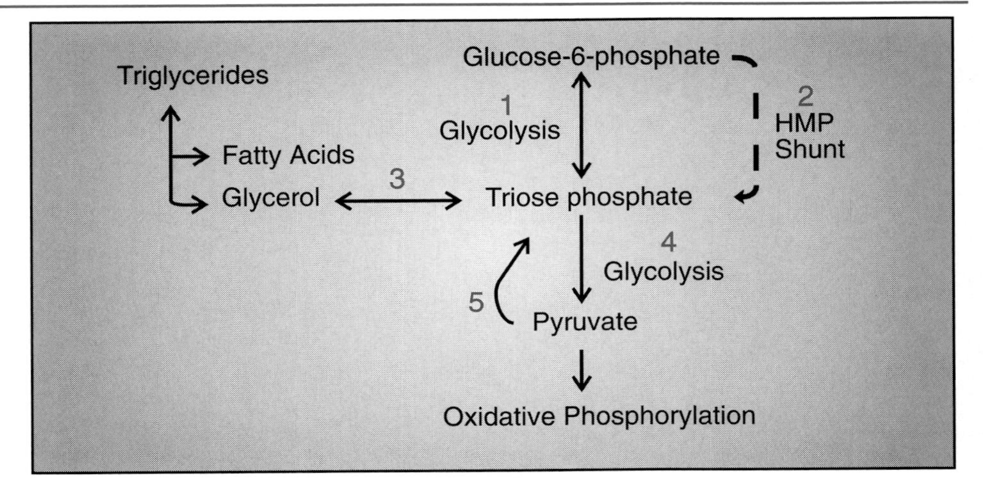

FIG. 7.6. Illustration of four pathways for the formation of triose phosphate. From glycolysis (1), from the HMP shunt (2), from glycerol (3), and from pyruvate by a separate enzymatic path (5). Triose phosphate may continue on the glycolytic pathway to pyruvate (4), be converted back to glucose-6-P (glucogenesis, reversal of 1), or be converted to glycerol (reversal of 3). Glycolysis is the major pathway for the formation of triose phosphate.

HORMONAL REGULATION OF PLASMA GLUCOSE[2]

Hormones regulate the plasma glucose concentration by affecting one or more of the various metabolic pathways illustrated in Figures 7.1 and 7.3. A perturbation in glucose concentration normally elicits the secretion of a hormone that rapidly counteracts the effect. The specific actions of the various hormones in this regulatory process are shown in Table 7.1.

TABLE 7.1
EFFECT OF HORMONES ON PLASMA GLUCOSE CONCENTRATION

ON PLASMA GLUCOSE CONCENTRATION			
DECREASE	INCREASE	HORMONAL ACTION	SIGNAL FOR RELEASE
Insulin		Primary; glycogenesis, glucose cell entry, triglyceride synthesis	↑ Plasma glucose
	Glucagon	Primary; glycogenolysis, gluconeogenesis	↓ Plasma glucose
	Epinephrine	Backup for glucagon, glycogenolysis	Nerve stimulation, "emergency" response
	Cortisol	Secondary; gluconeogenesis, antagonism to insulin action	ACTH
	Growth hormone	Secondary, gluconeogenesis, antagonism to insulin action	GH releasing factor
Somatostatin, a minor regulator		Tertiary; inhibits release of GH, insulin, and glucagon to varying degrees	Food in gastrointestinal tract

Insulin, secreted by pancreatic beta cells (islets) in response to an elevation in the plasma glucose concentration, is the only hormone that lowers the circulating glucose level (Table 7.1). Circulating insulin rapidly binds to receptors on cell surfaces, increases glucose entry into cells, particularly those of muscle, fat and liver, and alters the following metabolic pathways: (1) enhances the conversion of glucose to glycogen in the liver, (2) inhibits lipolysis and accelerates triglyceride synthesis in adipose cells, (3) promotes glycolysis by accelerating peripheral utilization of glucose, and (4) stimulates the synthesis of amino acids from pyruvate.

The adequate secretion of insulin and its unimpeded attachment to cells are obligatory for the regulation of the plasma glucose concentration. A relative excess of insulin in plasma produces a lowering of the plasma glucose concentration, whereas a relative deficiency is associated with a glucose elevation.

The glandular secretion of insulin is governed by the plasma glucose concentration; an increase is the signal for secretion, whereas a decrease inhibits its release.

Some amino acids, such as leucine and arginine, lower the plasma glucose concentration by stimulating insulin secretion. Sulfonylureas (tolbutamide, chlorpropamide, and others) also induce insulin secretion and are used as oral antidiabetic agents.

Glucagon, the principal counter-regulatory hormone for glucose, is secreted by the alpha cells of the pancreas in response to a fall in plasma glucose concentration. It raises the plasma glucose level by promoting (1) the conversion of liver glycogen to glucose and (2) the formation of glucose from amino acids (Table 7.1). Glucagon is the principal hormone for producing a rapid increase in plasma glucose concentration. Its secretion is suppressed by a hyperglycemia.

Epinephrine, secreted by the adrenal medulla, produces a rapid increase in plasma glucose concentration but simultaneously increases the heart rate and blood pressure and has other physiologic effects as well. This hormone serves as a back-up for glucagon (Table 7.1) and does not play an important role in the homeostasis of plasma glucose as long as the secretion of glucagon is adequate. Epinephrine secretion is triggered by nerve stimulation in response to a physical or emotional stress. Like glucagon, it induces rapid conversion of liver glycogen to glucose.

Growth hormone (GH), as its name implies, stimulates the growth of soft tissue, bone, and cartilage as its primary function. Its effect on the plasma glucose concentration is secondary to its tissue actions, some of which are antagonistic to those of insulin (Table 7.1). These actions are (1) inhibition of glucose entry into peripheral cells, (2) stimulation of liver glycogenolysis, and (3) acceleration of fatty acid catabolism. A prolonged excess of GH (acromegaly) is usually accompanied by a mild hyperglycemia and a diabetic-like glucose tolerance test result. A severe deficiency of GH may cause hypoglycemia, particularly after fasting for 16 to 24 hrs.

Cortisol, like other 11-oxysteroids secreted by the adrenal cortex, raises the plasma glucose concentration primarily by promoting gluconeogenesis (conversion of amino acids to glucose). The effect is not immediate and the hyperglycemia is mild. Cortisol, like GH, has some metabolic effects that are antagonistic to those of insulin. Many patients receive steroids as part of their therapy, and side effects may cause a mild hyperglycemia or an abnormal glucose tolerance test result. Excess cortisol secretion, whether caused by a pituitary adenoma secreting adrenocorticotropic hormone (ACTH) or by an overproduction caused by an adrenocortical neoplasm (Cushing's syndrome), is accompanied by a hyperglycemia, whereas a deficiency (Addison's disease) is usually characterized by a lowered plasma glucose concentration.

Somatostatin, a polypeptide hormone formed in the pancreatic delta cells and elsewhere, inhibits to varying degrees the release of insulin and glucagon (see Chap. 13). Somatostatin is secreted in response to the presence of food in the gastrointestinal tract and plays only a minor, indirect role in modifying plasma glucose concentration.

Thyroxine, the thyroid hormone (see Chap. 13), promotes the conversion of liver glycogen to glucose, which, over a period of time, can seriously deplete glycogen stores. It accelerates the absorption of glucose from the intestines during meals, but its role in regulating the plasma glucose concentration is insignificant.

DETERMINATION OF SERUM (PLASMA) OR WHOLE BLOOD GLUCOSE CONCENTRATION

Until about 35 years ago, most of the quantitative tests for glucose determination depended on the oxidation of glucose by hot alkaline copper solutions or solutions of potassium ferricyanide. These tests were supplanted by the *o*-toluidine test and later by enzymatic methods using either glucose oxidase or hexokinase. Burrin and Price have reviewed the various methods for measuring the glucose concentration in body fluids.[3]

Several glucose test methods are discussed in the following sections. No matter which method is used, one must take precautions in sample collection to prevent glucose utilization by leukocytes. The glucose loss on standing in a warm room may be as high as 10 mg/dL (0.6 mmol/L)/hr. The decrease in serum glucose concentration is negligible if the blood sample is kept cool and the serum separated from the clot within 0.5 hr of drawing. Otherwise, addition to the collection tube of 2 mg sodium fluoride per mL blood to be collected prevents glycolysis for 24 hr without interfering with the glucose determination.

Enzymatic Methods

Enzymatic methods are the most popular procedures for the determination of plasma glucose because of their high specificity, rapidity of assay, use of small sample quantities (10 μL or less), and ease of automation. The 1991 Comprehensive Chemistry Survey of the College of American Pathologists did not list a single laboratory among the 5700 participants that used a nonenzymatic (solely chemical) method for the determination of glucose; all laboratories employed an enzymatic assay, primarily with hexokinase or glucose oxidase.

Hexokinase Method[4]

Fewer substances (medication or physiologic constituents of plasma) interfere in the hexokinase method than in the glucose oxidase assay. Although hexokinase also phosphorylates mannose and fructose, these sugars are normally not present in sufficiently high concentrations in serum to interfere. *Principle.* Glucose is phosphorylated by hexokinase, in the presence of ATP, as shown in reaction (1). In reaction (2), the G-6-P is converted by a second enzyme, G-6-P dehydrogenase (GPD), in the presence of $NADP^+$, to 6-phosphogluconate. NADPH is produced in the reaction. The absorbance of NADPH is measured at 340 nm; the increase in absorbance is proportional to the amount of glucose originally present.

$$\text{(1) Glucose + ATP} \xrightarrow[\text{Mg}^{2+}]{\text{Hexokinase}} \text{G-6-P + ADP}$$

$$\text{(2) G-6-P + NADP}^+ \xrightarrow{\text{GPD}} \text{6-phosphogluconate + NADPH + H}^+$$

Note: $NADP^+$ is required when yeast is the source of GPD, but NAD^+ must be used when GPD is obtained from bacteria *(Leuconostoc mesenteroides)*. The hexokinase preparation should be free from contamination with isomerases because they could slow down the hexokinase reaction.
Reference Values. The usually accepted reference values for "fasting" serum glucose are from 65 to 100 mg/dL (3.6 to 5.6 mmol/L). The pathologic conditions associated with hyperglycemia and hypoglycemia are reviewed in this chapter.

Glucose Oxidase Method[6]

Principle. Glucose oxidase is highly specific for the β-isomer of glucose and, in the presence of oxygen, converts it to gluconic acid and hydrogen peroxide, as shown in reaction (3). Solutions of glucose that have been standing contain an equilibrium mixture of the α and β forms. Some glucose oxidase mixtures contain the enzyme mutarotase, which converts α-glucose to the β form as the β form is oxidized by glucose oxidase.

$$\text{(3) } \beta\text{-glucose + H}_2\text{O + O}_2 \xrightarrow[\text{Oxidase}]{\text{Glucose}} \text{gluconic acid + H}_2\text{O}_2$$

The serum glucose concentration is proportional to the oxygen consumed in reaction (3) or to the H_2O_2 produced. Some instruments polarigraphically measure the rate of oxygen consumption by means of an oxygen electrode and present the glucose concentration as a digital readout. The glucose oxidase reaction may be coupled with the decomposition of H_2O_2 by catalase and the trapping of the liberated oxygen by oxidation of ethanol to acetaldehyde.

Other methods measure the H_2O_2 by coupling it with some variant of reaction (4). As illustrated in reaction (4), a colorless hydrogen donor, such as *o*-tolidine* or *o*-dianisidine, becomes intensely colored on losing two of its hydrogen atoms to an oxygen coming from the decomposition of H_2O_2.

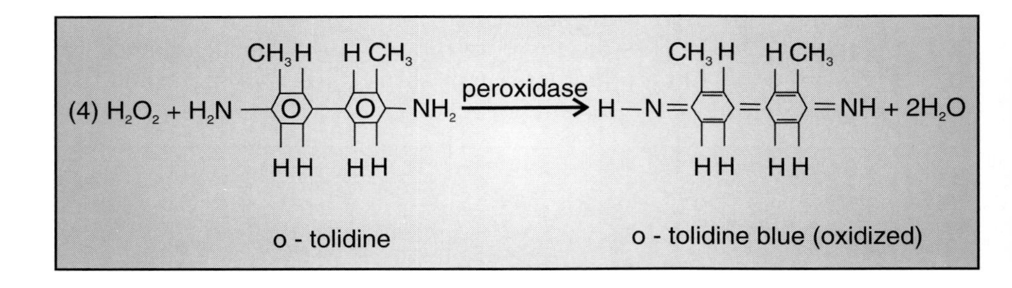

Unfortunately, various serum components, such as uric acid, ascorbic acid, and glutathione, compete with the chromogen as reducing agents and cause falsely low values by their interference with full color production.

The specificity of the glucose measurement is increased by substituting a different peroxide indicator reaction for reaction (4). Two different indicator systems, among the many variants, are still used today. In the first, 4-aminophenazone is oxidatively coupled with phenol[5] as H_2O_2 is decomposed, whereas in the other, 3-methyl-2-benzothiazolinone hydrazone (MBTH) is oxidatively coupled with N,N-dimethylaniline (DMA) to form an indamine dye.[6] The absorbance of a highly colored product is measured in each of these variants. Uric acid, ascorbic acid, or glutathione in the concentrations likely to be encountered in serum do not interfere.

Nonenzymatic Methods

Although nonenzymatic methods are now obsolete, the following are presented for historical reasons.

Nelson-Somogyi Method[3]

This method was accurate but labor intensive and difficult to automate. Glucose in a protein-free filtrate was oxidized with Cu^{2+} in hot alkaline solution to yield a Cu_2O precipitate. The Cu_2O was reoxidized by the addition of an acid solution of arsenomolybdate, with the concurrent reduction of molybdate to form a blue molybdate complex, the absorption of which was measured.

*Not to be confused with *o*-toluidine, an *o*-aminotoluene that condenses with glucose in hot acetic acid.

$$\text{Glucose} + 2\,Cu^{2+} + OH^- \longrightarrow Cu_2O + \text{glucose oxidation products}$$

$$Cu_2O + \text{arsenomolybdate } (Mo^{VI}) \longrightarrow 2Cu^{2+} + \text{arsenomolybdate blue } (Mo^{IV})$$

The concentration of glucose is proportional to the absorbance of the arsenomolybdate blue.

Ferricyanide method[3]

Serum glucose concentration was measured on the early AutoAnalyzers by oxidation in a hot alkaline solution of potassium ferricyanide as follows:

$$\text{Glucose} + \underset{\text{Ferricyanide (yellow)}}{[Fe^{III}(CN)_6]^{3-}} \xrightarrow[\text{heat}]{} \underset{\text{Ferrocyanide (colorless)}}{[Fe^{II}(CN)_6]^{4-}} + \text{glucose oxidation products}$$

The glucose concentration is proportional to the decrease in absorbance at 400 nm. The method gave falsely high results because several compounds in serum (creatinine, uric acid, ascorbic acid, and others) also were oxidized by the ferricyanide.

o-Toluidine Method[7]

Glucose in serum is condensed with *o*-toluidine by heating it in a concentrated acetic acid solution. The absorbance of the green chromogens at 630 nm is proportional to the glucose concentration.

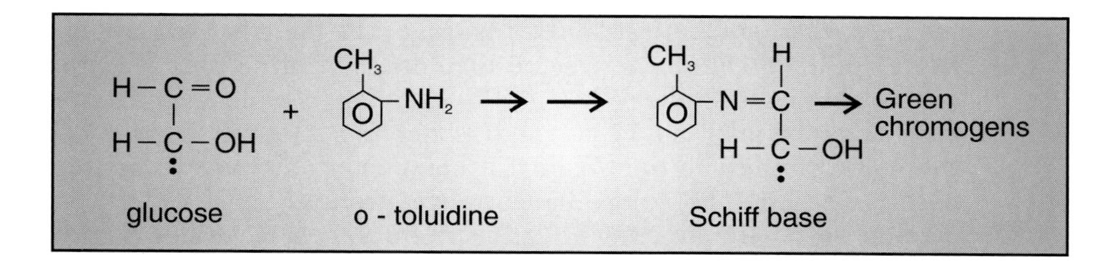

The acetic acid vapors are irritating to the eyes and skin; the hot acid solution causes rapid deterioration of the plastic tubing used in some of the automated methods. The enzymatic methods have replaced *o*-toluidine because of their rapidity and convenience.

Test Strips for Monitoring of Blood Glucose

Home monitoring of blood glucose concentrations has become widespread among patients with diabetes mellitus in whom frequent insulin adjustment is necessary.[8] Such patients are primarily individuals with insulin infusion pumps, gestational diabetes, or insulin-dependent diabetes mellitus that is difficult to control. The measurement of glucose excretion in the urine is not sufficiently sensitive or reliable for effective control of plasma glucose levels. Test strips for the estimation of glucose in a drop of whole blood are available from several commercial sources; these yield results that agree fairly well with those obtained by laboratory methods if the manufacturers' instructions are followed carefully. The test user must recognize that glucose values for *whole blood* are of necessity about 10% lower than those of *plasma* or *serum* because of the difference in water content; the glucose is distributed evenly between cell and plasma water. Test strips for the home monitoring of blood glucose are available (for example, Dextrostix, Miles; Chemstrips, Boehringer Mannheim). Each strip is impregnated with a glucose oxidase-catalase mixture and a hydrogen donor to give a color that is related to the glucose concentration. The intensity of color may be matched with a color chart, but can be obtained more accurately by reading in a small reflectance meter. The timing of the reaction and the subsequent washing are crucial; as with all laboratory tests, reliable results are obtained only when the directions are followed precisely and good quality control is practiced.

Determination of Urine Glucose Concentration

Qualitative or Semiquantitative

Copper Reduction. As described in the review of methodology for blood glucose testing, a hot alkaline solution of cupric sulfate oxidizes all reducing sugars (glucose, fructose, galactose, maltose, lactose, xylulose, arabinose, and ribose) and forms a brick-red to yellow precipitate of Cu_2O. This outcome is reached whether Clinitest tablets (Ames) or the Benedict method is used as the source of the alkaline Cu reagent. Five drops of urine plus ten drops of water are placed in a test tube. When a Clinitest tablet is added, the mixture begins to boil. The tube must not be agitated. About 15 s after the reaction ceases, check the color of the solution against a color chart. A blue color (of Cu^{2+}) is a negative result. The color change ranges from greenish (1^+) to yellow (4^+), depending on the amount of Cu_2O produced. A 1^+ reaction corresponds to about 50 mg/dL of reducing sugar; the 4^+ reaction indicates 2 or more g/dL. The test is adequate for making a semiquantitative estimate, but it is not specific for glucose.

Glucose Oxidase Impregnated Strips. Dipsticks manufactured by Ames, Boehringer Mannheim, and others contain glucose oxidase, peroxidase, and a chromogen. The strips are dipped into the urine and checked for color at the appropriate time (10 s for Labstix, 30 s for Chemstrip). These glucose oxidase tests are more sensitive than the Cu reduction tablets and are specific for glucose. False-positive tests are rare, but can occur if the glassware is contaminated with sodium hypochlorite (bleaching solution) or if the reaction is allowed to proceed too long before the result is read. False-negative results occur more commonly because ascorbic acid and urates inhibit the reaction; they might be a real problem for subjects with a high vitamin C intake or urate excretion. The sticks or strips should be checked again after 2 min to look for a delayed reaction. If present, confirm by Cu reduction tablets or prepare a Somogyi Zn $(OH)_2$ filtrate[3] of the urine and test again with the sticks. Glucose, if present, should be detected easily because the Zn $(OH)_2$ removes the inhibitors.

Normal subjects excrete less than 100 mg of glucose in a 24-hr period (less than 60 mg/L urine).

Quantitative

A quantitative estimation of the daily glucose excretion is frequently desired. Urine glucose can be quantitated satisfactorily by the *o*-toluidine, hexokinase, or glucose oxidase method that measures the rate of oxygen consumption in the same manner as that described for serum. Glucose oxidase colorimetric methods depending on catalase decomposition of H_2O_2 require prior removal of interferents (preparation of a Zn $(OH)_2$ filtrate or filtration through a mixed-bed resin).

No matter which glucose method is used for quantitation, the urine should be diluted with water if the glucose concentration exceeds 400 mg/dL. First use a semiquantitative method to determine whether dilution is required and, if so, to what extent.

Glucose in Other Body Fluids

The glucose concentration in cerebrospinal or other body fluids may be measured by the same methods and procedures as described for serum.

Cerebrospinal Fluid (CSF)

The concentration of glucose in CSF is about 60 to 75% of that in plasma and is in equilibrium with it; equilibration is slow and requires about 2 hr for changes in plasma glucose to show up in the CSF. Serum glucose should be measured at the same time as the glucose concentration of the CSF.

The determination of the glucose concentration is useful in the diagnosis of meningitis. The many bacteria and leukocytes in the infected CSF rapidly metabolize glucose and lower its concentration. The concentration of glucose in normal individuals varies from 45 to 70 mg/dL (2.5 to 3.9 mmol/L) but is usually less than 30 mg/dL (1.7 mmol/L) in various types of bacterial meningitis. A viral meningitis usually does not affect the CSF glucose concentration.

Exudates and Transudates

Exudates are the fluids that accumulate when the membranes are injured by inflammatory or infectious processes. The affected membranes allow larger molecules to pass through, and hence the protein concentration of exudates is higher than that in transudates.

The most frequent clinical chemistry analysis performed on transudates and exudates is the total protein assay ordered to differentiate between an inflammation and an infection, but glucose is sometimes ordered. The glucose concentration of transudates is approximately that of plasma. When membranes are injured by an infection, bacteria and leukocytes utilize glucose and lower its concentration in the fluid. A glucose concentration in exudate that is at least 25 mg/dL (1.4 mmol/L) lower than that in plasma is consistent with the presence of an infection.

KETONE BODIES IN URINE

Ketones are not present in the urine of healthy individuals eating a mixed diet. They may be present in the urine of persons with uncontrolled diabetes, subjects who have been without food for several days, or persons on a high-fat, low-carbohydrate diet. Infants develop ketonuria much earlier than adults when deprived of food.

The most common method for detecting ketones in urine uses a reaction between sodium nitroprusside (nitroferricyanide) and acetoacetate or acetone under alkaline conditions; a lavender color is produced. β-hydroxybutyrate does not react with the reagent.

Impregnated Strips or Sticks

By comparison with a color chart, the concentration of acetoacetate + acetone is expressed as negative, small, moderate, or large. The color chart correlates approximately with the following concentrations of acetoacetate: small = 10 mg/dL, moderate = 30 mg/dL, large = 80 mg/dL.

Reagent Tablets

Acetest (Ames) tablets may be used instead of a dipstick or strip. Crush a tablet of Acetest and place 1 drop of urine on it. Compare the color with that in a color chart.

IDENTIFICATION OF REDUCING SUGARS IN URINE

The reducing sugars that may be found in urine as a result of disease or a genetic defect are the disaccharide lactose, some common hexoses—glucose, fructose, galactose—and a few pentoses—xylulose and arabinose. Table sugar (sucrose) is not a reducing sugar. Reducing sugars in the urine are found by testing for copper reduction with a Clinitest tablet as described earlier in this chapter.

Principle. The separation of urinary sugars is accomplished by thin-layer chromatography (TLC) on plastic sheets or glass plates coated with fine cellulose particles. Individual sugars are identified by their relative migration distances and by the colors formed when heated with certain reagents.[9]

Interpretation. The relative separation of the sugars on TLC cellulose is similar to that by conventional paper chromatography. In the solvent system described, glucose migrates about 0.28 as far as the solvent front (an R_f of 0.28). The other sugars have the following R_fs: xylose, 0.44; fructose, 0.33; galactose, 0.24; lactose, 0.13; and glucuronic acid, 0.04. The aniline phosphate location reagent gives red-brown colors with pentoses, brown with glucose, galactose, and lactose, and yellow-blue with glucuronic acid. The naphthoresorcinol location reagent turns red when fructose or ketoses are present, but does not react with xylose, glucose, galactose, or lactose. Glucuronic acid produces a blue spot. If present, sucrose migrates closely to galactose but turns red with the naphthoresorcinol reagent; it does not react with the aniline phosphate.

Glucose and galactose may not separate cleanly if both are present. In this event, add some glucose oxidase powder to a small portion of the urine, adjust pH to 6 to 7, let stand 20 min, and test with a glucose oxidase dipstick. When the glucose has been completely oxidized, repeat the chromatographic separation, and test with the aniline phosphate location reagent. A brown spot with the R_f of glucose-galactose signifies the presence of galactose.

DISORDERS OF CARBOHYDRATE METABOLISM

The various disorders in carbohydrate metabolism may be grouped into several categories that depend primarily on laboratory findings. They are associated with: (1) a raised plasma glucose concentration (hyperglycemia); (2) a decreased plasma glucose concentration (hypoglycemia); and (3) a normal or lowered plasma glucose concentration, but with the excretion of a nonglucose reducing sugar in the urine.

Diseases Associated with Hyperglycemia

Diabetes Mellitus[10,11]

Hyperglycemia and "glucose intolerance" are common manifestations of several types of hormonal disturbances or imbalances, the most important of which is diabetes mellitus. This disorder is the seventh leading cause of death in the United States, is responsible for 12% of all new cases of blindness per year, and is involved in 25% of all cases of end-stage renal disease.[12] Diabetes mellitus is not a single, well-defined disease entity, but is a syndrome, a cluster of symptoms, caused by any of the several endocrine gland disorders that characteristically have the following features: an increased plasma glucose concentration, an absolute or relative insulin deficiency, and the probable occurrence of complications over a period of time. Glucosuria may or may not be present.

The National Diabetes Data Group[13] tentatively classified the various causes of hyperglycemia and glucose intolerance shown in Table 7.2. Thus, the three distinguishable types of diabetes mellitus[13] are insulin-dependent, noninsulin-dependent, and gestational. The other disorders listed in Table 7.2 are temporary or permanent conditions that exhibit diabetic-like glucose tolerance test results or are secondary to various situations, such as severe pancreatic or liver disease and drug intoxication.

Insulin-Dependent Diabetes Mellitus (IDDM or Type 1). This severe form of diabetes usually becomes evident during the stress of puberty. Individuals with IDDM have an absolute deficiency of insulin because of destruction or degeneration of their pancreatic islet beta cells. IDDM seems to be an autoimmune disease in which antibodies destroy the insulin-producing beta cells; replacement therapy with daily insulin injections or pancreas implants is a necessity. In the absence of insulin, various metabolic pathways are altered and lead to the following abnormalities: (1) *hyperglycemia;* (2) excretion of glucose into the urine *(glucosuria),* when the glomerular filtration rate of glucose exceeds the reabsorption capacity of the renal tubules; (3) *ketosis and ketoacidosis* as acetoacetate and β-hydroxybutyrate are produced in large quantities from the accelerated catabolism of fatty acids (Fig. 7.3), thus resulting in the high production and excretion of the ketoacids that exhaust the blood buffer system and cause acidosis; (4) the *positive test for ketones* in both blood and urine in ketoacidosis; (5) excretion into the urine of large amounts of

TABLE 7.2
CLASSIFICATION OF DIABETES MELLITUS AND OTHER CATEGORIES OF GLUCOSE INTOLERANCE (ADAPTED FROM THE NATIONAL DIABETES DATA GROUP)[13]

DISEASE	ABBREVIATION	CHARACTERISTICS
Idiopathic diabetes mellitus		
a. Insulin-dependent	IDDM or Type 1	Absolute insulin deficiency
b. Noninsulin-dependent	NIDDM or Type 2	Relative insulin deficiency
Gestational diabetes	GDM	Precipitated by pregnancy
Impaired glucose tolerance	IGT	Abnormal OGTT
Previous abnormality of glucose tolerance	Prev AGT	
Potential abnormality of glucose tolerance	Pot AGT	
Secondary diabetes		Hyperglycemia caused by excess GH, ACTH, or cortisol, or by severe pancreatic or liver disease

glucose and ketoacids, thereby producing an osmotic diuresis, with much loss of water and electrolytes; (6) the fat mobilization and increased cholesterol synthesis that lead to hyperlipidemia and hypercholesterolemia.

Prompt therapy with insulin saves the lives of nearly all IDDM patients, but most of them develop serious complications within 10 to 15 years. The end stages may include one or more of the following: stroke, myocardial infarction, loss of eyesight, renal failure, or neurologic defects.

The periodic or prolonged occurrence of hyperglycemia is believed to be partially responsible for some of these complications. Glucose slowly forms an addition product with some proteins in a nonenzymatic reaction by condensing with an exposed amino group to form a labile aldimine (Schiff base), which slowly and irreversibly rearranges into a stable ketoamine structure (Amadori rearrangement). The cross-linkage of ketoamine proteins over time may be responsible for some of the membrane or collagen defects in these patients.

Glycohemoglobin (GHb)* is the best studied glycated protein. Glucose slowly combines with the NH_2 group of valine, the N-terminal amino acid of the hemoglobin beta chains; it reacts to a lesser extent with other exposed amino groups (ϵ-group of lysine). The reaction rate varies directly with the glucose concentration. The first reaction, the formation of an aldimine, is reversible; but the second reaction, the rearrangement into a ketoamine, is not, as shown here:

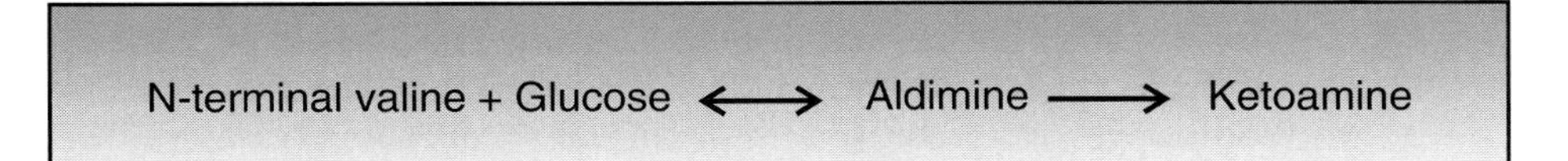

GHb consists primarily of three closely related ketoamines, collectively referred to as HbA_1 or individually as HbA_{1a}, HbA_{1b}, and HbA_{1c}; the major ketoamine is HbA_{1c}. The aldimine is called *pre*-HbA_{1c} because the glucose can become uncoupled. GHb, HbA_{1c}, or HbA_1 reflects the mean plasma glucose concentration for the preceding 6-week period, which is the average age of the circulating red cell (life span of approximately 4 months).[10,14] In a normal individual, HbA_{1c} comprises from 3.0 to 6.0% (mean 4.5%) of total hemoglobin; the figures for HbA_1 vary from 5 to 8%, with a mean of 5.6%. The percentage of HbA_{1c} may rise as high as 15 or 20% in diabetic persons whose disease is difficult to control.

Determination of Glycohemoglobin. The determination of the stable, glucose ketoamine complexes of hemoglobin (GHb) provides an assessment of the time-averaged control of the plasma glucose concentration for the preceding 4 to 8 weeks. Among the many different techniques for measuring GHb, those using affinity chromatography† have several distinct advantages for a hospital laboratory: (1) the procedure is not sensitive to temperature or pH fluctuations,

**Glycated* protein denotes the nonenzymatic condensation product of glucose with a protein. When the reaction occurs enzymatically, the product is called a *glycosylated* protein.

†Available from Isolab, Akron, OH 44321, or from Helena Laboratories, Beaumont, TX 77704-0752

(2) abnormal hemoglobins and nonglycated hemoglobins do not affect results, and (3) pre-HbA_{1c} is washed out and does not interfere.

Principle. The stationary phase of a small column consists of an insoluble, inert matrix (usually cellulose or agarose) that is covalently bound to a boronic acid derivative (commonly *m*-amino-phenylboronic acid). When a blood hemolysate passes through the column, the boronic acid chain forms a five-membered ring complex with the adjacent hydroxy groups in the *cis* configuration of any GHb containing glucose. The column is washed with two different buffers and the eluates are collected separately. The first buffer washes out all nonglycated hemoglobin, as well as pre-HbA_{1c}. The second buffer (usually sorbitol) displaces GHb from the column. The absorbance of hemoglobin in both tubes is measured at 415 nm. GHb is calculated as percent of the total hemoglobin.

Some physicians prefer to know the time-averaged glucose concentration for a shorter (more recent) period to control the blood glucose concentration of their diabetic patients more closely. The quantitation of protein-bound fructosamine[15,16] in serum provides a time-averaged glucose concentration of 1 to 3 weeks because it is derived from the glycation of all serum proteins. Abbott Laboratories and Roche Diagnostics Systems have developed fructosamine assays for their respective instruments.

Noninsulin Dependent Diabetes Mellitus (NIDDM or Type 2). This milder form of diabetes is characterized by adult onset, functional pancreatic beta cells, and insulin resistance (an insufficient peripheral response to a given concentration of plasma insulin). The insulin resistance may be associated with an imbalance of counterregulatory hormones (GH, glucagon, cortisol), with the production of antibodies against either insulin or cell receptors for insulin or with receptor or postreceptor defects. NIDDM can frequently be controlled by reducing the weight of obese patients, or by oral therapy with sulfonylureas; other patients require insulin treatment. NIDDM accounts for about 90% of all cases of diabetes mellitus. Inadequate control of the plasma glucose concentration can also lead to the long-term complications described for IDDM.

Gestational Diabetes (GDM). This type of diabetes is usually first diagnosed during the latter half of pregnancy because of some laboratory test abnormality (glucosuria, hyperglycemia, or abnormal glucose tolerance test). The increased secretion of the placental hormone, placental lactogen (hPL), inhibits the action of insulin at some postreceptor site. When known diabetic women become pregnant, the disease becomes more severe in the latter part of pregnancy for the same reason. Hyperglycemia must be prevented in GDM because of danger to the fetus. The newborn of a mother with GDM is larger than normal (about 500 g heavier), has delayed lung maturation and hyperplasia of the pancreatic beta cells, and is more likely to have some malformation.

Some patients with GDM develop NIDDM some years later, but in others, the abnormal glucose tolerance response may return to normal after childbirth and never reappear.

Diagnosis of Diabetes Mellitus. The acute state of IDDM is readily diagnosed from the history, symptoms (hunger, thirst, frequent urination, weight loss), and the following abnormal laboratory tests: plasma glucose ($>$ 200 mg/dL, 11.1 mmol/L), glucosuria (usually 4 + or $>$ 2 g/day), and presence of ketone bodies in urine and serum. The asymptomatic patient presents a diagnostic problem, but the tests described in the following paragraphs are helpful.

Fasting Plasma Glucose. The Diabetes Data Group[13] proposed that a fasting (overnight) plasma glucose concentration exceeding 140 mg/dL (7.8 mmol/L) on 2 separate occasions should be accepted as a criterion for diabetes. Others[17] have proposed the higher limit of 150 mg/dL. Milder cases of hyperglycemia require some type of carbohydrate-loading test for detection.

Postprandial Plasma Glucose. The simplest loading test is the measurement of plasma glucose concentration 2 hr after the patient consumes a meal containing approximately 100 g of carbohydrate mixed with other foodstuffs. A plasma glucose value exceeding 200 mg/dL is indicative of diabetes, whereas a value below 120 mg/dL is considered normal. Concentrations between 120 and 200 mg/dL are equivocal and require further study.

Oral Glucose Tolerance Test (OGTT). The patient should be placed on a diet containing adequate calories, protein, and at least 150 g carbohydrate per day for 3 days; drugs that may influence the results (steroids, estrogen, propranolol, phenytoin, thiazides) should be withheld for 3 days before the test.
Procedure. A blood sample is drawn from the patient after an overnight fast; 1.75 g glucose/kg body weight, with 75 g maximum, are dissolved in a palatable liquid and ingested by the patient within a 5-min period. Blood samples are drawn for serum glucose analysis 1, 2, and 3 hr after glucose ingestion.
Interpretation. According to the National Diabetes Data Group,[13] the OGTT is deemed positive in the following situations:

1. For *diabetes mellitus,* if the plasma glucose concentration is $>$ 200 mg/dL at 2 hr and 1 other time period in the test.
2. For *impaired glucose tolerance,* if the plasma glucose is between 140 and 200 mg/dL at 2 hr and at least 1 other value $>$ 200 mg/dL.
3. For *gestational diabetes in pregnancy,* if 2 or more values exceed the following: (a) zero time—105 mg/dL; (b) 1 hr—190 mg/dL; 2 hr—165 mg/dL; 3 hr—145 mg/dL.

Figure 7.7 illustrates normal and abnormal responses to the OGTT. Many factors not connected with diabetes can affect the OGTT result. Cellular response to insulin and the body's tolerance for glucose diminish with time after age 50. The OGTT response may also be altered by diet, drug therapy, anxiety, stress, degree of physical activity, and rate of intestinal absorption of glucose.

Other Hormonal Disorders

As reviewed previously in this chapter, chronic overproduction of GH, cortisol, or ACTH is accompanied by mild hyperglycemia. The overproduction may be the result of hyperplasia or an adenoma of the gland. A malignant growth of the adrenal medulla (pheochromocytoma) or of the alpha islet cells of the pancreas, which secrete epinephrine and glucagon, respectively, may be accompanied by sharp increases in the plasma glucose level during these periods of hormone spurts.

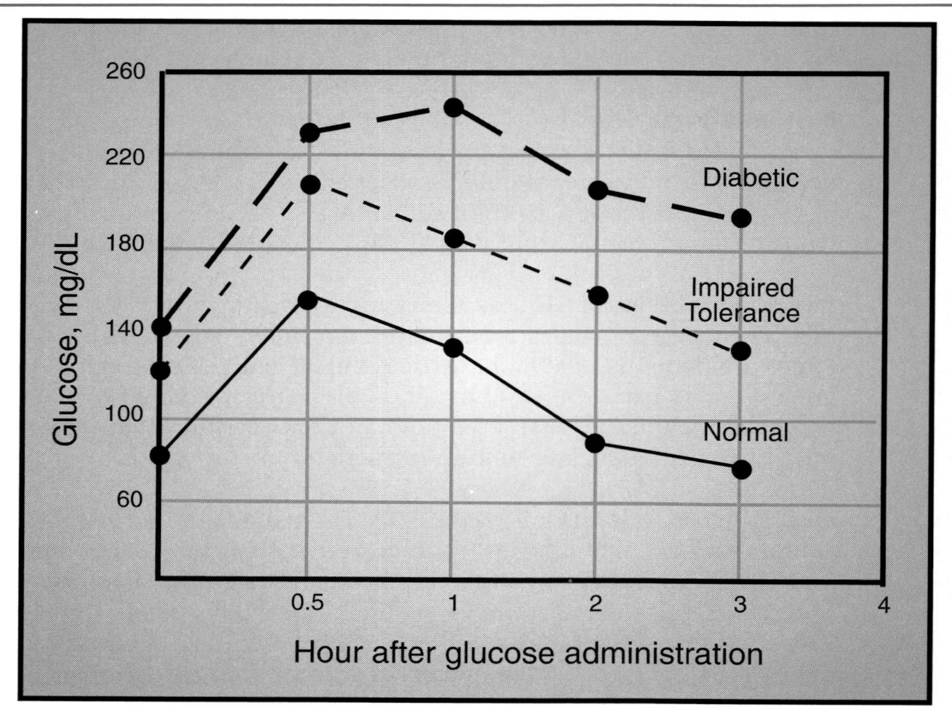

FIG. 7.7. Illustration of normal, diabetic, and impaired responses to an oral glucose tolerance test.

Disorders Associated with Hypoglycemia[2]

Any condition in which the plasma glucose concentration falls below the lower limits of normal by more than 2 standard deviations (below 60 mg/dL, 3.3 mmol/L) is called hypoglycemia. Because the brain depends on an adequate supply of glucose for its energy, the clinical symptoms of hypoglycemia resemble those of cerebral anoxia and may include one or more of the following: faintness, weakness, dizziness, tremors, anxiety, hunger, palpitation of the heart, or "cold sweat"; mental confusion and motor incoordination may even be experienced. Adults may lose consciousness when the plasma glucose concentration falls below 40 mg/dL (2.2 mmol/L) but the rapidity of fall is also a factor. Widespread convulsions may accompany or even precede coma. Newborn infants are less sensitive to a decreased concentration of plasma glucose and may not go into convulsions unless the plasma glucose concentration falls below 25 to 30 mg/dL (1.4 to 1.7 mmol/L).

The lower the plasma glucose level, the deeper is the coma for both adults and infants. The cortical centers are the first to be affected because they have the highest energy requirement in the brain, but with time, the lower centers are also affected. Irreversible brain damage or death may occur if the hypoglycemia and coma persist too long. If the length of time in coma is less than 20 min, intravenous injection of glucose usually restores consciousness immediately, with no permanent brain damage.

A technologist or supervisor who finds a patient's serum glucose concentration at a hypoglycemic level must contact the attending physician at once by telephone and transmit the result because of the limited time for appropriate action.

The following are the principal causes of hypoglycemia.

Hormonal

1. Insulin excess
   a. Overdosage of insulin in a diabetic person (too much insulin injected or failure of the diabetic person to eat after usual dosage).
   b. Excessive secretion of insulin by the pancreas as a result of:
      (1) Pancreatic hyperplasia.
      (2) Islet cell tumor of the pancreas (insulinoma).
      (3) Leucine.
      (4) Sulfonylureas.
2. Deficiency of glucagon (from alpha cells of pancreatic islets) or catecholamines (from adrenal medulla).
3. Primary deficiency of cortisol (Addison's disease) or secondary to ACTH deficiency. Fasting produces hypoglycemia.
4. Anterior pituitary (GH, ACTH) insufficiency. Hypoglycemia may appear upon fasting.

Hepatic

1. Depleted liver glycogen stores resulting from
   a. Prolonged fasting or starvation.
   b. Severe hepatocellular damage.
   c. Acute drug toxicity.
2. Failure to release liver glycogen (genetic defects)
   a. Type 1 glycogen storage disease (von Gierke's disease), a deficiency of glucose-6-phosphatase in liver.
   b. Rarer glycogen storage diseases (types 3, 6, and 8), in which glycogen-splitting enzymes are deficient in the liver.

Hereditary Disorders (Enzymatic Defects) with Reducing Sugars in the Urine

1. Galactosemia.[18] Galactose cannot be metabolized because galactose-1-P uridyl transferase is lacking or greatly reduced in activity. As a result, galactose-1-P piles up in cells and cannot be converted to glucose. Because galactose is a component of lactose, the sugar in milk, the disease is serious in infants. To survive, they must receive artificial milk containing no lactose. The diagnosis must be made early to avoid the crippling effects of galactose, which are hypoglycemia, hepatosplenomegaly (enlarged liver and spleen), cirrhosis of the liver, cataracts, and mental retardation. The infants fail to thrive and may die unless lactose is removed from the diet. The diagnosis is made by identifying galactose in the urine and confirmed by finding a deficiency of the enzyme galactose-1-P uridyl transferase in erythrocytes. The method for detection of urine sugars is given at the end of this chapter. Because galactosemia is rare, it

does not pay for a small laboratory to set up the enzyme test unless an active screening program is in progress.

2. Hereditary fructose intolerance.[19] In this rare genetic disorder, fructose-1-P accumulates in cells because the enzyme aldolase, which converts fructose-1-P to triosephosphate, is lacking. The ingestion of fruit or sucrose (a disaccharide composed of glucose and fructose) produces vomiting, hypoglycemia, failure to thrive, and hepatomegaly. Children can survive with this defect if they avoid fruit and products containing cane sugar. The laboratory findings are fructose in the urine and hypoglycemia after administration of a test meal containing fructose.
3. Fructose-1,6-diphosphatase deficiency.[19] Fructose-1,6-diphosphatase is essential for gluconeogenesis in the conversion pathway of pyruvate to glucose. Newborns and infants with this rare genetic disorder soon develop a fasting hypoglycemia, severe lactic acidosis, and hepatomegaly. Prognosis is poor.
4. Several harmless hereditary conditions with normal plasma glucose levels but in which reducing sugars (sugars that reduce Cu^{2+} to Cu_2O in a hot alkaline solution) appear in the urine. These conditions are essential fructosuria and essential pentosuria in which fructose and xylulose, respectively, are the sugars excreted. In both conditions, the lack of a specific enzyme causes an increase in concentration of the particular sugar and its resultant excretion in urine. The diagnosis of fructosuria and pentosuria is missed when one relies solely on the use of a glucose oxidase strip for testing.
5. Lactose intolerance (lactase deficiency syndrome).[20] Lactose, a disaccharide, is the principal carbohydrate in milk. It is split into equimolar quantities of glucose and galactose by the intestinal enzyme lactase. The liberated hexoses are absorbed rapidly in the intestine. A lactase deficiency can occur as a genetic defect (primary) or as an acquired deficiency secondary to several intestinal disorders. The symptoms are the same: milk intolerance, abdominal cramps and distention, and gas and diarrhea following the ingestion of milk products. Ingested lactose passes intact into the colon, where certain anaerobes ferment the lactose to yield lactic acid, short-chain fatty acids, and hydrogen gas. The host has abdominal distention, cramps, and diarrhea. Most of the hydrogen is passed as flatus, but some is absorbed into the blood, reaches the lungs, and is exhaled. The diagnosis of lactose intolerance (primary or secondary) can be established by finding significant amounts of hydrogen in a sample of the individual's expired air. A breath sample is collected and analyzed for hydrogen by means of a gas chromatograph.[21,22]

BLOOD LACTATE AND PYRUVATE

Lactic Acidosis

Blood lactate and pyruvate ions exist at low concentrations in an equilibrium mixture under normal circumstances. The liver usually converts a lactate excess during exercise back to glucose (Pathway No. 3, Fig. 7.1), but this process requires oxygen. The blood lactate concentration may be increased greatly by a severe oxygen deficiency because of (1) blockage of glucose formation from pyruvate, (2) inability of the TCA cycle to oxidize all the pyruvate to CO_2 and water (Fig. 7.1 and 7.2), and (3) an increase in hepatic glycolysis. The $[H^+]$ rises concomitantly with the lactate and creates the condition known as *lactic acidosis*. This situation prevails in conditions of severe anoxia and/or greatly reduced hepatic blood flow, as in shock, terminal illnesses, heart failure, and after ingestion of some drugs.[23] Lactic acidosis can be life-threatening if treatment is delayed too long. Laboratory testing reveals an anion gap in electrolyte measurement, a high blood lactate concentration, and a low pH.

Elevated blood lactate and pyruvate in a newborn could also be caused by an inborn error of pyruvate metabolism. A deficiency of pyruvate dehydrogenase, the enzyme complex that catalyzes the oxidation of pyruvate to CO_2 and acetyl CoA, has been defined. Deficiencies in various components of pyruvate carboxylase, the enzyme system that catalyzes the formation of oxaloacetate from CO_2 and pyruvate, have also been observed.[24] These disorders are inherited in an autosomal recessive manner. Diagnosis is often difficult because of the complexity of the enzyme assays (fresh muscle tissue is often most informative) and the lack of availability. Treatment has not been successful in these fatal diseases.

Determination of Blood Lactate

A measured volume of heparinized blood is immediately transferred to a tube containing dilute metaphosphoric or perchloric acid to precipitate proteins and prevent glycolysis. After centrifugation, the lactate in the supernate is determined spectrophotometrically by addition of the enzyme lactate dehydrogenase (E.C. 1.1.1.27) and NAD^+, as indicated by the following reaction:

$$\text{Lactate} + NAD^+ \xrightarrow{LD} \text{Pyruvate} + NADH + H^+$$

where LD signifies the enzyme lactate dehydrogenase. The reaction goes to completion (formation of pyruvate and NADH) at pH 9.0 to 9.6 by trapping the pyruvate as a hydrazone as fast as it is formed. The NADH formation is proportional to the lactate and is measured spectrophotometrically at 340 nm. Kits for performing the test are available.

Determination of Blood Pyruvate

The same metaphosphoric or perchloric acid precipitate is often used for the quantitation of pyruvate. The reaction is the reverse of that used to determine lactate.

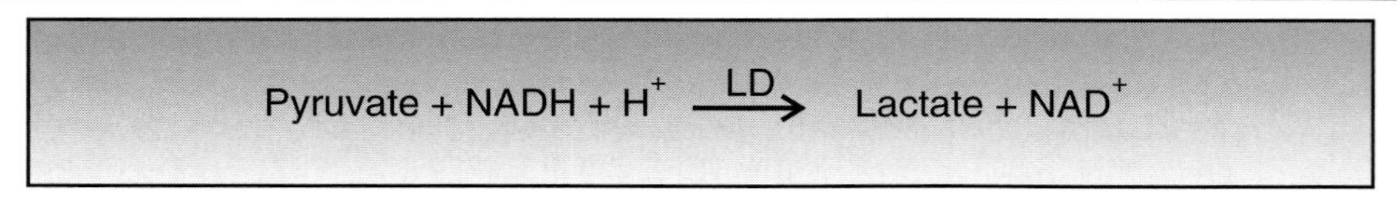

The equilibrium constant strongly favors the production of lactate at pH 7.5. The disappearance of NADH is measured spectrophotometrically at 340 nm as NADH is converted to NAD^+.

Both pyruvate and lactate are unstable in blood. Samples should be obtained while the patient is fasting and completely at rest. Venous samples should be obtained without the use of a tourniquet to avoid an artificial rise in pyruvate and lactate. If a tourniquet must be used to locate a vein, one should perform the venipuncture, remove the tourniquet, and allow the blood to circulate for 2 min before actually drawing the blood sample. Drawing a sample from small children can be a problem, as crying and kicking artificially raise the pyruvate and lactate. If an acceptable sample cannot be obtained from a venipuncture, a sample obtained from an arterial line may be more appropriate.

Reference Ranges

Lactate concentration in fasting venous blood is 0.5 to 1.7 mmol/L (5 to 15 mg/dL). Arterial values are slightly lower—up to 1.3 mmol/L (12 mg/dL). The pyruvate concentration in venous blood in fasting individuals is 0.03 to 0.10 mmol/L (0.3 to 0.9 mg/dL). Arterial blood is slightly lower—0.02 to 0.08 mmol/L (0.2 to 0.7 mg/dL). The usual ratio of blood lactate to pyruvate is 10 to 20:1.

Lactic acidosis becomes clinically evident when the blood lactate concentration exceeds 7 mmol/L. In severe cases, the level may surpass 20 mmol/L. The ratio of blood lactate to pyruvate most often increases in severe lactic acidosis; however, patients with pyruvate dehydrogenase complex deficiencies often produce pyruvate in excess of the normal ratio.

REFERENCES

1. Stryer, L.: Biochemistry. 3rd Edition. New York, W.H. Freeman, 1988.
2. Cryer, P.E.: Glucose homeostasis and hypoglycemia. *In* William's Textbook of Endocrinology. 8th Edition. Edited by J.D. Wilson and D.W. Foster. Philadelphia, W.B. Saunders, 1992.
3. Burrin, J.M., and Price, C.P.: Measurement of blood glucose. Ann. Clin. Biochem., *22*:327, 1985.
4. Neese, J.W.: Glucose, direct hexokinase method. *In* Selected Methods of Clinical Chemistry. Vol. 9. Edited by W.R. Faulkner and S. Meites. Washington, D.C., American Association for Clinical Chemistry, 1982.
5. Trinder, P.: Determination of glucose in blood using glucose oxidase with an alternative oxygen acceptor. Ann. Clin. Biochem., *6*:24, 1969.
6. Gochman, N., and Schmitz, J.M.: Application of a new peroxide indicator reaction to the specific, automated determination of glucose with glucose oxidase. Clin. Chem., *18*:943, 1972.
7. Passey, R.B., Gillum, R.L., and Baron, M.L.: Glucose, *o*-toluidine method. *In* Standard Methods of Clinical Chemistry. Vol. 9. Edited by W.R. Faulkner and S. Meites. Washington, D.C., American Association for Clinical Chemistry, 1982.

8. Bergman, M., and Felig, P.: Self-monitoring of blood glucose levels in diabetes. Arch. Intern. Med., *144*:2029, 1984.
9. Menzies, I.S., and Seakins, J.W.T.: Sugars. *In* Chromatographic and Electrophoretic Techniques. 3rd Edition. Vol. 1. Edited by I. Smith. New York, Interscience, 1969.
10. Foster, D.W.: Diabetes mellitus. *In* Harrison's Principles of Internal Medicine. Vol. 2. 12th Edition. Edited by J.D. Wilson, et al. Philadelphia, W.B. Saunders, 1991.
11. Unger, R.H., and Foster, D.W.: Diabetes Mellitus. *In* William's Textbook of Endocrinology. 8th Edition. Edited by J.D. Wilson and D.W. Foster. Philadelphia, W.B. Saunders, 1992.
12. National Diabetes Data Group. Diabetes in America, diabetes data compiled 1984, Bethesda, MD: National Institute of Arthritis, Diabetes, Digestive and Kidney Disease, National Institutes of Health, 1985.
13. National Diabetes Data Group: Classification and diagnosis of diabetes mellitus and other categories of glucose intolerance. Diabetes, *28*:1039, 1979.
14. Baynes, J.W., Bunn, H.F., Goldstein, D.E., et al.: National Diabetes Data Group: report of the expert committee on glucosylated hemoglobin. Diabetes Care, *7*:602, 1984.
15. Baker, J.R., et al.: Serum fructosamine concentration as measure of blood glucose control in type 1 (insulin dependent) diabetes mellitus. Br. Med. J., *290*:352, 1985.
16. Armbruster, D.A.: Fructosamine: structure, analysis, and clinical usefulness. Clin. Chem., *33*:2153, 1987.
17. Ito, C., Mito, K., and Hara, H.: Review of criteria for diagnosis of diabetes mellitus based on results of follow-up study, Diabetes, *32*:343, 1983.
18. Segal, S.: Disorders of galactose metabolism. *In* Metabolic Basis of Inherited Disease. Vol. 1. 6th Edition. Edited by C.R. Scriver, et al. New York, McGraw-Hill, 1989.
19. Gitzelmann, R., Steinmann, B., and van den Berghe, G.: Disorders of fructose metabolism. *In* Metabolic Basis of Inherited Disease, Vol. 1. 6th Edition. Edited by C.R. Scriver, et al. New York, McGraw-Hill, 1989.
20. Greenberger, N.J., and Isselbacher, K.J.: Disaccharidase deficiency syndrome. *In* Harrison's Principles of Internal Medicine. Vol. 2. 12th Edition. Edited by J.D. Wilson, et al. Philadelphia, W.B. Saunders, 1991.
21. Rosado, J.L., Torres, L., and Solomons, N.W.: Chromatograph with external column evaluated for determining breath hydrogen in clinical studies. Clin. Chem., *30*:1838, 1984.
22. Joseph, F., Jr., and Rosenberg, A.J.: Identifying lactose malabsorbers through breath hydrogen measurements. Lab. Med., *17*:85, 1986.
23. Levinsky, N.G.: Lactic acidosis. *In* Harrison's Principles of Internal Medicine. Vol. 1. 12th Edition. Edited by J.D. Wilson, et al. Philadelphia, W.B. Saunders, 1991.
24. Robinson, B.H.: Lactic acidemia. *In* The Metabolic Basis of Inherited Disease. 6th Edition. Edited by C.R. Scriver, et al. New York, McGraw-Hill. 1989.

REVIEW QUESTIONS

1. If a person eats a large, balanced meal containing more calories than needed for immediate energy demands, how does the body handle the carbohydrate?
2. What hormones are involved in glucose homeostasis, and how do they act?
3. What are the distinguishing features of the two types of diabetes (IDDM and NIDDM)?
4. How does measurement of blood glycated hemoglobin assist the physician in controlling the plasma glucose concentration of a diabetic patient?
5. What laboratory tests would you recommend for ascertaining whether a patient has (a) IDDM or (b) NIDDM ? Start with the simplest tests first.
6. What are some of the causes of hypoglycemia, and how would you test for them?
7. What are the principles of the hexokinase method for the determination of plasma glucose concentration?

8 Lipid Metabolism

OUTLINE

OBJECTIVES

After reading this chapter, the student will be able to:

1. Recognize the different lipid structures and know their principal functions.
2. Describe the body sources of cholesterol, its main functions in the body, and its limited homeostatic control.
3. Describe the composition and functions of the different classes of lipoproteins.
4. Describe how lipids are transported in both the endogenous and the exogenous pathways.
5. Describe the different classes of apoproteins and know which classes are cofactors for enzymes active in lipid metabolism.
6. Know the features of the different classes of familial hyperlipemias and the lipoprotein electrophoretic pattern typical of each class.
7. Describe the analytical principles involved in the measurement of plasma cholesterol, triglycerides, and HDL cholesterol.
8. Calculate the concentration of serum LDL cholesterol when given the concentration of total and HDL cholesterol and triglycerides.

Lipid is a term applied to naturally occurring substances that are soluble in nonpolar solvents, such as hexane or chloroform, and insoluble in water. The main lipid classes of interest in clinical chemistry are fatty acids, triglycerides, phospholipids, sterols, sphingolipids, and cholesterol. Their functions are described in the following sections.

LIPID CONSTITUENTS

Fatty Acids[1]

Fatty acids are straight-chain hydrocarbons with a terminal carboxyl group. They are frequently identified by the number of carbon atoms and the number of double bonds, as in the diunsaturated linoleic acid (18:2). The location of the double bond in the n- or omega (ω) numbering system designates the number of carbon atoms from the terminal methyl. Thus, linoleic acid is designated as 18:2n-6 and is called an ω-6 fatty acid.

Fatty acids exist mostly as esters of glycerol in both triglycerides (Fig. 8.1) and some phospholipids (Fig. 8.2) and as esters of high-molecular-weight alcohols (cholesterol [see Fig. 8.5] and sphingosine [see Fig. 8.4]). The fatty acids of triglycerides are mostly C_{16} or C_{18}, and in phospholipids, they are C_{18} to C_{22}. Those esterified with sphingosine (sphingomyelin, cerebrosides, gangliosides) are usually C_{24}. The fatty acids of plant and animal tissues, including those of humans, usually contain an even number of carbon atoms because they are synthesized from 2-carbon units (acetyl CoA). The fatty acids may be saturated (no double bonds) or unsaturated (one or more double bonds) (Table 8.1). In general, the fatty acids derived from plant seeds or fish lipids are

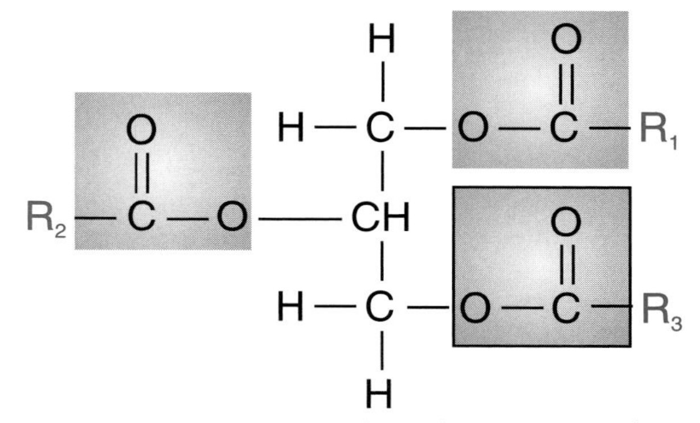

FIG. 8.1. Structure of triglyceride. R_1, R_2, and R_3 represent long-chain fatty acids.

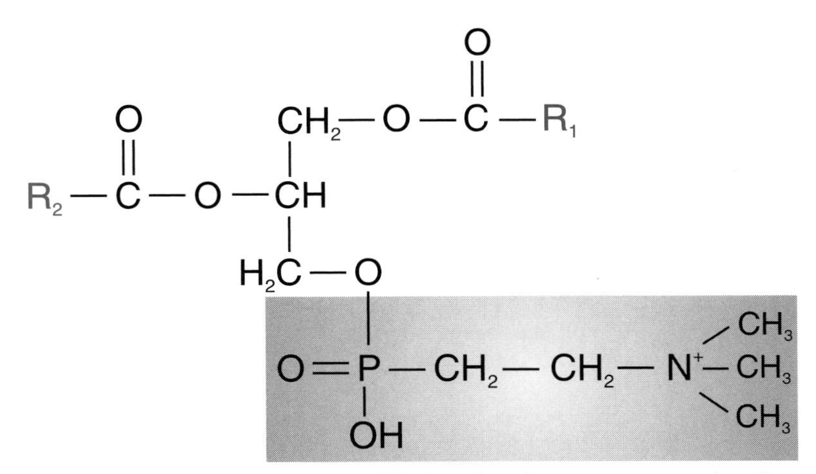

FIG. 8.2. Structure of phosphatidylcholine (lecithin), a common phospholipid. R_1 and R_2 represent long-chain fatty acids in ester linkage to the glycerol moiety. R_2 is unsaturated and usually has two or more double bonds. The colored portion of the molecule shows where choline is linked to phosphate.

TABLE 8.1
COMMON DIETARY FATTY ACIDS

NUMBER OF CARBONS AND DOUBLE BONDS	COMMON NAME	OMEGA (ω-) SERIES*	COMMON SOURCE
16:0	Palmitic		Most plant and animal fats
18:0	Stearic		
18:1	Oleic	ω-9	Most plant and animal fats
18:2	Linoleic	ω-6	Many seed oils (safflower,corn, soy, cottonseed)
18:3	γ-Linolenic	ω-6	In above seed oils
18:3	α-Linolenic	ω-3	Oils of soybean, linseed, wheat germ, chia
20:4	Arachidonic	ω-6	
20:5	Eicosapentaenoic (EPA)	ω-3	Fat of cold-water fish (herring, salmon, sardines, mackerel)
22:6	Docosahexaenoic (DHA)	ω-3	

*The ω-series starts numbering from the terminal methyl, in contrast to the Δ system, which starts numbering from the terminal carboxyl.

more unsaturated and contain much more of the essential fatty acids, linoleic (18:2) and linolenic (18:3) acids, that humans require and cannot synthesize. Essential fatty acids appear mostly in the 2-position of phospholipids and are necessary for the structural integrity of cell or organelle membranes. Arachidonic acid (Fig. 8.3) is also essential for body function, but can be formed from linoleic and linolenic acids by desaturation and chain elongation and need not be present in the diet. An enzyme elongates the fatty acid chain by attaching acetyl CoA to the carboxyl end; the chain lengthens by two carbon atoms each time. Oxidation of fatty acids involves the formation of a fatty acid-CoA conjugate (acyl CoA), with transfer by a carrier molecule (carnitine) into mitochondria. Acetyl CoA is split off from the acyl CoA molecule and is oxidized to CO_2 and water in the tricarboxylic acid (TCA) cycle (Chap. 7). This process of β-oxidation is repeated sequentially until the entire chain is used for the generation of adenosine triphosphate (ATP). Disorders of β-oxidation are reviewed in Chapter 16.

Eicosanoids and Prostaglandins

The highly unsaturated arachidonic acid (20:4, ω-6) and the ω-3 acids, EPA (20:5) and DHA (22:6) (Table 8.1), are preferentially selected for the 2-position

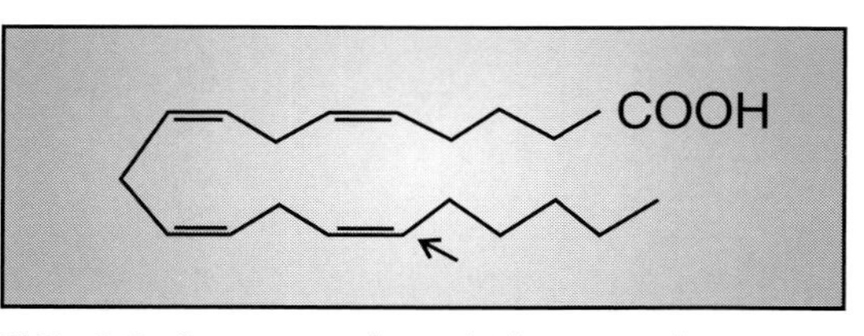

FIG. 8.3. Structure of arachidonic acid (C20:4), showing the location of the four double bonds. The arrow points to the ω-6 carbon atom.

of membrane phospholipids of all tissues. In stressful situations, a phospholipase liberates these polyunsaturated fatty acids from the membrane phospholipids. Intracellular enzymes convert those C_{20} or C_{22} fatty acids into 3 families of eicosanoids designated as prostaglandins, thromboxanes, and leukotrienes. The structures are modifications of the parent molecule, arachidonic acid, shown in Figure 8.3.

These fatty acid derivatives have potent, hormone-like biologic activity. They are formed in tissues at the site of action, have short half-lives, and cannot be stored.

Each of the prostaglandin and leukotriene families contains three or four members that differ slightly in structure and have different physiologic activities. Thus, one type of prostaglandin may contract smooth muscle and be a vasoconstrictor, whereas a second type may have the opposite effect and a third type may inhibit blood platelet aggregation and induce the formation of cyclic AMP. Thromboxane (TX), like prostaglandin (PG), is derived from a 20:4, ω-6 fatty acid (arachidonic acid). The potent TXA_2 is formed in blood platelets and increases blood coagulability by aggregating blood platelets. This characteristic is a fine, defensive reaction in bleeding situations, but frequent generation of TXA_2 increases the danger of myocardial infarction in patients at risk. The third family of eicosanoids, the leukotrienes (LTs), are formed mainly from ω-3 fatty acids (20:5, 22:6) and compete with PG and TX for formation. LTs are formed primarily in segmented white blood cells, particularly in neutrophils; they influence the white blood cell inflammatory and immune response.

Triglycerides[1]

Most of the fatty acids in the body are components of triglycerides (Fig. 8.1) and are stored in the depots (adipose tissue) as fat. Adipose cells convert fatty acids into triglycerides by esterification with glycerol-3-phosphate, a compound that arises from glucose catabolism (see Fig. 7.6). Cells must contain glucose for triglyceride formation. Glucose is absent during periods of fasting, starvation, or uncontrolled diabetes mellitus, and in these conditions, hydrolysis of triglycerides and withdrawal of their fatty acids from the depots predominate. Excess carbohydrate ingested during a meal may be stored temporarily as triglycerides after conversion of glucose to fatty acids. The hormone insulin promotes the synthesis of triglycerides by adipose cells, whereas its deficiency accelerates triglyceride hydrolysis.

The first step in the catabolism of triglycerides begins with their hydrolysis. The fatty acids appear in plasma as nonesterified (free) fatty acids bound to albumin as a carrier. Various tissues, but muscle cells in particular, use fatty acids for energy purposes by means of β-oxidation as previously described. When the body is flooded by the catabolism of fatty acids, as in insulin deficiency (diabetes mellitus) or prolonged starvation, the acetyl CoA piles up and condenses into acetoacetyl CoA, an obligatory intermediate in the formation of ketone bodies (Fig. 7.4). The acetoacetyl CoA may also condense with another molecule of acetyl CoA to form 3-hydroxy-3-methylglutaryl CoA (HMG-CoA). The reaction can proceed to the formation of acetoacetate by splitting off acetyl CoA or to the synthesis of cholesterol by a series of enzymatic reactions.

The serum triglyceride concentration is rarely requested as a solitary test; it is usually requested in combination with total cholesterol and, perhaps, HDL and LDL cholesterol (cholesterol carried by high-density and low-density lipoproteins, respectively). As explained later in this chapter, this combination of tests aids in estimating the degree of risk for coronary artery disease or in elucidating a hyperlipoproteinemia. Hence, all the methods for measuring lipids or lipoproteins are covered in the latter part of this chapter.

Phospholipids[1]

The principal phospholipids are composed of a diglyceride esterified with phosphoric acid, which, in turn, is bound as an ester to a nitrogen-containing base (choline, ethanolamine) or to serine (Fig. 8.2). Phosphatidyl derivatives of ethanolamine, serine, and inositol are sometimes collectively referred to as cephalins. In phosphatidyl choline (lecithin), the base end (choline) carries a positive charge and is polar, whereas the fatty acid segments are nonpolar. Phosphatidyl choline emulsifies easily in water and helps to form stable, colloidal suspensions of fat in water; the molecules orient in such a way that the polar head is in the water phase and the nonpolar end is in the lipid phase. Phosphatidyl choline is the principal lipid component of a surfactant secreted by the lung alveoli and is essential for proper distention of the alveoli on breathing (see Chap. 15). Phospholipids are essential components of cell membranes because of their ability to align themselves between water and lipid phases. Phosphoethanolamine, a constituent of blood platelets, is a necessary participant in the clotting process. Phospholipids in lipoproteins also supply the fatty acids necessary for the esterification of cholesterol. The phospholipids play a role in mitochondrial metabolism, in blood coagulation, and in lipid transport as part of lipoproteins, and are important structural components of membranes. Membrane phospholipids contribute the highly unsaturated fatty acid for the formation of prostaglandin and other eicosanoids. In most cases, however, little clinically useful information is gained by measuring the concentration of plasma phospholipids.

Sphingolipids[1]

The sphingolipids are all compounds containing the long-chain, dihydroxyamino alcohol sphingosine (Fig. 8.4). All the sphingolipids bind a fatty acid in amide linkage to the amino group and are also known as ceramides because they are cerebral lipids containing an amide group. In the ceramide sphingomyelin, phosphorylcholine is bound to a hydroxyl group (Fig. 8.4); sphingomyelin is also classified as a phospholipid because the molecule contains phosphate. Sphingomyelin is an essential component of cell membranes, particularly of red blood cells, and of the nerve sheath. Sphingomyelin accumulates in the liver and spleen of patients suffering from a rare lipid

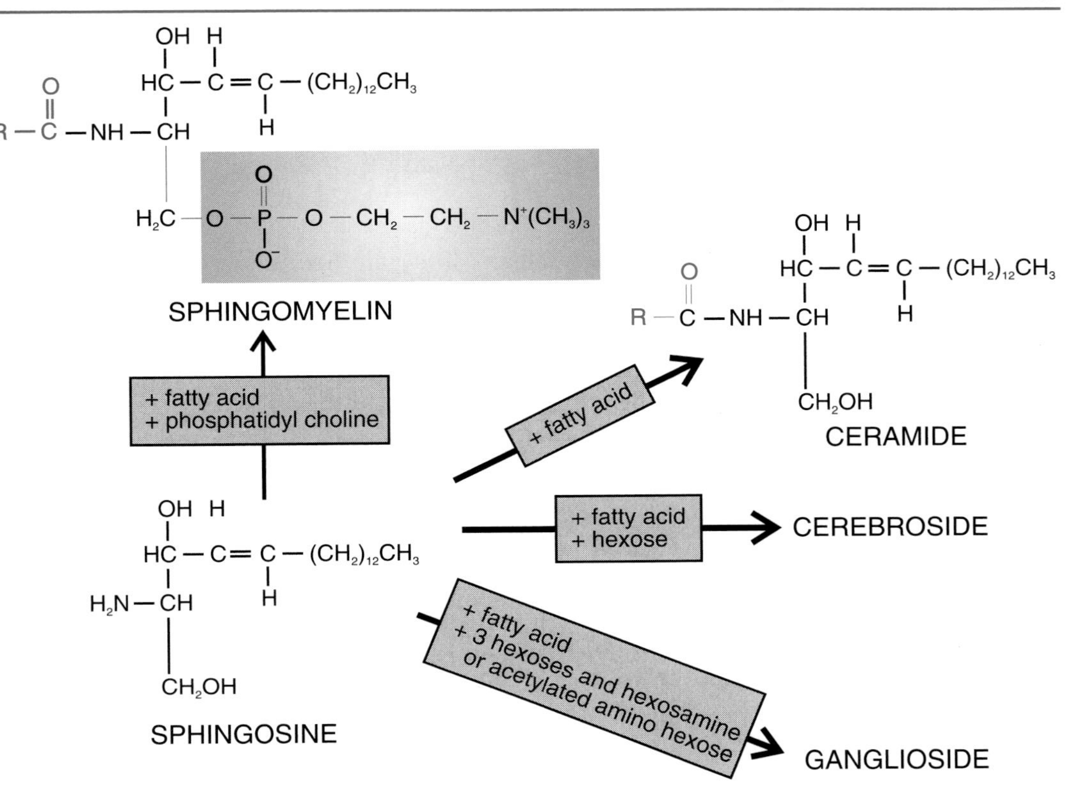

FIG. 8.4. Structures of sphingosine and various sphingolipids.

storage disease (Niemann-Pick disease); an enzyme capable of splitting off phosphorylcholine from sphingomyelin is lacking.

A ceramide becomes a cerebroside when a hydroxyl group binds a hexose (glucose or galactose) in a glycosidic linkage. The ceramides become gangliosides when the carbohydrate chain is composed of hexosamine or an acetylated amino sugar bound to three hexoses (Fig. 8.4). Cerebrosides and gangliosides are also called glycolipids because of the carbohydrate moieties in a lipid structure. Cerebrosides are present as a high percentage of the lipid matter in brain and nerve tissue, particularly in the myelin sheath. Gangliosides are abundant in the gray matter of the brain and in nerve ganglia; they are also present in cell membranes in which the protruding carbohydrate chain serves as a determinant or recognition center for certain immunologic reactions (for example, blood group reactions). Genetic lipid storage diseases have been discovered that show deficiency of an enzyme that splits off a particular portion of the carbohydrate side chain of cerebrosides and gangliosides, with accumulation of these lipids in various tissues. These diseases (lipidoses) are rare. Little clinical information is obtained from measurement of cerebrosides or gangliosides in tissues except in research studies or to confirm a diagnosis at autopsy; identification of the enzyme defect is more convincing. For pregnant patients at risk for lipidoses, cell cultures are grown from amniotic fluid or chorionic villus samples and a search is made for a specific enzyme deficiency in splitting the carbohydrate chain of glycolipids (see Chap. 15). Enzyme assays of biopsy samples may be performed for diagnostic purposes at large medical centers.

Cholesterol[1,2]

Cholesterol, the principal body sterol, is a complex alcohol formed of four fused rings and a side chain (Fig. 8.5); pure cholesterol is a solid at body temperature. Cholesterol is present in all cells as a structural component of membranes and also serves as a precursor for the formation of steroid hormones by the gonads and the adrenal cortex. Although not soluble in water, both cholesterol and triglycerides are normal constituents of plasma, in which they are suspended in solution by a shell of proteins and phospholipids known as lipoprotein particles.

Approximately 70% of plasma cholesterol exists in an acyl ester form. The esterification takes place almost exclusively in a high-density lipoprotein (HDL) complex. An enzyme, lecithin:cholesterol acyltransferase (LCAT), catalyzes the transfer of an unsaturated fatty acid from the R_2-position of phosphatidyl choline (lecithin) (Fig. 8.2) to form an ester with the hydroxyl group of cholesterol. A similar enzyme, acyl:cholesterol acyltransferase (ACAT), is present in liver and other tissues, where it catalyzes the esterification of cholesterol within cells.

Most of the cholesterol in the body is synthesized from acetyl CoA, but we also ingest some when we eat meat, dairy products, or eggs. Plants do not contain cholesterol, although they do have closely related sterols.

Cholesterol is catabolized in hepatic cells by oxidation to bile acids (cholic and chenodeoxycholic acids, Fig. 8.6) that conjugate with glycine or taurine before secretion into bile. These bile acids and conjugates are emulsifying agents that are essential for the digestion and absorption of fats (see Chap. 11,

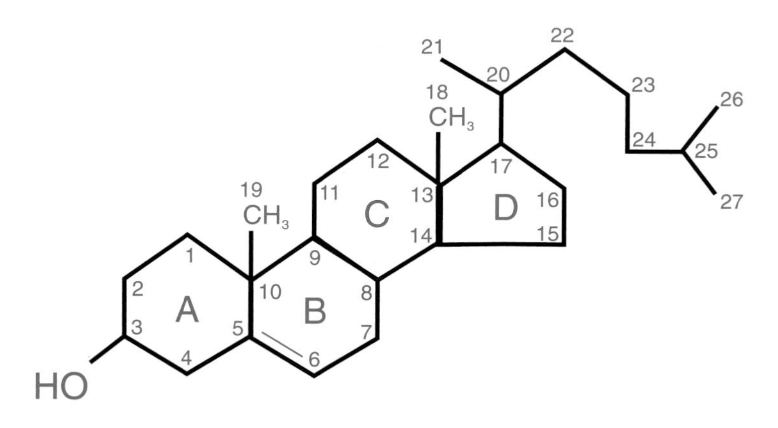

FIG. 8.5. Structure of cholesterol. A cholesterol ester is formed when the hydroxyl group is esterified.

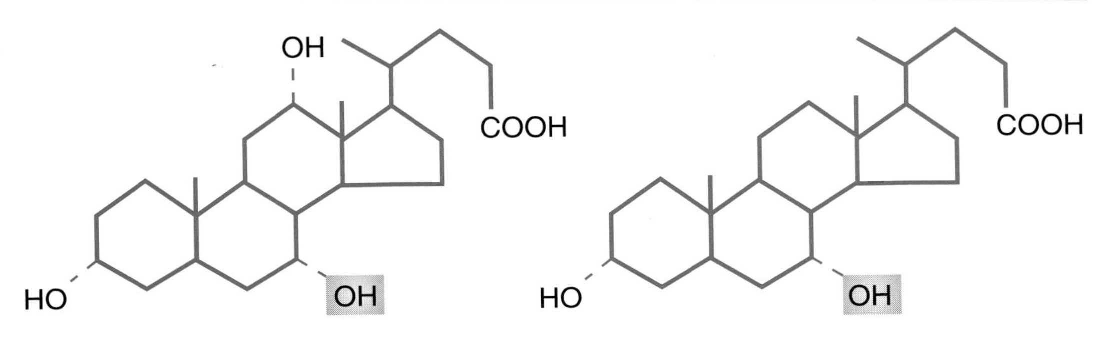

FIG. 8.6. Primary bile acids formed in the liver from cholesterol. Bacterial action in the intestines removes a hydroxyl group (colored yellow) from a portion of cholic and chenodeoxycholic acids to form secondary bile acids, namely deoxycholic and lithocholic acids, respectively.

Figs. 11.3 and 11.4). Some cholesterol is also secreted as such into the bile. Both bile acids and biliary cholesterol are reabsorbed to some extent in the intestines by an enterohepatic circulation (see Chap. 11). Thus, the liver is the site of cholesterol disposal or degradation, as well as its major site of synthesis.

A negative feedback mechanism controls to a limited extent the rate of synthesis of cholesterol. When the diet is high in cholesterol, the increased amount of cholesterol brought to the liver decreases the receptor-mediated hepatic uptake of cholesterol and inhibits the rate-limiting enzyme (β-hydroxy-β-methylglutaryl CoA reductase) essential for the synthesis of mevalonic acid, a step in the synthesis of cholesterol. Furthermore, the reabsorption of bile acids and cholesterol in the enterohepatic circulation is decreased, so more cholesterol is excreted in the form of bile acids and free cholesterol. These mechanisms keep the balance of cholesterol relatively constant when the dietary intake does not exceed 300 mg/day. Typical American diets, however, contain approximately 600 mg/day, an amount that promotes the accumulation of cholesterol and the expression of cholesterol-related diseases.

Serum cholesterol concentration can rise to high levels in some pathologic states. An elevated cholesterol concentration has been implicated as one of several risk factors leading to coronary artery disease (atherosclerosis or myocardial infarction); thus, the measurement of serum cholesterol is a fairly common laboratory procedure.

The role of cholesterol in various clinical situations and methods for its measurement are reviewed later in this chapter.

EXOGENOUS AND ENDOGENOUS PATHWAYS

The body lipids are derived from two sources that require separate metabolic pathways. The first source is fats, oils, and tissue lipids in the diet. After ingestion, the dietary lipids are hydrolyzed in the intestines and absorbed and transported to various tissues. This route is the *exogenous pathway*, dealing with lipids from the outside. The liver, however, readily synthesizes saturated and monounsaturated fatty acids from acetyl CoA and converts them to triglycerides that are distributed to tissues. Cholesterol is also synthesized in the liver from acetyl CoA units. This internal synthesis and distribution of lipids is the *endogenous pathway*. Both pathways require a means for solubilization and transportation of water-insoluble lipids through the bloodstream; lipoproteins are the particles that transport and distribute the lipids. Elucidation of the exogenous and endogenous pathways is continued in the section following the description of the lipoproteins and their composition (Fig. 8.9).

LIPOPROTEINS AND APOLIPOPROTEINS

Lipoproteins are lipid-filled particles that have an outer membrane consisting of a monolayer of special proteins called apolipoproteins interspersed with the polar lipids (phospholipids and nonesterified cholesterol). The polar lipids are aligned with their charged heads facing outward and their hydrophobic tails pointing inward (Fig. 8.7). The outer membrane surrounds a central core of neutral lipids (triglycerides and cholesteryl esters).

Classes of Lipoproteins

The five different classes of lipoproteins have distinctive physical properties, structure, and functions, and each differs in size, density, and electro-

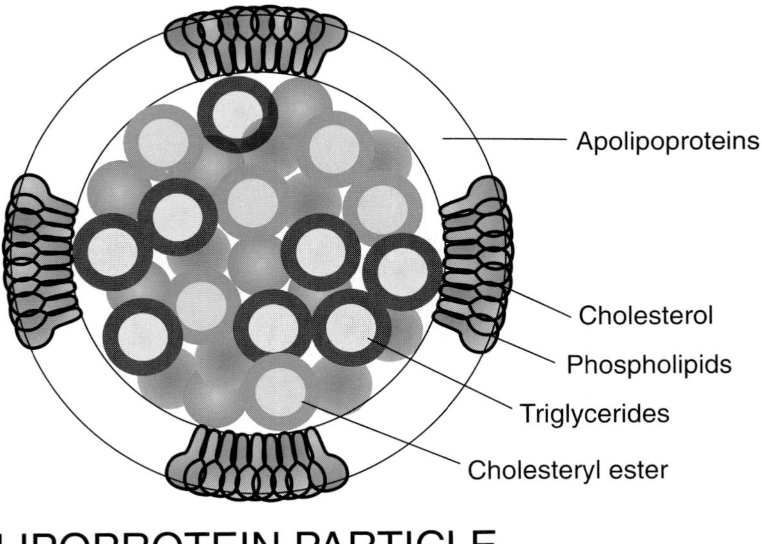

FIG. 8.7. Schematic representation of a lipoprotein particle in cross section illustrates the intermingling of apolipoproteins, phospholipids, and cholesterol in the outer hydrophilic membrane.

TABLE 8.2
LIPOPROTEINS: COMPOSITION, ORIGIN, AND FUNCTION

			COMPOSITION			
LIPOPROTEIN AND (DENSITY)	ELECTRO-PHORETIC MOBILITY	APOPROTEIN CONTENT	MAJOR LIPID (%)	APOLIPO-PROTEIN (%)	TISSUE OF ORIGIN	PRIMARY FUNCTION
Chylomicron (< 1.006)	At origin	AI, AII CI, CII, CIII B48	TG: 90	2	Intestine	Transports dietary TG to muscle or depots.
VLDL (< 1.006)	Pre-β	B100 CI, CII, CIII EII, EIII, EIV	TG: 55	5–8	Liver (primarily) and intestine	Transports endogenous TG to muscle or depots.
IDL (1.006–1.019)	Between β and pre-β	B100 CI, CII, CIII EII, EIII, EIV	TG: 30 Chol: 30	15	Intravascular	Intermediate between VLDL and LDL.
LDL (1.019–1.063)	β	B100 Traces of C and E.	CE: 45	20–24	Intravascular	Transports cholesterol to tissues.
HDL (1.063–1.21)	α	AI, AII, AIV CI, CII, CIII EII, EIII, EIV D (trace)	PL: 30	50	Liver and intestine	Transports cholesterol from tissues to liver; HDL complex is cholesterol esterification site.

TG = triglycerides; CE = cholesteryl esters; PL = phospholipids.

electrophoretic mobility (Table 8.2). Each class of lipoproteins has a specific set of apolipoproteins in the membrane and a different mixture of lipids in the core. A schematic lipoprotein is illustrated in Figure 8.8.

The most commonly used names for the lipoprotein classes are derived from their relative densities upon ultracentrifugation; a less widely used nomenclature system is derived from the relative positions after lipoprotein electrophoresis (Table 8.2).

The lipoprotein classes, arranged in order of increasing density, are listed as follows. The structure and function of the lipoproteins are described in more detail in a subsequent section.

Chylomicrons

Chylomicrons are the largest and least dense of all the lipoproteins. They arise in the intestine and transport ingested triglycerides to adipose tissue and muscle cells.

Very Low-Density Lipoprotein (VLDL)

VLDL is a lipoprotein made in the liver and is designed primarily to transport triglycerides synthesized by the liver to muscle and adipose cells.

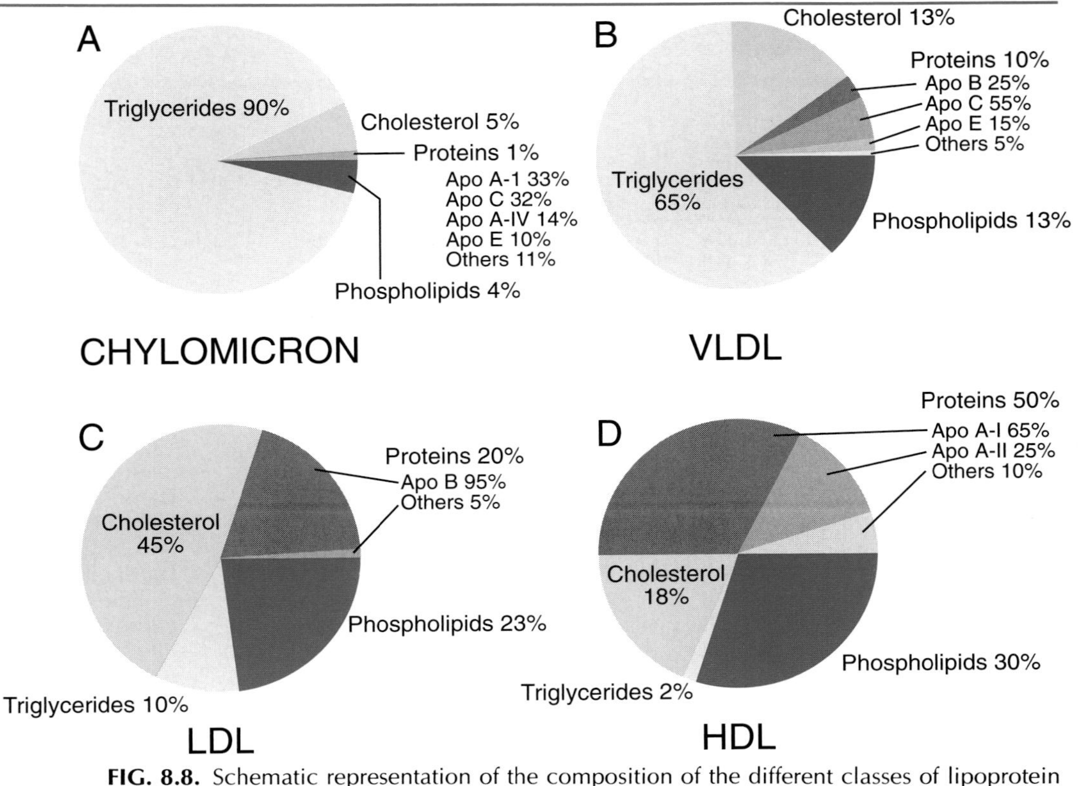

FIG. 8.8. Schematic representation of the composition of the different classes of lipoprotein particles. A, B, C, and D illustrate the typical apolipoprotein and lipid composition of chylomicrons, VLDL, LDL, and HDL, respectively. (Courtesy of the American Association for Clinical Chemistry, copyright owner of Behring Diagnostic's Wall Chart of Apolipoproteins.)

Intermediate-Density Lipoprotein (IDL)

IDL is a transitory remnant of VLDL circulating in plasma after about half of VLDL triglycerides have been transferred to adipose or muscle cells. Most of the IDL undergoes further delipidation, transfers to HDL all its apolipoproteins except Apo B, and thus becomes LDL. A small percentage of IDL binds to liver cells, where it is degraded.

Low-Density Lipoprotein (LDL)

LDL, rich in cholesterol, arises in plasma from IDL. LDL delivers cholesterol either to the liver for bile acid formation or to other tissues for use as a structural component of new cell membranes, as a precursor of steroid hormones (in appropriate endocrine glands), or for storage as cholesteryl esters.

High-Density Lipoprotein (HDL)

HDL has a complicated life cycle and undergoes growth and change after its initial formation. HDL particles are made both by liver and by intestinal mucosa cells. A newly formed (nascent) HDL particle forms a complex with some apolipoproteins, and LCAT esterifies cholesterol by transferring to it a fatty acid from lecithin. All the plasma lipoproteins join temporarily to an HDL cholesterol ester transfer complex for esterification of some of the cholesterol in their membranes. This complex is the source of all cholesterol esters in plasma. HDL also transfers some apolipoproteins back and forth to other lipoproteins at various stages in their life cycles (see a subsequent section on apolipoproteins).

Clinically Important Aberrant Lipoproteins

Floating β-lipoprotein

This abnormal lipoprotein appears as a broad beta band in plasma electrophoresis patterns of individuals with remnant hyperlipoproteinemia (type III). The abnormal lipoprotein is an IDL that occurs in individuals without the E3 isoform of apolipoprotein E. The lack of apolipoprotein E activity decreases the hepatic clearance of VLDL and chylomicron remnant particles, thereby resulting in hyperlipidemia. Individuals with remnant hyperlipidemia often have the apolipoprotein E2/E2 phenotype.

Lipoprotein Little A Antigen, Lp(a)

This abnormal LDL lipoprotein, also known as sinking pre-β-lipoprotein, is associated with a higher risk for atherosclerosis. It can be detected in the laboratory by immunochemical measurement of an unusual apolipoprotein, Apo Lp(a), a glycoprotein associated with apolipoprotein B-100.

Lipoprotein X

Lipoprotein X is the only lipoprotein to migrate toward the cathode when plasma is subjected to lipoprotein electrophoresis in *agar;* it migrates toward the anode as a β-globulin on other support media. Lipoprotein X contains an unusually high concentration of phospholipids and nonesterified cholesterol and is found only in the plasma of patients with biliary cirrhosis or severe biliary obstruction.

Apolipoprotein Families[3,4]

Apolipoproteins are the proteins in the monolayer surface of lipoproteins. They consist of five families; some families have several subgroups. Many of the apolipoproteins do not remain with particular lipoprotein particles, but may be transferred to particles of a different class as lipid metabolism proceeds during the transport process. Some of the apolipoproteins are cofactors of enzymes involved in the initial phases of lipid catabolism. Other apolipoproteins serve as binders to cell receptors on specific tissues in addition to their structural role. The liver is the site of synthesis of most of the apolipoproteins, whereas the intestine primarily generates those for chylomicrons.

Apolipoprotein A (Apo A)

Apo A is found primarily in HDL; the measurement of its concentration in plasma parallels the concentration of HDL. Apo A comprises about 90% of the protein in HDL (Fig. 8.8). A small amount of Apo A appears in newly formed chylomicrons, but later is transferred to HDL. Apo AI, the major member of the Apo A family, is a prerequisite for the activation of LCAT, the enzyme that esterifies cholesterol. Apo AII contributes to the structural integrity of HDL, whereas Apo AIV may also activate LCAT and play a role in reverse cholesterol transport (the movement of cholesterol from cells to liver for disposal). When chylomicrons have been converted to remnants by progressive delipidation (loss of triglycerides through hydrolysis and subsequent escape of fatty acids), they transfer their Apo A and Apo C to HDL.

Apolipoprotein B (Apo B)

Apo B is the major protein constituent of LDL; measurement of its plasma concentration is the best direct way to estimate LDL concentration. The Apo B group consists of two apolipoproteins of quite different molecular weights. The larger one, Apo B100, is synthesized by the liver, whereas the smaller one, Apo B48, is made by the intestines. As shown in Figure 8.8, Apo B100 comprises 95% of the protein in LDL and 25% of the protein in VLDL; the chylomicron contains little apolipoprotein, of which Apo B48 comprises about 10%. Apo B is completely absent from HDL.

Apo B100 in LDL, in cooperation with Apo E, functions as a binder to receptors in specific extrahepatic tissue cells. Although VLDL also contains Apo B100, its high triglyceride content (65%) masks its binding centers so that VLDL does not bind to extrahepatic tissues; after delipidation, VLDL remnants (IDL) can bind to hepatocytes as a first step in their engulfment (endocytosis).

Apolipoprotein C (Apo C)

The Apo C family interacts with enzymes that act on lipids during their transportation by lipoproteins. Apo CII activates lipoprotein lipase, the catalyst for the extrahepatic hydrolysis of triglycerides transported by chylomicrons and VLDL. The functions of Apo CI and Apo CIII are not clear. Apo CI activates the enzyme LCAT in vitro, but its role in the body is unknown. Apo CIII seems to modulate the uptake of triglyceride-rich lipoprotein remnants by hepatic cells.

Apolipoprotein D (Apo D)

Little is known about the function of Apo D, but it appears to be involved in the transfer of cholesteryl esters from an HDL cholesteryl ester transfer complex to VLDL, LDL, and HDL.

Apolipoprotein E (Apo E)

Apo E exist in several isoforms that differ only by charge on the molecule; they can be separated by electrophoresis. Apo E is associated with HDL at its formation, and later is shared with chylomicrons and VLDL during transport. Apo E is the membrane protein that recognizes and binds to hepatocytes as the first step in endocytosis, the internalization of the Apo E-containing lipoprotein by the hepatic cell. Individuals with remnant hyperlipidemia often demonstrate the Apo E2/E2 phenotype (a less active form of Apo E) and poor hepatic clearance of IDL particles. Apo E, in conjunction with Apo B100 (Apo B/E), is the necessary LDL binder to extrahepatic tissues for delivery of its cholesterol load.

Lipid Transport and Functions of Lipoproteins[3,4]

The Exogenous Pathway[7]

Dietary lipids, mainly triglycerides, are hydrolyzed in the gastrointestinal tract and their products are absorbed by intestinal mucosa cells. In the endoplasmic reticulum of these cells, the fatty acids, monoglycerides, and some of the cholesterol are esterified to yield a new set of conjugated lipids. The triglycerides and cholesteryl esters are enclosed in a monolayer containing Apo B48, Apo AI, Apo AII, phospholipid, and nonesterified cholesterol; this is the newly formed chylomicron that is secreted into lymph. The chylomicrons flow in the lymph channels until they enter the blood circulatory system via the thoracic duct. During the passage in lymph, the chylomicron acquires Apo C and Apo E from HDL.

Fates of Triglycerides and Cholesterol. The chylomicron adheres to a binding site on capillary walls that is mostly in muscle and adipose tissue. The enzyme, lipoprotein lipase, that is in the vicinity is activated by Apo CII and starts the initial process of delipidation; triglycerides are hydrolyzed to free fatty acids (FFA) and monoglycerides that migrate to adjacent muscle or adipose cells. There, they are either oxidized for energy purposes (muscle) or reconverted to triglycerides (adipocyte) and stored as depot fat. When the chylomicron meets an HDL-cholesteryl ester complex, Apo AI activates LCAT, which esterifies cholesterol in the membrane; the ester shifts to the chylomicron core. The chylomicron remnant, with its rich load of cholesteryl esters and small remaining load of triglycerides, transfers its Apo A to HDL and receives ApoE. It continues to circulate until it reaches the liver, where Apo E recognizes and binds to a hepatocyte. The chylomicron remnant is quickly internalized by endocytosis and migrates to a lysosome; lysosomal enzymes soon hydrolyze the chylomicron lipids and proteins to their basic components. The fatty acids, amino acids, and cholesterol enter their respective pools. The fatty acids usually are converted to triglycerides and transported from the liver by VLDL, but the cholesterol pool is handled in one of three possible ways as described in the previous section on cholesterol. Chylomicrons disappear from the circulation about 10 hours after a high-fat meal.

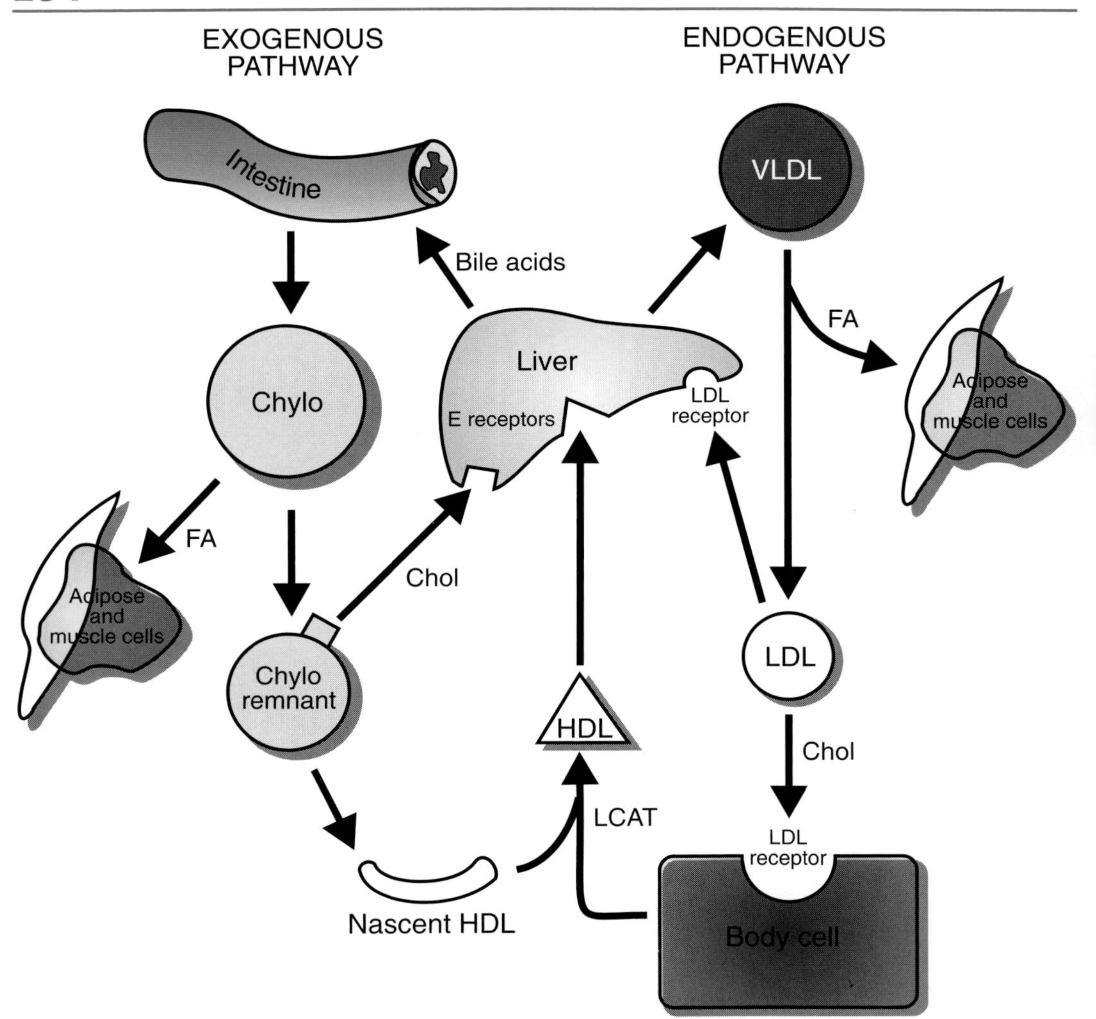

FIG. 8.9. Lipid transport by the exogenous and endogenous pathways. The chylomicrons transport triglycerides from the intestine to adipose and muscle cells. Hydrolysis of triglycerides by lipoprotein lipase during transport enables the liberated fatty acids to migrate to these cells. In a similar fashion, the triglycerides synthesized in the liver are transported to adipose and muscle cells by VLDL. VLDL is transformed into LDL as it sheds its load of triglycerides. The major portion of the LDL cholesterol is delivered to body cells; the remainder enters the liver as the LDL remnant is internalized by endocytosis. Chylo = chylomicrons; Chol = cholesterol; VLDL = very low-density lipoproteins; LDL = low-density lipoproteins; HDL = high-density lipoproteins; LCAT = lecithin:cholesterol acyltransferase. (Modified from Diagnostic Medicine, April, 1985.)

In summary, chylomicrons transfer the bulk of dietary triglycerides to adipose and parenchymal cells as FFA for either energy use or for storage as triglycerides. Dietary cholesterol is transferred almost entirely to hepatic cells; its disposition depends upon many homeostatic factors (see previous section on cholesterol).

The Endogenous Pathway[7]

Hepatic cells continuously synthesize fatty acids from the pool of acetyl CoA that comes from the catabolism of triglycerides, carbohydrate, and protein in the liver. Some cholesterol is also synthesized, starting with acetyl CoA units. Under normal, healthy conditions, the liver cannot store either triglycerides or cholesterol to any appreciable extent.

The liver generates a lipoprotein (VLDL) to transport these lipids from the liver to tissues throughout the body. The hepatocyte encloses newly synthesized triglycerides in the core and cholesterol in the membrane of a newly formed VLDL, similar to that of the chylomicron in the intestine. The protein portion of the VLDL membrane is made up almost exclusively of Apo B100, with a little Apo CII and Apo E. VLDL is primarily a transporter of triglycerides; its cholesterol load is actually transported to tissues by LDL. VLDL attaches to the HDL-cholesteryl ester transfer complex described in the previous section on high-density lipoproteins. The complex esterifies cholesterol in the VLDL membrane; the ester migrates to the core. As shown in Figure 8.8, VLDL contains about five times more triglycerides than cholesterol. HDL transfers some Apo CII to VLDL, which then leaves the complex.

Conversion to IDL. The Apo CII-enriched VLDL travels in the blood until it binds to an adipocyte. The Apo CII activates the nearby lipoprotein lipase that hydrolyzes about one-half of the triglycerides in the VLDL particle; the FFA migrate into an adipocyte for storage as triglyceride. Some Apo E is transferred from an HDL to the partially delipidated VLDL that then detaches from the adipocyte. This VLDL remnant is called IDL, an intermediate, transitory lipoprotein that contains about equal quantities of triglycerides and cholesterol.

Conversion to LDL. As the IDL circulate, they continue to interact with lipoprotein lipase until most of the triglycerides have been transferred to adipocytes or muscle cells. They transfer to HDL all of their apoproteins except Apo B100 and a small amount of Apo E. The delipidation and subsequent transfer of apoproteins converts IDL into a cholesterol-rich, triglyceride-poor LDL particle. The small percentage of IDL particles that escape this fate eventually bind to hepatocytes, where they are engulfed and carried into the hepatic cells by endocytosis. Their lipids and apoproteins are hydrolyzed to the basic constituents that enter their respective liver pools.

The LDL now delivers cholesterol to tissues. The Apo B/E binders recognize and attach to appropriate tissue cells for delivery of the LDL cholesterol. The LDL is internalized by endocytosis and degraded by lysosomal enzymes. The now nonesterified cholesterol may be used in membranes of dividing cells, may be converted to steroid hormones by adrenal or gonadal cells, or may be esterified and stored in the cell as an ester. The LDL that does not bind to peripheral cells eventually binds to liver cells, where they are internalized and degraded. The cholesteryl esters in the core are hydrolyzed and the cholesterol enters the liver pool.

In summary, VLDL transports endogenous triglycerides to peripheral tissues for catabolism or storage. The endogenous cholesterol remains in the VLDL remnants as the latter are transformed successively to IDL and then to LDL. The cholesterol-rich LDL transports these cholesteryl esters to peripheral tissues for use or deposit. All remaining LDL are finally internalized by the liver and degraded; its cholesterol enters the hepatic pool.

HYPERLIPOPROTEINEMIAS AND HYPERLIPIDEMIAS[5–7]

A *hyperlipoproteinemia,* by definition, is an increased concentration of one or more lipoproteins in plasma. Because the lipoproteins transport plasma lipids, a *lipoproteinemia* is also a *hyperlipidemia.* The two main plasma lipids of clinical interest are cholesterol (total) and triglycerides. Hence, a hyperlipoproteinemia may also be a *hypercholesterolemia,* a *hypertriglyceridemia,* or a mixture that may be designated as *hyperlipidemia.*

Several decades ago, Fredrickson and Levy classified inherited lipoprotein disorders according to the lipoprotein electrophoresis pattern, plasma concentrations of cholesterol and triglycerides, and a description of the plasma on standing, that is, whether a creamy layer was evident and whether the plasma was clear or turbid. Six different phenotypes (patterns of test results) emerged from consideration of these criteria. The Fredrickson system of classification is oriented to laboratory results rather than to disease states. The Fredrickson hyperlipidemia phenotypes do not designate specific diseases or the mechanism of disease, and patients may change phenotypes over time. Recently, reclassification of the hyperlipidemia states by inherited disease and pathophysiologic mechanism is receiving greater emphasis and was made possible by advances in lipid and lipoprotein research. Both the pathophysiologic and Fredrickson classifications are in current use in clinics and are presented in this text.

Detection of an elevated level of serum cholesterol or triglyceride in a patient typically prompts the following two questions: Is this a primary or secondary hyperlipidemia? Was a proper *fasting* serum sample used for the analysis? The term *primary hyperlipoproteinemia (or primary hyperlipidemia)* is used for familial conditions with one or more gene defects. A gene defect in this instance means a defect in an apolipoprotein, a receptor site, or any of the enzymes used by the lipid transport system. *Secondary hyperlipidemia* is a consequence of an existing disease, dietary factors, drugs, or a combination of these factors. Obesity and alcohol consumption frequently exacerbate hyperlipidemia of both primary and secondary disease states.

Primary Hyperlipoproteinemias

Familial Hypercholesterolemia

Individuals with this disease usually develop atherosclerosis at an early age and succumb prematurely to coronary artery disease. Approximately 75% of patients, particularly older individuals, exhibit tendon xanthomas. The levels of plasma cholesterol and LDL cholesterol are greatly elevated, but the level of triglycerides remains normal, and the lipoprotein electrophoresis pattern type IIa is observed in both homozygotes and heterozygotes with the disorder. The plasma LDL concentration is elevated (Fig. 8.10) because of a defective Apo E receptor that fails to bind LDL efficiently. The LDL escapes degradation in the liver and continues to circulate with its load of cholesterol.

Polygenic Hypercholesterolemia

As the term polygenic suggests, many subtle genetic deficiencies in the enzymes and proteins of cholesterol metabolism are thought to contribute to the mild elevation of plasma cholesterol characteristic of this disease state (type IIa or type IIb by lipoprotein electrophoresis, Fig. 8.10). The combination of polygenic susceptibility and regular consumption of high-fat, high-cholesterol foods is thought to explain most elevated cholesterol levels observed in most clinics. Unlike familial hypercholesterolemia, these patients seldom have tendon xanthomas.

Familial Combined Hypercholesterolemia

An elevation of both plasma cholesterol and triglyceride occurs in this disorder in patients older than 30 years of age. The condition is associated with an excessive production of Apo B and either VLDL or LDL particles (products of the endogenous pathway). Patients display various lipoprotein electrophoresis patterns (types IIa, IIb, IV, and occasionally V, Fig. 8.10), depending on secondary influences.

Familial Endogenous Hypertriglyceridemia

Moderate elevation of plasma triglyceride without an elevation of cholesterol in obese patients with hyperglycemia and hyperuricemia is typical of this condition. The expression of disease varies between families, but the net result is an excess level of plasma VLDL (lipoprotein electrophoresis pattern type IV and type V, Fig. 8.10), which is exacerbated by weight gain, estrogens, alcohol, or high-carbohydrate diets.

Lipoprotein Lipase and Apolipoprotein CII Deficiencies

The lack of lipoprotein lipase activity caused by either a direct genetic lesion or the lack of apolipoprotein CII activity (a lipoprotein lipase activator) results in profoundly lipemic plasma stemming from the buildup of chylomicrons, in either the absence or the presence of VLDL (lipoprotein patterns type I and V, Fig. 8.10). Milder forms of this disorder require secondary factors to aggravate symptoms of skin xanthomas, enlarged liver, lipemia retinalis, and acute pancreatitis.

Remnant Hyperlipidemia

This disease state is associated with the loss of an apolipoprotein E-mediated binding of VLDL and chylomicron remnant particles by the liver. Conse-

Type Hyper-lipoproteinemia	Appearance of Plasma	Electrophoretic Pattern	Lipoprotein Abnormality	Lipid Abnormality
Normal Control	Clear	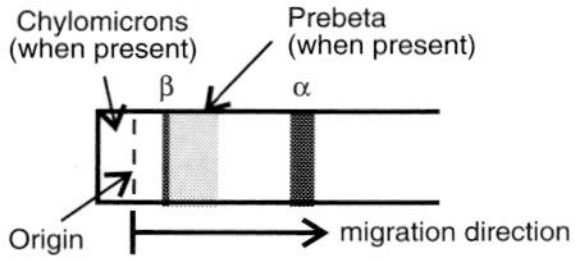 Chylomicrons (when present); Prebeta (when present); β; α; Origin; migration direction	None	None
Type I	Creamy layer over clear plasma	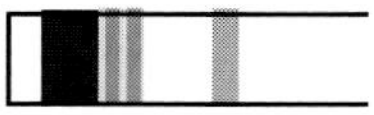	Chylo $\uparrow\uparrow\uparrow$	C $\rightarrow$ N TG$\uparrow\uparrow\uparrow\uparrow$
Type II A	Clear	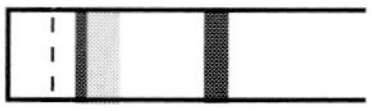	LDL $\uparrow\uparrow\uparrow$	C$\uparrow\uparrow\uparrow$ TG $\rightarrow$ N
Type II B	Clear or slightly turbid		LDL $\uparrow\uparrow$ VLDL $\uparrow$	C$\uparrow\uparrow$ TG$\uparrow$
Type III	Turbid	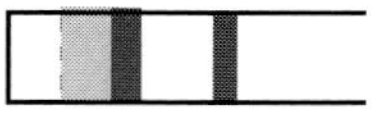	IDL $\uparrow$ Abnormal LDL	C$\uparrow\uparrow$ TG$\uparrow\uparrow$
Type IV	Turbid to Milky	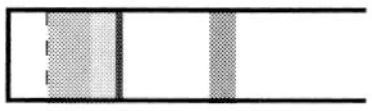	VLDL $\uparrow\uparrow$	C$\uparrow$ or N TG$\uparrow\uparrow$
Type V	Creamy layer over turbid suspension	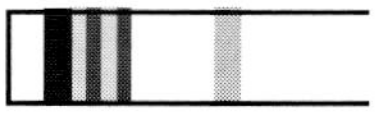	Chylo $\uparrow\uparrow\uparrow$ VLDL $\uparrow\uparrow\uparrow$	C$\uparrow$ TG$\uparrow\uparrow\uparrow$

FIG. 8.10. Fredrickson classification of hyperlipidemia phenotypes by lipoprotein electrophoresis.

quently, degradation of those particles is greatly diminished (see earlier section on floating β-lipoprotein). The presence of elevated levels of remnant lipoprotein particles with IDL density is typical of lipoprotein electrophoresis pattern type III (Table 8.3).

Familial Hyper-α-Lipoproteinemia

This rare condition has no clinical features that would induce an individual to seek a physician. It is a biochemical abnormality that does no harm and may even reduce the risk of coronary artery disease. Both HDL and HDL cholesterol

TABLE 8.3
HYPERLIPEMIC DISORDERS

ELEVATED LIPOPROTEIN CLASS	PRIMARY HYPERLIPOPROTEINEMIAS		SECONDARY DISORDERS*
	TYPE	NAME	
Chylomicrons (Exogenous hyperlipemia)	I	Familial lipoprotein lipase deficiency CII apolipoprotein deficiency	Dysglobulinemias Systemic lupus erythematosus
LDL (Hypercholesterolemia)	IIa	Familial hypercholesterolemia Familial multiple lipoprotein-type hyperlipemia	Nephrotic syndrome Hypothyroidism Cushing's syndrome†
LDL-VLDL (Combined hyperlipemia)	IIb	Familial multiple lipoprotein-type hyperlipemia	Nephrotic syndrome Hypothyroidism Cushing's syndrome†
β-VLDL (Remnant hyperlipemia)	III	Familial dys-β-lipoproteinemia	Hypothyroidism Systemic lupus erythematosus
VLDL (Endogenous hyperlipemia)	IV	Familial hypertriglyceridemia (mild) Sporadic hypertriglyceridemia Familial an-α-lipoproteinemia (Tangier disease)	{ Diabetic hyperlipemia Glycogenosis, type 1 Lipodystrophies Hypopituitarism Nephrotic syndrome Uremia Alcoholism
VLDL + chylomicrons (Mixed hyperlipemia)	V	Familial hypertriglyceridemia (severe) Familial lipoprotein lipase deficiency CII apolipoprotein deficiency	

Adapted from Havel, R.J.: Approach to the patient with hyperlipidemia. Med. Clin. North Am., *66*:319, 1982.
*Secondary hyperlipemias may mimic several different classes of hyperlipoproteinemias, depending on the stage and severity of the underlying disease. Other factors (stress or alcoholism) may contribute to the problem.
†Usually mild and induced by glucocorticoid use or stress.

are elevated in this condition, whereas the plasma concentration of cholesterol may be normal or slightly elevated. The triglyceride concentration is within normal limits. The ratio of LDL cholesterol to HDL cholesterol is reduced below the usual ratio of 3:1, thus possibly resulting in a reduced risk for coronary artery disease.

Secondary Hyperlipoproteinemias

As mentioned earlier, many secondary hyperlipoproteinemias mimic some of the familial hyperlipoproteinemia patterns (Table 8.3). No intrinsic or permanent defects occur, however, in any of the apolipoproteins, binding sites, or enzymes in these secondary disturbances. The abnormality disappears if the primary cause (disease, metabolic process, excess hormones) is returned to normal. The degree of hyperlipidemia is affected by many factors and can be modified by decreasing the amount of cholesterol and saturated fat in the diet and by modifying life-style to include more exercise, fewer calories, and less alcohol intake. Because the risk of coronary artery disease is increased by hyperlipidemia, vigorous steps are often taken to reduce both cholesterol and triglycerides by prescribing appropriate diet and drugs, as well as by treating other coronary artery disease risk factors, such as hypertension.

The lipoprotein electrophoretic patterns of secondary hyperlipoproteinemia are subject to change as a disease becomes more severe. For example, hypothyroidism may have electrophoretic patterns resembling those of type IIa, IIb, or III, depending on the stage and severity of the disease. Table 8.3 lists a group of secondary causes of hyperlipidemia.

HYPOLIPOPROTEINEMIAS

Familial A-β-lipoproteinemia

A-β-lipoproteinemia (the absence of plasma β-lipoproteins) was the first familial hypolipoproteinemia to be discovered. It is rare but severe. Chylomicrons, VLDL, and LDL are absent because of a failure to synthesize Apo B. Individuals with this defect have retarded growth and progressive degeneration of the central nervous system.

These individuals have poor absorption of dietary fat, of the fat-soluble vitamins (vitamins A, D, E, and K), and of the essential fatty acids, linoleic and linolenic acids. The human body cannot synthesize from acetyl CoA units a fatty acid that has more than one double bond; multiple unsaturated fatty acids are obtained from polyunsaturated seed oils or oils from cold-water fish (salmon, for example). Nerve tissue requires polyunsaturated fatty acids in its structure, and prostaglandins are synthesized from arachidonic acid, a polyunsaturate (Fig. 8.3). The severity of this familial lipoprotein disorder may be ascribed in part to the lack of essential nutrients caused by the failure in lipid absorption.

Hypo-β-lipoproteinemia

Familial hypo-β-lipoproteinemia is characterized by a low LDL and a low plasma total cholesterol concentration, as well as by a low LDL cholesterol. Individuals with this disorder appear normal in all other respects and get a bonus from their low LDL cholesterol concentration. They are less likely to suffer from coronary artery disease than are their "normal" counterparts who have a much higher LDL cholesterol.

Familial An-α-lipoproteinemia

Familial an-α-lipoproteinemia, initially called Tangier disease, is a rare disorder in which HDL (α-lipoprotein) is virtually absent, as demonstrated by lipoprotein electrophoresis. Individuals with this disorder cannot synthesize Apo AI or Apo AII. Cholesteryl esters accumulate in all tissues of the reticuloendothelial system; the tonsils become enlarged and have an orange-yellow color. The plasma cholesterol concentration is usually reduced because of the absence of HDL cholesterol, whereas the triglyceride concentration may be normal or raised moderately, depending on such secondary factors as diet and life-style. Despite the absent HDL (and HDL cholesterol), an increased incidence of coronary artery disease does not occur in this group.

LABORATORY TESTS FOR LIPOPROTEIN AND LIPID DISORDERS

The definitive separation of lipoprotein classes is performed by ultracentrifugation, but this technique is neither feasible for most laboratories nor desirable in most workups for possible lipid disorders.

As mentioned in an earlier section, gene defects in the synthesis of lipoproteins (primary hyperlipoproteinemias) are much rarer causes of hyperlipidemia than are the multitude of secondary hyperlipoproteinemias accompanying a host of abnormal metabolic, hormonal, or pathologic events. Poor dietary or life-style habits may also aggravate the underlying conditions. Hence, the most productive and economical laboratory investigation of hyperlipidemias is as follows:

1. Properly collect a fasting blood sample. The patient must be without food or alcohol for at least 12 hr before the blood is drawn to insure that all chylomicrons arising from the last source of food have been cleared.
2. Place the separated plasma (EDTA anticoagulant) in a refrigerator and let stand overnight. Then check for the presence of a cream layer (chylomicrons), and note whether the plasma is clear or turbid (see Fig. 8.10).
3. Thoroughly mix the sample, take aliquots, and measure the concentrations of plasma total cholesterol and triglycerides. If an estimate of the risk for coronary artery disease is desired, also measure the HDL cholesterol concentration. (See subsequent section on LDL cholesterol for the formula for calculating its concentration from the measured concentrations of triglycerides and total and HDL cholesterol.)
4. Sometimes a lipoprotein electrophoretic pattern may aid in making finer distinctions.

For definitive diagnosis of some cases of primary hyperlipoproteinemia, ultracentrifugation and subsequent lipoprotein electrophoresis on these fractions may be necessary. Immunochemical techniques on plasma or lipoprotein fractions may be required to quantitate specific apolipoproteins. This type of work is usually carried out at large medical centers or in lipid specialty laboratories, but the current trend is for clinical laboratories to quantitate specific apolipoproteins as these tests become readily available.

Determination of Total Cholesterol

Enzymatic Methods[8]

Enzymatic methods for the hydrolysis of cholesterol esters and oxidation of cholesterol are simpler, faster to perform, and use less corrosive chemicals than the more tedious chemical analyses. The serum can be used directly, with little interference from serum constituents. The required volume of serum varies from 5 to 100 μL, depending on the system. Enzymes and reagents are available in commercial kit form, and the tests can be performed manually or adapted to various types of automated instruments. Oxidation of cholesterol by cholesterol oxidase yields H_2O_2, which can be quantitated in a manner similar to that in a glucose assay.

Principle. Cholesteryl esters in serum are hydrolyzed by cholesterol ester hydrolase. All the cholesterol is then oxidized by cholesterol oxidase to the corresponding ketone, with a shift in the location of a double bond. The H_2O_2 generated by the oxidation is decomposed by horseradish peroxidase in the presence of 4-aminoantipyrine (4-aminophenazone) and phenol to yield a quinoneimine dye. The absorbance of the dye, measured at 500 nm, is proportional to the cholesterol concentration. All the enzymes and chromogen are contained in a single, buffered reagent. An endpoint method at 37° C is recommended. As an alternative, one can use an O_2 electrode for measuring the O_2 released by the decomposition of H_2O_2.

The following reactions take place*

Hydrolysis 1. Cholesteryl esters $\xrightarrow{CE}$ Cholesterol + Fatty Acids

Oxidation 2. Cholesterol + O_2 $\xrightarrow{C\ Ox}$ Cholest-4-ene-3-one + H_2O_2

3. 2 H_2O_2 + 4-aminophenazone $\xrightarrow{Per}$ Quinoneimine dye + 4 H_2O

4. Measure absorbance at 500 nm

Alternate 2A. $2H_2O_2 \xrightarrow{Per} 2H_2O + O_2$

3A. Measure O_2 consumption with an oxygen electrode.

Reference Values. The plasma cholesterol concentration is low at birth, but increases about 40% by the third day of life to reach a value of about 125 mg/dL (3.2 mmol/L). The concentration slowly increases with age. The cholesterol values of women are slightly lower than those of men until the menopause; values in women then exceed those in men. The 1988 report of the Cholesterol Education Program (NCEP)[9] defined the levels of plasma cholesterol deemed desirable, tolerable, or a serious risk factor for developing coronary artery disease. As shown in Table 8.4, *desirable* concentrations are below 200 mg/dL (5.2 mmol/L), those between 200 and 239 mg/dL (5.2 to 6.2 mmol/L) are *borderline,* and those exceeding 240 mg/dL (6.2 mmol/L) are *high cholesterol concentrations that should be reduced* by dietary means and/or appropriate drug therapy. These designations apply to all individuals over 20 years of age irrespective of sex or age bracket. In view of these recommendations, a table of reference values by age and gender is unnecessary. See Reference no. 8A for such detailed information.

Conversion to SI: Multiply mg/dL by 0.0259 to obtain mmol/L.

Increased Concentration. Total serum cholesterol concentration is increased in hypothyroidism, uncontrolled diabetes mellitus, nephrotic syndrome, extrahepatic obstruction of the bile ducts, and various types of hyperlipidemias, especially in xanthomatosis. Cholesterol rises in late pregnancy but returns to

*Abbreviations: CE = cholesterol esterase; Per = peroxidase; C Ox = cholesterol oxidase.

TABLE 8.4
RECOMMENDED AND HIGH-RISK PLASMA LIPID CONCENTRATIONS

	DESIRABLE		BORDERLINE		HIGH RISK	
CONSTITUENT	(mg/dL)	(mmol/L)	(mg/dL)	(mmol/L)	(mg/dL)	(mmol/L)
Cholesterol*						
Total	< 200	< 5.2	200–239	5.2–6.2	> 240	> 6.2
HDL	> 45	> 1.2	35–45	0.91–1.2	< 35	< 1.2
LDL	< 130	< 3.4	130–159	3.4–4.1	> 159	> 4.1
Triglyceride†	< 250	< 2.8	250–500	2.8–5.7	> 500	> 5.7

*National Cholesterol Education Program.[9]
†Consensus Conference: Treatment of hypertriglyceridemia.[18]

normal levels within a month after delivery. The NCEP considers a long-time, elevated cholesterol concentration ($>$ 240 mg/dL) a serious risk factor for the development of coronary artery disease.

Decreased Concentration. Hypocholesterolemia is usually present in hyperthyroidism, hepatocellular disease, anemias, starvation, and certain genetic defects. In a rare genetic disease, a-β-lipoproteinemia, affected individuals have no β-lipoproteins (LDL), and the serum cholesterol concentration is low.

Determination of HDL Cholesterol

The serum concentration of *total* cholesterol is of limited value in assessing the risk of coronary artery disease because cholesterol is transported primarily by two lipoprotein classes with different functions. HDL transports cholesterol *from tissues to the liver for catabolism* (conversion to bile salts), whereas LDL transports cholesterol *from sites of origin to deposition in tissues*, including blood vessels. The HDL competes with LDL for binding to tissue receptors and may thus reduce cholesterol accumulation in blood vessel walls. The concentration of HDL cholesterol appears to be *inversely* related to the risk of cardiovascular disease,[10] but a word of caution must be inserted. The normal range of HDL cholesterol is narrow, and the precision of the analysis is poorer than that for total cholesterol. This combination of factors provides a high relative error in the estimation of risk of cardiovascular disease. Commercial kits are available for the HDL cholesterol test.

Principle. Polyvalent anions complex with lipoproteins containing Apo B and precipitate. Thus, VLDL and LDL are precipitated by the addition of a reagent containing heparin-$MgCl_2$, dextran sulfate-$MgCl_2$, or phosphotungstate-$MgCl_2$. After centrifugation, HDL is the only lipoprotein remaining in the supernate, and its cholesterol is determined by an enzymatic method in the same manner as that described for total cholesterol.

The apolipoproteins of HDL (Apo AI and AII) and apolipoprotein B of LDL can be measured directly by enzyme-linked immunoassay (ELISA) or by immunonephelometry (INA). These assays may eventually supplant those of HDL and LDL cholesterol.

Determination of LDL Cholesterol

The concentration of LDL cholesterol is difficult to measure directly, but it may be calculated from the measurements of total and HDL cholesterol and triglycerides in plasma, as follows:

1. LDL chol = Total chol − (HDL chol + VLDL chol)
2. When the concentration of triglycerides (TG) is less than 400 mg/dL, the usual ratio of VLDL chol:TG = 1:5. Hence, VLDL chol = TG/5.
3. By substitution in Equation 1, LDL chol = Total chol − (HDL chol + TG/5).

The risk for coronary artery disease increases with a rising plasma concentration of LDL cholesterol but decreases with an increasing concentration of HDL cholesterol. For this reason, some laboratories calculate the ratio of LDL cholesterol:HDL cholesterol for estimating the degree of risk; a ratio of less than 3:1 is considered desirable.

Reference Values. Reference values are not provided for HDL and LDL cholesterol for the same reasons as for total cholesterol. See Table 8.4 for desirable and high-risk values. The desirable concentrations are > 45 mg/dL (> 2.0 mmol/L) for HDL cholesterol and < 130 mg/dL (< 3.4 mmol/L) for LDL cholesterol.

Increased Concentration. LDL cholesterol is elevated in primary hyperlipoproteinemia types IIa and IIb and in many secondary hypercholesterolemias (see section on total cholesterol). HDL cholesterol is elevated in familial hyper-α-lipoproteinemia. Life-style affects the concentration of both of these constituents, as described earlier for lipoproteins and cholesterol.

Decreased Concentration. Familial hypolipoproteinemias were reviewed in an earlier section. LDL cholesterol is low or absent in situations in which LDL is low or absent, including a-β-lipoproteinemia and hypo-β-lipoproteinemia. In like manner, no measurable HDL cholesterol is present in familial an-α-lipoproteinemia because of the failure to synthesize HDL.

Plasma Triglycerides[11–16]

All laboratories today quantify triglycerides by measuring the glycerol moiety of the molecule after hydrolysis. They perform the analysis by utilizing enzymes in every step of the procedure from initial hydrolysis to liberate the glycerol to formation of the end product that is measured spectrophotometrically, fluorometrically, or colorimetrically, depending on the series of coupled enzymatic reactions and the end product. For example, the disappearance of NADH is measured spectrophotometrically in a widely used method (Method

A[12]), whereas it is estimated fluorometrically in another (Method B[13]); two other popular methods measure the absorbance of different highly colored dyes by colorimetry (Methods C[14] and D[11,15]). Although triglycerides can be assayed by using only chemical reactions (no enzymes) that produce a highly colored or fluorescent compound, the methods are labor intensive because they require an initial isopropanol extraction in the presence of an alumina adsorbent mixture to remove phospholipids and other interferents.[17] None of the 5000 laboratories participating in the 1991 College of American Pathologists (CAP) Comprehensive Chemistry Survey employed a strictly chemical procedure. By contrast, enzymatic methods need no prior extraction or purification steps and are completely automated.

Enzymatic Methods

All the laboratories listed in the 1991 CAP Comprehensive Chemistry Survey employed enzymes for the hydrolysis of triglycerides and measured the glycerol moiety or its equivalent by a variety of coupled enzymatic reactions.*

Method A.[12] In method A, the analyte adenosine diphosphate (ADP), formed in Reaction 2, is measured by coupling it with Reactions 3 and 4. The constituent measured in Reaction 4 is the disappearance of NADH at 340 nm. Each μmole of NADH utilized is equivalent to 1 μmole of original triglyceride.

Reaction 1 $TG + 3\ H_2O \xrightarrow{Lipase} Glycerol + 3\ FA$

Reaction 2 $Glycerol + ATP \xrightarrow{GK} Glycerol\text{-}3\text{-}P + ADP$

Reaction 3 $ADP + Phosphoenolpyruvate \xrightarrow{PK} ATP + Pyruvate$

Reaction 4 $Pyruvate + NADH + H^+ \xrightarrow{LD} Lactate + NAD^+$

Method B.[13] Method B is the same as Method A except that fluorometry is substituted for spectrophotometry for estimating the disappearance of NADH in Reaction 4. The excitation wavelength is 355 nm, whereas the emitted light is measured at 460 nm.

Most laboratories in the CAP survey, however, preferred alternative enzymatic reactions that produced a highly colored dye with an absorbance that could be measured in the visible light range. The two most popular variants are methods C and D, shown as follows.

Method C.[14] In this method, Reactions 1 and 2 are coupled with Reactions 5 and 6. The NADH generated in Reaction 5 reduces a tetrazolium salt to a formazan dye in Reaction 6. The absorbance is measured between 500 and 590 nm; the triglyceride concentration is proportional to the absorbance.

*The following abbreviations are used: ADP = adenosine diphosphate; ATP = adenosine triphosphate; NADH and NAD^+ = the reduced and oxidized forms, respectively, of nicotinamide adenine dinucleotide; DA = diaphorase; FA = fatty acid; GK = glycerol kinase; Glycerol-3-P = glycerol-3-phosphate; GPDH = glycerol-3-phosphate dehydrogenase; GPO = L-alpha-glycerol phosphate oxidase; LD = lactate dehydrogenase; PK = pyruvate kinase; PO = peroxidase; TG = triglyceride; tetrazolium salt = 2-p-iodophenyl-3-p-nitrophenyl-5-phenyltetrazolium (oxidized); formazan dye-reduced form of the tetrazolium salt.

Reaction 5 Glycerol-3-P + NAD^+ $\xrightarrow{GPDH}$ Dihydroxyacetone phosphate + NADH + H^+

Reaction 6 NADH + H^+ + tetrazolium salt $\xrightarrow{DA}$ Formazan dye + NAD^+

Method D. Method D also utilizes Reactions 1 and 2. Glycerol-3-P, the analyte selected from Reaction 2, is enzymatically converted to dihydroxyacetone phosphate plus H_2O_2 by L-alpha-glycerol phosphate oxidase (GPO), as shown in Reaction 7. In a complicated reaction,[14] a red quinoneimine dye is formed as H_2O_2, in the presence of peroxidase, oxidizes the substrate shown in Reaction 8. The absorbance of the dye is measured at 505 nm and is proportional to the triglyceride concentration. In all these methods, a glycerol blank is advisable.

Reaction 7 Glycerol-3-P + O_2 $\xrightarrow{GPO}$ Dihydroxyacetone phosphate + H_2O_2

Reaction 8 H_2O_2 + 4-chlorophenol + 4-aminophenazone $\xrightarrow{PO}$ HCl + 2 H_2O + 4-(*p*-benzoquinoneimino)-phenazone

Bilirubin interferes by reacting with the hydrogen peroxide, thereby giving a lower result for triglyceride because less dye is formed. Naito[11] and colleagues minimized the bilirubin interference by adding potassium ferrocyanide to the reagent mixture in Reaction 8. The bilirubin reacts with potassium ferrocyanide instead of the hydrogen peroxide. Bilirubin is reduced to an inactive compound as the ferrocyanide is oxidized to ferricyanide.

Reference Values. As with cholesterol, the reference values for triglycerides in Western industrialized nations are skewed toward the high side and do not necessarily reflect a healthy or desirable state. The reference range is broad and varies with age and gender. The Consensus Conference on Triglyceridemia[18] concluded that a mild elevation (less than 250 mg/dL or 2.8 mmol/L) in plasma triglyceride concentration *alone* generally does not increase the risk for coronary artery disease as long as the cholesterol concentration remains within safe, recommended limits. The Consensus Conference also concluded that triglyceride levels between 250 and 500 mg/dL (2.8 and 5.7 mmol/L) should be considered as possible markers for the presence of some genetic form of hyperlipoproteinemia that might require diet therapy. Patients with concentrations greater than 500 mg/dL (5.7 mmol/L) should be treated vigorously because of the increased risk for pancreatitis (Table 8.4).

Increased Concentration. The concentration of serum triglycerides is moderately elevated after a meal containing fat and may rise as high as 260 mg/dL (3 mmol/L) after a meal containing 50 g fat (for example, a large hamburger and French fries). The peak of the triglyceride elevation occurs in 4 to 5 hr postprandial.

The plasma triglyceride concentration is high in the hyperlipemias in which the concentrations of chylomicrons and/or VLDL (type I and type V) are greatly elevated, is less so in type IV, and is only moderately raised in types IIb and III (Fig. 8.10). Hypertriglyceridemia commonly occurs secondary to the following pathologic conditions: hypothyroidism, nephrotic syndrome, acute alcoholism, obstructive liver disease, acute pancreatitis, uncontrolled diabetes, glycogen storage disease (type I), and sometimes insulin-treated diabetes (Table 8.3). The serum or plasma is usually turbid or milky when the triglycerides are elevated. The plasma triglycerides are mildly increased in familial an-α-lipoproteinemia (deficiency of HDL).

Conversion to SI: Multiply mg/dL by 0.0113 to convert to mmol/L; for calculation purposes, the triglyceride is assumed to be triolein, which has a molecular weight of 885.

Decreased Concentration. The plasma triglyceride concentration is low in the rare disease a-β-lipoproteinemia (absence of LDL) and seldom exceeds 15 mg/dL (0.17 mmol/L).

Colorimetric and Fluorometric Methods[17]

Principle. Serum triglycerides are extracted with isopropanol in the presence of an alumina adsorbent mixture that removes phospholipids, monoglycerides, and diglycerides, glucose, bilirubin, and other interfering substances. Triglycerides in the extract are saponified to glycerol and soaps of the fatty acids or transesterified to yield glycerol and ethyl esters of the fatty acids as shown in Reactions 9a and 9b. The glycerol is oxidized to formaldehyde by means of periodate, yielding 2 moles of formaldehyde per mole of glycerol (Reaction 10). The formaldehyde is determined by the Hantzsch condensation of formaldehyde with NH_3 and acetylacetone (Reaction 11). The resulting diacetyl dihydrolutidine is yellow and is also fluorescent when activated with light at 400 nm. It may be determined colorimetrically as well as fluorometrically; the fluorometric method is more sensitive.

Reaction 9a Triglyceride + 3 KOH → glycerol + 3 K-soaps (K-salt of fatty acid)

Reaction 9b Triglyceride + 3 Na ethoxide → glycerol + 3 ethyl esters of FA

Reaction 10

$$\underset{\text{Glycerol}}{H_2C(OH){-}CH(OH){-}CH_2(OH)} + \underset{\text{Periodate}}{2\,HIO_4} \rightarrow \underset{\text{Formaldehyde}}{2\,H{-}\overset{H}{\overset{|}{C}}{=}O} + \underset{\text{Formic Acid}}{H{-}COOH} + \underset{\text{Iodate}}{2\,HIO_3} + H_2O$$

Reaction 11 Hantzsch Reaction:

$$\underset{\text{Formaldehyde}}{H{-}\overset{H}{\overset{|}{C}}{=}O} + \underset{\text{Acetylacetone}}{2\,CH_3{-}\overset{O}{\overset{\|}{C}}{-}CH_2{-}\overset{O}{\overset{\|}{C}}{-}CH_3} + \underset{\text{Ammonia}}{NH_3} \rightarrow \text{3.5-diacetyl-1.4-dihydrolutidine} + 3\,H_2O$$

(Product structure: 1,4-dihydropyridine ring with CH_2 (H₂) at position 4, $CH_3{-}C(=O){-}$ groups at positions 3 and 5, CH_3 groups at positions 2 and 6, and N–H at position 1.)

3.5-diacetyl-1.4-dihydrolutidine

Reaction 12	Colorimetric:	Measure the absorbance at 412 nm.
	Fluorometric:	Measure the emitted light at 485 nm when the exciting wavelength is 400 nm.

Note: Reaction 1 (hydrolysis reaction) of the enzymatic method may be substituted for Reactions 9a and 9b.

Lipoprotein Electrophoresis[19]

The principal reason for requesting a lipoprotein electrophoresis is to ascertain the type of hyperlipoproteinemia in patients with hyperlipidemia. As pointed out previously, in many cases, a classification can be made by correlating the analytical values for total cholesterol and triglycerides with the appearance of the serum after standing overnight in a refrigerator. An electrophoresis pattern does help in difficult situations.

Principle. Lipoproteins carry an electric charge at pH 8.6 and can be separated by electrophoresis on agarose gel in a manner similar, to that for serum proteins. Chylomicrons are uncharged and remain at the origin. After electrophoretic separation, the lipoproteins are stained with a fat stain (for example, Fast Red 7B); the support medium is destained and then dried. The

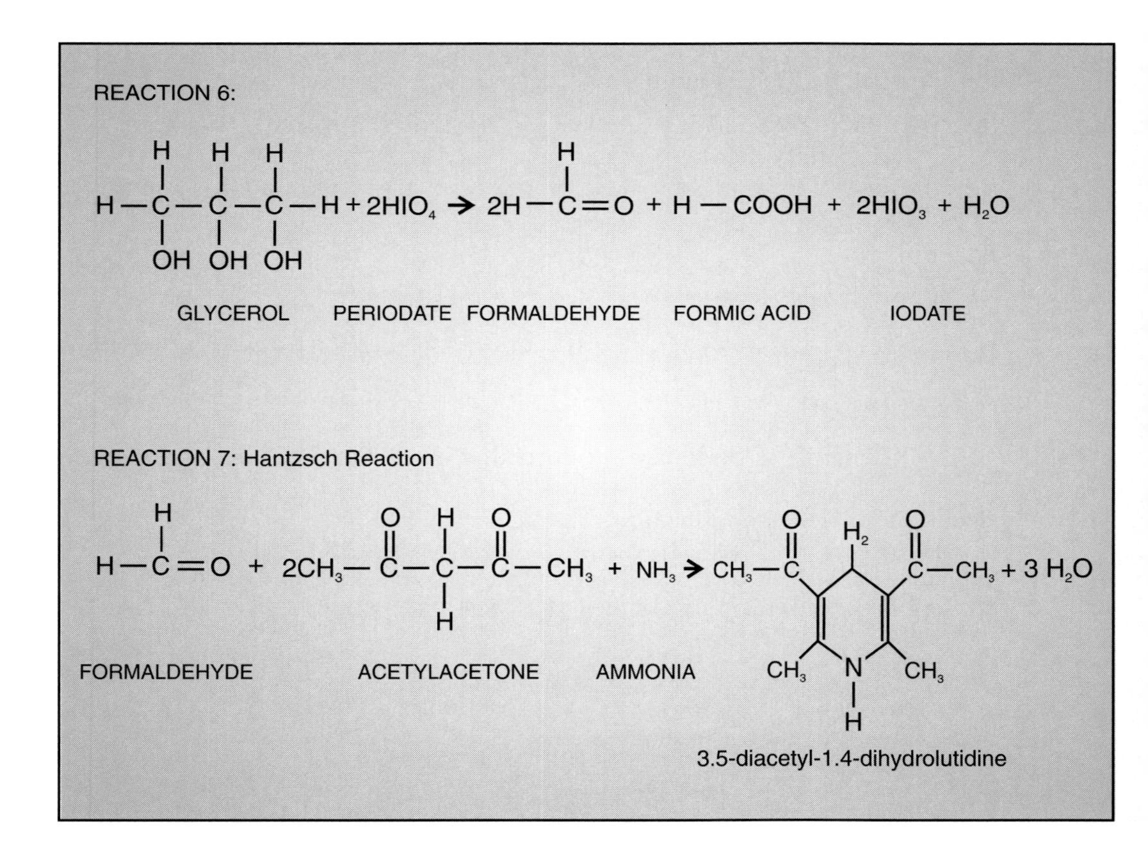

lipoprotein bands and chylomicrons are visible because of the fat stain dissolved in them. Abnormal bands may be seen by visual inspection or quantitated from a densitometric scan of the membrane.

Note: Serum samples should be obtained after an overnight fast to avoid postprandial hyperlipidemia. The lipid-protein bond is labile and easily ruptured by freezing. Samples should be assayed as soon as possible and kept in a refrigerator, not a freezer. On removal from the refrigerator, samples should be allowed to stand at room temperature for at least 30 min to allow for warm-up and reformation of any lipid-protein bond. Observe for turbidity and cream layer. Mix by inversion before taking sample for electrophoresis.

Interpretation. The interpretation is made from a combination of three sets of data: abnormal bands on the electrophoresis strip; the appearance of the serum, that is, whether clear, turbid, or milky, or whether a cream layer forms on standing; and analytical values for serum cholesterol and triglycerides. The interpretation is made from the correlation of these data with the criteria outlined in Figure 8.10.

REFERENCES

1. Mayes, P.A.: Lipids of physiologic significance. *In* Harper's Biochemistry. 23rd Edition. Edited by R.K. Murray, et al. Norwalk, Appleton and Lange, 1993.
2. Mayes, P.A.: Cholesterol synthesis, transport and excretion. *In* Harper's Biochemistry. 23rd Edition. Edited by R.K. Murray, et al. Norwalk, Appleton and Lange, 1993.
3. Mayes, P.A.: Lipid transport and storage. *In* Harper's Biochemistry. 23rd Edition. Edited by R.K. Murray, et al. Norwalk, Appleton and Lange, 1993.
4. Havel, R.J.: Introduction: structure and metabolism of plasma lipoproteins. *In* The Metabolic Basis of Inherited Disease. Vol. 1. 6th Edition. Edited by C.R. Scriver, et al. New York, McGraw-Hill, 1989.
5. Kane, J.P., and Havel, R.J.: Disorders of the biogenesis and secretion of the lipoproteins containing the B apolipoproteins. *In* The Metabolic Basis of Inherited Disease. Vol. 1. 6th Edition. Edited by C.R. Scriver, et al. New York, McGraw-Hill, 1989.
6. Brunzell, J.D.: Familial lipoprotein lipase deficiency and other causes of chylomicronemia syndrome. *In* The Metabolic Basis of Inherited Disease. Vol. 1. 6th Edition. Edited by C.R. Scriver, et al. New York, McGraw-Hill, 1989.
7. Havel, R.J.: Approach to the patient with hyperlipidemia. Med. Clin. North Am., *66*:319, 1982.
8. Allain, C.C., Poon, L.S., Chan, C.S.G., et al.: Enzymatic determination of total serum cholesterol. Clin. Chem., *20*:470, 1974.

8A. Lipid Research Clinics Population Studies Data Book, Vol. 1, The prevalence study. Washington, D.C., Dept. of Health and Human Studies, July 1980, pp. 28–81. NIH publication no. 80-1527.

9. National Cholesterol Education Program: Report of the panel on detection, evaluation and treatment of high blood cholesterol in adults. Arch. Intern. Med., *148*:36, 1988.
10. Lipid Research Clinics Program. The Lipid Research Clinics Prevention Trial Results. I. Reduction in incidence of coronary heart disease. JAMA, *251*:351, 1984. II. The relationship of reduction in incidence of coronary heart disease to cholesterol lowering. JAMA, *251*:351, 1984.
11. Naito, H.K., Gatautis, V.J., and Galen, R.S.: The measurement of serum triacylglycerol on the Hitachi 705 chemistry analyzer. Clin. Chem., *31*:948, 1985.
12. Bucolo, G., and David, H.: Quantitative determination of serum triglycerides by the use of enzymes. Clin. Chem., *19*:476, 1973.
13. Rietz, E.B., and Guilbault, G.G.: Fluorometric estimation of triglycerides in serum by a modification of the method of Bucolo and David. Clin. Chem., *23*:286, 1977.

14. Megraw, R.E., Dunn, D.E., and Biggs, H.G.: Manual and continuous flow colorimetry of triglycerides by a fully enzymatic method. Clin. Chem., *25*:273, 1979.
15. Nagele, U., Hagele, E.O., Sauer, G., et al.: Reagent for the enzymatic determination of serum total triglycerides with improved lipolytic efficiency. J. Clin. Chem. Clin. Biochem., *22*:165, 1984.
16. Klotzsch, S.G., and McNamara, J.R.: Triglyceride measurements: A review of methods and interferences. Clin. Chem., *36*:1605, 1990.
17. Soloni, F.G.: Simplified manual micromethod for determination of serum triglycerides. Clin. Chem., *17*:529, 1971.
18. Consensus Conference: Treatment of hypertriglyceridemia. JAMA, *251*:1196, 1984.
19. Noble, R.P.: Electrophoretic separation of plasma lipoproteins in agarose gel. J. Lipid Res., *9*:593, 1968.

REVIEW QUESTIONS

1. What are the principal lipid constituents of the body and their distinguishing features?
2. What body mechanisms tend to maintain a relatively constant cholesterol concentration? Are these mechanisms always successful?
3. What are the composition and function of the different classes of lipoproteins?
4. What is an apolipoprotein and what are its functions? Which apolipoproteins are cofactors for enzymes active in lipid metabolism?
5. How are lipids transported in the exogenous pathway? In the endogenous pathway?
6. What are the distinguishing features of the five types of familial hyperlipoproteinemias?
7. What is the difference between a primary and secondary hyperlipemia?
8. What principles are involved in the determination of serum total cholesterol? Of serum HDL cholesterol?

Proteins 9

OUTLINE

- **I. AMINO ACIDS**
  - A. Structure
  - B. Properties
- **II. PROTEINS**
  - A. Formation
  - B. Plasma (serum) proteins
    - 1. Functions
    - 2. Classification
    - 3. Electrophoresis
    - 4. Electrophoretic separation
      - a. Albumin
      - b. α_1-globulin
      - c. α_2-globulin
      - d. β-globulin
      - e. γ-globulin
    - 5. Immunoglobulins
      - a. IgG
      - b. IgA
      - c. IgM
      - d. IgD and IgE
      - e. Monoclonal and polyclonal disorders
        - (1) Light-chain disease
        - (2) Heavy-chain disease
    - 6. Determination of serum total proteins
      - a. Biuret method
        - (1) Reference values
        - (2) Increased concentration
        - (3) Decreased concentration
      - b. Refractive index method
      - c. Fractionation by electrophoresis
        - (1) Fraction 1: albumin
          - (a) Reference values
          - (b) Increased concentration
          - (c) Decreased concentration
        - (2) Fraction 2: α_1-globulin
        - (3) Fraction 3: α_2-globulin
        - (4) Fraction 4: β-globulin
        - (5) Fraction 5: γ-globulin
    - 7. Determination of individual serum proteins
      - a. Serum albumin
      - b. Prealbumin
    - 8. Determination of acute phase reactants
  - C. Cerebrospinal fluid proteins
    - 1. Determination by turbidity
      - a. Increased concentration
      - b. Decreased concentration
    - 2. Fractionation by electrophoresis
    - 3. CSF/serum protein ratios
  - D. Urine protein
    - 1. Principle
    - 2. Reference values
    - 3. Increased excretion
    - 4. Urinary albumin
  - E. Transudates and exudates

OBJECTIVES

After reading this chapter, the student will be able to:

1. Understand the importance of the three-dimensional structure of proteins.
2. List five specific serum proteins and their physiologic functions.
3. Understand the clinical utility of specific protein measurement.
4. Describe the structure and function of the immunoglobulins IgG, IgM, and IgA.
5. Interpret serum protein electrophoresis patterns.
6. Describe the significance of cerebrospinal fluid/serum protein ratios.
7. Understand the term oligoclonal banding.
8. Understand the term acute-phase reactant.

AMINO ACIDS[1]

Structure

Amino acids are the modules or building blocks from which all polypeptides and proteins are synthesized. Features common to all amino acids are a terminal carboxyl group (-COOH) and an adjacent amino group (-NH_2) on the alpha carbon atom (Fig. 9.1). The side chain, -R, is unique for each of the 20 naturally occurring amino acids. Proteins contain many, if not all, of these 20 amino acids in varying combinations. Humans can synthesize 11 of the amino acids, but 9 must be provided by dietary intake and hence are termed essential. The essential amino acids are histidine, isoleucine, leucine, lysine, methionine, phenylalanine, threonine, tryptophan, and valine. Our amino acids are derived primarily from dietary protein of vegetable or animal origin; the most common sources are grains, beans, meat, fish, poultry, dairy products, and eggs.

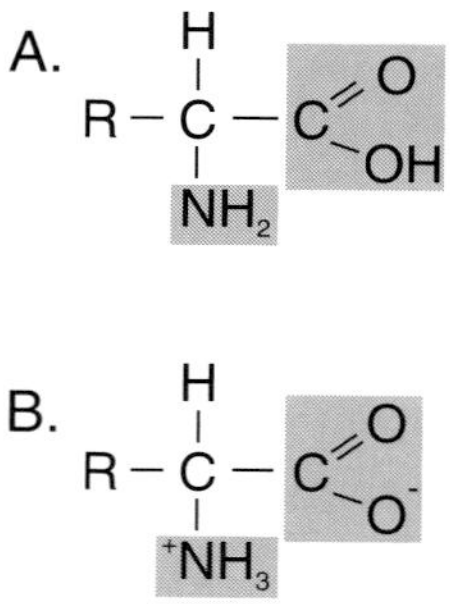

A. $R-C(H)(NH_2)-C(=O)OH$

B. $R-C(H)(^{+}NH_3)-C(=O)O^{-}$

C.

Amino Acid	Side Chain, R	Group
Serine	$HO-CH_2-$	Hydrophilic
Phenylalanine	$C_6H_5-CH_2-$	Hydrophobic
Glutamic	$^{-}OOCCH_2CH_2-$	Acidic
Lysine	$H_3^{+}N-CH_2-CH_2-CH_2-CH_2-$	Basic
Cysteine	$H-S-CH_2-$	Sulfhydryl (Hydrophilic)

FIG. 9.1. General structure of amino acids. All contain the terminal carboxyl and adjacent α-amino group. The uncharged form appears in A and its ionized form in plasma is shown in B. One example of the different groups of amino acids is illustrated in C.

Properties

Amino acids are amphoteric, meaning that they contain both a proton-accepting group (-NH_2) and a proton-donating group (-COOH). Figure 9.1 illustrates the ionized form and charged terminal groups of amino acids in plasma. The R-groups may also contribute charges and versatility to the combining actions of amino acids. Glutamic and aspartic acids are dicarboxylic; thus, peptides containing them are acidic and have a negative charge contributed by the second carboxyl group (Fig. 9.1). Lysine, arginine, and histidine are basic amino acids, and peptides with these amino acids carry a positive charge derived from the second amino group (lysine and arginine) or from the imino group of histidine. The R-group of cysteine contains a terminal sulfhydryl group (-SH) that can react with another sulfhydryl to form a disulfide bridge (-S-S-). Many of the R-groups are hydrophobic (nonpolar), whereas all those with a hydroxyl group are hydrophilic (polar, but not charged). An example of each group is shown in Figure 9.1.

PROTEINS

Formation

Proteins are built by the sequential condensation of a terminal amino group of one amino acid with the carboxyl group of another amino acid to form peptide bonds (Fig. 9.2). The synthesis takes place on cell ribosomes by means of information stored in coded form in the genes (DNA) and transmitted by messenger RNA (see Chap. 16).

A protein assumes a three-dimensional structure that is determined by the interaction between the various R-groups and charges on the amino acids in the polypeptide chain and the interlinked chains of a multichain protein. The interactions may include covalent bonding by two SH-groups to form a disulfide bridge (-S-S-); attraction of unlike charged groups to each other; grouping together of polar or nonpolar groups; hydrogen bonding that arises when two electronegative atoms, such as N or O, share an H atom between

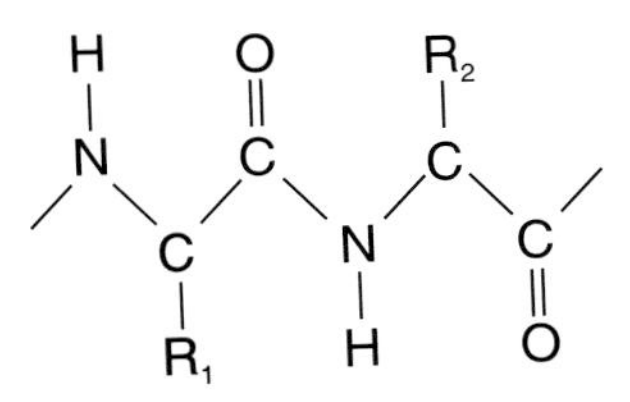

FIG. 9.2. Linkage of amino acids by means of peptide bonds to form a polypeptide chain. The R side chains may carry positive charges, negative charges, or none, depending on the polarity of the group.

them; and attraction among weak, electrostatic, intermolecular forces (van der Waals forces). Except for structural proteins, most proteins assume a globular shape in which the hydrophobic R-groups are brought together within the folded protein, and the hydrophilic amino acids are grouped on the outer surface. The function of a protein depends on the sequence of amino acids, which determines its shape and conformation.

The code provides directions for the sequential step-by-step assembly of amino acids into polypeptide chains for every cell protein. A protein may consist of 1 or more intertwined polypeptide chains, some of which may contain 100 or more amino acids. For example, hemoglobin, the principal protein in the red blood cell, is composed of 2 sets of 2 different polypeptide chains.

Plasma (Serum) Proteins

Proteins are present in all body fluids, but the protein concentration is normally high (> 3g/dL) only in blood plasma, lymphatic fluid, and some exudates. Protein concentration in the cerebrospinal fluid of normal subjects is < 45 mg/dL, whereas the urine contains only a trace.

Concentration of total protein in serum usually ranges from 6.0 to 8.2 g/dL and is about 0.3 g/dL higher for plasma because of the presence of fibrinogen. The protein concentration cannot be expressed in mmol/L because serum is a mixture of so many proteins with differing molecular weights. The molecular weight may vary from 40,000 to 50,000 daltons for some of the smaller globulins to about 1 million daltons for the macroglobulins. Some of the lipoproteins may be as large as 10 to 20 million daltons. These proteins are large molecules that cannot pass through the membranes lining the blood vessels. Because water, other small molecules, and most ions can diffuse in both directions through the capillary walls, the presence of the proteins in the vascular bed draws water into the system and creates an osmotic pressure, usually referred to as oncotic pressure.

Functions

The plasma proteins have many functions, which may be summarized as follows:

1. They affect the distribution of extracellular fluid between the vascular bed and the interstitial fluid by means of the oncotic pressure they generate. This property is characteristic of all proteins, but the most important protein in this respect is albumin because of its relatively small size (69,000 daltons) and high concentration (about 60% of the total plasma proteins).
2. They serve as carriers for various cations and some compounds that are relatively insoluble in water, such as bilirubin, fatty acids, steroid hormones, and lipids.
3. Some function as antibodies to provide a defense system for the body

against foreign proteins, viruses, and bacteria. This role is reserved for the gamma globulins, as is reviewed later in this chapter.

4. Some form part of the endocrine system. Many of the hormones are proteins or polypeptides, which reach their target organ by circulating in plasma.
5. Some protect against damage to the vascular system by forming a complex blood-clotting system.
6. They provide tissues with a source of nutrients for building materials, or calories. Proteins can be catabolized for energy purposes or reused for structural or other purposes by contributing to the amino acid pool.
7. Some function as enzymes. Most plasma enzymes are derived from intracellular sources and appear in plasma during the course of ordinary breakdown and replacement of cells. Because this process is greatly increased during tissue necrosis, the plasma enzyme pattern may provide clues concerning a specific tissue or organ disorder in some disease states.

Classification

Proteins may be classified into different groups according to function (immunoglobulins, clotting factors, or enzymes), composition (glycoproteins, mucoproteins, lipoproteins), relative migration distance in an electrical field, or sedimentation rate when subjected to a centrifugal force. Although all these types of nomenclature are used, the most widely used general classification system is to name the protein groups after the five main categories or bands obtained on cellulose acetate or paper when separation takes place in an electrical field.

Electrophoresis

The technique for separating proteins by means of an electrical current is called electrophoresis and is described in greater detail in Chapter 3. At pH 8.6, all the serum proteins carry a negative charge to a greater or lesser extent. If a small serum sample is placed upon a wet (pH 8.6 buffer) support medium, such as cellulose acetate or agarose gel interposed between 2 electrodes immersed in pH 8.6 buffer, and subjected to an electrical charge of several hundred volts, the proteins move toward the positive pole (anode). The migration distance varies directly with the charge carried by the protein, but is modified to a slight extent by buffer flow.

The buffer flow (electro-osmotic or endosmotic flow) is caused by the negative charge of the support medium when the capillaries or pores become wet, thereby leaving the liquid medium (buffer) with a small positive charge. When the current is turned on, a flow or movement of buffer occurs toward the negative electrode (cathode), and this movement passively carries with it, to a small extent, the proteins that tend to migrate toward the anode because of their negative charge.

After separation on cellulose acetate or agarose gel by electrophoresis, protein bands are visualized by staining with a Coomassie blue type of stain.

Densitometry is used to quantitate the individual protein bands, but a visual inspection of the membrane or gel is advisable for interpretation.

Electrophoretic Separation

In a normal serum, five fractions are present: albumin, alpha one (α_1) globulin, alpha two (α_2) globulin, beta (β) globulin, and gamma (γ) globulin, starting with the greatest migration toward the anode (Fig. 9.8). Each fraction is composed of many individual proteins. Because of endosmotic flow, the gamma band is usually displaced from the point of application slightly toward the cathode. The slowest of the beta fraction (fibrinogen, if plasma is used) is usually at the point of application, and all faster fractions move toward the anode.

Albumin

Serum albumin consists of a single species of protein and is the most prominent band in an electropherogram. Serum albumin constitutes approximately 60% of the total serum protein and migrates the farthest toward the anode. It is synthesized exclusively in the liver and functions as a regulator of blood oncotic pressure, as a carrier for many cations and water-insoluble substances (calcium, bilirubin, fatty acids, therapeutic drugs), and as a pool of amino acids for caloric or synthetic purposes. It has a half-life in plasma of about 17 days; thus, its plasma concentration would fall about 3% per day if synthesis were completely halted. Hepatocellular damage usually results in a decrease in the serum albumin concentration, but the change is relatively slow. The concentration of serum albumin is decreased in the following situations:

1. Decreased synthesis, whether caused by damaged hepatic cells, deficient protein intake, as in malnutrition or starvation, or impaired digestion or absorption of protein products (sprue).
2. Extensive protein loss, whether through the kidneys, as in the nephrotic syndrome, through the skin following extensive burns or severe skin lesions, as in exfoliative dermatitis, or through the gastrointestinal tract, as in protein-losing enteropathies (protein-losing intestinal diseases).
3. Shift to ascitic fluid, which may happen in chronic liver disease with cirrhosis. Ascites, the accumulation of fluid in the peritoneal cavity, is frequently caused by portal hypertension (increased portal vein blood pressure in the liver) and a lowered plasma oncotic pressure (lowered albumin concentration), by obstruction to hepatic vein flow, or by a combination of factors occurring in hepatocellular disease.

The remaining four bands in serum electrophoresis are globulins. The globulins are less soluble than albumin in various salt solutions and formerly were separated as a single group from albumin by precipitation with a concentrated Na_2SO_4 solution. Much more clinical information is obtained by an electrophoretic separation, however.

α_1-Globulin

The second fastest band (the band adjacent to albumin but closer to the cathode) is the α_1-globulin fraction. It is a mixture of many proteins, some of which have been identified and characterized. Many in this group are combined with carbohydrate and are called glycoproteins; others exist in combination with lipids and are named α_1-lipoproteins. Some of the prominent individual proteins in the α_1-band are α_1-antitrypsin, α_1-acid glycoprotein, and thyroxine-binding globulin. In general, the α_1-globulin band increases as a nonspecific response to inflammation (acute-phase reactants) arising from various causes, such as infection, trauma, and neoplasms. The α_1-globulins are synthesized in the liver.

α_2-Globulin

The third protein band visible on cellulose acetate electrophoresis is composed of the α_2-globulins. This fraction also contains some α_2-lipoproteins. Some of the well-known proteins in this fraction include haptoglobin (a hemoglobin-binding protein), α_2-macroglobulin, and ceruloplasmin (a copper-binding protein that has oxidase activity). The mucoproteins in the α_2-fraction are also increased in inflammatory conditions.

β-Globulin

The fourth band is the β-globulin fraction, which contains the β-lipoproteins, the iron-transporting protein (transferrin), fibrinogen, and other lesser known proteins. The newer high-resolution agarose gels actually separate the β-lipoproteins out from the others, thereby giving two bands in the β-region. Any condition that increases the β-lipoproteins causes this band to become more prominent because the lipoproteins are predominant in this fraction.

γ-Globulin

The slowest band is the γ-globulin fraction, which contains the immunoglobulins, or circulating antibodies, essential for defense against foreign proteins of all sorts. These proteins are synthesized by cells of the reticuloendothelial system—plasma cells and lymphocytes.

Immunoglobulins

The five known classes of immunoglobulins are IgG, IgA, IgM, IgD, and IgE.[2] These proteins are related structurally in that each member of a group, except IgM and secretory IgA, is a monomer composed of two light and two heavy polypeptide chains linked together by disulfide bonds. IgM is a pentamer consisting of five such monomeric units, and secretory IgA is a dimer composed of two monomers. The same two light chains, designated as kappa (κ) and lambda (λ) chains, are used in all the five classes of immunoglobulins, but the heavy chains are unique and distinctive for each class. Both light chains are the same for any particular molecule, but each class of immunoglobulin contains a mixture of κ and λ light chains. Each monomer has two sites for binding antigen, one at each end of a polypeptide chain where the light and heavy chains lie side by side (Fig. 9.3).

IgG

The IgG class (Fig. 9.4) is present in the highest concentration in serum and comprises about 75 to 80% of the γ-globulin fraction. Most of the circulating

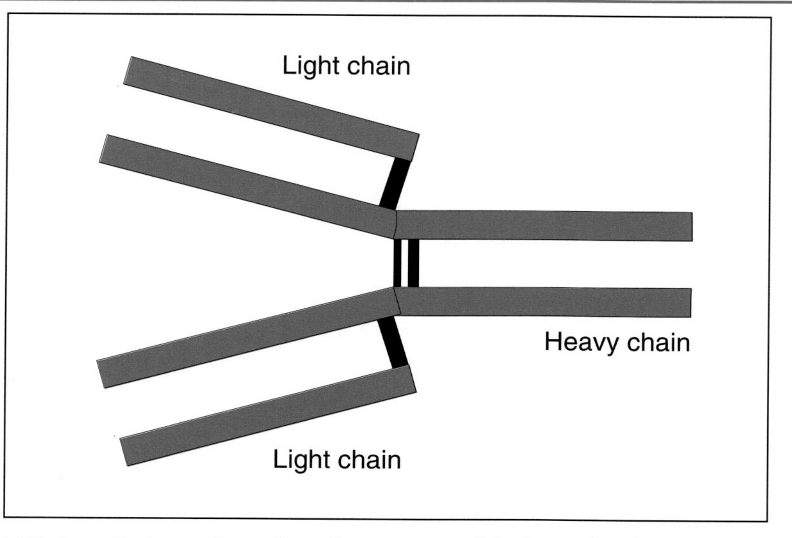

FIG. 9.3. Basic configuration of an immunoglobulin molecule. (Modified from Turgeon, M.L.: Fundamentals of Immunohematology. Philadelphia, Lea & Febiger 1989.)

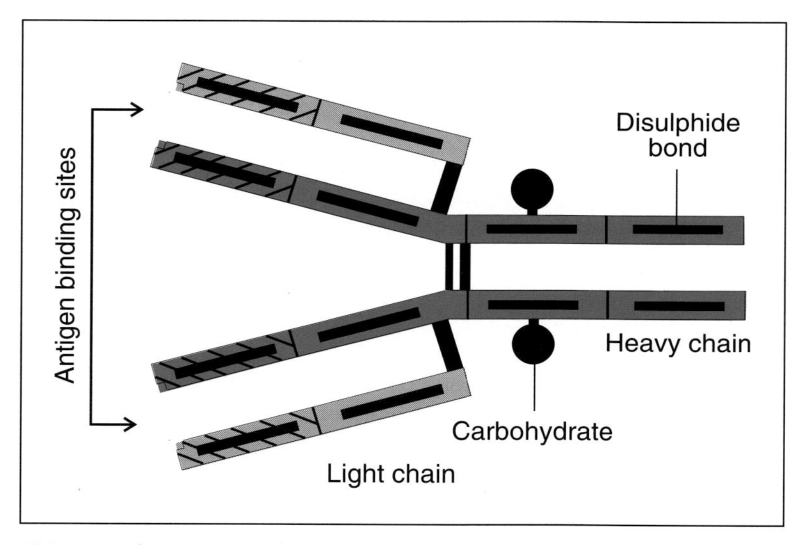

FIG. 9.4. The structure of an IgG molecule. (Modified from Turgeon, M.L.: Fundamentals of Immunohematology. Philadelphia, Lea & Febiger 1989.)

antibodies belong to this class. IgG is composed of 4 subclasses, IgG_1 (70%), IgG_2 (20%), IgG_3 (6%), and IgG_4 (4%), which differ somewhat in amino acid composition. IgG_1 predominates and is the maternal immunoglobulin found in the highest concentration in the newborn. Production of all the IgGs is stimulated by the invasion of bacteria or viruses; the antibodies are particularly effective in coating bacterial toxins. IgG_1 and IgG_3 bind most avidly to and activate complement (the lytic proteinases and esterases that actually destroy

TABLE 9.1
CLASSES OF IMMUNOGLOBULINS

IMMUNO-GLOBULIN	MOLECULAR FORMULA	PHYSICAL STATE	VALENCY (NO. OF ANTIGEN BINDING SITES)	MOLECULAR WEIGHT (DALTONS)	ADULT SERUM CONCENTRATION (mg/dL)
IgG	$\gamma_2 \kappa_2$ $\gamma_2 \lambda_2$	Monomer	2	160,000	1,100
IgA	$\alpha_2 \kappa_2$ $\alpha_2 \lambda_2$	Monomer	2	170,000	230
	$\alpha_4 \kappa_4$ $\alpha_4 \lambda_4$	Dimer	4	385,000	
IgM	$\mu_{10} \kappa_{10}$ $\mu_{10} \lambda_{10}$	Pentamer	10	900,000	190
IgD	$\delta_2 \kappa_2$ $\delta_2 \lambda_2$	Monomer	2	184,000	4
IgE	$\epsilon_2 \kappa_2$ $\epsilon_2 \lambda_2$	Monomer	2	188,000	0.03

attacked bacteria and cells by disrupting areas of the cell membrane). The attachment of IgG molecules to a bacterial surface, however, does not kill the invader. IgG molecules serve instead as places for attachment of phagocytic cells, which perform the killing, or for binding and activating complement, which through its enzymatic actions facilitates the destruction. Activated complement also attracts macrophages to the scene of action by chemotaxis and thereby stimulates phagocytosis.

The IgG class of immunoglobulins consists of humoral antibodies that also appear in all the extravascular compartments. They cross the placenta and confer passive immunity upon the fetus; 4 to 6 months must pass before a newborn has matured sufficiently to make an adequate amount of IgG antibodies. Some of the properties, composition, and serum concentrations of each of the immunoglobulin classes are shown in Table 9.1. They are all glycoproteins that contain from 3 to 12% carbohydrate.

IgA

The IgA class of immunoglobulins (Fig. 9.5) protect mucous membrane surfaces from bacterial and viral attack. They are synthesized in plasma cells below the mucosal layer. Humoral IgA, like IgG, is a monomer, but about 20% of IgA is a dimer that appears in various fluids and secretions (colostrum, milk, saliva, tears, and sweat). The two monomers are joined by a small glycopeptide called the J chain. The dimer, secretory IgA, helps to maintain the integrity of mucosal surfaces and transfers some passive immunity to nursing infants. IgA comprises about 10 to 15% of the circulating immunoglobulins.

IgM

The IgM class (Fig. 9.6) is the largest in molecular size because each molecule is composed of five monomeric subunits, each of which is joined to a central J piece (a glycopeptide, as in IgA). A monomeric subunit resembles IgG or humoral IgA in that it consists of two light chains (κ or λ) and two heavy chains

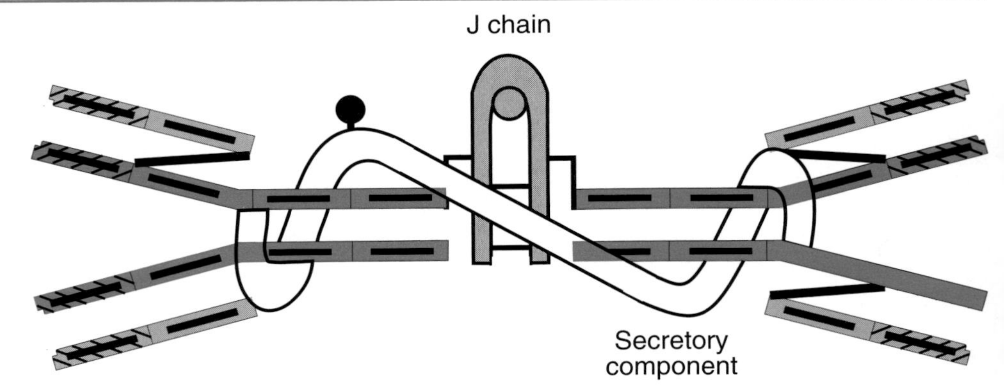

FIG. 9.5. The structure of a secretory IgA molecule. (Modified from Turgeon, M.L.: Fundamentals of Immunohematology. Philadelphia, Lea & Febiger 1989.)

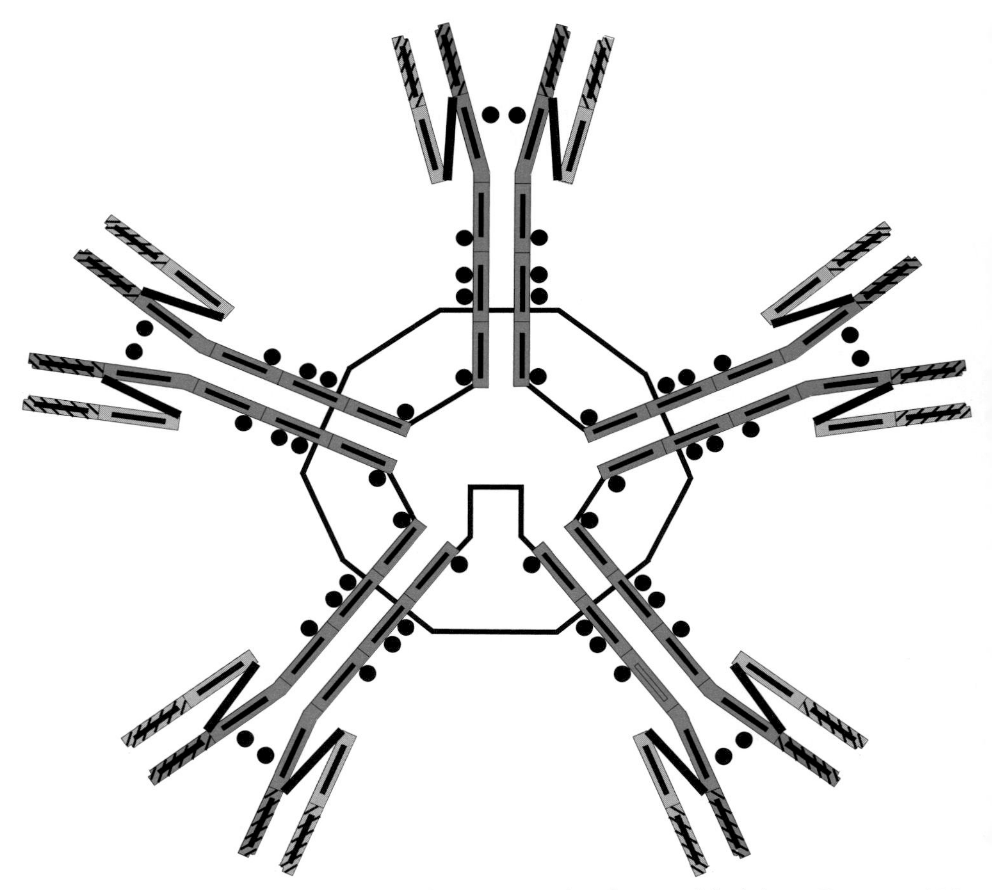

FIG. 9.6. The pentameric structure of an IgM molecule. (Modified from Turgeon, M.L.: Fundamentals of Immunohematology. Philadelphia, Lea & Febiger 1989.)

TABLE 9.2
REFERENCE VALUES* FOR SERUM IMMUNOGLOBULINS IN VARIOUS AGE GROUPS (mg/dL)

AGE	IgG RANGE	IgA RANGE	IgM RANGE
Cord serum	775–1665	0.05–9	4–25
0.5–3 months	305–835	3.5–6.3	16–125
3–6	145–970	5–86	19–114
6–12	425–1120	15–90	45–215
1–2 years	360–1185	14–113	38–232
2–3	500–1245	24–131	51–197
3–6	575–1355	36–199	53–207
6–9	670–1505	30–271	52–220
9–12	645–1570	63–282	67–268
12–16	695–1520	85–242	47–247

*Modified from Hicks, J.M., and Boeckx, R.L.: Pediatric Clinical Chemistry. Philadelphia, W.B. Saunders, 1984.

that are unique to its class (Table 9.1). Although IgM synthesis begins late in fetal life, it is the first of the immunoglobulins to be formed. It reaches adult levels at about age 6 months. IgM in an adult constitutes about 5 to 10% of the circulating immunoglobulins, and its increase is rapidly stimulated by the presence of foreign particulate bodies, such as bacteria. Like IgG, IgM can bind and activate complement. The change in concentration of the immunoglobulins with age is shown in both Table 9.2 and Fig. 9.7.

IgD and IgE

Because only IgG, IgA, and IgM are present to any appreciable extent in normal serum, they are the immunoglobulins usually measured in the laboratory. IgD and IgE are normally present in such small amounts that they can be measured only by sensitive methods (Table 9.1) IgE binds so firmly and selectively to mast cells that little of this antibody circulates in plasma. IgD and IgE were first detected in the serum of patients with multiple myeloma, a disease in which one class of immunoglobulins is greatly elevated.

Monoclonal and Polyclonal Disorders

The plasma concentration of the gamma globulins as a group usually rises in chronic infection with an increase in IgG, IgA, and IgM. In some diseases, however, the increase may be in only one class of the immunoglobulins. These disorders are called monoclonal diseases because a single class of immunoglobulin is synthesized in excess by a proliferation of plasma cells. The cell proliferation is believed to arise from a single clone that results in a mass of cells all producing the same class of immunoglobulin. An abnormal (monoclonal) immunoglobulin is called a paraprotein (Figs. 9.8 and 9.9). The most common of the monoclonal diseases is *multiple myeloma (myelomatosis)*, a malignant proliferation of plasma cells that results in an abnormally high concentration of serum immunoglobulins, usually IgG or IgA. *Waldenström's macroglobulinemia* is a less common monoclonal malignant disorder characterized by a high serum concentration of IgM. *Cryoglobulinemia* is also a monoclonal disease in which plasma or serum proteins (IgM) precipitate when

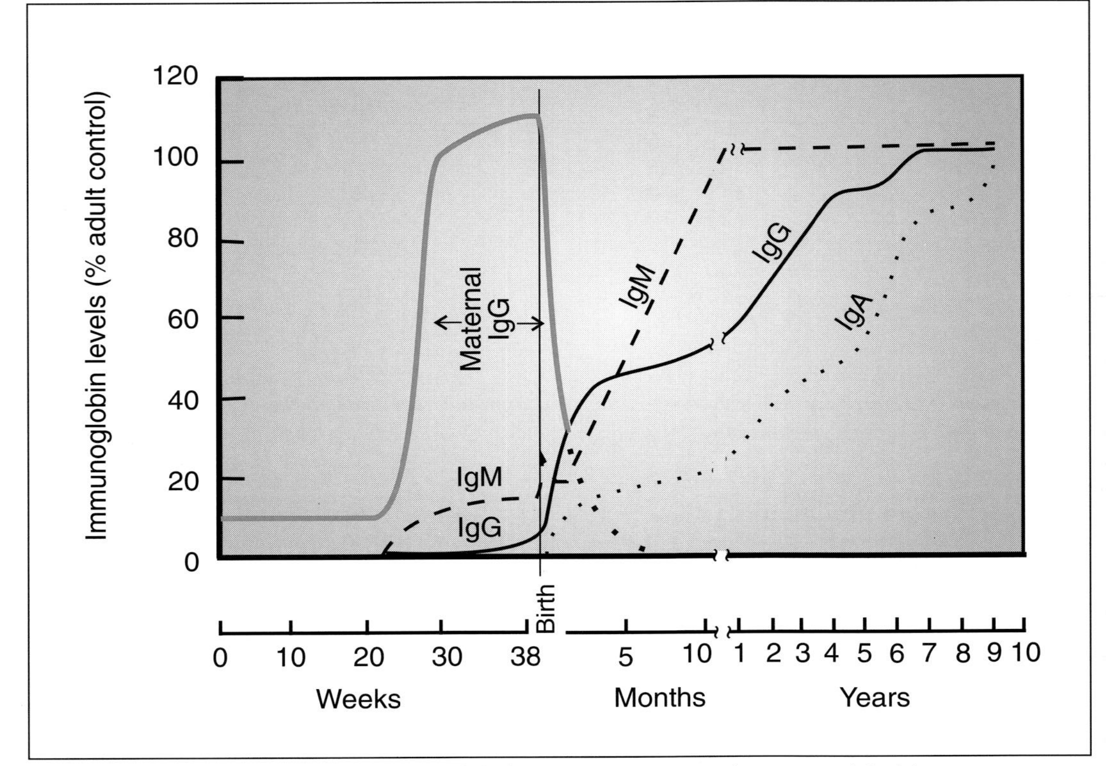

FIG. 9.7. Changes in immunoglobulin concentrations with age. (Modified from Bauer, J.D.: Clinical Laboratory Methods. 9th Edition. St. Louis, CV Mosby 1982.

cooled below body temperature. Some cases of *lymphomatous disease* may be classified as monoclonal as well.

Diseases accompanied by an increased concentration of two or more classes of immunoglobulins are called polyclonal disorders. Polyclonal gammopathies are often seen in viral hepatitis, sarcoidosis, rheumatoid arthritis, chronic infection, and other disease states.

Light-Chain Disease. When immunoglobulins are synthesized in the plasma cells, the heavy-chain polypeptides are usually assembled on a different ribosome from those of the light chain. The process operates synchronously as two light chains are attached to two heavy chains to form the finished immunoglobulin. In multiple myeloma, more light chains than heavy chains are produced and enter the bloodstream; they are of relatively low molecular weight and hence pass through the glomerular membrane and appear in the urine. These protein chains of low molecular weight are known as *Bence Jones proteins* and have peculiar solubility properties that were first reported by Dr. Henry Bence Jones. The Bence Jones proteins precipitate when the urine is heated from 45 to 60° C and redissolve when the heating is continued above 80° C. This test is not reliable because too many false-negative results appear when the protein concentration is low. The presence of other protein obscures

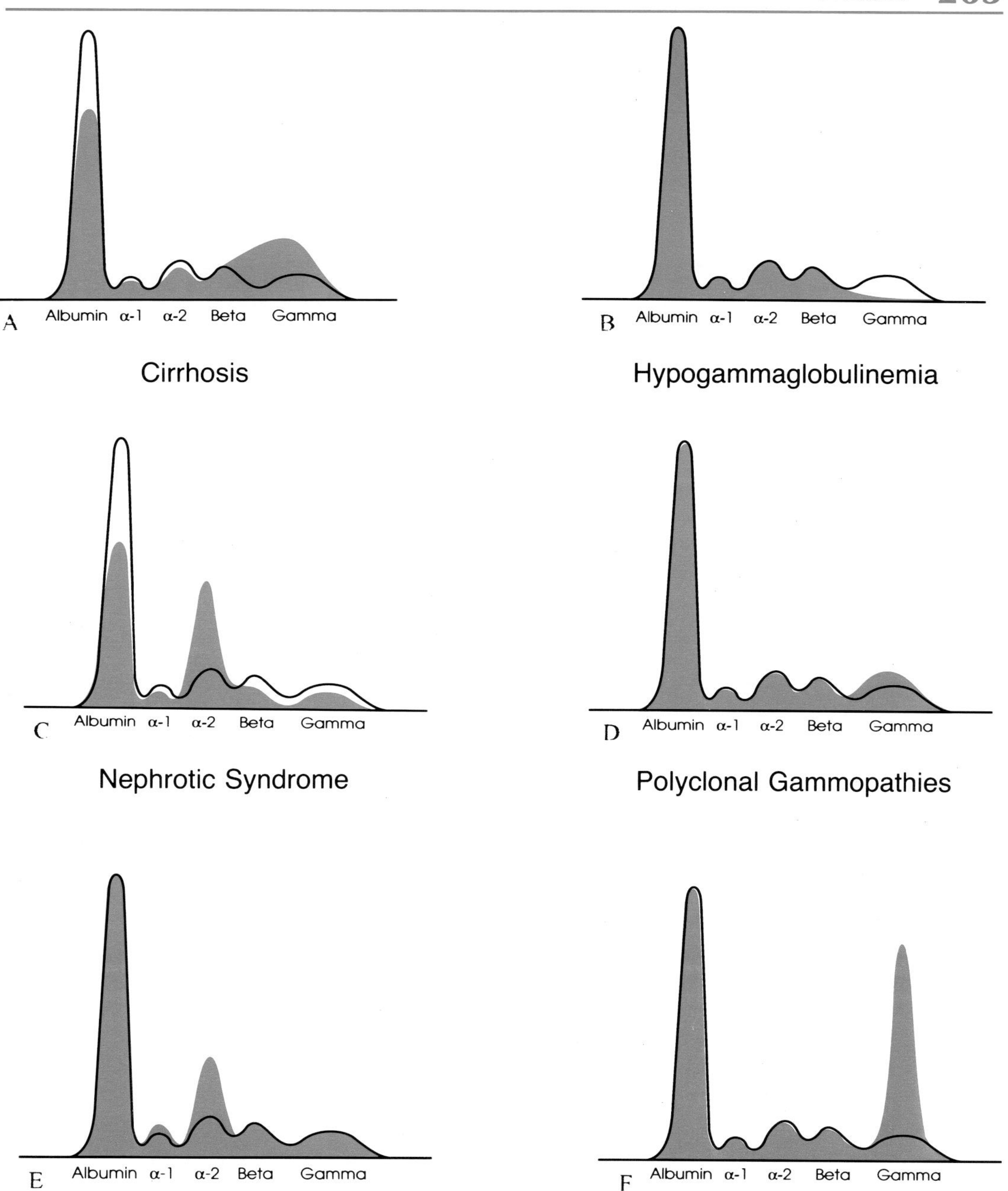

FIG. 9.8. Typical serum electrophoretic patterns found in disease (blue) and in the normal state (black). A, Cirrhosis of the liver. B, Hypogammaglobulinemia. C, Nephrotic syndrome. D, Polyclonal gammopathy. E, Acute inflammation. F, Monoclonal gammopathy (Illustration courtesy of Helena Laboratories.)

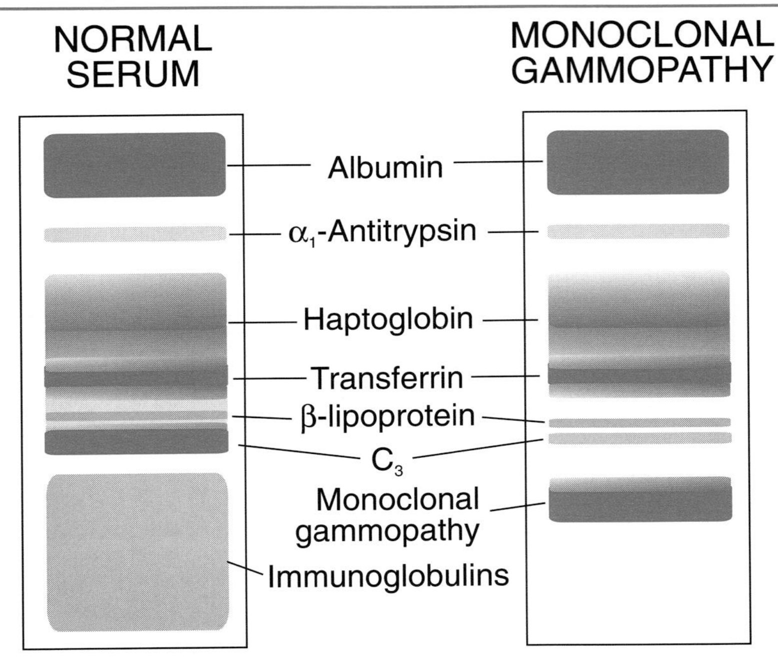

FIG. 9.9. Protein electrophoresis gels from normal serum and a patient's serum with a monoclonal gammopathy. (Illustration adapted from Killingsworth, L.M.: High Resolution Protein Electrophoresis; A Clinical Overview with Case Studies. Beaumont, TX, Helena Laboratories, 1985.[14])

the test. A myelomatosis with Bence Jones protein in the urine is now called "light-chain disease." The chain may be of either the κ or the λ type. This finding can be overlooked in a routine urinalysis because the urine dipstick test for protein is not sensitive to κ or λ chains. The presence of light chains in the urine in multiple myeloma is often the first indicator of glomerular damage caused by a buildup of amyloid protein in the kidney, and not necessarily by the light chains themselves.

A more definitive test for Bence Jones protein is double diffusion in agar gels containing antibodies to κ and λ chains (see Chap. 4). A precipitation line in the gel occurs when the antigen (Bence Jones protein in the urine sample) diffuses out of the sample well in sufficient quantity to react and precipitate with the specific antibody. This test not only reacts positively with Bence Jones protein, but identifies the type of chain. Immunoelectrophoresis and immunofixation are being used with increasing frequency in the clinical laboratory to identify light chains in urine. Immunonephelometric methods are also available for rapid quantitation.

A paraprotein in urine can also be detected by protein electrophoresis if the urine is concentrated to a protein concentration of 2 to 6 g/dL. The urine protein(s) are separated by electrophoresis and stained in the same manner as serum or cerebrospinal fluid proteins. The presence of a Bence Jones protein is demonstrated as a distinct band on the membrane; when a graph of the absorbance is made with a densitometer, it appears as a sharp spike somewhere in the globulin region.

Heavy-Chain Disease. The production of more heavy chains than light chains also can occur, but this reverse situation happens much less frequently. Because the heavy chains are of a sufficiently high molecular weight to be retained by the intact glomerular membrane, a kidney lesion, in addition to heavy-chain overproduction, is required before the proteins are present in the urine.

Determination of Serum Total Proteins[3]

The serum total protein, as its name implies, represents the sum total of numerous different proteins, many of which vary independently of each other. The total protein concentration must be measured when performing an electrophoresis to calculate the concentration of each protein fraction from its percentage. Aside from this situation, the determination of total protein supplies limited information except in conditions relating to changes in plasma or fluid volume, such as shock, dehydration, possible overhydration, and hemorrhage. The need for fluids is revealed by an elevated serum protein concentration that shows hemoconcentration. Measurement of the total serum protein level is also useful when determining the calcium concentration so that one can more accurately assess free or available calcium, which is not protein bound.

Biuret Method

Many colorimetric or dye-binding methods are used for the measurement of protein. One of the most common is the biuret method in which Cu^{2+} reacts in alkaline solution with the peptide linkages of proteins to form a violet-colored complex. The intensity of the color produced is proportional to the protein concentration. A structure containing at least two peptide linkages $(\text{-CO-NH-})_2$ is required to form the complex; hence, free amino acids do not react. Alkaline copper reagent forms a similar complex with biuret ($NH_2\text{-CO-NH-CO-}NH_2$), from which the reaction takes its name.

Reference Values. The concentration of serum total proteins in normal adults ranges from 6.0 to 8.2 g/dL. The values in plasma are about 0.2 to 0.4 g/dL higher because of the presence of fibrinogen. The values are about 0.5 g/dL higher in an ambulatory person than in one at bed rest because the erect position produces a pooling of fluid (extravasation of protein-free fluid from the blood vessels to interstitial fluid) in the lower extremities. The serum concentrations in infants for the first 3 or 4 months of life are about 1.0 g/dL lower than those in adults.

Increased Concentration. The total protein concentration of serum is usually increased in patients with dehydration, monoclonal disease (multiple myeloma, macroglobulinemia, cryoglobulinemia), and some chronic polyclonal diseases (liver cirrhosis, sarcoidosis, systemic lupus erythematosus, chronic infections).

Decreased Concentration. The serum total protein concentration is decreased in inadvertent overhydration; in conditions involving protein loss through the kidneys (nephrotic syndrome), skin (severe burns), or gut (protein-losing enteropathies); or in failure of protein synthesis (starvation, protein malnutrition, severe nonviral liver cell damage).

Refractive Index Method

The refractive index of an aqueous solution increases directly with an increase in solute mass concentration. Because the mass concentration of serum proteins is much higher than that of all the other serum constituents, measurement of the refractive index of a drop of serum provides a rapid way of estimating serum protein concentration. The refractometer should have a calibrated scale that gives a direct readout of protein concentration.[4]

Fractionation by Electrophoresis

An electrophoretic separation of serum proteins aids in the diagnosis of several clinical entities. The protein profiles (electrophoretic tracings) usually present distinctive patterns in the following diseases (Fig. 9.8): nephrotic syndrome, hepatocellular disease, hypogammaglobulinemia, and monoclonal paraprotein disease (multiple myeloma, Waldenström's macroglobulinemia).

The principles involved in the electrophoretic separation of proteins are reviewed in a previous section. Because procedural details vary with instrumentation and the particular composition of the cellulose acetate or agarose used as the support medium, none are given here. Several convenient and economical systems are available commercially (Beckman, Helena Laboratories, Corning, Gelman) that permit the separation of 8 microsamples (less than 0.5 μL each) simultaneously on a small membrane. The newer agarose gels have allowed for higher resolution and are preferable at this time to cellulose acetate.

The total protein (TP) concentration is always measured in performing an electrophoresis because it allows the concentration of the various fractions to be expressed as g/dL when TP is multiplied by the percent of that fraction.

$$\text{Protein fraction, g/dL} = \frac{\text{absorbance of specific protein band}}{\text{sum of absorbance of all bands}} \times \text{TP (g/dL)}$$

Reference Values. On separation by electrophoresis, the following concentrations in g/dL and percentages are usually found:

FRACTION	CONCENTRATION	PERCENT OF TOTAL
Albumin	3.5–5.2	50–65
α_1-globulin	0.1–0.4	2–6
α_2-globulin	0.5–1.0	6–13
β-globulin	0.6–1.2	8–15
γ-globulin	0.7–1.6	10–20

The actual concentrations are more important clinically than are the percentages because the concentrations give more information concerning an excess or deficit of a protein class. Electrophoretic patterns are illustrated in Figure 9.8. Separated bands of abnormal serums are shown in Figure 9.9

Fraction 1: Albumin.

Reference Values. The reference range of serum albumin varies from 3.5 to 5.2 g/dL. For about the first 4 months of life, the serum albumin concentration of most infants is about 10 to 12% lower than that of adults (Table 9.2). The discrepancy may be a little larger for premature infants.

Increased Concentration. Conditions of hyperalbuminemia are rarely seen except in the presence of acute dehydration or shock. An increase in albumin concentration is only temporary because interstitial water is drawn into the vascular bed by increased osmotic forces.

Decreased Concentration. The conditions that produce a lowered total protein concentration also lower the serum albumin concentration, as reviewed in the total protein section.

Fraction 2: α_1-*globulin.*

The α_1-globulin fraction is frequently increased when the serum albumin concentration falls, particularly in infections and inflammatory diseases. Its concentration is lowered in acute hepatitis and in familial α_1-antitrypsin deficiency, which is a cause of pulmonary emphysema.

Fraction 3: α_2-*globulin.*

The concentration of this fraction is greatly increased in the nephrotic syndrome owing to a large increase in α_2-macroglobulin, a protein too large to pass through the lesions in the glomerular basement membrane that occur in this disease. The α_2-fraction is frequently increased in inflammatory conditions, such as rheumatoid arthritis and lupus erythematosus, or after a myocardial infarct. In rare cases of multiple myeloma, the paraprotein may migrate in the α_2-band. The serum concentration of α_2-globulins is decreased in acute hepatocellular disease.

Fraction 4: β-*globulin.*

The β-globulin fraction is usually elevated in hyperlipemias of various types. Iron deficiency anemia may also result in an elevation in the β-region caused by transferrin. Sometimes the paraprotein of multiple myeloma migrates with the β-fraction and appears as a sharp spike in the electrophoretic pattern.

Fraction 5: γ-*globulin.*

As mentioned earlier, the paraproteins of the monoclonal diseases are abnormal γ-globulins. Each is unique and has an amino acid composition different from that of the normal immunoglobulins. Hence, they may migrate farther toward the anode than usual if amino acids with a greater negative charge should be substituted in the molecule. Most of the paraproteins migrate in the γ-fraction, some as a slow γ and some as a fast γ; a small percentage of the paraproteins may migrate in the β-band, and only rarely are they found in the α_2-fraction. In a monoclonal disease, such as multiple myeloma, the abnormal clones, whether of IgG or IgA, suppress the synthesis of the normal immunoglobulins. The *monoclonal* diseases exhibit a sharp spike in the electrophoretic pattern. The immunoglobulins may be elevated as a group in various *polyclonal* diseases and appear as a broad, dark band on the cellulose acetate or agarose membrane after staining, that is, the densitometer tracing appears as a broad hump. This pattern is seen in viral hepatitis, sarcoidosis, rheumatoid arthritis, chronic infections, and some leukemias and lymphomas. Some typical densitometry patterns are shown in Figure 9.8, and a portion of a membrane photograph is in Figure 9.9. Changes found in various diseases are summarized in Table 9.3.

Decreased serum γ-globulin concentration (hypogammaglobulinemia) occurs when the synthesis of these proteins by lymphocytes is impaired; the condition may be congenital (an inherited defect in the γ-globulin synthesizing system) or acquired. The acquired type may be caused by drugs to which the patient is sensitive; the γ-globulin concentration can be restored to normal by withdrawal of the offending drug. The body defense mechanism is impaired in severe hypogammaglobulinemia, but can be partially restored by the periodic intramuscular injection of γ-globulins. The γ-globulin concentration may be decreased in the terminal stages of Hodgkin's disease, a malignant disease involving the lymph nodes; in the early stages, the γ-globulin concentration may be moderately elevated.

Determination of Individual Serum Proteins

Serum Albumin

The determination of serum albumin, together with total protein, is frequently requested as part of a liver profile. The test is performed easily by using an anionic dye that binds tightly to albumin. Bromcresol green[3] and bromcresol purple[5] are widely used.

TABLE 9.3
CHANGES IN TOTAL SERUM PROTEIN AND ELECTROPHORETIC FRACTIONS IN VARIOUS DISEASES*

DISEASE	TOTAL PROTEIN	ALBUMIN	GLOBULIN			
			α_1	α_2	β	γ
Rheumatoid arthritis	N	↓	N ↑	↑	N	↑
Lupus erythematosus	N ↑	↓	N	↑	N	↑
Acute glomerular nephritis	N	N ↓	N ↑	↑	N	N ↑
Nephrotic syndrome	↓	↓	N ↓	↑	N ↑	↓
Hepatitis						
Acute	N ↓	↓	N ↑	↓	N	↑
Chronic (cirrhosis)†	N ↓	↓	N	↓	↑	↑
Biliary obstruction	N	N	N	↑	↑	N
Acute infection	N	N ↓	↑	N ↑	N	N
Chronic infection	N ↓	↓	↑	↑	N	↑
Malnutrition	↓	↓	N	N	N	↓
Multiple myeloma‡	↑	↓	N	N ↑	N ↑	N ↑
Hypogammaglobulinemia	↓	N	N	N	N	↓

*A decrease in concentration is denoted by ↓, an increase by ↑, and a normal value by N. These patterns, with a few exceptions, are not sufficiently specific to be diagnostic of a particular disease, but are compatible with the diseases listed.

†Area between β- and γ-bands is "bridged" because of increase in IgA and IgM.

‡The abnormal globulin may migrate with the velocity of an α_2-, β-, or γ-globulin, but it appears most frequently in the γ-band.

Prealbumin

Prealbumin is also known as thyroxin-binding prealbumin or transthyretin and migrates electrophoretically slightly faster than does albumin. Its main function is transport of thyroxin and vitamin A. The concentration of prealbumin is relatively low in serum (reference range = 10 to 40 mg/dL) compared to that of serum albumin. Consequently, prealbumin is usually not visible upon serum protein electrophoresis. Because prealbumin is a low-molecular-weight protein, its concentration in cerebrospinal fluid is relatively high. One can visualize prealbumin clearly in cerebrospinal fluid protein electrophoresis with agarose gel.

Prealbumin is made in the liver, and has a relatively short half-life of 2 days. This characteristic makes it a sensitive indicator of changes affecting its synthesis or catabolism, as in protein malnutrition, liver disease, and acute inflammation. The primary clinical utility of prealbumin measurement is nutritional assessment. Laboratory analysis methods include radial immunodiffusion, immunodiffusion, and nephelometry (see Chap. 4).

Acute Phase Reactants

Acute phase proteins are used clinically as markers for tissue necrosis and inflammation. These proteins migrate differently on electrophoresis and have various physiologic functions, but they all show an increased concentration in blood in response to inflammation. The acute phase proteins include ceruloplasmin, fibrinogen, α_1-acid glycoprotein, α_1-antitrypsin, haptoglobin, and C-reactive protein (CRP).[6]

The concentration of CRP can increase several hundredfold with inflammation, thus making CRP the most sensitive and clinically useful protein for this purpose. Used in conjunction with fever, white cell count, and red cell sedimentation rate, CRP is used in diagnosis of overt infection and in monitoring of an inflammatory process. In spite of the magnitude of the increase in CRP, one cannot visualize much change in the electrophoresis pattern (CRP migrates in the γ-region). The protein is quantitated most frequently by rate immunonephelometry (Beckman), enzyme immunoassay, and radial immunodiffusion. Normally, CRP is present in plasma at a concentration less than 0.8 mg/dL.

Physiologically, CRP activates the classic complement system in a manner similar to that of an antigen/antibody complex. CRP and complement then stimulate phagocytosis, or the ingestion of bacteria. In this way, CRP aids in the elimination of toxic substances (endotoxins) produced by the inflammatory process. The serum concentrations of many other individual proteins are sometimes requested as diagnostic aids. A partial list appears in Table 9.4. (The protein hormones are reviewed in Chapter 13, transferrin and ferritin in Chapter 12, and α-fetoprotein in Chapter 15.) All these special proteins are estimated by some type of immunoassay. Many are measured by immunonephelometry,[7] but those in low concentration may require the use of radioimmunoassay or immunofluorescence.[8] Commercial kits are generally available.

TABLE 9.4
INDIVIDUAL SERUM PROTEINS AS DIAGNOSTIC AIDS

SERUM PROTEIN	CLINICAL SIGNIFICANCE	SERUM REFERENCE RANGE	FUNCTION
Prealbumin	↓ Inflammation, malignancy, hepatocellular disease, malnutrition	10–40 mg/dL	Transports thyroxine, vitamin A
α_1-acid glycoprotein	↑ Acute phase reactant, inflammation, rheumatoid arthritis, systemic lupus erythematosus	50–150 mg/dL	Undocumented; ? steroid transport
α-antitrypsin	↑↓ Acute phase reactant, congenital deficiency associated with pulmonary emphysema	75–200 mg/dL	Protease inhibitor
Carcinoembryonic antigen (CEA)	↑ Cancer of colon, pancreas, lung, or stomach	< 3.0 ng/mL, nonsmoker < 5.0 ng/mL, smoker	
α-fetoprotein	↑ Fetal neural tube defects (see Chap. 15), liver tumors	10–60 ng/mL, 16th wk of pregnancy	Primary fetal plasma protein
Ferritin	↓ Iron deficiency anemia	12–125 ng/mL, woman 30–250 ng/mL, man	Iron storage
Haptoglobin	↑↓ Acute phase reaction, chronic intravascular hemolysis	25–200 mg/dL	Binds free hemoglobin in plasma
β_2-microglobulin	↑ Renal failure, inflammation, neoplasm, kidney transplant rejection	0.10–0.26 mg/dL	
Transferrin	Differential diagnosis of anemia	220–400 mg/dL	Transports iron
C-reactive protein (CRP)	↑ Acute phase reactant	< 0.8 mg/dL	Activates classic complement pathway
Ceruloplasmin	↑ Wilson's disease, pregnancy, oral contraceptives	18–45 mg/dL	Binds copper; is a ferroxidase

Cerebrospinal Fluid Proteins

Cerebrospinal fluid (CSF) is a clear, colorless liquid that circulates in the brain over the cerebral hemispheres and downward over the spinal cord. CSF is usually obtained by lumbar puncture of the spinal column; only a limited amount can be removed without causing a headache. A CSF specimen is precious because of the difficulty and inconvenience in obtaining a good specimen and because several laboratories (microbiology, hematology, and chemistry) usually must work with the same limited specimen (5 to 10 mL in an adult). Every effort should be made to conserve material and to save all surplus fluid in the refrigerator for at least 2 weeks in case further testing is required.

The commonly requested chemistry tests on CSF are protein and glucose. The protein concentration of the CSF is low in relation to that in plasma, ranging between 15 to 45 mg/dL or about 0.4% of the plasma concentration. It rises in various types of infection of the meninges, the membranes covering the brain and spinal cord. Such infections are known as meningitis. CSF protein concentration also rises in inflammatory lesions of the brain, after trauma, or in conditions that produce an elevated pressure (tumor, brain abscess). In addition to the analysis for protein, other laboratory observations provide useful information, for example, the presence of blood or other chromogens, numerous white cells, or bacteria.

Determination by Turbidity

CSF protein is frequently determined turbidimetrically; a sulfosalicylic acid–Na_2SO_4 reagent produces a fine suspension of protein when added to CSF. The protein concentration is proportional to the turbidity that is measured by the decrease in light transmittance. A more sensitive Coomassie blue method is also widely used.[9]

Increased Concentration. The CSF protein concentration is elevated in various types of meningitis, neurosyphilis, some cases of encephalitis, brain tumors and abscesses, and frequently after cerebral hemorrhage. Lesions that inflame the meninges or increase CSF pressure usually increase the CSF protein concentration because of a compromised blood-brain barrier. An increase in CSF protein may also be seen in multiple sclerosis and other demyelinating diseases because of an increase in synthesis of IgG within the central nervous system.

Decreased Concentration. Decreased concentration is of no clinical significance.

Fractionation By Electrophoresis

The principal reason for protein fractionation of CSF is to evaluate the γ-globulin region. Patients with multiple sclerosis or other demyelinating diseases may or may not show an obvious elevation of CSF protein. The most frequent finding on electrophoresis is the appearance of oligoclonal IgG bands (Fig. 9.10). Oligoclonal banding (several dark bands in the γ-region) is present in 85 to 95% of patients with clinically diagnosed multiple sclerosis.

When the total protein in CSF is within normal limits, the fluid must be concentrated at least 100-fold to carry out a successful electrophoresis on cellulose acetate or agarose. Concentration may be performed by ultrafiltration at reduced pressure.[10] The addition of 40 mg of finely ground polyacrylamide gel (20,000 daltons exclusion, Gelman Instrument Co.) to 200 μL of CSF quickly results in a 20- to 40-fold concentration.[11]

With concentrated CSF, electrophoresis is carried out exactly as for serum except that two or more applications of the concentrate may be necessary to apply the proper amount of protein to the membrane. Normally, γ-globulin in CSF is less than 11% of the total protein, but in multiple sclerosis, it usually

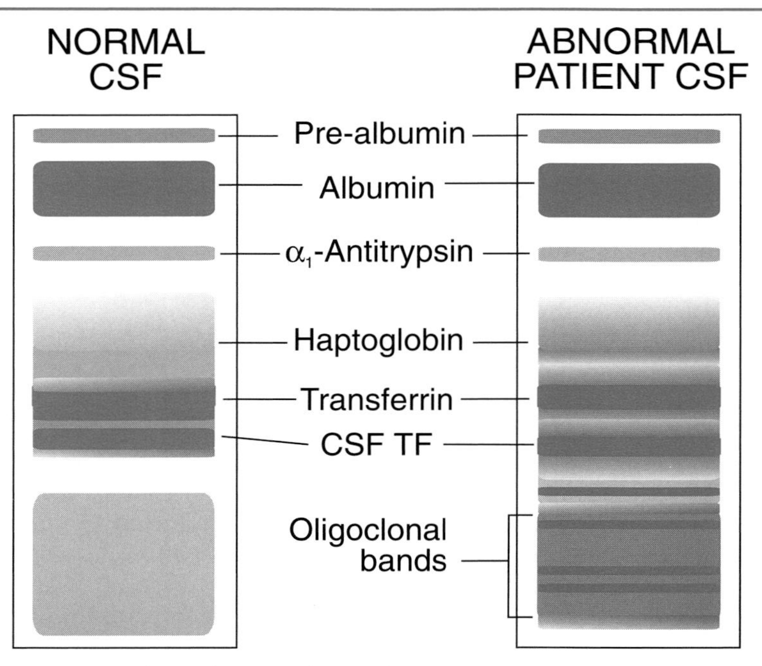

FIG. 9.10. Electrophoresis of normal CSF (note the prominent prealbumin band) and abnormal CSF with oligoclonal banding. (Illustration adapted from Killingsworth, L.M.: High Resolution Protein Electrophoresis; A Clinical Overview with Case Studies. Beaumont, TX, Helena Laboratories, 1985.[14])

exceeds 18%. A normal pattern is shown in Figure 9.10. Note the prominent prealbumin band.

CSF/Serum Protein Ratios

The determination of whether an increase in CSF protein is the result of diffusion of protein across the blood-brain barrier or of synthesis of protein within the CSF is often helpful for diagnostic purposes. CSF/serum protein ratios help to evaluate the integrity of the blood-brain barrier.

A CSF/serum albumin ratio can be calculated by dividing the CSF albumin (in mg/dL) by the serum albumin (in g/dL). A ratio of less than 9 is considered normal. Because albumin is not made within the CSF, any increase indicates a compromised blood-brain barrier. A ratio of greater than 14 is considered mild breakdown, and a ratio of greater than 100 is considered total breakdown of the blood-brain barrier.

An increase in production of IgG within the central nervous system can be documented by calculating the CSF immunoglobulin index, as follows:

$$\text{Index} = \frac{\text{IgG (CSF) mg/dL}}{\text{IgG (serum) g/dL}} \div \frac{\text{Albumin (CSF) mg/dL}}{\text{Albumin (serum) g/dL}}$$

This is equivalent to:

$$\text{Index} = \frac{\text{IgG (CSF)}}{\text{IgG (serum)}} \times \frac{\text{Albumin (serum)}}{\text{Albumin (CSF)}}$$

A reference range of 0.3 to 0.77 has been established for this index.[12] A value of greater than 0.77 is consistent with increased synthesis of IgG within the central nervous system and would be compatible with multiple sclerosis.

Urine Protein

A semiquantitative method of determining urine protein using dipsticks is described in Chapter 6. To evaluate some renal diseases, one must know how much protein is lost in the urine in a 24-hr period; a quantitative method is used to do so.

Principle. Trichloroacetic acid precipitates the protein in urine as a fine suspension. The turbidity of the suspension is proportional to the amount of protein precipitated.

Reference Values. A normal urine may contain 0.05 to 0.1 g in a 24-hr period, even though this amount may not be detected by routine methods.

Increased excretion. Protein may be lost in large quantities in nephrotic syndrome because lesions occur in the basement membrane of the glomerulus; lesser amounts of protein are excreted in other diseases producing renal lesions. Bence Jones protein, the light chains of immunoglobulins, appears in the urine in many cases of multiple myeloma. The amount excreted depends to a large extent on the stage and severity of the disease. Some individuals have an orthostatic proteinuria, protein appearing in the urine when they stand erect or walk for any period of time. This condition is benign.

Urinary Albumin

Normal albumin excretion is < 15 mg/day.[13] Routine screening tests for proteinuria may only give a positive reading at levels of albumin excretion greater than 250 mg/day. The range of intermediate proteinuria between 30 to 300 mg albumin/day is above normal, yet is not recognized as overt proteinuria. This condition is known as microalbuminuria, which refers to small amounts of albumin and not to a small albumin molecule. Albuminuria is a consequence of increased permeability and damage to the glomerulus, thereby making it a good measure and prognostic indicator of diabetic nephropathy. Laboratory methods of analysis include nephelometry, enzyme immunoassay, and radioimmunoassay.

Transudates and Exudates

Sometimes fluids from surfaces between membranes or from body cavities are brought to the laboratory as specimens for analysis. These include pleural, ascitic, pericardial, and peritoneal fluids. (Amniotic fluid is treated in a separate section in Chapter 15.)

A fluid that accumulates as an ultrafiltrate is called a transudate. The capillary pores of the membranes usually pass protein molecules of 200,000 to 300,000 daltons in molecular weight, but hold back the larger molecules. The fluid is low in protein relative to plasma, but the protein concentration still

amounts to 2 or 3 g/dL. Transudates accumulate because of increased hydrostatic pressure in the capillaries, as in congestive heart failure, or because of a decreased oncotic pressure, as in the nephrotic syndrome. Ascitic fluid accumulates in the peritoneal cavity in chronic severe liver disease because of both factors: an increase in portal vein pressure caused by fibrotic scarring (cirrhosis) and a low plasma oncotic pressure accompanying the hypoalbuminemia.

Exudates are the fluids that accumulate when the membranes are injured by inflammatory or infectious processes. The affected membranes allow larger molecules to pass through, and hence, the protein concentration of exudates is higher than that of transudates (> 3 g/dL).

Glucose concentration can also be helpful in differentiating a transudate from an exudate. The glucose concentration of a transudate is approximately that of plasma. In exudates, however, bacteria and leukocytes utilize glucose and lower its concentration. An exudate glucose concentration that is at least 25 mg/dL (1.4 mmol/L) lower than the plasma glucose is consistent with an infectious process.

REFERENCES

1. Rodwell, V.W.: Amino acids and peptides. *In* Harper's Review of Biochemistry. 20th Edition. Edited by D.W. Martin Jr., P.A. Mayes, V.W. Rodwell, and D.K. Granner. Los Altos, CA, Lange Medical Publications, 1985.
2. Goodman, J.W.: Immunoglobulins. *In* Basic and Clinical Immunology. Edited by D.P. Stites, J.D. Stobo, and J.V. Wells. Norwalk, Appleton and Lange, 1987.
3. Peters, T., and Biarnonta, G.T.: Protein (total protein) in serum, urine, and cerebrospinal fluid; albumin in serum. *In* Selected Methods of Clinical Chemistry. Vol. 9. Edited by W.R. Faulkner and S. Meites. Washington, D.C., American Association for Clinical Chemistry, 1982.
4. Barry, K.G., McLaurin, A.W., and Parnell, B.L.: A practical temperature-compensated hand refractometer (the TS meter): its clinical use in estimating total serum proteins. J. Lab. Clin. Med., *55*:803, 1960.
5. Pinnell, A.E., and Northam, B.E.: New automated dye-binding method for serum albumin determination with bromcresol purple. Clin. Chem., *24*:80, 1978.
6. Bienvenu, J.: Acute Phase Proteins in the Neonate. Brea, CA, Beckman Instruments, 1987.
7. Savory, J.: Nephelometric immunoassay techniques. *In* Ligand Assay. Edited by J. Langan and J.J. Clapp. New York, Masson Publishing, 1981.
8. Ullman, E.F.: Recent advances in fluorescence immunoassay techniques. *In* Ligand Assay. Edited by J. Langan and J.J. Clapp. New York, Masson Publishing, 1981.
9. Johnson, J.A., and Lott, J.A.: Standardization of the Coomassie Blue method for cerebrospinal fluid proteins. Clin. Chem., *24*:1931, 1978.
10. Kaplan, A., and Johnstone, M.: Concentration of cerebrospinal fluid proteins and their fractionation by cellulose acetate electrophoresis. Clin. Chem., *12*:717, 1966.
11. Schubert, E.T.: Protein electrophoresis of concentrates from micro samples of pediatric cerebrospinal fluid. Clin. Chem., *32*:1416, 1986.
12. Grant, G.H., Silverman, L.M., and Christenson, R.H.: Amino acids and proteins. *In* Fundamentals of Clinical Chemistry. 3rd Edition. Edited by N.W. Tietz. Philadelphia, W.B. Saunders, 1987.
13. Winer, R.L.: Microalbuminuria: A Discussion of Disease Involvement. Brea, CA, Beckman Instruments, 1989.
14. Killingsworth, L.M.: High Resolution Protein Electrophoresis; A Clinical Overview with Case Studies. Beaumont, TX, Helena Laboratories, 1985.

REVIEW QUESTIONS

1. What are the features of amino acids that determine whether they are basic, acidic, or neutral?
2. What are five principal functions of plasma proteins?
3. What plasma protein is present in the highest concentration, and what are its functions?
4. What are the structures and functions of the different classes of immunoglobulins?
5. Separation of serum proteins by electrophoresis depends on what principles?
6. What are the characteristic serum protein electrophoresis patterns in patients with (a) nephrotic syndrome, (b) acute infectious hepatitis, (c) multiple myeloma, and (d) hypogammaglobulinemia?
7. What principles are involved in the determination of total protein by the biuret reaction? Of albumin by the bromcresol green method?
8. What are the functions of (a) prealbumin and (b) acute phase reactants, and how may their concentrations in serum be measured?
9. What is the origin of cerebrospinal fluid proteins, and what is the rationale for measuring same?
10. In what circumstances is the concentration of urine protein elevated?

Enzymes 10

OUTLINE

IV. CLINICAL USAGE OF ENZYMES *(continued)*

E. Diagnosis of acute pancreatitis
   1. Serum amylase
      a. Increased activity
      b. Decreased activity
   2. Serum amylase isoenzymes
      a. Principle
   3. Urine amylase
      a. Procedure
      b. Calculation
      c. Reference values
      d. Amylase creatinine clearance ratio (ACCR)
      e. Increased excretion
      f. Decreased excretion
   4. Serum lipase
      a. Increased concentration

F. Diagnosis of metastasizing cancer of the prostate
   1. Increased concentration
   2. Decreased concentration
   3. Acid phosphatase isoenzyme

G. Investigation of genetic disorders
   1. Pseudocholinesterase
   2. Glucose-6-phosphate dehydrogenase (GPD)
   3. Galactose-1-phosphate uridyltransferase

H. Enzymes as reagents or labels

OBJECTIVES

After reading this chapter, the student will be able to:

1. Describe the different factors affecting the velocity of an enzyme reaction.
2. Explain what is meant by zero order kinetics.
3. Describe and select parameters for optimal measurement of enzyme activity.
4. Explain the differences between enzyme activity and enzyme mass.
5. Discuss why the measurement of a number of serum (and other body fluid) enzymes is clinically useful.
6. Discuss which enzymes and/or isoenzymes are useful in the diagnosis of myocardial infarction, liver disease, and acute pancreatitis.
7. Define the term "isoenzyme" and give an example of the use of an isoenzyme in diagnosis.
8. Identify the rate-limiting components in enzyme activity measurements.
9. Calculate enzyme activity in U/L given a molar extinction coefficient and change in absorbance.

NATURE OF ENZYMES

Biologic Catalysts

Most chemical reactions, particularly those for the oxidation or transformation of organic compounds, proceed imperceptibly, or not at all at 37° C because they require elevated temperatures for their action. Living cells cannot be subjected to high temperatures because of the delicate nature of protoplasm;

most protein solutions coagulate at temperatures above 56° C. Biologic systems, however, have developed numerous catalysts that enable the necessary metabolic reactions to occur at body temperature. The biologic catalysts speed up reaction rates and are called *enzymes*.

Enzymes are present in all body cells, and each functions in a specific reaction. Enzymes catalyze all essential reactions (oxidation, reduction, hydrolysis, esterification, synthesis, molecular interconversions) that supply the energy and/or chemical changes necessary for vital activities (muscle contraction, nerve conduction, respiration, digestion, growth, reproduction, maintenance of body temperature). As catalysts, enzymes are not used up in the process and do not change the equilibrium point of the reaction; they merely accelerate the rate for reaching equilibrium.

All known enzymes are proteins and are synthesized in the body in the same manner as all other proteins; the synthesis of each particular enzyme is under the control of a specific gene. Many genetic defects are manifested by a deficiency of the enzyme protein or an absence of specific enzyme activity, and can be demonstrated by appropriate laboratory tests.

Enzyme Action

The properties, mode of action, and kinetics of enzymes are treated more extensively in many biochemistry textbooks; supplementary readings in enzymology books also are helpful.[1,2]

The first step in the action of an enzyme is the formation of an enzyme-substrate complex; that is, the binding of the substrate molecule to the large protein enzyme (Fig. 10.1). Some enzymes require the presence of a coenzyme, a nonprotein molecule, before they can function. Most coenzymes are small, heat-stable compounds that are usually derivatives of particular members of the vitamin B family, such as thiamine phosphate (B_1), pyridoxal (B_6), nucleotides of riboflavin (B_2), nucleotides of nicotinamide, and others. Enzyme assays requiring nicotinamide adenine dinucleotide (NAD) or its phosphate as

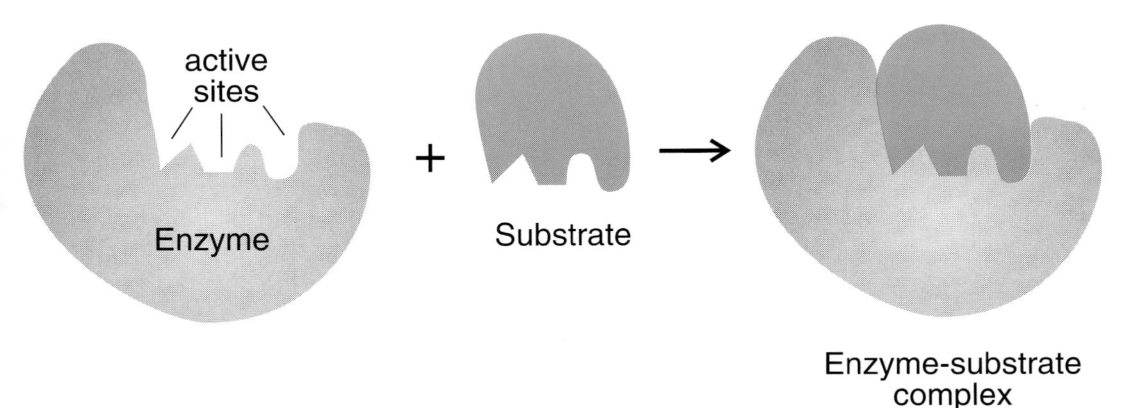

FIG. 10.1. Binding of substrate to enzyme in a precise lock-and-key fashion. The specificity of binding depends on the active site on the enzyme molecule.

part of the system are commonly used in the clinical chemistry laboratory because the formation or disappearance of the reduced coenzyme is easily followed by measuring the absorbance change at 340 nm. The reduced coenzyme (NADH, NADPH) has a strong absorbance at 340 nm, whereas the oxidized form (NAD^+, $NADP^+$) has none. The great sensitivity of this measurement allows the use of small serum samples.

Enzymes have an active site, the actual place where particular bonds in the substrate are strained and ruptured. The active site has a specific three-dimensional configuration, depending on the amino acid sequence of the enzyme; the sequence determines the distribution of charges around the site. Part or all of the substrate fits into the active site in the same manner as a key matches a lock, a process that confers specificity upon the enzyme. Certain enzymes require the presence of a particular ion to induce the necessary configuration for proper binding to the substrate; such inorganic ions are called *activators*. The most common activators are calcium, magnesium, and zinc, but other cations, such as iron, manganese, and copper, are required by certain enzymes. Anions may be required by some enzymes for full activity; chloride ion is essential for amylase action.

After the enzymatic action (rupture of a particular bond in the substrate) has occurred, the products of the reaction are no longer bound and the enzyme is recycled as the products diffuse away. The free enzyme can bind another substrate molecule and repeat its action.

Enzyme Inhibition

Because the catalytic action of an enzyme depends on its specific configuration as a protein, many substances can inhibit this activity by changing the configuration of the enzyme (binding of heavy metal cation, drugs, or other charged molecules to the protein).

Some inhibitors closely resemble the binding portion of the substrate in shape and charge and therefore can bind to the enzyme. These molecules are competitive inhibitors because the substrate and inhibitor compete for the same binding site; the addition of more substrate results by mass action in the displacement of inhibitor molecules by substrate molecules, with a consequent increase in enzymatic activity. Competitive inhibitors of common metabolic reactions are also called *antimetabolites*. The noncompetitive types of inhibitors alter the configuration of the enzyme in such a way as to reduce or abolish access of the substrate to the active site, thereby causing a decrease in enzyme activity. The addition of more substrate has no effect on this type of enzyme inhibition because it cannot reverse the structural alteration that has occurred.

FACTORS AFFECTING THE VELOCITY OF ENZYME REACTIONS

The velocity of enzyme reactions is affected by various factors, both general and specific, such as concentration of reactants, pH, temperature, ionic strength, presence of specific ions (activators), and presence of inhibitors. Any substance that alters the configuration of an enzyme affects its activity.

Any assay of enzyme activity must be carried out under optimum conditions for the substrate concentration, pH, buffer selection and concentration, coenzymes, activators, temperature, and time of incubation. Variations in reference values are frequently a reflection of differences in these assay conditions from laboratory to laboratory; reference ranges suggested by this text are method dependent.

Enzyme Concentration

When the concentration of substrate is high and essentially constant during an enzyme reaction, the velocity of the reaction is directly proportional to the enzyme concentration. The reaction proceeds more rapidly when more enzyme molecules are present to bind the abundant substrate, a mass action effect.

Substrate Concentration

When the amount of enzyme in a reaction mixture is fixed and the reaction rate is plotted against increasing amounts of substrate, the type of curve illustrated in Figure 10.2 is obtained. The reaction rate varies directly with substrate concentration when its concentration is low, as illustrated by the linear segment *ab* of the curve. The direct proportionality between reaction rate and substrate concentration is known as a *first order reaction.* At higher substrate concentrations, fewer binding sites are available for attachment, so the increase in reaction velocity for each added increment of substrate is progressively diminished (segment *bc*, Fig. 10.2). The reaction velocity becomes independent of the substrate concentration when the binding sites are saturated (at point d); the reaction proceeds at maximum velocity and follows Michaelis-Menten kinetics. At maximum velocity, the *rate of increase* of the velocity falls to zero, and this reaction state is called a *zero order reaction.* Most enzyme tests used in the clinical chemistry laboratory follow Michaelis-Menten kinetics and are performed under the conditions of zero order kinetics, that is, a large excess of substrate is present so that the amount of enzyme activity is the only rate-limiting factor in the assay. The term *enzyme kinetics* is used in connection with the study of the rates of enzyme reactions.

pH

Enzymes, being proteins, carry a net charge on the molecule that is sensitive to alterations in pH; the binding of substrate to enzyme may be greatly affected by the environmental pH. In general, the optimum pH range for an enzyme is relatively small, and the curve produced by plotting enzyme activity versus pH is usually steep, with a rapid falling off after the optimum point. Also, the range of permissible pH has a distinct limit before protein denaturation of the enzyme affects activity. The effect of pH on the relative activities of acid and alkaline

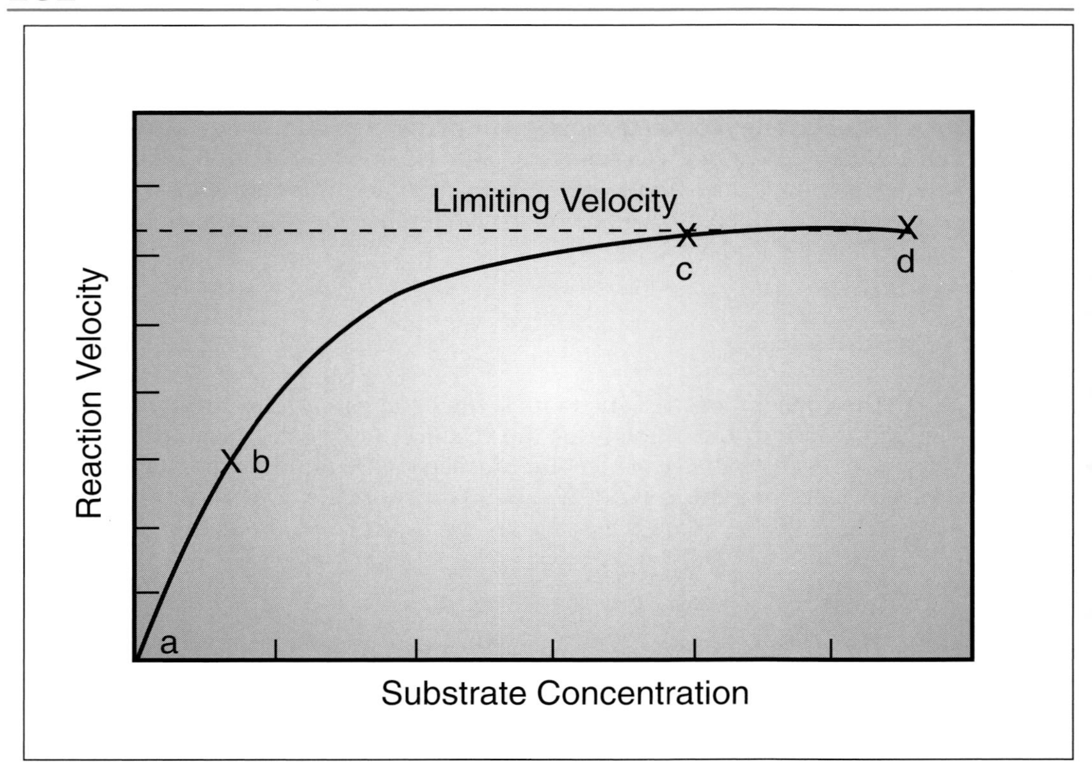

FIG. 10.2. Relationship between substrate concentration and enzyme reaction velocity.

phosphatase is illustrated in Figure 10.3. Note that the optimum pH range is narrower for alkaline phosphatase and that the falloff is steeper.

Temperature

The reaction velocities of most chemical reactions increase with temperature and approximately double for each 10° C rise. Again, the protein nature of enzymes places strict limits on the permissible temperature range because most proteins become denatured and insoluble within minutes of being subjected to temperatures of 60° C and above. Most of the body enzymes have a temperature optimum close to 37 to 38° C and have progressively less activity as the temperature rises above 42 to 45° C. All enzyme activity ceases when the protein is completely denatured. Low temperatures also decrease enzyme activity, and enzymes may be completely inactive at temperatures of 0° C and below. The inactivity at low temperatures is reversible, so many enzymes in tissues or extracts may be preserved for months by storing at −20 or −70° C. Measurement of enzymes at a variety of temperatures is inconvenient for the laboratory. Usually the waterbaths and instruments are set to conduct enzyme measurements at 25, 30, or 37° C. Unfortunately, agreement is not universal on the exact temperature to use, and thus, temperature is one of the factors responsible for variation in results from laboratory to laboratory. The incuba-

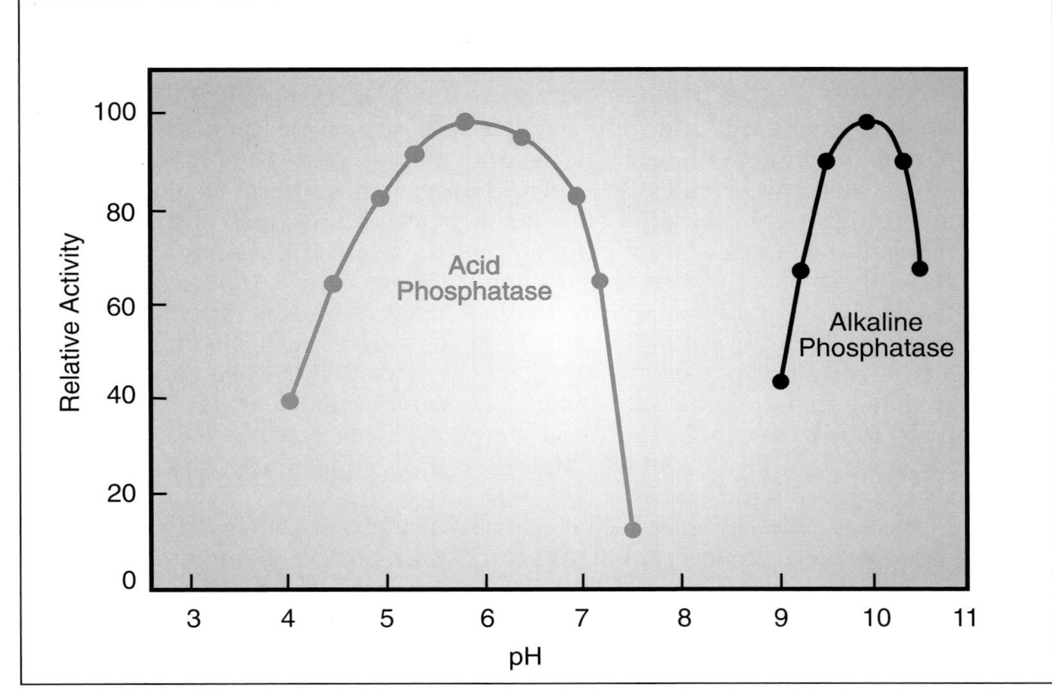

FIG. 10.3. Relative reaction velocities of alkaline phosphatase and acid phosphatase at various pH levels.

tion temperature must be controlled to an accuracy of ±0.1° C because of appreciable change of enzyme activity with temperature.

Electrolyte Environment

Many enzymes are sensitive to certain buffer ions in the incubation mixture; hence, different buffers must be tested before deciding on optimum conditions for measuring activities. Certain cations at a particular concentration range may be necessary for complete activation of the enzyme, and the total ionic strength of the incubation mixture may also affect activity.

Inhibitors

For some enzymes, a product of the reaction may inhibit enzyme activity. Inorganic phosphate, a reaction product formed during the hydrolysis of organic phosphate esters by alkaline phosphatase, is a competitive inhibitor of this enzyme. Also, various drugs administered to patients may be competitive or noncompetitive inhibitors of certain serum enzymes.

MEASUREMENT OF ENZYME ACTIVITY

The body contains hundreds of different enzymes, most of which are located and function primarily on or within cells, that is, attached to cell membranes, nuclei, and various organelles, such as mitochondria, ribosomes, lysosomes, and endoplasmic reticulum. Other enzymes may exist free in the cytoplasm.

Enzymes also are found in low concentration in such body fluids as plasma, cerebrospinal fluid, urine, and exudates. These enzymes come primarily from the disintegration of cells during the normal process of breakdown and replacement (wear and tear), but certain enzymes appear in these fluids in much higher concentration after injury or death of tissue cells. Some of the intracellular enzymes from injured or necrotic cells diffuse into the plasma or enter the blood by way of the lymph. Cytoplasmic enzymes diffuse more readily from injured cells than do mitochondrial or other structurally bound enzymes; both types are found in serum after tissue necrosis. In addition, the altered membrane permeability that may occur with inflammation may be sufficient to permit the diffusion of some cytoplasmic enzymes from cells into body fluids. Serum is the usual fluid taken for enzyme measurement for diagnostic purposes, but all the other body fluids may be chosen under appropriate circumstances.

The catalytic activities of enzymes are usually measured rather than their mass or molar concentration. With the introduction of immunoassay techniques, the concentration of a few isoenzymes (for example, prostatic) may be expressed as μg of isoenzyme per L of serum. The prostatic-specific isoenzyme became the first enzyme to be routinely measured by mass rather than by catalytic activity.[3] Creatine kinase MB isoenzyme, used in evaluating cardiac patients, is also often expressed in mass units today. Common analytical measurement of enzyme activity may involve the decrease in substrate concentration with time, the increase in a reaction product, or the increase or decrease of a cofactor, such as NADH. If the substrate is present in a concentration sufficiently high to saturate the enzyme (occupy all binding sites), the concentration of the reaction product depends directly on the concentration of the enzyme and the time period of the reaction (zero order kinetics).

The activity of an enzyme may be measured by two different approaches: fixed-time (two-point) assay and multipoint continuous monitoring (kinetic assay).

Fixed-Time (Two-Point) Assay

The first enzyme tests commonly used in the clinical laboratory (amylase, lipase, alkaline, and acid phosphatase) used a fixed time for the reaction and expressed the enzyme activity as the amount of substrate transformed by a specified volume of serum under the particular conditions of the test. Many of the incubation times were too long, requiring 30 to 60 min of incubation. As a consequence, some enzyme inactivation (partial denaturation), product inhibition, and suboptimal concentration of substrate occurred after a time. The rate of enzyme activity decreased with time after an early period of linearity.

The greater stability and sensitivity of modern spectrophotometers have improved the performance of two-point assays by making shorter incubation times possible. In such systems, however, the limits of enzyme activity in which zero order kinetics are followed must be established. Preliminary experiments with different time intervals and with various aliquots of serum sample are essential for the selection of the conditions that produce linearity.

In some clinical situations, the activities of particular serum enzymes may be so great that the substrate concentration in the two-point assay may become exhausted before the test is completed or so reduced in concentration that the binding sites of the enzyme do not remain saturated. In either case, the measured activity is no longer directly proportional to enzyme concentration or to time, and a falsely low value is obtained. The test must be repeated using a much shorter time period, a smaller serum sample, or a serum dilution.

Zero order kinetics are also not followed if the enzyme in a serum to be tested has a lag phase, a time period during which the enzyme activity builds up to its linear zero order rate. A lag phase may be produced by the enzyme binding some metabolic substrate in serum or by an impurity in the added substrate that is slowly broken down. The lag phase is a direct result of transitory, competitive inhibition and may be avoided by instituting a short preincubation period before adding the desired substrate. A lag phase may also appear in linked enzyme reactions when the indicator reaction takes a little time to reach a maximum rate. Lag phase caused by impurities in the substrate can be eliminated only by substrate purification or by obtaining another source of substrate that does not contain the impurity. Figure 10.4 illustrates an enzyme reaction with a lag phase (segment *ab* in the curve). In this example, the true enzyme activity in a 5-min, 2-point assay is from minute 2 to minute 7, with an activity of segment *bd*. The activity measured from time 0 to 5 min would be falsely low because of the lag phase. Also, in this example, the substrate no longer saturates the enzyme after 11 min, so zero order kinetics are followed only in segment *be*. A measurement of activity in a 15-min period after the lag phase would be falsely low, as indicated by the segment *fg*.

Multipoint Continuous Monitoring (Kinetic Assays)

The *rate of the enzyme reaction as a function of time* is measured by incubating serum under specific reaction conditions and measuring the rate of change in substrate, cofactor, or product concentration. Because the reaction time is usually short (from seconds to a few minutes), enzyme inactivation is improbable. Furthermore, continuous monitoring permits multiple readings for the determination of the rate. If the substrate concentration falls below the optimum because of high enzyme activity, a warning is provided by the falling reaction rate; no such warning is obtained in a fixed-time reaction. Continuous monitoring is used most commonly with enzymes in which changes in NADH or NADPH are measured, but can also be used to determine other enzyme activities (for example, alkaline phosphatase) if a colored product is generated from a noncolored substrate. Many assays require the introduction of accessory purified enzymes to produce a coupled system that uses or generates NADH, the substance for which change in concentration is measured by a decreasing or an increasing absorbance at 340 nm. The accessory enzymes must be added in excess so that the rate limiting factor is the concentration (activity) of the enzyme of interest. A preincubation period is usually necessary so that

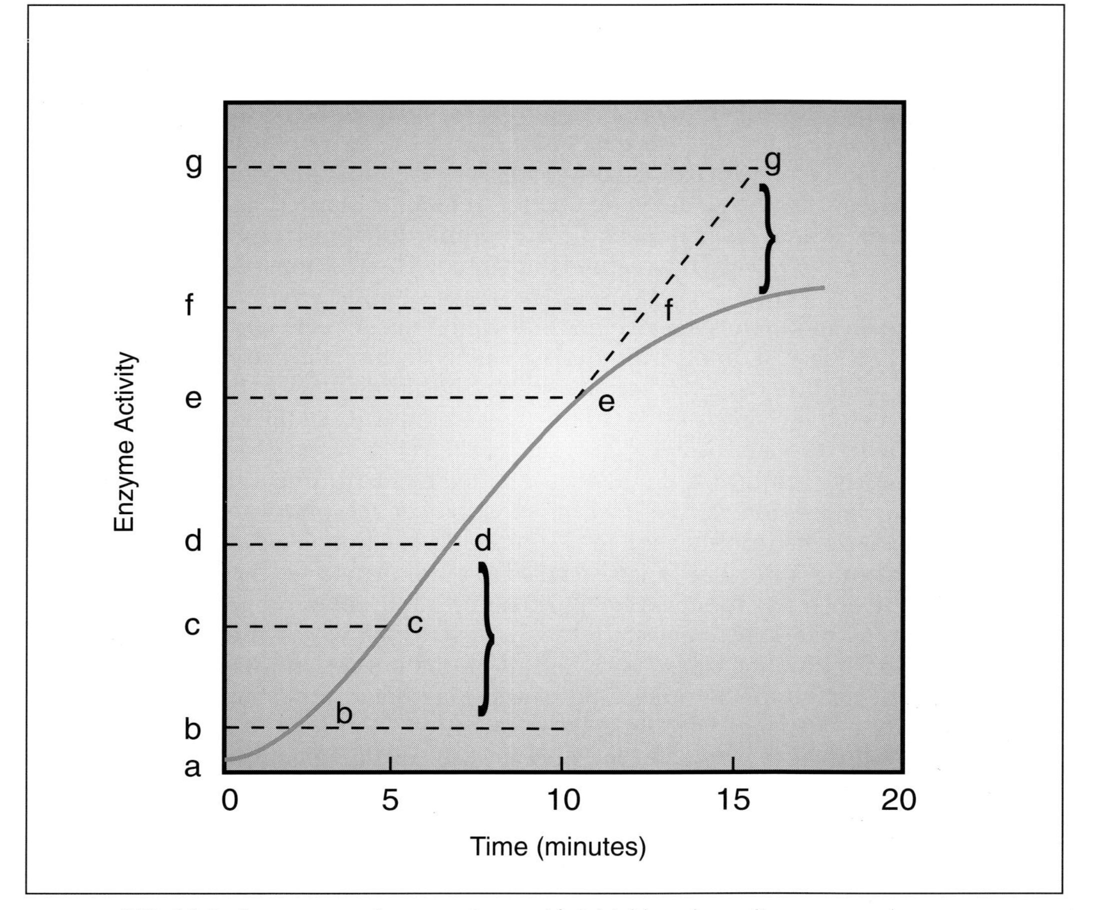

FIG. 10.4. Enzyme reaction over time, with initial lag phase, linear measurement range, and final exhaustion of substrate.

extraneous reactions using or producing NADH are completed before measurements are started on the enzyme of interest. Also, some serum enzymes have a lag phase, as previously described in the section on two-point assay, and allowance for this lag period must be made before initiating measurements of the reaction rate. Some precautions also must be taken with continuous monitoring methods to eliminate possible errors. Inspection of the chart recording of absorbance change with time allows one to check for linearity of the enzyme action and to see whether substrate is sufficient for serum samples with high enzyme activity. Some instruments provide a built-in check for linearity of the absorbance readings.

The International Unit (U) of enzyme activity was defined by the International Commission on Biochemical Nomenclature (1972) as the amount of an enzyme that converts 1 μmole of substrate per minute in an assay system. The activity is usually expressed in terms of International Units per liter of serum (U/L). The Système Internationale introduced a confusing note, however, by designating the second as the basic unit of time; the SI unit of enzyme activity is called a katal and is defined as 1 mole of substrate transformed per second.

$$1\ \text{U} = 10^{-6}\ \text{mol/60 sec} = 16.7\ \text{nkat}$$
$$1\ \text{nkat/L} = 0.06\ \text{U/L}$$

We express enzyme units as U/L in the text of this book. The absorbance change during an enzyme reaction can be converted into U/L if the molar absorptivity (ϵ) of the substrate or product is known. It accomplished by the following:

$$\frac{\Delta A}{\text{min}} \times \frac{1}{\epsilon} \times \frac{\text{TV (mL)}}{\text{SV (mL)}} \times 10^6 = \text{enzyme activity in U/L}$$

where ΔA = change in absorbance; ϵ = molar absorptivity in liter $\text{mol}^{-1}\ \text{cm}^{-1}$ (6220 for NADH at 340 nm); TV = total reaction volume; SV = sample volume; and 10^6 = conversion from mol to μmol.

Because the activity of an enzyme depends on a host of factors, such as pH, temperature, the concentrations of substrate, cofactors (if any), ionic strength, salts, and buffers, the optimum conditions for a test must be defined and followed. Normal values for a particular enzyme test may vary from laboratory to laboratory because of small or large differences in methodology, even though all the laboratories may be expressing the results in International units. The historical nomenclature of enzymes was confusing because some names were arbitrarily chosen and others were constructed by adding the suffix "ase" to the substrate acted upon (for example, proteinase, esterase). The Enzyme Commission of the International Union of Biochemistry (IUB) has provided a systematic nomenclature* and code number for all enzymes.[2]

Because the measurement of an enzyme level uses its catalytic activity, its assay becomes a sensitive means of diagnosing disease. Tiny changes can be detected because each enzyme molecule is recycled over and over again as it acts on the substrate. Although a few enzyme tests have been used in the clinical laboratories for 50 or 60 years, the past 3 decades have seen a

*No official international abbreviations exist for the enzymes. The following nonstandard abbreviations are used in this book:

ACP, acid phosphatase
ALD, aldolase
ALT (GPT), alanine aminotransferase
ALP, alkaline phosphatase
AST (GOT), aspartate aminotransferase
CK (CPK), creatine phosphotransferase (phosphokinase), creatine kinase
CK-BB (CK_1), the BB (brain) isoenzyme of CK
CK-MB (CK_2), the MB (heart) isoenzyme of CK
CK-MM (CK_3), the MM (muscle) isoenzyme of CK
G-1-P and G-6-P, glucose-1 and -6 phosphate, respectively
GD, glutamate dehydrogenase
GGT, gamma glutamyltransferase
GPD, glucose-6-phosphate dehydrogenase
LD (LDH), lactate dehydrogenase
LD_1, the most rapidly migrating LD isoenzyme (heart)
LD_2, the second most rapidly migrating LD isoenzyme
MD, malic dehydrogenase
MP, maltose phosphorylase
5NT, 5′-nucleotidase
OCT, ornithine carbamoyl transferase
P_i, inorganic phosphate

tremendous increase in their use and scope. The discovery in 1954 that the serum aspartate aminotransferase (AST, previously called glutamic-oxaloacetate transaminase [GOT]) activity in serum increased significantly within 24 hr after an acute myocardial infarction initiated the modern era of clinical enzymology. Many different enzymes were investigated in an attempt to find early indicators of specific tissue injury. The serum patterns of several different enzymes obtained after a suspected organ injury are usually more revealing than is the activity of a single enzyme; this finding has led to the multiplicity of enzyme tests.

Some enzymes exist in multimolecular forms (isoenzymes) with similar catalytic activity but different physical and immunologic properties. Current theory ascribes the formation of isoenzymes to differences in the combination of polypeptide chains comprising the enzyme. Unique patterns or combina-

TABLE 10.1
SOME SERUM ENZYMES OF CLINICAL INTEREST

COMMON NAME (ABBREVIATION)	SYSTEMATIC NAME (IUB)	EC* CODE NUMBER	DIAGNOSTIC PURPOSE
Acid phosphatase (ACP)	Orthophosphoric monoester phosphohydrolase	3.1.3.2	Metastasizing cancer of the prostate
Alanine aminotransferase (ALT)‡	L-alanine:2-oxoglutarate aminotransferase	2.6.1.2	Liver disease
Aldolase (ALD)	Fructose 1,6-diphosphate: D-glyceraldehyde-3-phosphate lyase	4.1.2.13	Muscle disorders
Alkaline phosphatase (ALP)	Orthophosphoric monoester phosphohydrolase	3.1.3.1	Bone and liver disorders
Amylase	1,4-α-D-Glucan glucanohydrolase	3.2.1.1	Acute pancreatitis
Aspartate aminotransferase (AST)†	L-aspartate:2-oxoglutarate aminotransferase	2.6.1.1	Myocardial infarction, liver disease, muscle disease
Creatine kinase (CK)	ATP:creatine phosphotransferase	2.7.3.2	Myocardial infarction, muscle disease
γ-glutamyl transpeptidase (GGT)	γ-glutamyl transferase	2.3.2.2	Liver disease
Lactate dehydrogenase (LDH or LD)	L-lactate:NAD^+ oxidoreductase	1.1.1.27	Myocardial infarction, liver disease, malignancies
Lipase	Triacylglycerol acylhydrolase	3.1.1.3	Acute pancreatitis
5′-nucleotidase (5NT)	5′-ribonucleotide phosphohydrolase	3.1.3.5	Liver disease
Ornithine-carbamoyl transferase (OCT)	Carbamoyl phosphate:L-ornithine carbamoyltransferase	2.1.3.3	Liver disease
Pseudocholinesterase (SChE)	Acylcholine acylhydrolase	3.1.1.8	Exposure to organophosphate insecticides

*Enzyme commission
†Old name, Glutamic oxaloacetic transaminase (GOT)
‡Old name, Glutamic pyruvic transaminase (GPT)

tions of these enzyme variants may occur in some organs. The isoenzymes in serum can usually be separated on the basis of electrophoretic mobility, by differences in adsorption properties (column chromatography), or by reaction with specific antibodies (immunoassay). Serum isoenzyme patterns are used for diagnostic purposes because they may reveal injury to a specific organ or tissue. The use of isoenzymes (CK, LD and ACP isoenzymes) in clinical enzymology is elaborated upon later in this chapter.

Some of the enzyme tests currently being performed widely in clinical chemistry laboratories are described in the following paragraphs and summarized in Table 10.1. This table lists not only the common or trivial name of the enzyme but also its common abbreviation, its systematic name, the Enzyme Commission (EC) code number, and some of the situations in which the tests are used for clinical diagnosis or patient management.

CLINICAL USAGE OF ENZYMES

Diagnosis of Myocardial Infarction

A myocardial infarct is a necrotic area in the heart caused by a deficient blood flow to that area as the result of a clot in a coronary vessel and/or narrowing of the vessel lumen by atherosclerotic plaques. When the cardiac cells in the necrotic area die, their intracellular enzymes diffuse out of the cell into tissue fluid and end up in plasma. Because a definitive diagnosis of myocardial infarction is not always possible by electrocardiogram, appropriate enzyme tests are extremely helpful for this purpose.[1,4,5]

The enzyme tests that have proved most helpful in the diagnosis of myocardial infarction are listed in Table 10.2. These are creatine kinase

TABLE 10.2
ENZYMES IN THE DIAGNOSIS OF MYOCARDIAL INFARCTION

ENZYME			ENZYME ACTIVITY INCREASE		
NAME	ABBREVIATION	ADULT UPPER LIMIT OF NORMAL (ULN)‡ (Units/L)	USUAL RISE (× ULN)	TIME FOR MAXIMUM RISE* (Hours)	TIME FOR RETURN TO NORMAL (Days)
Creatine kinase	CK	100	5–8	24	2–3
Isoenzyme MB	CK-MB	6	5–15	24	2–3
Aspartate aminotransferase	AST	25	3–5	24–48	4–6
Lactate dehydrogenase	LD	290	2–4	48–72	7–12
Isoenzyme #1	LD_1	100	2–4†	48–72†	6–12
Isoenzyme #2	LD_2	115	1.1–3†	48–72†	6–10

*Time after occurrence of infarct as manifested by chest pain.
†The LD-isoenzyme indication of a possible MI is an abnormal increase in the total amount of LD and an increase in LD_1 so that the ratio LD_1/LD_2 is greater than 1 (the *flipped* LD ratio). The flipped ratio is usually not maintained as long as an abnormally elevated LD_1.
‡All upper limits of normal are method dependent.

(CK), aspartate aminotransferase (AST), lactate dehydrogenase (LD), isoenzyme CK-MB, and the isoenzymes of LD. All the isoenzymes of LD must be separated to obtain the absolute values of LD_1 and LD_2, as well as the ratio, LD_1:LD_2.

AST is mentioned for historical reasons, as it is no longer of clinical importance in a workup for myocardial infarction because of its nonspecificity. As shown in Table 10.2 and Figure 10.5, some of the enzyme activities increase early after an infarction (CK and CK-MB), some appear a little later (AST), and some increase even later and remain elevated for prolonged periods (LD, LD_1, and LD_2). Each enzyme has its own particular time course when the serum activity of the enzyme is plotted against time after the myocardial infarct (Fig. 10.5). Because the laboratory has no control over when the patient may elect to see the physician or when the enzyme tests are ordered, some tests must be available that can help to diagnose a myocardial infarction in a time period that may vary from 4 hours to 10 days.

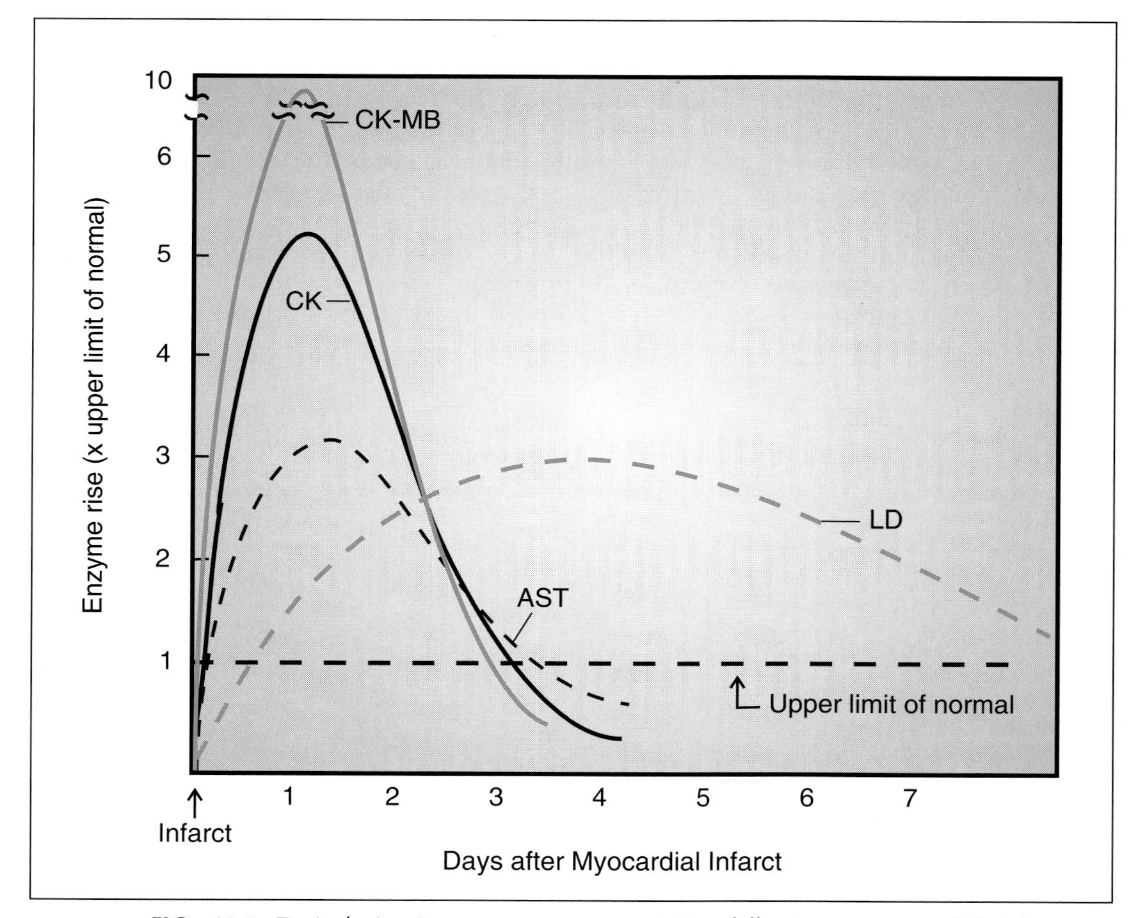

FIG. 10.5. Typical rise in serum enzyme activities following a myocardial infarction. Although AST increases following MI, it has been replaced clinically by CK, CK-MB, and sometimes LD.

Creatine Kinase (CK)

CK is an enzyme that catalyzes the reversible transfer of phosphate from adenosine triphosphate (ATP) to creatine, as shown in Reaction 1. This reaction makes possible the storage of high-energy phosphate in a more stable form than in ATP, as reviewed in Chapter 6.

Reaction 1: Creatine + ATP $\xrightarrow{CK}$ Creatine phosphate + ADP

CK is present in high concentration in skeletal muscle, cardiac muscle, thyroid, prostate, and brain; it is present only in small amounts in liver, kidney, lung, and other tissues. Hence, an increase in serum CK activity must be ascribed primarily to damage to striated muscle (skeletal or cardiac) and, in rare cases, to brain. Differentiation of these various diseases frequently can be made upon clinical grounds, but in some situations such differentiation is not possible. Measurement of the CK-MB isoenzyme helps to solve the problem.

As seen in Figure 10.5, the serum CK activity begins to rise 3 to 6 hr after a myocardial infarction and reaches a maximum around 24 hr. The maximum rise in serum for a patient is usually about 5 to 8 times the upper limit of normal (× ULN), but in severe cases, the increase may reach 10 to 20 × ULN. In general, the higher the rise, the worse the prognosis. The CK activity in serum may return to normal levels in 3 or 4 days; a subsequent increase indicates an extension of the infarct.

Serum CK may be determined by using a series of enzymatic reactions coupled with Reaction 1; those based on the formation of creatine phosphate and adenosine diphosphate (ADP) in Reaction 1 are deemed to proceed in the "forward" direction (at pH 9.0), and those that measure the ATP produced from the hydrolysis of creatine phosphate are considered to move in the "reverse" direction (at pH 6.7). The reverse-direction methods are preferred because they proceed much more rapidly. Prepared reagents or kits for CK are readily available from many manufacturers (Boehringer Mannheim, Roche).

Principle. The activity of CK is followed by using the ATP produced from creatine phosphate in Reaction 1 to form glucose-6-phosphate (Reaction 2); the glucose-6-phosphate is dehydrogenated in Reaction 3, and the rate of formation of NADPH is measured at 340 nm. CK activity is proportional to the increase in absorbance.

The reagent mixture usually contains N-acetylcysteine to prevent oxidation of the CK sulfhydryl groups and adenosine monophosphate (AMP) compounds to inhibit the possible reaction of ADP with myokinase, which reduces assay interference in hemolyzed specimens.

Reaction 2: ATP + glucose $\xrightarrow{HK}$ ADP + glucose-6-phosphate

Reaction 3: Glucose-6-phosphate + $NADP^+$ $\xrightarrow{GPD}$ 6-Phosphogluconate + NADPH + H^+
where HK is hexokinase and GPD is glucose-6-phosphate dehydrogenase.

Increased Activity. Because CK is located primarily in skeletal muscle, myocardium, and brain, the serum activity increases after damage to these tissues and is not usually affected by the pathologic conditions in other organs. CK activity is usually higher in men than in women because of the greater muscle mass in men.

1. *In damage to heart tissue.* As described previously in this chapter, a sharp but transient rise in CK activity follows a myocardial infarction; the degree of increase varies with the extent of the tissue damage. CK is the first enzyme to appear in serum in higher concentration after myocardial infarction and is probably the first to return to normal levels if no further coronary damage occurs. The serum CK may be increased in some cases of coronary insufficiency without myocardial infarction.[1,4-6] The simultaneous determination of the CK-MB isoenzyme and LD isoenzymes helps to make the diagnosis, as described later.
2. *In damage to skeletal muscle.* The serum CK activity may rise to high levels following injury to skeletal muscles. Some of the causes may be trauma, muscular dystrophy, massage of chest during a heart attack, an intramuscular injection, or even strenuous exercise. The serum activity parallels the amount of muscle tissue involved. In prolonged shock, the CK enzyme also leaks from ischemic muscle cells and appears in the serum.
3. *In brain damage.* The CK levels in serum are increased in brain injury only in instances of some damage to the blood-brain barrier; the rise is in the BB fraction. Damage to the blood-brain barrier may be caused by trauma, infection, stroke, or severe oxygen deficiency.

Creatine Kinase Isoenzymes

Enzymes are proteins assembled from polypeptide subunits. CK is a dimer that consists of two subunits, M and B; the M subunit predominates in skeletal muscle and the B subunit predominates in brain. Three isoenzymes separated by electrophoresis behave as dimers: CK-BB (CK_1), composed of two B chains; CK-MB (CK_2), composed of one M and one B chain; and CK-MM (CK_3), composed of two M chains. The BB isoenzyme is the most negatively charged and migrates farthest toward the positive pole (anode) with the mobility of prealbumin. MB is of intermediate negative charge and is the middle band (mobility of β-globulin). MM has the least negative charge and migrates slightly toward the cathode. The migration is similar to that of γ-globulin proteins (Fig. 10.6).

Tissue Distribution

The MM isoenzyme is found primarily in skeletal and cardiac muscles, but activity is low in lung and kidney. Cardiac muscle cells contain a mixture of the CK-MM and CK-MB isoenzymes. The major portion is MM, but the MB content is considerable and may comprise from 15 to 20% of the total CK activity. By contrast, skeletal muscle CK consists of approximately 99% MM fraction and only about 1% of MB. The BB isoenzyme is present in brain tissue, in gastrointestinal and genitourinary tracts (colon, prostate, uterus), and, with lower activity, in thyroid and lung.

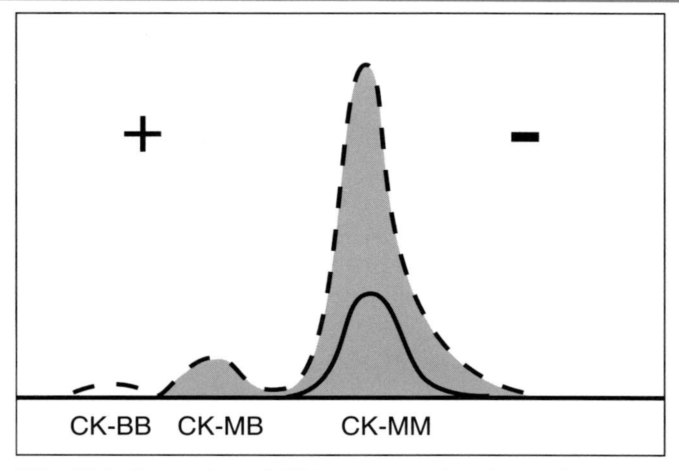

FIG. 10.6. Separation of CK isoenzymes by electrophoresis. A normal pattern (solid black line) shows only CK-MM. The presence of elevated levels of CK-MB and CK-MM seen following an MI is illustrated by the solid blue. CK-BB is rarely seen; its position is noted by a dotted line.

Serum CK Isoenzymes

The predominant CK isoenzyme in the serum of normal individuals is the MM fraction, which comprises 94 to 98% of the total. The MB isoenzyme may comprise as much as 6% of the total, but is usually only 2 to 4% (1 to 4 U/L). In normal serum, the BB isoenzyme is undetectable by electrophoretic methods but may increase appreciably in women immediately post partum and in patients with cardiovascular accidents (stroke), acute renal disease, adenocarcinomas of the prostate or other tissues, severe hypoxia (lack of oxygen), and brain injury that damages the blood-brain barrier.

The most important use for CK isoenzymes is in the diagnosis of myocardial infarction (MI). Following a moderate to severe MI, the MB isoenzyme rises rapidly, reaches a maximum within 24 hr, and then falls rapidly (Fig. 10.5). Its relative increase in serum is greater than that for total CK, but it returns to normal values a little earlier than does total CK. After a small MI, the MB isoenzyme may become elevated even though the total CK remains within normal limits. When dealing with a possible MI, one must consider the percentage increase in CK-MB, as well as its activity in units because trauma to skeletal muscle results in an increase in both total CK and the MB isoenzyme. The MB activity as percent of total activity, however, is less than 3% in muscle trauma and greater than 6% in MI. The assay for CK-MB is an important diagnostic aid in patients suspected of having had an MI. Further information on CK isoenzymes in the diagnosis or assessment of MI appears in several recent reviews.[5,7,8]

CK-MB activity can be measured after separating the CK isoenzymes by electrophoresis or ion exchange chromatography. It can also be quantified by various immunotechniques using enzyme or radioisotopic (RIA) labels.

Electrophoretic Separation of CK Isoenzymes. Serum CK-MB activity can be determined by enzymatic assay (fluorometric or colorimetric) after electrophoretic separation of the isoenzymes on agarose gel (Fig. 10.6). The separated isoenzyme bands can be visualized under fluorescent or visible light after incubation with substrate and then quantified by means of a scanning densitometer or fluorometer. The BB isoenzyme migrates the farthest toward the anode, whereas the MM is almost stationary; MB has an intermediate anodic mobility. The concentration of the BB isoenzyme in normal individuals, however, is present in only trace amounts and cannot be visualized on an electropherogram, but the increased concentration that accompanies certain pathologic states (cerebrovascular accidents, some adenocarcinomas, and other conditions) appears as a fast anodic band. In a small percentage of cases, the appearance of atypical isoenzyme bands may be caused by increased CK-BB coming from damaged heart muscle,[8] macro CK type I (IgG linked to BB isoenzyme),[9] or macro CK type II (mitochondrial CK).[10] These possibilities require further study.

Immunotechniques for Determination of CK Isoenzymes. Antibodies specific for the B (or M) subunits of CK bind to half the MB and all the BB (or MM) isoenzymes. This principle underlies all the immunoprecipitation and immunoinhibition assays. Some of the early immunoinhibition tests made the unwarranted assumption that serum BB activity is too low to be detected, but this deficiency can be corrected by the use of an appropriate blank to eliminate the occasional elevated BB fraction encountered in 1 to 2% of patients in critical care units. The technology is changing rapidly in attempts to improve specificity and sensitivity. The advances include new methods of immunoprecipitation, the use of antibodies bound to a solid phase to facilitate separation, labeling of antibodies with an enzyme or a radioactive tracer (usually ^{125}I) to increase sensitivity, substituting monoclonal antibodies for the polyclonals, or using multiple antibodies, as in the sandwich technique (adding an anti-IgG antibody of a different species to react with and precipitate an anti-M or anti-B antibody already bound to its specific isoenzyme subunit), to increase specificity. The production of a monoclonal antibody specific for the MB isoenzyme is a new development that appears to be advantageous.[11]

Radioimmunoassay (RIA) methods use ^{125}I-labeled antibodies (to either the M or the B subunit) and depend on a conventional competitive binding assay; they are more costly than other methods and no more selective. All radiometric methods quantify the MB isoenzyme in terms of mass (ng/mL) instead of activity (U/mL).

Reference Values. The CK-MB band in normal serum is usually not visible when separated by electrophoresis. Normal serum may have as much as 6 U/L or 2 to 30 μg/L of MB protein. CK-MB is usually less than 6% of the total CK.

Increased Activity. The MB isoenzyme starts to increase within 4 hr after an MI and reaches a maximum within 24 hr. The maximum rise may range from 10 U to as high as 400 U, depending on the severity of the infarct.

The MB isoenzyme is always elevated within 48 hr of an MI, but it can also be elevated with coronary insufficiency. When taken together with the LD isoenzymes, the finding of a flipped ratio ($LD_1 > LD_2$) plus an elevated MB isoenzyme is a positive indication of an MI. A diagnosis is always made with more assurance when the pattern of key enzyme changes is taken into consideration (see Fig. 10.5).

The BB isoenzyme is not detectable by electrophoresis in the serum of healthy individuals, but traces can be found by RIA. The BB fraction rises to electrophoretically measurable levels in the conditions mentioned previously in this chapter.

Lactate Dehydrogenase (LD)

Lactate dehydrogenase reversibly catalyzes the oxidation of lactate to pyruvate by transferring hydrogen from lactate to the cofactor NAD^+, according to Reaction 4.

Reaction 4: $CH_3-CH(OH)-C(=O)-O^- + NAD^+ \overset{LD}{\longleftrightarrow} CH_3-C(=O)-C(=O)-O^- + NADH + H^+$

Lactate → Pyruvate

LD is distributed widely in tissues and is present in high concentration in liver, cardiac muscle, kidney, skeletal muscle, erythrocytes, and other tissues.[1,4]

Principle. In the conditions of this assay, the enzyme LD converts lactate to pyruvate while reducing NAD^+ to NADH. The rate of increase in absorbance of NADH at 340 nm is proportional to LD activity. The backwards assay (pyruvate → lactate) can also be used to measure LD and is the principle of the Kodak reagent system for automated analyzers. In this case, the decrease in absorbance as NADH is oxidized to NAD^+ is monitored.

Reference Values. The forward reference range of serum LD activity varies from 125 to 290 U/L. The backwards assay proceeds much faster and has reference ranges of 297 to 537 U/L.

Increased Concentration. The serum LD activity is increased in many various disorders because it is so widely distributed in tissues.[1,4] The principal clinical uses of the LD test are the following:

1. *In myocardial infarction.* Serum LD activity increases after MI, but the rise occurs later than that for CK and is of less intensity (Table 10.2). Its value in the diagnosis of MI lies in the prolongation of its increased

activity; it may remain elevated for 7 to 10 days, long after the CK level has returned to normal (Fig. 10.5). The isoenzymes of LD also aid in the diagnosis of MI and are described in the following sections.

2. *In other diseases.* Serum LD activity is increased in liver disease, but other enzymes are more sensitive and specific for liver disorders (see Chap. 11). The serum activity is also increased following muscle trauma, renal infarct, hemolytic diseases, and pernicious anemia. Hemolyzed blood specimens have artifactually elevated LD activities owing to LD enzymes coming from the ruptured red blood cells; the same is true if the serum is allowed to stand too long upon the clot (see Table 10.3).

Lactate Dehydrogenase Isoenzymes

The LD enzyme is composed of four subunits. The two different polypeptide chains are an M type typical of skeletal muscle and an H type from cardiac muscle. Five different isoenzymes are possible from this combination: HHHH (LD_1, from heart muscle, red blood cells, or kidney), MHHH (LD_2, also in heart), MMHH (LD_3, in lung and other tissues), MMMH (LD_4, in many tissues), and MMMM (LD_5, primarily in skeletal muscle and liver). LD_1 migrates farthest toward the anode and is called the fastest fraction; the others migrate more slowly, with LD_5 moving slightly toward the cathode as does γ-globulin (Fig. 10.7).

Principle.[12] The LD isoenzymes in serum are separated by electrophoresis on agarose in a manner similar to that for serum proteins. The membrane containing the separated isoenzyme bands is then overlaid on an agarose gel containing lactate substrate, buffer, NAD^+, nitroblue tetrazolium, and phenazine methosulfate. As each LD isoenzyme band oxidizes lactate to pyruvate, the NADH formed transfers electrons to phenazine methosulfate, which reduces nitroblue tetrazolium to an insoluble formazan dye upon the isoenzyme band. The concentration of lactate substrate affects the measurement of the individual isoenzymes. A lower lactate concentration yields a relatively high LD_1 fraction, whereas a higher substrate concentration yields a relatively higher LD_5 fraction. After washing and drying the membrane, the intensity of the dye deposited on the isoenzyme bands is quantitated by densitometry.

TABLE 10.3
CHANGES IN LD ISOENZYME PATTERNS SEEN IN DISEASE

ELEVATED LD_1 AND LD_2	ELEVATED LD_3	ELEVATED LD_4 AND LD_5
Myocardial infarct ($LD_1 > LD_2$)	Pulmonary embolism	Liver disease
In vivo or *in vitro* hemolysis	Malignant disorders	Arthritis and joint effusions
Sickle cell anemia	Rapid destruction of lymphocytes	Skeletal muscle infections
Megaloblastic anemia	Infectious mononucleosis	Prostate cancer
Tumors	Splenic necrosis	Malignant disorders
Renal infarct		Septic shock
Muscular dystrophy		

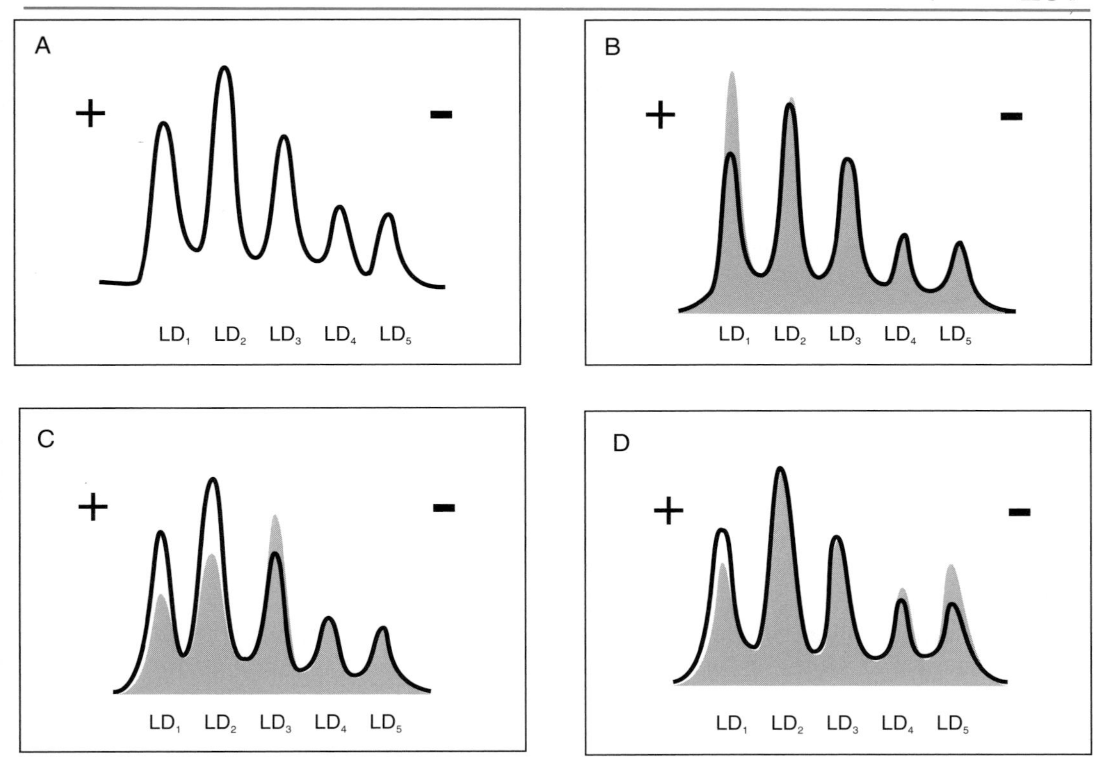

FIG. 10.7. Separation of LD isoenzymes by electrophoresis in normal (A) and disease states. The light blue indicates the pattern seen in myocardial infarction (B), pulmonary embolism (C), and liver disease (D).

Reference Values. The serum LD composition of normal individuals is as follows:

LD_1	100 U/L
LD_2	115 U/L
LD_3	65 U/L
LD_4	40 U/L
LD_5	35 U/L

Clinical Applications. The LD isoenzyme determination may be useful in the following clinical situations (Table 10.3):

1. *In myocardial infarction.* The LD isoenzymes from heart (LD_1 and LD_2) increase in serum within 24 to 48 hr after an infarct. LD_1 and LD_2 are the fastest LD isoenzymes. The ULN for LD_1 is 100 U/L or 34% of the total LD; the ULN for LD_2 is 115 U/L or 40% of the total. In an MI, the ratio of LD_1:LD_2 becomes *flipped,* that is, greater than 1. Because this isoenzyme pattern (the flipped ratio) may occur after a renal infarct and in hemolytic situations, these possibilities must be ruled out on other grounds. If the time interval is right, a CK isoenzyme pattern can resolve the problem.

The combination of an elevated CK-MB isoenzyme and a flipped LD isoenzyme ratio in a patient suspected of having had an MI makes the diagnosis certain. This combination never occurs in coronary insufficiency without an MI.

2. *In liver disease.* See Chapter 11 for the role of LD isoenzyme in helping to distinguish between viral hepatitis and infectious mononucleosis.
3. *In pulmonary infarction.* In this condition, LD_3 frequently becomes a prominent band.

Diagnosis of Hepatic Disease

Serum Aspartate Aminotransferase (AST)

Aspartate aminotransferase (AST) is found in practically every tissue of the body, including red blood cells, and is in particularly high concentration in cardiac muscle and liver, intermediate in skeletal muscle and kidney, and in much lower concentrations in the other tissues. The discovery in 1954 that the serum AST (glutamine-oxaloacetic transaminase [GOT]) concentration increased shortly after the occurrence of an MI greatly accelerated the search for enzymes in serum as indicators of specific tissue damage. Although AST increases following an MI, its use in following MI patients is limited by its nonspecificity. The measurement of serum AST levels is most helpful for the diagnosis and monitoring of hepatocellular disease.

Principle.[13] The enzyme AST reversibly transfers an amino group from aspartate to α-ketoglutarate (oxoglutarate) according to Reaction 5 and forms oxaloacetate in the process. This reaction is coupled with that of malate dehydrogenase, in which the oxaloacetate is reduced to malate as NADH is simultaneously oxidized to NAD^+ (Reaction 6); the decrease in absorbance at 340 nm is measured. Pyridoxal phosphate binds to the enzyme and serves as a coenzyme for the transamination reaction.

Reaction 5: $\text{Aspartate} + \alpha\text{-ketoglutarate} \xrightarrow{\text{AST}} \text{glutamate} + \text{oxaloacetate}$

Reaction 6: $\text{Oxaloacetate} + \text{NADH} + \text{H}^+ \xrightarrow{\text{MD}} \text{malate} + \text{NAD}^+$

Preincubation of the serum sample with lactate dehydrogenase before adding the α-ketoglutarate substrate is necessary for the removal of any endogenous pyruvate that may be present. LD also uses NADH in converting pyruvate to lactate, but the supply of endogenous pyruvate is soon exhausted. Blood specimens should be drawn in a manner that avoids hemolysis. Erythrocyte AST falsely elevates the serum AST activity if the blood is hemolyzed.

Reference Values. The normal concentration of serum AST is 6 to 25 U/L.

Increased Activity. The serum activity of AST is increased with intracellular liver damage (hepatitis, cirrhosis, hepatotoxins), after MI, in trauma to or diseases affecting skeletal muscle, after a renal infarct, and in various hemolytic conditions. Prolonged myocardial ischemia may be accompanied by a rise in serum AST. Congestive heart failure is also associated with an increased serum activity of AST because of the hepatic ischemia and anoxia that are produced.

Serum Alanine Aminotransferase (ALT)

The enzyme that transfers an amino group from the amino acid alanine to α-ketoglutarate is alanine aminotransferase (ALT).

The concentration of ALT in tissues is not nearly as great as that of AST. ALT is present in moderately high concentration in liver, but is low in cardiac and skeletal muscles and other tissues. Its use for clinical purposes is primarily for the diagnosis of intracellular hepatic disease (see Chap. 11). ALT is the more specific transaminase for the liver, and is elevated in the presence of hepatocellular necrosis. Measurement can be helpful in distinguishing hepatocellular jaundice from obstructive jaundice, as ALT shows marked elevations in hepatocellular jaundice and only mild elevations in obstructive jaundice.

Principle.[13] ALT reversibly transfers the amino group from alanine to α-ketoglutarate, thereby forming pyruvate and glutamate (Reaction 7). The rate of formation of pyruvate is determined by coupling the ALT reaction with that of LD, which converts the pyruvate to lactate (Reaction 8). The decrease in absorbance at 340 nm is measured as NADH is oxidized to NAD^+.

Reaction 7: $\text{Alanine} + \alpha\text{-ketoglutarate} \xrightarrow{\text{ALT}} \text{glutamate} + \text{pyruvate}$

Reaction 8: $\text{Pyruvate} + \text{NADH} + H^+ \xrightarrow{\text{LD}} \text{lactate} + NAD^+$

Notes:

1. The blood specimen should be drawn and handled carefully to avoid hemolysis. Hemolysis falsely elevates the ALT.
2. Serum ALT is relatively unstable and should be kept at 4° C for short periods or stored frozen at −20° C or lower for longer periods.

Reference Values. The reference range for serum ALT is 3 to 30 U/L.

Increased Activity. The serum activity of ALT is increased in various intracellular hepatic disorders.

Gamma Glutamyltransferase (GGT)

This enzyme catalyzes the transfer of a glutamyl group from one peptide to another. It is present in numerous tissues, including kidney, liver, pancreas, and prostate. Elevation of GGT can be seen in individuals taking drugs, such as phenobarbital, that induce the microsomal drug metabolizing system. GGT is also sensitive to alcohol; chronic alcoholic persons often have elevated GGT, regardless of liver damage.

The most clinically useful application of GGT is diagnosis of diseases of the liver and biliary tract. Its degree of elevation is only moderate in hepatocellular liver disease, but is marked in obstructive liver disease. GGT helps to determine whether an elevated alkaline phosphatase is coming from bone or liver.

Ornithine Carbamoyltransferase (OCT)

Physiologically, OCT is found only in the liver, as a component of the urea cycle. It catalyzes the transfer of a carbamyl group from ornithine to citrulline. As a liver function test, its specificity is excellent. Elevations are quite marked in hepatocellular disease. A genetic deficiency of the enzyme causes severe and often fatal hyperammonemia in the newborn period (see Chap. 16). Unfortunately, the assay of this enzyme is not widely available in most clinical laboratories. The methods do not lend themselves readily to automation, and other enzymes offer the advantage of more rapid turnaround times.

Alkaline Phosphatase

Alkaline phosphatase is reviewed in the following section, but in association with liver disease, alkaline phosphatase is most useful in diagnosis of cholestasis or obstructive liver disease.

Diagnosis of Bone Disease

Alkaline Phosphatase (ALP)

Alkaline phosphatases are a group of enzymes that split off a terminal phosphate group from an organic phosphate ester in alkaline solution. Their optimum pH is usually around pH 10, but this number varies with the particular substrate and isoenzyme. ALP is widely distributed in the body and is present in high concentration in bone (osteoblasts, the cells of growing bone), intestinal mucosa, and renal tubule cells and in lower concentration in the liver (highest in the biliary tree), leukocytes, and placenta.

Alkaline Phosphatase Isoenzymes[14]

Isoenzymes of ALP separated by acrylamide or agarose gel electrophoresis have been identified in the liver, bone, intestinal mucosa, placenta, and bile. Physical and chemical properties are sometimes used for distinguishing between the various isoenzymes. For example, the placental enzyme is heat stable at 65° C, but the others are not. At 55° C, the liver isoenzyme is moderately heat stable, but the bone enzyme is heat labile. This property is sometimes used to differentiate between bone and liver disease when the total ALP is elevated. The intestinal ALP isoenzyme is inhibited by L-phenylalanine. Normal serum ALP consists of a mixture of isoenzymes derived primarily from liver, intestines, and bone; the isoenzyme from liver predominates except under conditions of rapid skeletal growth.

Principle.[14] ALP catalyzes the hydrolysis of *p*-nitrophenylphosphate to *p*-nitrophenylate ion and phosphate. The substrate is colorless, but the *p*-nitrophenylate resonates to a quinoid form in alkaline solution and strongly absorbs light at 404 nm. The reaction is followed by continuous monitoring. A buffer, such as 2-amino-2-methyl-1-propanol, is used to act as a phosphate acceptor (transphosphorylation) and to speed up the reaction. Mg^{2+} and Zn^{2+} are activators of ALP.

Reference Values.
Adults: 20 to 105 U/L
Infants and children: 0 to 3 months = 70 to 220 U/L
3 months to 10 years = 50 to 260 U/L
10 years to puberty = 60 to 295 U/L

The increased normal values for growing children reflect the increased osteoblastic activity that occurs during periods of rapid skeletal growth.

Increased Activity. Serum ALP activity is raised in all *bone disorders* accompanied by *increased osteoblastic activity*. Such disorders include Paget's disease (osteitis deformans), osteoblastic tumors with metastases, hyperparathyroidism when mobilization of Ca and P from bone exists, rickets, and osteomalacia. The activity of the hepatic enzyme, 5′-nucleotidase (5NT), is normal under these circumstances. Serum ALP activity is also increased in liver disease, particularly in *disorders of the hepatic biliary tree* (see Chap.11) and during the *third trimester of pregnancy* owing to the contribution of a placental isoenzyme of ALP that is absorbed into the maternal bloodstream. Transient but extreme elevations in ALP are occasionally seen in children, usually between the ages of 2 months to 2 years.[15] In the absence of clinical symptoms of bone or liver disease, these children should merely be followed to ensure that their ALP value returns to normal within 6 months. The transient ALP elevation has not been linked to disease and seems to be predominantly from bone.

The predominant serum ALP isoenzyme is the bone fraction in bone diseases, the liver and (sometimes) bile fractions in liver ailments, and the placental type in late pregnancy. Atypical isoenzymes may appear in some malignant conditions.

Decreased Activity. Low levels of ALP are found in *hypophosphatasia*, a rare congenital defect in dwarfs resulting from depressed osteoblastic activity, in hypothyroidism, and in pernicious anemia. Deficiencies of thyroid hormone in hypothyroidism and of vitamin B_{12} in pernicious anemia are responsible for the lowered serum ALP activity. Hypophosphatasia is also known as phosphoethanolaminuria, named for the presence of phosphoethanolamine in the urine. Patients show premature tooth loss and rickets-like bone lesions. They are most often identified by their dentist during childhood.

5′-Nucleotidase

The enzyme 5′-nucleotidase (5NT) is useful in interpreting an elevated ALP result of unknown origin.[4] 5NT is a phosphatase enzyme that splits off the phosphate from 5′-adenosine monophosphate. Its serum concentration is elevated in liver disorders, but not in bone diseases. A high ALP activity with a normal 5NT activity indicates that the ALP is coming from a tissue other than liver and presumably from bone, if such is the alternative. If both activities are elevated, the high ALP can be ascribed to a liver problem and not to bone.

Diagnosis of Muscle Disorders[4]

The three serum enzymes used most frequently for this purpose, in order of their general usefulness, are creatine kinase (CK), aspartate aminotransferase (AST), and aldolase (ALD). These enzymes are elevated in all types of progressive muscular dystrophy, but the greatest relative rise in the CK enzyme occurs in the Duchenne type of muscular dystrophy. The CK serum activity may be increased by as much as 50 × ULN early in the disease, whereas the activity of AST may be 10 × ULN and that of ALD may be 6 × ULN. The activities of the enzymes in serum become progressively lower with the duration of the disease because of the decreased muscle mass.

Raised serum enzyme levels are detectable in virtually all neurogenic muscle atrophies and are temporarily raised following muscle trauma, surgery when muscles are cut, and intramuscular injections when long-lasting preparations are used.

Methods for detecting CK and AST are described in previous sections of this chapter. The enzyme aldolase, listed in Table 10.1, converts fructose-1,6-diphosphate into two triose phosphate esters, dihydroxyacetone phosphate and glyceraldehyde-3-phosphate.[16] This process is an early step in the glycolysis of glucose.

Diagnosis of Acute Pancreatitis[4,5]

The diagnosis of acute pancreatitis is difficult to establish without the assistance of laboratory tests. The patient usually complains of an intense pain in the upper abdomen that could be caused by several different disorders. The two tests most commonly used for diagnostic purposes are those of serum amylase and serum lipase. The measurement of the amount of amylase excreted into the urine per hour also provides useful information.

Serum Amylase

Starch is the storage form of carbohydrate in plants, and glycogen is the storage form in animals. Both are polymers of glucose molecules that are linked together in a chain as condensation takes place between the hydroxyl groups on carbons 1 and 4 of adjacent glucose molecules. Water is split off, and the linkage is called an α-1,4-glycosidic bond. An occasional branch point in the starch or glycogen molecule is produced by forming an α-1,6-glycosidic bond.

In humans, the α-amylases are enzymes that randomly split the α-1,4-glycosidic bonds on a starch chain or hydrolyze related polysaccharides containing 3 or more glucose units linked by such bonds. The products of amylase action are a mixture of glucose, maltose, and dextrins (small polymers containing the branch points).

Amylases are secreted by the salivary and pancreatic glands into their respective juices, which enter the gastrointestinal tract. These enzymes are important for the digestion of ingested starches, but the amylase from the pancreas plays the major role, because the salivary amylase soon becomes inactive in the acidic condition prevailing in the stomach. Several isoenzymes of both pancreatic and salivary amylase exist. The amylase that is normally present in serum is derived from both pancreas and salivary glands.[17] The activity of serum amylase rises after an obstruction to the flow of fluid from either the salivary or the pancreatic glands, but the elevation is usually much greater when the outflow from the pancreatic gland is blocked. Acute pancreatitis is caused by blockage of the pancreatic ducts, by direct injury to the pancreatic tissue by toxins, inflammation, or trauma, or by impaired blood flow to the pancreas. The inflammation and autodigestion by pancreatic enzymes that accompany pancreatic injury usually result in an obstruction to the flow of pancreatic juice into the intestine. High levels of amylase activity may be found in pleural fluid in some cases of acute pancreatitis.

Increased Activity. Serum amylase activity is raised considerably in acute pancreatitis and obstruction of the pancreatic ducts, and mildly in obstruction of the parotid (salivary) gland. The rise in serum amylase activity after obstruction of the pancreatic ducts, whether by a stone, inflammation, or compression of the common bile duct by a cancerous growth of the head of the pancreas, is rapid and temporary. It usually reaches a maximum value, which may be from 6 to 10 × ULN, in about 24 hr, with a return to normal in 2 or 3 days. The increase in serum amylase activity caused by a stone in the parotid duct or by the disease *mumps* usually is less than 4 × ULN.

Because serum amylase is rapidly cleared by the kidney, measurement of urinary amylase is a valuable adjunct to the serum test and is described in a subsequent section. Some types of renal damage may be accompanied by a mildly elevated level of serum amylase because of impaired excretion.

Decreased Activity. A decreased concentration of serum amylase may be found in acute or chronic hepatocellular damage, but this is not a sensitive liver function test.

Several different approaches are used to determine serum amylase activity.[18] The amyloclastic methods measure the disappearance of starch substrate; saccharogenic methods measure the reducing sugars (glucose and maltose) that are produced as a result of enzymatic action; and dye methods (chromolytic) measure the absorbance of soluble dye that is split from an insoluble amylose-dye substrate. Several ultraviolet enzymatic methods for amylase that have fewer interferences have more recently been developed and are technically suited for automated analyzers. Commercial kits are available that are based on coupling *p*-nitrophenol in an α-glycosidic linkage to a glucose or maltose polymer (Behring, Boehringer Mannheim). Amylase splits off the *p*-nitrophenol, the absorbance of which can be continuously monitored at 405 nm. Most laboratories prefer to use well-defined polymers of glucose or maltose, rather than of starch, as a substrate.

Boehringer Mannheim uses a maltoheptaose (G_7, containing 7 glucose units as a polymer) bonded to *p*-nitrophenol (PNP) as a substrate. The reaction mixture also contains the enzyme α-glucosidase, which cannot split an oligosaccharide containing more than 4 glucose molecules in the chain. The following reactions occur in the serum amylase determination:

Reaction 9: $G_7PNP + 5H_2O \xrightarrow{\alpha\text{-amylase}} G_4PNP + G_3PNP + G_2PNP + \text{glucose}_{5,4,3}$

Reaction 10: $(G_4\text{-}G_3\text{-}G_2)PNP \xrightarrow{\alpha\text{-glucosidase}} X\text{-glucose} + PNP$

The absorbance at 405 nm is proportional to the amylase activity.

Serum Amylase Isoenzymes

Several isoenzymes of salivary amylase and several different isoenzymes from pancreas have been demonstrated in serum. Not all isoenzymes are present in the serum of a normal person; the most prominent peaks are P_2 (the intermediate migrating pancreas band) and S_1 (the slowest migrating salivary band). The salivary isoenzymes migrate farther toward the anode than those from the pancreas. Patients with acute pancreatitis usually exhibit a prominent P_3 isoenzyme band after electrophoretic separation of their serum. The P_3 band does not appear in patients hospitalized for nonpancreatic gastrointestinal problems.[19] An elevated P_3 isoenzyme concentration frequently occurs, however, in patients with severe kidney disease, presumably because of impaired renal clearance of the isoenzyme.

Principle. Serum is separated by electrophoresis using cellulose acetate as the support medium. After separation, the membrane is laid on an agarose gel containing a starch-dye substrate and incubated for 20 min. Dark-colored bands appear in areas to which the isoenzymes have migrated. The membrane is air-dried, and the absorbance of the various bands is read in a recording densitometer.

Amylase isoenzymes may also be determined by the Beckman Amylase DS assay performed before and after preferentially inhibiting the salivary amylase by means of a wheat germ inhibitor.[20] Many kits (Boehringer Mannheim) also use monoclonal antibodies against the salivary amylase to inhibit activity from salivary isoforms.[21] The Boehringer Mannheim pancreatic amylase method uses two monoclonal antibodies to inhibit more than 97% of the salivary isoenzyme activity, leaving the pancreatic isoenzyme activity unaffected. The pancreatic activity is then measured as previously described. This technique has proved more reliable in measuring true pancreatic amylase than has the wheat germ inhibition method.

Urine Amylase

Because amylase has a molecular weight of less than 50,000 daltons, it is readily excreted by the kidney. Accordingly, the urinary excretion of amylase is high in patients with acute pancreatitis. The increased excretion of amylase persists longer than the elevation in serum amylase activity and can help to establish the diagnosis of acute pancreatitis. The urinary amylase may be elevated for 7 to 10 days, whereas the serum amylase returns to normal in 2 or 3 days after an attack.

Procedure. Collect an accurately timed urine specimen for 1 to 2 hr and measure the enzyme activity in the same manner as for serum.

Calculation. Calculate the enzyme activity/mL urine in the same manner as for serum. Multiply this value times the urine volume (mL)/hr.

Reference Values. The normal excretion of amylase is 2 to 19 U/hr by the Beckman DS Assay.

Amylase Creatinine Clearance Ratio (ACCR). The clearance of amylase can be determined from a simultaneous urine and serum sample. This procedure may be preferable if an accurately timed urine cannot be collected.

$$\text{ACCR (\%)} = \frac{\text{Urine amylase (U/L)} \times \text{serum creatinine (mg/dL)}}{\text{Serum amylase (U/L)} \times \text{urine creatinine (mg/dL)}} \times 100$$

Increased Excretion. The amount of amylase excreted per hour may become high in acute pancreatitis, as much as 5 to 10 × ULN. The elevation may persist for as long as a week after the serum amylase has returned to normal. Other conditions that increase the serum amylase also result in an elevated urinary amylase excretion.

Decreased Excretion. The test for urinary amylase is meaningless in acute or chronic renal disease because, with a decreased glomerular filtration rate (as shown by a decreased creatinine clearance), the clearance of amylase is also decreased. Severe damage to hepatic cells may be accompanied by a lowered level of serum amylase. The reporting of results as a ratio of amylase clearance to creatinine clearance may be misleading if renal or hepatic injury is present.

Serum Lipase

Lipase is an enzyme that hydrolyzes emulsified triglycerides. The ester bonds at carbon atoms 1 and 3 of glycerol are preferentially split, thereby liberating 2 moles of long-chain fatty acids and 1 mole of 2-acylmonoglyceride per mole of triglyceride attacked. The reaction with triolein follows:

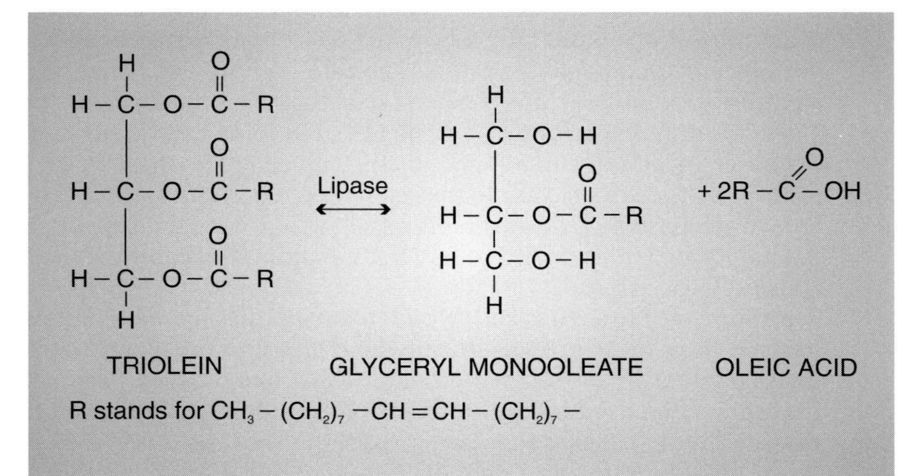

The pancreas is the principal organ for the production of lipase, which is secreted in the pancreatic juice along with the other digestive enzymes. Some lipases may be found in gastric and intestinal mucosa, but they play no significant role in the digestion of fat.

Serum normally contains a low activity of lipase, but the tissue source of this enzyme has not been clearly identified. Like serum amylase, the activity of serum lipase is rapidly elevated in acute pancreatitis, but remains elevated for a longer time than does amylase. The classic method for assaying serum lipase (hydrolysis of an olive oil emulsion and titration of the liberated fatty acids) was not easy to perform technically and was not as reproducible as the serum amylase test; consequently, it was not frequently used. The determination of serum lipase by a specific turbidimetric method[22] is much simpler to perform. A modification of this method is now available on the DuPont *aca*, and similar spectrophotometric methods are available from Boehringer Mannheim and Kodak (Ektachem analyzer). The presence of colipase in the Boehringer Mannheim and Kodak methodologies provides better correlation with the reference methods. The inclusion of colipase eliminates activity of lipoprotein lipase (postheparin clearing factor), thus giving the assay higher specificity for pancreatic lipase.

Increased Concentration. Serum lipase activity is elevated in acute pancreatitis and may reach 10 to 40 × ULN, depending on the method of assay and the time at which the sample is collected. The serum lipase activity usually reaches its maximum at 72 to 96 hr after an attack of acute pancreatitis and declines more slowly than does the amylase activity.

Diagnosis of Metastasizing Cancer of the Prostate

Cancer of the prostate is a common disease that primarily affects elderly men. Diagnosis in its early stages is difficult because of the lack of signs and symptoms; usually the diagnosis is made only after metastasis has occurred.

The prostate gland is rich in acid phosphatase (ACP); ACP is present in lower concentration in many tissues (spleen, kidney, liver, bone, blood platelets, and others). ACP acts optimally below pH 6.0 to split off a phosphate group from various organic phosphate esters. The ACP activity of normal serum is derived from some of the previously named tissues, but primarily from blood platelets. Normal serum has a low activity of ACP, but in metastasizing carcinoma of the prostate, its activity increases greatly and may rise to 3 to 15 × ULN. The carcinoma must metastasize, that is invade blood capillaries, lymph channels, and other tissues, before the elevation in the serum level of ACP occurs; a discrete prostatic cancer that has not penetrated beyond the capsule does *not* cause this rise in serum ACP.

Because erythrocytes and blood platelets also contain an ACP, the ACP derived from these sources during the clotting of the blood specimen must be distinguished from the ACP coming from the prostate. Two different techniques may be used to help to identify the serum ACP derived from prostatic tissue. The first technique uses a substrate that the ACP from prostate splits more readily than does the ACP from platelets and erythrocytes; sodium thymolphthalein monophosphate and α-naphthylphosphate are such substrates. The second technique measures the ACP activity before and after adding tartrate to the mixture. Tartrate greatly inhibits the ACP from prostate, but is much less inhibitory for the ACP from erythrocytes or platelets. A combination of both techniques is considered most satisfactory when α-naphthylphosphate is used as the substrate.

Increased Concentration. Increased activity of the tartrate-inhibitable ACP is characteristically found in metastasizing carcinoma of the prostate. Occasionally, elevated activities of the tartrate-inhibitable ACP may be found in Gaucher's disease or some bone diseases (Paget's disease or female breast cancer that has metastasized to bone), and these elevations in ACP obviously are not derived from the prostate. Massage of the prostate increases ACP activity for 1 or 2 days.

Decreased Concentration. No physiologic significance is attached to a low serum ACP activity.

ACP Isoenzymes

Many ACP isoenzymes can be separated electrophoretically; several originate in the prostate gland. The different prostatic isoenzymes vary in their content of sialic acid (an acetylated amino-sugar acid), and because this group is charged, the isoenzymes migrate to different spots in an electric field. Various immunotechniques, such as RIA, counterimmunoelectrophoresis, and enzyme immunoassay based on a monoclonal antibody against a prostate-specific acid phosphatase isoenzyme (PAP), are used in an attempt to detect a prostatic carcinoma before it is palpable (detectable by touch in a rectal examination). These immunoassays are about 1000-fold more sensitive than conventional activity measurements (See Chap. 4). The sensitivity of PAP is estimated between 31 to 69% for detection of early prostatic carcinoma, but the false-positive rate (prostatic nodular hyperplasia) is about 10%.[3]

More recently, PAP has been supplemented, and in some cases replaced, by prostate specific antigen (PSA).[23,24] PSA, a glycoprotein produced only by prostate tissue, was first isolated by Wang in 1979.[25] The specificy of this protein for prostate makes it an excellent tumor marker. Although PSA is tissue specific, it is not cancer specific, as it is found in benign, normal, and malignant prostates. An estimated 3 to 21% of patients with benign prostatic hypertrophy have modest increases (> 10 μg/L) of PSA. Sensitivities for PSA in tumor detection are reported to range from 73 to 84%. The overlap of prostate cancer with benign prostatic hypertrophy has encouraged many laboratories to offer both tests as screening tools for men following surgical removal of tumor tissue to look for recurrence. Screening is also useful in the older male population, especially when symptoms (frequent urination) occur.

Methods that employ some type of immunoassay are commercially available from Yang Laboratories (RIA), Hybritech (enzyme immunoassay [EIA]), Abbott (microparticle capture EIA), and Ciba-Corning (chemiluminescence immunoassay).

Investigation of Genetic Disorders

Most genetic diseases are characterized by a deficiency in at least one enzyme system. In certain situations, the deficiency can be demonstrated by finding an absent or greatly decreased enzyme activity in particular tissues or cells. Carriers of the trait usually demonstrate a moderately lowered concentration of the enzyme. This type of enzyme analysis is usually carried out in special laboratories dedicated to the study of genetic disorders because of the special techniques involved. In the following paragraphs, a few of the enzymes measured for this purpose are mentioned, but details of the methodology are not given.

Pseudocholinesterase

Patients undergoing surgery frequently are injected with the short-acting muscle relaxant succinylcholine. An enzyme in plasma, pseudocholinesterase, rapidly hydrolyzes the succinylcholine so that its action in the body is shortlived. Some people have a genetic deficiency of this enzyme,[4] and when injected with succinylcholine, they fail to inactivate the drug and may be subjected to its action for as long as 2 to 3 hr instead of the usual 2 min. The respiratory muscles may be so relaxed by the drug that breathing is inadequate and the patient's life is endangered. This situation can be avoided by screening all patients beforehand for decreased activity of pseudocholinesterase, using succinylcholine as the substrate.

Measurement of the enzyme (in red cells instead of serum) is also useful in following patients who have been exposed to organophosphates, an insecticide commonly used in certain farm areas. Enzyme inhibition is proportional to the amount of exposure to the insecticide.

Glucose-6-phosphate Dehydrogenase (GPD)

The enzyme GPD catalyzes the oxidation of glucose-6-phosphate as the first step in the pentose phosphate pathway. The enzyme is present in many types of cells, but it is particularly important in the red blood cell because the NADPH generated is a necessary ingredient for other enzyme systems in the erythrocyte to prevent the accumulation of methemoglobin. Some people have a genetic deficiency of GPD, and those who have 30% or less of the normal amount of the enzyme are susceptible to acute hemolysis of the red blood cells if exposed to certain drugs, such as antimalaria drugs (primaquine), some sulfonamides, quinine, and others. A hemolytic crisis can also be precipitated by the ingestion of fava beans or inhalation of pollen in some sensitive individuals. Screening methods are available for detecting a deficiency of GPD in red cells. An excellent review of the subject, with a review of various methodologies, has been written by Luzzatto and Mehta.[26]

Galactose-1-phosphate Uridyltransferase

The disease galactosemia is caused by hereditary deficiency of the enzyme galactose-1-phosphate uridyl transferase, which is necessary for the conversion of ingested galactose to glucose. This conversion must take place before galactose can be used for energy. In the absence of the enzyme, continued ingestion of galactose (a component of lactose, the sugar in milk) causes a buildup of galactose-1-phosphate in cells. This buildup produces the typical symptoms of galactosemia: cataract formation, enlarged liver and spleen, and mental retardation. If the deficiency is discovered early enough, the infant can be saved by removing all sources of galactose from the diet (substituting an artificial milk devoid of lactose for natural milk).

Infants suspected of having galactosemia can be checked by analyzing their red blood cells for the presence of the enzyme galactose-1-phosphate uridyl transferase. The subject is well reviewed by Segal.[27]

Enzymes as Reagents or Labels

In addition to the measurement of serum enzyme activity as an aid to the diagnosis and management of disease, enzymes are widely used in the clinical chemistry laboratory as reagents highly specific for particular chemical constituents of serum or as reagents in coupled reactions to yield a product that can be measured conveniently. The following are examples:

1. *Hexokinase* and *glucose-6-phosphate dehydrogenase* in the measurement of CK activity. These enzymes are used as coupled enzymes to generate NADPH (See Reactions 2 and 3, this chapter).
2. *Glucose oxidase* in the determination of serum glucose concentration. Glucose oxidase reacts specifically with glucose + O_2 to produce gluconic acid + H_2O_2. The glucose concentration may be determined from the rate of O_2 consumption or from the amount of H_2O_2 produced.
3. *Uricase* in the determination of uric acid. The enzyme converts uric acid into allantoin, CO_2, and H_2O_2. The urate concentration can be measured by the decrease in absorbance at 293 nm or by quantitation of the H_2O_2 produced.

Much more recently, enzymes have been introduced into the clinical chemistry laboratory as a means of labeling antigens or antibodies in various immunoassays for determining the serum concentration of drugs, hormones, or other compounds of interest. This role is elaborated further in Chapter 4. The use of an enzyme label provides a means for amplification of a chemical reaction; it greatly increases the sensitivity of detection of constituents in serum by several orders of magnitude. Enzyme labeling approaches RIA in sensitivity and specificity, but is performed without the use of radioactive compounds or counting equipment.

REFERENCES

1. Wilkinson, J.H.: The Principles and Practices of Diagnostic Enzymology. Chicago, Year Book Medical Publishers, 1976.
2. Dixon, M., and Webb, E.C.: Enzymes. 3rd Edition. New York, Academic Press, 1979.
3. Killian, C.S., et al.: Prognostic importance of prostate-specific antigen for monitoring patients with stages B2 to D1 prostate cancer. Cancer Res., *45*:886, 1985.
4. Moss, D.W.: Isoenzymes. London and New York, Chapman and Hall, 1982.
5. Lott, J.A.: Serum enzyme determinations in the diagnosis of acute myocardial infarction. Hum. Pathol., *15*:706, 1984.
6. Szasz, G., Gruber, W., and Bernt, E.: Creatine kinase in serum. Clin. Chem., *22*:650, 1976.
7. Wu, A.H.B.: CK-MB assay methods: a comparison. Lab. Manager, *23*:44, 1985.

8. Loshon, C.A., Rittenhouse, S.E., Bowers, G.N., Jr., and McComb, R.B.: Unusual findings related to atypical creatine kinases in two hospitalized patients. Clin. Chem., *32*:207, 1986.
9. Lang, H., and Wurzburg, U.: Creatine kinase: an enzyme of many forms. Clin. Chem., *28*:1439, 1982.
10. James, G., and Harrison, R.L.: Creatine kinase isoenzymes of mitochondrial origin in human serum. Clin. Chem., *25*:943, 1979.
11. Vaidya, H.C., Maynard, Y., Dietzler, D.N., and Ladenson, J.H.: Direct measurement of creatine kinase-MB activity in serum after extraction with a monoclonal antibody specific to the MB isoenzyme. Clin. Chem., *32*:657, 1986.
12. A modification of Rosalki, S.B.: Standardization of isoenzyme assays with special reference to lactate dehydrogenase isoenzyme electrophoresis. Clin. Biochem., *7*:29, 1974.
13. Bergmeyer, H.U., Scheibe, P., and Wahlefeld, A.W.: Optimization of methods for aspartate aminotransferase and alanine aminotransferase. Clin. Chem., *24*:58, 1978.
14. McComb, R.B., Bowers, G.N., Jr., and Posen, S.: Alkaline Phosphatase. New York, Plenum Press, 1979.
15. Posen, S., Lee, C., Vines, R., et al.: Transient hyperphosphatasia of infancy—an insufficiently recognized syndrome. Clin. Chem., *23*:292, 1977.
16. Pinto, P.V.C., Kaplan, A., and Van Dreal, P.A.: Aldolase: II. Spectrophotometric determination using an ultraviolet procedure. Clin. Chem., *15*:349, 1969.
17. Goldberg, D.M.: Enzymes and isoenzymes in the evaluation of diseases of the pancreas. *In* Clinical and Analytical Concepts in Enzymology. Edited by H.A. Homburger. Skokie, IL, College of American Pathologists, 1983.
18. Kaufman, R.A., and Tietz, N.W.: Recent advances in measurement of amylase activity—a comparative study. Clin. Chem., *26*:846, 1980.
19. Legaz, M.E., and Kenny, M.A.: Electrophoretic amylase fractionation as an aid in diagnosis of pancreatic disease. Clin. Chem., *22*:57, 1976.
20. Huang, W.Y., and Tietz, N.W.: Determination of amylase isoenzymes in serum by use of a selective inhibitor. Clin. Chem., *23*:1525, 1982.
21. Zaninotto, M., Bertorelle, R., Secchiero, S., et al.: Assay of pancreatic amylase with use of monoclonal antibodies evaluated. Clin. Chem., *34*:2552, 1988.
22. Hoffmann, G.E., and Weiss, L.: Specific serum pancreatic lipase determination, with use of purified colipase. Clin. Chem., *26*:1732, 1980.
23. Griffiths, J.C.: The laboratory diagnosis of prostatic adenocarcinoma. CRC Crit. Rev. Clin. Lab. Sci., *19*:187, 1983.
24. Armbruster, D.A.: Prostate-specific antigen: Biochemistry, analytical methods, and clinical application. Clin. Chem., *39(2)*:181, 1993.
25. Wang, M.C., Valenzuela, L.A., Murphy, G.P., and Chu, T.M.: Purification of a human prostate specific antigen. Invest. Urol., *17*:159, 1979.
26. Luzzatto, L., and Mehta, A.: Glucose-6-phosphate dehydrogenase deficiency. *In* The Metabolic Basis of Inherited Disease. 6th Edition. Edited by C.R. Scriver, et al. New York, McGraw-Hill, 1989.
27. Segal, S.: Disorders of galactose metabolism. *In* Metabolic Basis of Inherited Disease. 6th Edition. Edited by C.R. Scriver. New York, McGraw-Hill, 1989.

REVIEW QUESTIONS

1. How do enzymes act, and what parameters affect the velocity of the reaction?
2. What is meant by zero order kinetics?
3. If you were working on a new method for measuring the serum concentration of enzyme X, how would you determine optimal conditions?
4. What is the difference between an enzyme and each of its isoenzymes (if it has any), and how do isoenzymes arise?
5. What enzymes and/or isoenzymes are useful in the diagnosis of (a) myocardial infarction, (b) acute hepatitis, (c) acute pancreatitis?

6. Why is the time factor (time of sample in relation to the onset of symptoms) so important when using enzymes or isoenzymes for the diagnosis of a possible myocardial infarction or acute hepatitis?

7. How may measurement of NADH (appearance or disappearance) be incorporated into the final step of some enzyme determinations?

8. Calculate the activity of AST in U/L given the following:
Δ Absorbance/min = 0.15
The molar absorptivity of NADH (in L/mol^{-1}/cm^{-1} at 340 nm) = 6220
Total volume of reaction mixture = 5.0 mL
Sample volume = 0.5 mL
(A) 241 U/L
(B) 42 U/L
(C) 0.241 U/L
(D) 109 U/L

9. How may enzymes be used as reagents? Give three examples.

11 The Liver and Tests of Hepatic Function

OUTLINE

I. PHYSIOLOGIC ROLE OF THE LIVER
- A. Metabolic functions
  - 1. Carbohydrate metabolism
  - 2. Protein synthesis
  - 3. Lipid metabolism
  - 4. Other syntheses or transformations
- B. Storage functions
- C. Excretory functions
- D. Protective functions
  - 1. Phagocytic action
  - 2. Detoxification
- E. Circulatory functions
- F. Function in blood coagulation

II. BILE PIGMENT METABOLISM
- A. Bilirubin
- B. Disorders of bile pigment metabolism
  - 1. Excessive load of bilirubin presented to liver
  - 2. Defective transport into the hepatocyte
  - 3. Congenital deficiency of the uridyldiphosphate glucuronyl (UDPG) transferase enzyme system
  - 4. Impairment in the esterification of bilirubin
    - a. Physiologic jaundice of the newborn
    - b. Hemolytic disease of the newborn (HDN) or Isoimmune hemolytic disease (IHD)
  - 5. Disturbances in excretion of bilirubin (cholestasis)
    - a. Intrahepatic
    - b. Posthepatic

III. HEPATIC DYSFUNCTION
- A. Viral hepatitis
  - 1. Hepatitis A virus (HAV)
  - 2. Hepatitis B virus (HBV)
  - 3. Non-A, Non-B hepatitis virus (NANBV)
    - a. Hepatitis C virus (HCV)
    - b. Hepatitis E virus (HEV)
    - c. Delta hepatitis virus (HDV)
- B. Alcoholic liver disease
- C. Hepatic drug toxicity
- D. Cholestasis
- E. Hepatic encephalopathy (hepatic coma)

IV. TESTS FOR DISCLOSING HEPATIC DYSFUNCTION
- A. Serum bilirubin
  - 1. Principles of analysis
    - a. Total bilirubin
    - b. Esterified (conjugated) bilirubin
    - c. Precautions
    - d. Reference values
    - e. Increased concentration
      - (1) Total bilirubin
      - (2) Esterified bilirubin
      - (3) Nonesterified bilirubin

IV. TESTS FOR DISCLOSING HEPATIC DYSFUNCTION *(continued)*

f. Decreased concentration

2. Direct spectrophotometric determination
   a. Principle

B. Urine bilirubin
   1. By tablet
   2. By dipstick

C. Urine urobilinogen
   1. By dipstick

D. Fecal urobilinogen

E. Dye tests of excretory function

F. Tests based on abnormalities of serum protein
   1. Albumin
   2. Gamma globulins
   3. Alpha globulins
   4. Clotting factors
      a. Interpretation

G. Selected enzyme tests
   1. Serum alkaline phosphatase (ALP)
      a. Reference values
      b. Increased activity
   2. Aminotransferases (AST and ALT)
      a. Reference values
      b. Increased activity

H. Less common enzyme tests for special situations
   1. γ-glutamyltransferase (GGT)
      a. Reference values
      b. Increased activity
   2. 5′-Nucleotidase (5NT)
      a. Reference values
      b. Increased activity

OBJECTIVES

After reading this chapter, the student will be able to:

1. Describe the many different functions of the liver and explain how their impairment can provide evidence of liver injury.
2. Describe how bilirubin is formed, its fate in the body, and its significance.
3. Identify some of the different disorders of bile pigment metabolism (jaundice) and classify as prehepatic, hepatic, or posthepatic.
4. Describe the types of viral hepatitis, understand their modes of transmission, and follow safety precautions when taking body fluids from these patients.
5. Describe the analytical principles for performing the following tests:
   a. bilirubin, total, esterified, and nonesterified.
   b. albumin and total protein.
   c. protein fractionation by electrophoresis.
   d. selected enzyme tests.
6. Describe the formation, physiologic role, and fate of bile acids.
7. Describe the role of the liver in maintaining cholesterol homeostasis.
8. Recognize typical abnormalities in the panel of liver performance (function) tests that are associated with common hepatic diseases.
9. Describe the role of the liver in the production of blood coagulation factors and how liver disease may affect the process.

The liver is an essential organ for performing many physiologic and metabolic functions necessary for life. It plays a central role in protein, carbohydrate, and lipid metabolism and in the intermediary metabolism of amino acids, simple sugars (hexose, pentose, triose), and fatty acids. The liver synthesizes blood coagulation factors and essentially all the plasma proteins except immunoglobulins; it is the principal storage center for carbohydrate (glycogen); it temporarily stores small amounts of protein and lipid; it produces and secretes bile into the intestine, a fluid necessary for the emulsification of lipids as a prerequisite for their digestion and absorption; and it protects the body from various foreign or dangerous materials by phagocytic action and by detoxification reactions. The liver is the only organ in the body of mammals that has the ability to regenerate; it can grow back after surgical removal of a large portion (as much as approximately 80%) of the organ.

Liver injury may impair to some degree the performance of one or more of the hepatic functions, depending on the site and extent of the damage. When the injury is accompanied by hepatic inflammation or necrosis (death of cells), the hepatocytes release into the circulation some of their intracellular enzymes. A combination of liver function tests, together with a few selected serum enzyme measurements, may provide information for ascertaining the status of the liver. The laboratory results may aid the physician in the diagnosis of liver disease, in prognosis during the course of the illness, and in confirmation of recovery.

PHYSIOLOGIC ROLE OF THE LIVER

The liver, a large organ, lies just under the diaphragm on the right side of the body. It has a tremendous blood supply and weighs about 1500 g. Like all organs, the liver receives oxygenated blood pumped from the heart in an artery that branches and subdivides to arterioles and finally perfuses the tissue as capillaries. The arterial exchange of O_2 for CO_2 between blood and tissue cells and the exchange of nutrients and waste products take place in the capillary beds. The capillary branches combine to form venules and then veins that finally return the blood from the organ to the heart.

The hepatic artery and hepatic vein serve this function for the liver, but the liver is unique in that it has a second great venous blood supply coursing through it. The portal vein, which brings blood from all parts of the gastrointestinal tract (stomach, small and large intestines, pancreas, and spleen), flows directly from these organs to the liver, where it terminates in capillary-like vessels called sinusoids; the blood from the sinusoids finally leaves the liver by merging with the blood of the hepatic vein. The great physiologic significance of the portal flow to the liver is that *all nutrients arising from the digestion of food in the gastrointestinal tract,* with the exception of fats, *pass through the liver* before entering the general circulation for transmission to the rest of the body. This nutrient-rich blood supply provides the liver with a high concentration of various substrates that enables it to carry out the many metabolic functions that characterize it as the "metabolic factory" of the body.

The liver is also unique in its fine anatomic structure or architecture. In essence, it is characterized by a series of plates of hepatic cells, one-layer thick, in contact with portal vein "capillaries" (sinusoids) on two sides, as well as with bile canaliculi that convey hepatic cell secretions (bile) to bile ducts. Thus, each hepatic cell (hepatocyte) has a large surface area in contact with both a nutrient intake system from the sinusoids and an outlet system, the bile canaliculi, which carry away the secretions and excretions from the hepatocytes.[1,2] The secretion arising from the hepatic cells is called bile. The bile canaliculi combine to form bile ducts that carry the bile secretion into the small intestine. The anatomic arrangement of the liver provides maximum efficiency for receiving materials to be used or processed, as well as for removing or excreting undesirable products. The flow of blood and bile in a liver lobule is shown schematically in Figure 11.1.

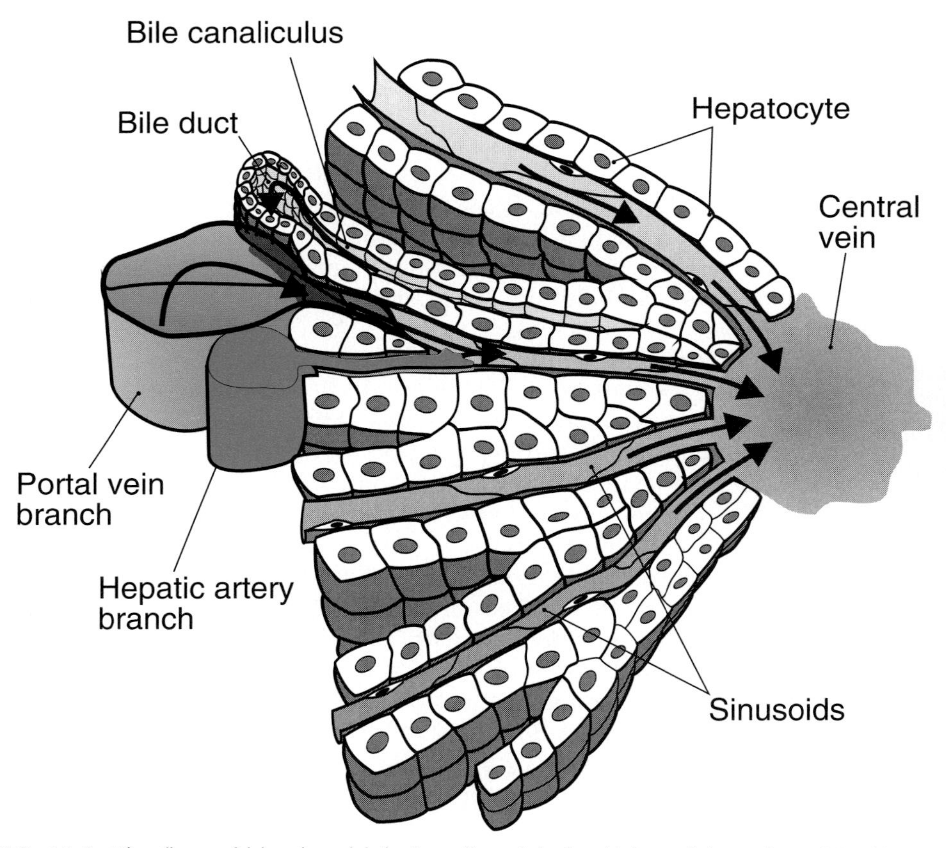

FIG. 11.1. The flow of blood and bile in a liver lobule. (Adapted from Fox, S.I.: Human Physiology. 3rd Edition. Dubuque, IA, W.C. Brown, 1990.)

Metabolic Functions

The liver carries out all the diverse reactions involved in both the anabolic and the catabolic phases of protein, carbohydrate, and lipid metabolism; some of these reactions are performed exclusively in this organ.

Carbohydrate Metabolism

The liver converts glucose to glycogen (glycogenesis) by one enzymatic pathway during periods of excess carbohydrate intake and converts glycogen to glucose (glycogenolysis) by a different pathway during times of high energy demand or lowered plasma glucose concentration. It converts lactate (arising from working muscle) back to glucose and, in times of a high and persistent glucose demand, synthesizes glucose from the catabolism of some amino acids (gluconeogenesis). The liver, in response to neural and hormonal signals, maintains a relatively constant plasma glucose concentration as it modifies the rates of the glycogenesis, glycogenolysis, and gluconeogenesis reactions (see Chap. 7).

Protein Synthesis

Almost all the plasma proteins, with the exception of the gamma globulins, are synthesized in the liver. The plasma proteins originating in the liver include albumin, apolipoproteins, acute phase reactants (α_1-antitrypsin, complement C_3, ceruloplasmin, C-reactive protein, haptoglobin), prealbumin, many specific carrier proteins, and many of the proteins involved in the blood coagulation process. These blood coagulation proteins include prothrombin; fibrinogen; Factors V, VII, and X from the extrinsic clotting system; and Factor IX from the intrinsic clotting system.

Lipid Metabolism

The liver is an active center of lipid anabolism and catabolism. It synthesizes the endogenous lipids (triglycerides, cholesterol, and phospholipids) described in Chapter 8, as well as the lipoproteins necessary to transport these lipids to tissues. Hepatocytes also catabolize lipids; fatty acids are broken down into a series of acetyl CoA units that may be utilized for the synthesis of other compounds or for the production of energy. Hepatic cells catabolize cholesterol by transforming it into the primary bile acids (cholic and chenodeoxycholic acids, Fig. 11.2), which are then conjugated in an amide linkage with either of the two amino acids glycine or taurine (Fig. 11.3). These conjugates are secreted into bile. Bile acid conjugates, together with phospholipids, form the outer membrane of micelles that enclose and solubilize cholesterol in hepatic cells and thus make possible its excretion. Bile acids and their conjugates act like detergents and, in the intestine, are essential emulsifiers of ingested lipids. Fat digestion and absorption do not occur when bile is excluded from the intestine because of bile duct obstruction or severe cholestasis.

Other Syntheses or Transformations

Other metabolic transformations too numerous to mention here occur in the liver. These include various transaminations (conversion of a keto acid to an amino acid, with concomitant production of a new keto acid), production of

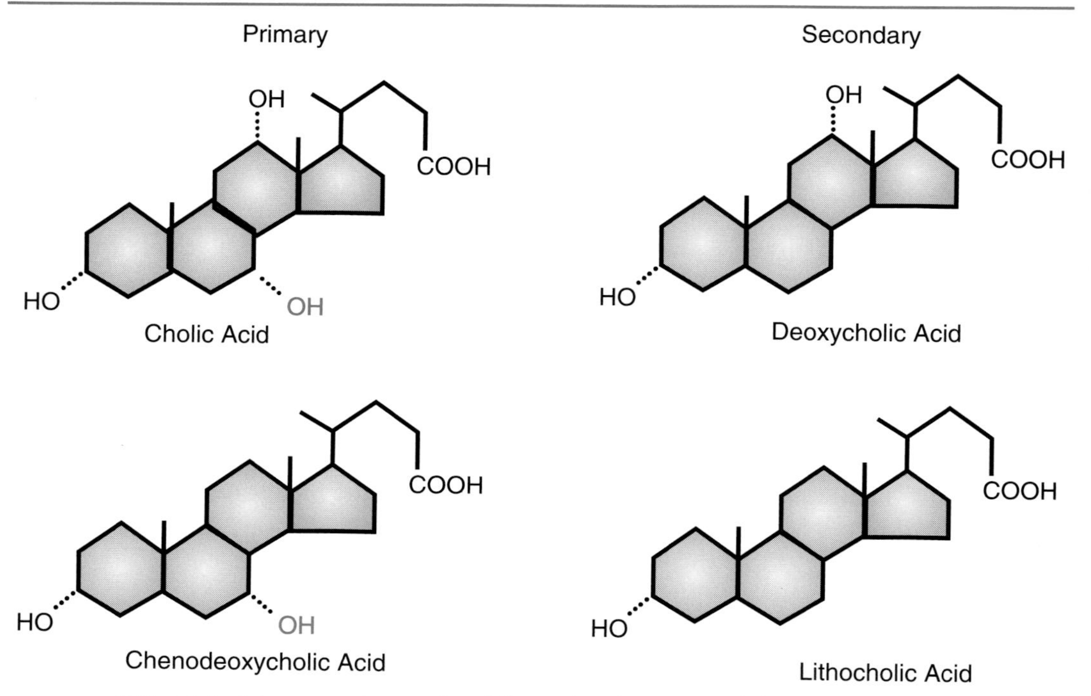

FIG. 11.2. Primary and secondary bile acids. The primary bile acids (cholic and chenodeoxycholic) are produced in the liver from the catabolism of cholesterol. The secondary bile acids (deoxycholic and lithocholic) are formed in the intestine as a result of bacterial removal of a hydroxyl group from cholic and chenodeoxycholic acids, respectively.

ketone bodies (acetoacetic acid and β-hydroxybutyric acid) from acetyl CoA, conversion of some amino acids to glucose and vice versa, conversion of lactic acid to glucose (or glycogen), synthesis of fatty acids from acetyl CoA, and many other reactions of importance.

Storage Functions

The liver is the primary storage site for glycogen and for vitamins A, D, and B_{12}. Significant amounts of iron are stored in the reticuloendothelial (R-E) cells of the liver, as well as in the bone marrow and spleen. Normally, the storage of lipids in the liver is limited.

Excretory Functions

The liver secretes into the bile canaliculi many different compounds to form a viscous fluid called bile. The bile contains bilirubin esters, bile acid conjugates, cholesterol, phospholipids, organic anions, soluble derivatives of steroid hormones and drugs, electrolytes, and many other materials. The bile is usually stored in the gallbladder until mealtimes; cholecystokinin-pancreozymin, a

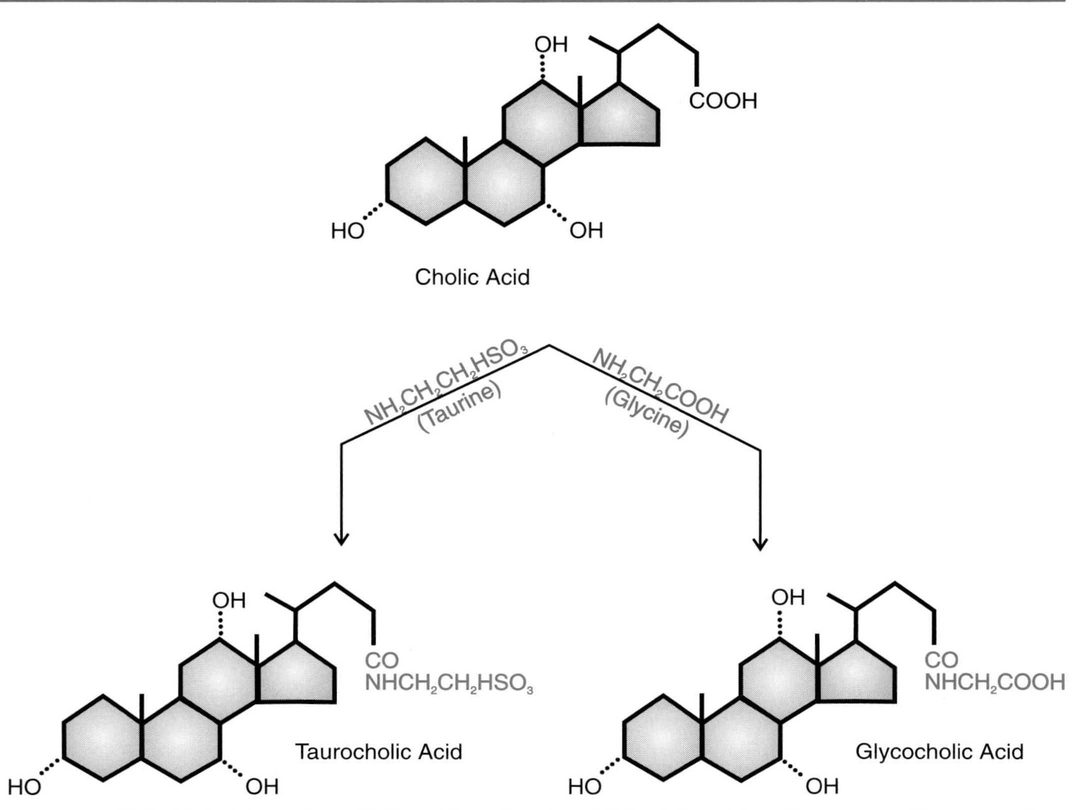

FIG. 11.3. Formation of bile acid conjugates. All the bile acids, primary as well as secondary, are conjugated with either glycine or taurine in the cells lining bile canaliculi and ductules. The reaction shown uses cholic acid as a model.

hormone produced by the intestinal mucosa, stimulates the gallbladder to contract and eject the bile into the small intestine (duodenum). Bile acids and their conjugates effectively emulsify all types of lipids as a necessary step for their digestion and absorption.

All the bile constituents in the intestine are subjected to hydrolytic action by enzymes, to partial degradation by bacteria, and to partial reabsorption before the residual intestinal contents are excreted as feces. Bacterial action removes a hydroxyl group from each of the primary bile acids, converting them into the secondary bile acids, deoxycholic and lithocholic acids (Fig. 11.2). Some of the cholesterol and bile acids (primary and secondary) are absorbed for recycling in a process called the *enterohepatic circulation* because the absorption products return directly to the liver via the portal vein.

Protective Functions

The liver helps to protect the body from various foreign or dangerous materials by phagocytic action and detoxification reactions.

Phagocytic Action

The liver contains many phagocytic cells, called Kupffer cells, which line the sinusoids and are active in removing foreign materials from the blood. The Kupffer cells are part of the reticuloendothelial system and act as scavengers.

Detoxification

Many noxious or comparatively insoluble compounds, including most therapeutic drugs, are converted to other forms that either are less toxic or become water-soluble and therefore excretable by the kidney. Conversion to a less toxic form many involve esterification, acetylation, methylation, oxidation, reduction, or other changes. Ammonia, a toxic substance arising in the large intestine through bacterial action upon amino acids, is carried to the liver by the portal vein and converted into the innocuous compound urea by the hepatocytes. Reaction with glucuronic acid is a common mechanism for converting insoluble materials into water-soluble compounds, glucuronides, that can be excreted by the kidney. The insoluble pigment bilirubin is esterified to a mono- or diglucuronide, which is then excreted into bile. More commonly, glucuronic acid is combined in ether linkage with hydroxyl groups of hormones and drugs. Likewise, many of the steroid hormones are converted into glucuronides and excreted through the kidney, thereby shortening their active stays in the body. A portion of these materials may be converted into sulfates and excreted.

Circulatory Functions

The liver plays a role in immunologic defense through its R-E system (Kupffer cells), helps to regulate blood volume by serving as a blood storage area, and is a means for mixing the blood from the portal system with that of the systemic circulation.

Function in Blood Coagulation

As mentioned under the protein synthesis section, certain protein clotting factors are synthesized solely by hepatic cells. These factors are fibrinogen, prothrombin, and Factors V, VII, IX, and X. Factors V, VII, IX, and X have short half-lives, turn over quite rapidly, and thus may quickly become limiting for the coagulation process in the presence of severe hepatic disease. Prothrombin and Factors VII, IX, and X require the presence of vitamin K, a fat-soluble vitamin, for their synthesis.

BILE PIGMENT METABOLISM

Bilirubin

Bilirubin, the principal pigment in bile, is derived from the breakdown of hemoglobin when senescent red blood cells are phagocytized. The erythrocytes are loaded with hemoglobin, a complex molecule containing four heme groups (ferroprotoporphyrin) attached to a protein, globin (see Fig. 11.4). The erythrocytes survive in the body for about 120 days before the aged cells are engulfed and digested by phagocytic cells. These scavenger cells of the R-E system that destroy the old red blood cells are located primarily in the spleen, liver, and bone marrow. The spleen is the main graveyard of the old red cells. The degradation of the hemoglobin molecule, shown schematically in Figure 11.4, leads to the formation of bilirubin. About 80% of the bilirubin formed

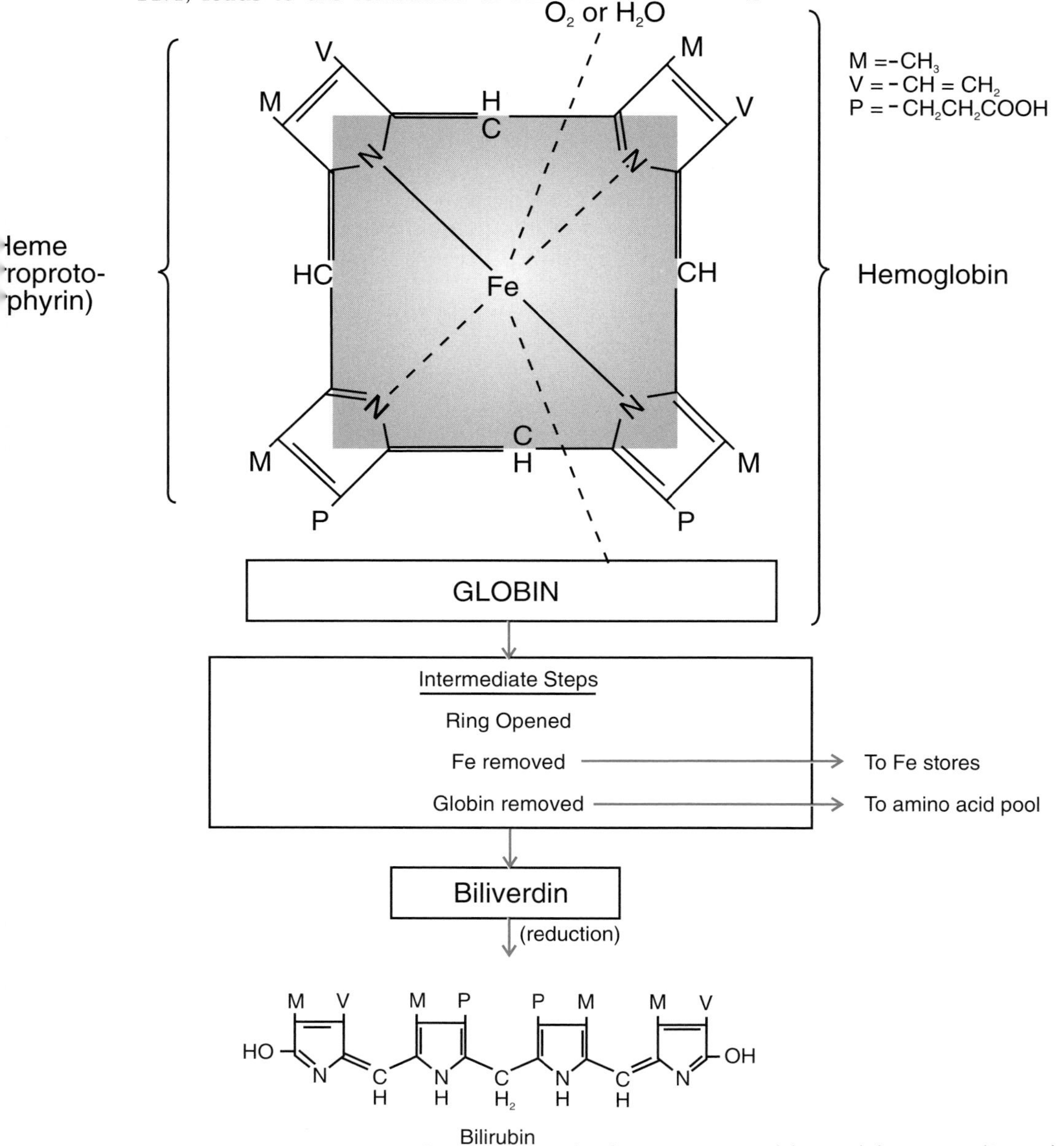

FIG. 11.4. Degradation of hemoglobin. The heme portion of hemoglobin is cyclic and contains four substituted pyrrole rings connected by methene groups; Fe^{2+} is bound inside the pyrrole chain. Hemoglobin is converted to bilirubin by a series of steps that include disruption of the cyclic pyrrole chain, removal of Fe and of globin, and reduction of one methene group.

daily is derived from the breakdown of the senescent erythrocytes; the remainder comes from the degradation in bone marrow of immature red blood cells and from the destruction of other heme-containing proteins (myoglobin, catalase, cytochromes). Thus, about 6 to 6.5 g of hemoglobin in aged red blood cells are broken down daily in an adult to form about 220 mg of bilirubin; another 50 or 60 mg of bilirubin originate from other sources. The stepwise breakdown of hemoglobin and the fate of the product, bilirubin, are shown schematically in Figure 11.5.

Step 1. The exact mechanism is not well understood, but the net effect is the splitting off of the protein globin, which may be reused or hydrolyzed to amino acids that join the amino acid pool for recycling. The porphyrin ring of the heme molecule is broken open, with loss of one of the methene groups connecting the four pyrrole rings. The resulting open-chain tetrapyrrole loses its iron, which becomes bound to a protein (ferritin), where it is stored until it is used again for the synthesis of new heme compounds in the bone marrow. The tetrapyrrole is reduced to form bilirubin, a reddish-yellow waste product that must be excreted.

Step 2. Bilirubin leaves the R-E cell and is solubilized in plasma by firmly binding to the protein albumin.

Step 3. Upon reaching the liver sinusoids, the bilirubin-albumin complex attaches to the hepatocyte membrane. The bilirubin is detached from its albumin carrier and is transported inside the hepatic cell network to microsomes in rough endoplasmic reticulum by a transport protein, ligandin.

Step 4. Esterification (conjugation) of bilirubin takes places in the endoplasmic reticulum (Fig. 11.6). An enzyme, uridyldiphosphate glucuronyl (UDPG) transferase, transfers a glucuronic acid molecule to one or both of the two propionic acid side chains in bilirubin, thereby converting bilirubin into a mono- or diglycuronide ester. The glucuronides are frequently referred to as conjugated bilirubin; in this book, they are called esterified bilirubin. A small amount of esterified bilirubin is regurgitated into plasma.

Step 5. The water-soluble bilirubin glucuronides (BG) are secreted from the hepatic cell by a transport mechanism into the bile canaliculi as a step in the formation of bile (Fig. 11.6).

Step 6. The BG, along with the rest of the bile, passes into larger channels, the bile ducts. Bile is produced continuously by the liver, but between periods of digestion of foods, it is stored temporarily in the gallbladder, where it is concentrated by the absorption of water. Under proper stimulation (food in the stomach or hormones that stimulate the flow of bile), bile from the gallbladder and that coming directly from the liver flow through the common bile duct into the small intestine.

Step 7. As BG approaches the colon (large intestine), it is exposed to the action of bacteria, the enzymes of which cleave off the glucuronic acid moieties and reduce (hydrogenate) the bilirubin molecule. The reduction products are known collectively as urobilinogen.

Step 8. A portion of the urobilinogen is absorbed into the portal blood, which flows to the liver. The healthy liver cells remove almost all the urobilinogen as it passes through and re-excrete it into the bile. The small portion of urobilinogen that escapes excretion by the hepatocyte (about 1% of the colon urobilinogen content) reaches the peripheral circulation, where it is excreted by the kidney into the urine (Step 10).

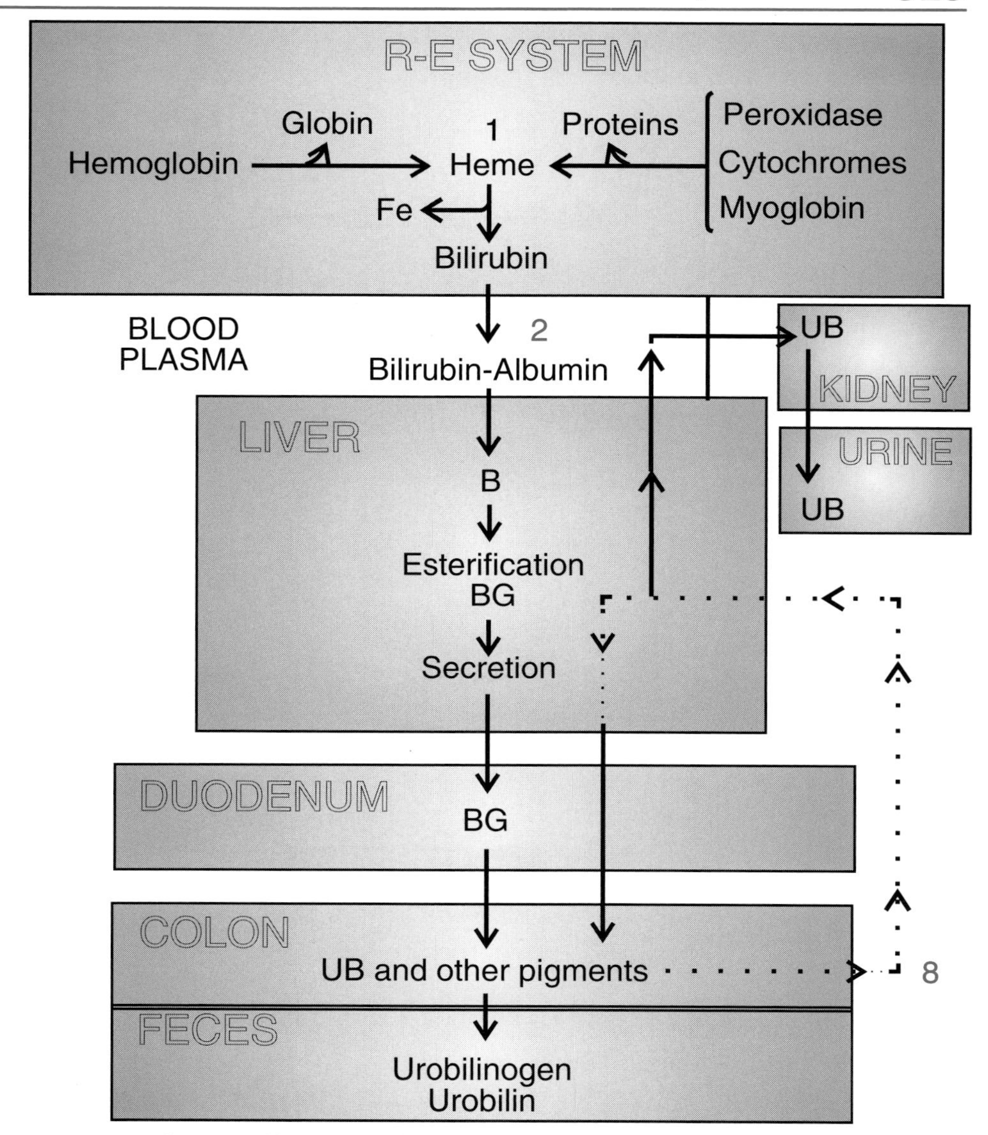

FIG. 11.5. Schematic illustration of formation and excretion of bile pigments. R–E = reticuloendothelial; B = bilirubin; BG = bilirubin diglucuronide; UB = urobilinogen. The uptake of bilirubin in the liver is from a sinusoid into a hepatocyte; esterification takes place on a microsome. The BG is excreted into bile canaliculi, which merge into bile ducts.

Step 9. The urobilinogen that is not absorbed in the colon becomes partially oxidized to urobilin and other brownish pigments that are excreted in the feces.

In summary, about 250 to 300 mg of bilirubin are produced daily in normal, healthy adults. A normally functioning liver is required to eliminate this amount of bilirubin from the body, a process requiring the prior esterification of bilirubin to form the water-soluble mono- and diglucuronides. Practically all the bilirubin is eliminated in the feces in the form of pyrrole fragments and pigments (dipyrroles, urobilinogen, urobilin); a small amount of the colorless

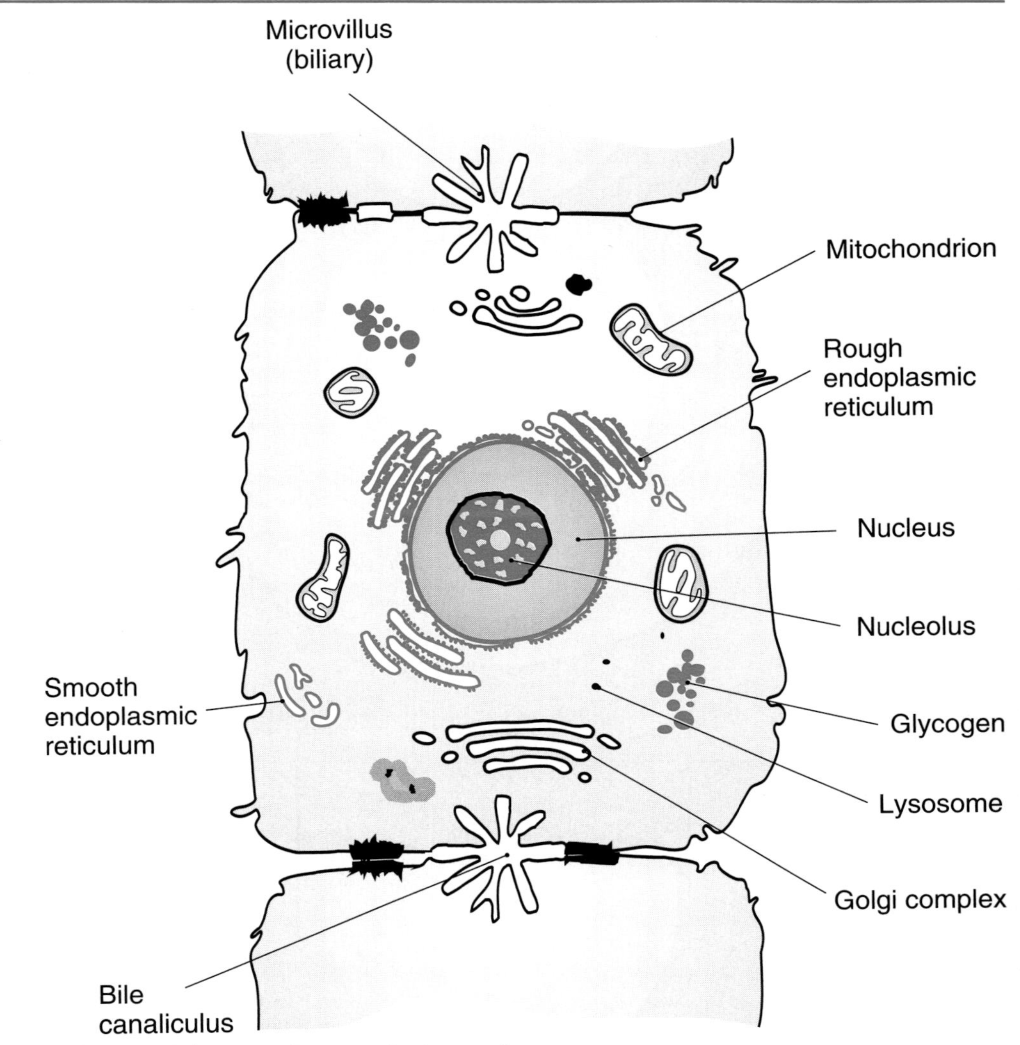

FIG. 11.6. Schematic diagram of a liver cell.

product urobilinogen is eliminated in the urine. A low concentration of bilirubin is found in normal plasma; most of this bilirubin is nonesterified but firmly bound to albumin. A small percentage of the total bilirubin exists in normal plasma as glucuronide. Tests are available for measuring the concentration of total bilirubin and esterified bilirubin in serum and of urobilinogen in urine.

Disorders of Bile Pigment Metabolism

Analytical tests for the measurement of concentrations of total and esterified bilirubin in serum have been available for a long time. The concentration of the nonesterified (unconjugated, indirect-reacting) bilirubin is obtained by difference. The concentration of total bilirubin in the serum of normal adults ranges from 0.1 to 1.0 mg/dL (1.7 to 17.0 μmol/L). The concentration of the esterified bilirubin may be as much as 0.3 mg/dL (5.0 μmol/L). When the bilirubin concentration in the blood rises, the pigment begins to be deposited in the sclera of the eyes and in the skin. This feature becomes evident to an experienced observer when the serum bilirubin reaches a concentration of 2.5 mg/dL (43 μmol/L) or greater. This yellowish pigmentation in the skin or sclera is known as jaundice or icterus; when visible jaundice is present, the serum bilirubin concentration must of necessity be above the upper limit of normal.

Because most liver diseases and several nonhepatic disorders are accompanied by jaundice, the differential diagnosis of jaundice plays an important role in elucidating hepatic dysfunction. An elevated level of serum bilirubin may be produced by one or more of the following four processes involved in the metabolism of bilirubin (Fig. 11.5).

Excessive Load of Bilirubin Presented to Liver (Fig. 11.5, Step 2)

This problem occurs in hemolytic disease or in any process of excessive erythrocyte destruction. With chronic hemolysis, more bilirubin than the liver can handle may be produced and presented to the liver for excretion. This problem of overproduction with retention is prehepatic in origin. The serum bilirubin elevation in chronic hemolytic anemia is mild, usually below 5 mg/dL (85 μmol/L), because the liver can excrete much more than the normal load of this pigment. The bilirubin fraction that is increased is the albumin-bound nonesterified bilirubin (Fig. 11.5, Step 2); the esterified fraction remains at normal levels. No bilirubin is detected in the urine because the esterified fraction has not risen.

During the catabolism of this increased load of bilirubin, more bilirubin is esterified in the liver (Fig. 11.5, Step 4) and excreted into the gut (Step 5), where it is converted into urobilinogen (Step 7). As a result of the formation of excessive amounts of urobilinogen and other pigments, fecal urobilinogen and urobilin are greatly increased (Step 9), whereas urine urobilinogen (Step 10) is moderately increased. These changes are shown in Table 11.1.

Defective Transport into the Hepatocyte

An active transport system involving ligandin is required for the transportation of bilirubin through the hepatic cell membrane to the microsomes (Step 3). A congenital malfunction of this transport system produces a harmless mild jaundice (serum bilirubin below 2 mg/dL or 34 μmol/L), usually detected accidentally during routine screening tests for blood chemistry or in monitoring of serum bilirubin concentrations during other illnesses. This transport defect is known as Gilbert's syndrome.

Congenital Deficiency of the UDPG Transferase Enzyme System

Some children are born with a defect in the transferase enzyme that leads to a severe jaundice of the type seen in hemolytic disease of the newborn. The disease, known as the Crigler-Najjar syndrome, is rare but severe; more than

TABLE 11.1
TYPICAL RESULTS OF LIVER FUNCTION TESTS IN VARIOUS TYPES OF HYPERBILIRUBINEMIA*

	TYPES OF JAUNDICE						
	PREHEPATIC		INTRAHEPATIC				EXTRAHEPATIC
COMMON LABORATORY TESTS	HDN†	Hemolytic	Acute Injury	Chronic Injury	Cholestasis	Cirrhosis	Obstruction (Cholestasis)
Bilirubin—Total	↑ ↑ ↑	↑	↑ ↑	N ↑	↑ ↑ ↑	↑	↑ ↑ ↑
Bilirubin—Esterified	N	N	↑ ↑	N ↑	↑ ↑ ↑	↑	↑ ↑ ↑
Urine Bilirubin	0	0	↑ ↑ ↑	↑	↑ ↑ ↑	↑	↑ ↑ ↑
Urine Urobilinogen	0	↑	↑	↑	N ↑	N ↑	↓ ↓
Fecal Urobilinogen	0	↑ ↑ ↑	↓	↓	↓	N ↓	↓ ↓
Albumin			N	↓	N	↓ ↓	N
γ-globulin			↑‡	↑ ↑‡	N	↑ ↑	N
Prothrombin Time			↑ ↑	↑	↑§	N	↑§
ALP			↑	↑	↑ ↑	N	↑ ↑ ↑
AST			↑ ↑ ↑ ↑	↑	↑ ↑	↑	↑ ↑
ALT			↑ ↑ ↑ ↑	↑	↑	N ↑	↑

*See text for situations where these tests are most useful.
†Hemolytic disease of the newborn
‡γ-globulin is increased in viral hepatitis but not in a hepatitis caused by drugs or chemical toxins.
§Reversed 24 hours after injection of soluble vitamin K preparations.

50% of the reported patients with this disease died within 1 year of birth, and about half of the survivors suffered from irreversible brain damage.

Impairment in the Esterification of Bilirubin

Physiologic Jaundice of the Newborn. The UDPG transferase enzyme system responsible for converting bilirubin into a glucuronide ester (Fig. 11.5, Step 4) is not fully developed at birth. In full-term infants, several days pass before the enzyme is produced in sufficient quantity to esterify the bilirubin presented to it. The serum bilirubin may rise as high as 8 mg/dL (137 μmol/L) in normal full-term infants by the third to the sixth day of life as a result of the immature enzyme system before falling to normal adult levels; this condition is known as physiologic jaundice of the newborn. This process is aggravated in premature infants, who must wait a longer time for generation of the esterifying enzyme. The serum bilirubin concentration may climb as high as 16 mg/dL (274 μmol/L) in premature infants in the absence of any disease process (Table 11.2).

Hemolytic Disease of the Newborn (HDN) or Isoimmune Hemolytic Disease (IHD). This isoimmune hemolytic disease is caused by an Rh, ABO, or other blood system incompatibility between mother and fetus. If the mother has been previously sensitized to a surface antigen present on fetal red blood cells (most commonly by a prior pregnancy or by a blood transfusion), maternal antibodies (IgG) cross the placenta and bind to the specific surface antigens; activated complement binds to the attached IgG and destroys the red cell membrane by lytic, enzymic action. The hemolytic process may continue for several weeks after birth because many red cells are coated with IgG antibodies before

TABLE 11.2
REFERENCE VALUES OF SERUM TOTAL BILIRUBIN IN FULL-TERM* AND PREMATURE INFANTS*

AGE	TERM		PREMATURE	
(DAYS)	(mg/dL)	(μmol/L)‡	(mg/dL)	(μmol/L)‡
At birth†	<2.0	<34	<2.0	<34
0–1	<6	<103	<8	<137
1–2	<8	<137	<12	<205
3–5	<10	<171	<16	<274
30	0.2–1.0	3.4–17.1		

*Infants without isoimmune hemolytic disease.
†In umbilical cord blood.
‡SI conversion factor for bilirubin: μmol/L = mg/dL × 17.1.

delivery. The hemolysis produces a large amount of bilirubin for excretion by an immature liver. The load is too large to handle effectively, so the plasma bilirubin concentration rises rapidly and the jaundice may be severe. Plasma albumin has a limited capacity for binding bilirubin, and when primary binding sites are saturated, the bilirubin becomes less tightly bound to secondary sites. When the blood circulates to the brain, the unbound bilirubin is partitioned between the lipid covering of brain cells and the plasma. It enters the brain cells and causes irreversible damage to the basal ganglia. Nuclear staining of brain cells by bilirubin is known as kernicterus. Many infants who survive kernicterus become spastic and may also suffer from mental retardation. Exchange transfusion can lower the bilirubin concentration and prevent this outcome.

The most severe IHD occurs when an Rh-D negative mother who is sensitized to the D antigen has an Rh-D positive fetus. Incompatibilities and sensitization to antigens of the Duffy, Kell, Kidd, or other blood type systems may also be severe, but occur less frequently. The ABO type of IHD may occur when the mother is type O and the baby is either Type A or Type B, because the mother has antibodies against A and B surface antigens. Fortunately, the antibodies against ABO system antigens are much weaker than those of the Rh type and do much less damage. The jaundice produced in the newborn is much less severe, and the plasma bilirubin rises more slowly and crests at lower levels. The need for an exchange transfusion is much less frequent in this type of IHD.

Historically and legally, the critical concentration of plasma bilirubin for possible brain damage in newborns was considered to be approximately 20 mg/dL (340 μmol/L), although this number may vary according to the concentration of plasma albumin, the drugs administered, and other factors, such as weight and well-being of the neonate. The critical level now tends to be lowered as information accumulates concerning bilirubin-binding capacity of albumin and the effects of various drugs on bilirubin binding. An exchange transfusion is usually performed before the critical level is reached; several transfusions may be required in some cases. Therapeutic exposure of neonates to ultraviolet light helps to reduce the plasma bilirubin concentration by converting it to an excretable form.

The laboratory has the crucial responsibility to perform the bilirubin analysis accurately and to report the result promptly in cases of IHD. The physician depends heavily on the laboratory values in deciding whether an exchange transfusion is necessary.

The rise in serum bilirubin levels in physiologic jaundice of the newborn and in IHD is in the nonesterified portion because of the immaturity of the transferase enzyme system. The first sign of production of the enzyme is an increase in the plasma concentration of the esterified fraction.

Disturbances in Excretion of Bilirubin (Cholestasis)

The largest percentage of patients with jaundice is found in this category. In this large group of disorders, the elevated concentration of serum bilirubin consists of both the esterified and the nonesterified fractions, because bilirubin can enter the hepatic cell and become esterified. Because of the impairment in the excretory transport system or of an obstruction in the biliary tree that impedes the excretion of bile into the intestine (Fig. 11.5, Steps 5 and 6), the concentration of esterified bilirubin builds up in the hepatocyte. Consequently, some of the esterified bilirubin is regurgitated into the bloodstream by way of the sinusoids. A second consequence of the increased amounts of bilirubin glucuronide in the hepatic cells is the slowdown of the transport system that transfers the bilirubin into the hepatocyte. The net result is hyperbilirubinemia, with the esterified bilirubin constituting 50 to 70% of the total. The test for urine bilirubin is usually positive in the early stages of the disease because of the elevated level of bilirubin glucuronide in plasma that can pass through the renal glomerular membrane (Table 11.1).

With respect to the bilirubin degradation products, less bilirubin is reaching the gut because of the obstruction, so fewer pigments are produced. The feces may vary in color from pale yellow or brown to clay-colored, depending on the completeness of the obstruction or impairment of the excretory transport process (Table 11.1). With complete cholestasis, the urinary urobilinogen falls to zero because no bile, and hence no bilirubin, reaches the intestine for later conversion to urobilinogen.

When the bile is completely excluded from the gut by an obstruction, a gross failure to digest and absorb ingested fats occurs because of the absence of bile salts to serve as emulsifiers. This problem produces a steatorrhea—pale, fat, bulky stools.

The various types of disorders responsible for faulty or limited excretion of bilirubin by the liver may be divided into two groups: (1) those of hepatic origin (intrahepatic) and (2) those originating beyond the liver (posthepatic).

Intrahepatic. The most common disorders are diffuse hepatocellular damage and liver cell destruction. These conditions may be caused by viral hepatitis, hepatitis produced by toxins (drugs, such chemicals as phosphorus or organic arsenicals, such chlorinated solvents as carbon tetrachloride, and chloroform), cirrhosis, or intrahepatic cholestasis (as in hepatic edema). In hepatocellular disease, the urinary excretion of urobilinogen may be increased despite a reduction in the amount of bilirubin that may reach the gut. This seeming paradox is explained by the *failure of the damaged hepatic cells to remove the* *reabsorbed urobilinogen* from the portal vein blood; the urobilinogen in the blood then becomes available for excretion by the kidney. In the Dubin-Johnson syndrome, a congenital condition, a disorder appears in the transport system for excreting the bilirubin diglucuronide into the bile canaliculi (Fig. 11.5, Step 5).

All intrahepatic problems must be treated by medical management; surgery would harm the patient.

Posthepatic. These disorders include all types of extrahepatic cholestasis or obstruction of the flow of bile into the intestine. The obstruction may be caused by stones in the bile ducts or gallbladder, carcinoma of the head of the pancreas, other tumors that obstruct the common bile duct, or strictures of the common bile duct. Problems related to obstruction by stones, benign tumors, and strictures are definitely alleviated or cured by surgery, whereas successful treatment of malignancies depends on the type of malignancy, its size and distribution, and the time of discovery (before or after metastasis). Because surgical intervention benefits patients with certain types of jaundice but harms others, the correct differential diagnosis of jaundice is extremely important to make. The laboratory tests that assist in making this distinction are reviewed later in this chapter.

HEPATIC DYSFUNCTION

Many factors or agents can diminish liver performance to the extent that some hepatic function tests are abnormal; the patient may or may not experience pain or feel sick. Some of the common situations are described in the following paragraphs.

Viral Hepatitis[3–6]

Viral hepatitis is a worldwide disease of serious proportions that is accompanied by acute hepatocellular injury. At least five different viruses are recognized as specific agents for causing hepatitis; they differ somewhat in their modes of transmission, incubation period, severity, and sequelae (Table 11.3). Although many other viruses, such as Epstein-Barr, cytomegalovirus, and others, may incidentally invade the liver at times and cause hepatocellular damage, these instances are rare compared to those of the five viruses under review.

The agents of viral hepatitis are the following:

Hepatitis A Virus (HAV)[3]

Hepatitis A, formerly called acute infectious hepatitis, is transmitted primarily by the enteric route, that is, by the ingestion of contaminated water or food supplies. The disease spreads rapidly where sanitation is poor and water supplies are unreliable. The feces of patients with hepatitis A are heavily laden with virus particles during the early stages of the disease. Close personal

TABLE 11.3
THE VARIOUS TYPES OF HEPATITIS AND SOME OF THEIR CHARACTERISTICS

			HEPATITIS NON-A, NON-B		
CHARACTERISTICS	HEPATITIS A	HEPATITIS B	HEPATITIS C	HEPATITIS D*	HEPATITIS E†
Principal transmission route	Enteric	Parenteral, sexual	Parenteral, sexual	Parenteral	Enteric
Incubation time (days)	15–45	40–180	15–160	?	14–60
Severity	Mild	Severe	Moderate to severe	Very severe	Mild
Carrier state	None	Sometimes	Usually	Frequently	None

*Hepatitis D occurs only in combination with hepatitis B; it requires the viral coat of hepatitis B.
†Hepatitis E occurs primarily in Asia, India, and Central America.

contact with such patients poses the danger of infection, especially when the practice of personal hygiene is at a minimum. Hepatitis A has a short incubation period (2 to 6 weeks) and practically all patients recover completely; there is no chronic carrier state. The disease is identified in the laboratory by finding IgM antiHAV antibodies in serum during the acute stage of the disease. See Table 11.3 for some of its main features.

Hepatitis B Virus (HBV)[3]

Hepatitis B, formerly called "serum" hepatitis, a designation no longer deemed appropriate, is transmitted by sexual intercourse and by the parenteral route (by needle sharing among drug users, by transfusion with contaminated blood or blood products, or by accidental puncture by a needle containing infectious blood). All body fluids of infected patients, including semen and urine, contain HBV virus; the disease is spreading rapidly among adolescents through sexual transmission. The usual incubation time is 2 to 3 months. Most patients recover completely, but 10% develop chronic hepatitis. Some recover but may become carriers of the live virus.

Three different markers in the blood are commonly used for ascertaining whether an individual has or has had hepatitis B and, if so, whether an infectivity persists (carrier state). HBV has a surface antigen (HBsAg) that appears in the blood early in the disease, before clinical symptoms are present and even before a detectable rise in serum aminotransferase activity. In typical cases, circulating HBsAg persists during the entire clinical course of the disease but usually decreases to an undetectable level from 2 to 6 months after the disappearance of symptoms. The presence of HBsAg in blood identifies the disease as hepatitis B.

The other two circulating markers of hepatitis B are the core antigen (HBcAg), derived from the surface of the nucleocapsid core of the B virion, and an antigen derived from the disintegration of the nucleocapsid (HBeAg). HBeAg is found only in patients who test positive for the HBsAg and is a marker for infectivity. IgM antibodies against HBcAg (IgM antiHBc) are detectable in patients with recent or present hepatitis B; in contrast, IgG antiHBc are the antibodies found circulating in patients who recovered some time ago from hepatitis B or who suffer from chronic hepatitis B at present.

Blood from a patient with hepatitis B usually has a high concentration of infective particles during the acute stage. Hospital laboratory workers must take precautions to avoid the disease (see Chap. 2).

Non-A, Non-B Hepatitis Virus (NANBV)

The transmission of hepatitis by blood transfusions, even when the donors tested negatively for HAV and HBV, was convincing evidence that other viral agents (NANBV) could cause acute hepatitis. We now know of at least three such agents.

Hepatitis C Virus (HCV).[4] Ninety percent of the transfusion-related cases of hepatitis in the United States prior to the development of virus-specific screening tests (July, 1990) are attributed to HCV.[3] The transmission of hepatitis by blood transfusion should decrease as blood bank donors are tested for HCV in addition to HAV and HBV. The screening test for HCV depends on the detection of circulating antibodies to HCV. The development of a measurable concentration of IgM antiHCV may take from 1 to 3 months, however. Thus, some individuals with HCV who are tested during this critical "window" period will escape detection. For this reason, blood banks cannot completely eliminate all HCV-infected blood donors. HCV, like HBV, is transmitted sexually, as well as by the parenteral route; approximately 170,000 new cases of HCV occur each year in the United States. About 50% of people infected with HCV develop chronic hepatitis, of which 20% deteriorate further to cirrhosis. The characteristics of HCV infection resemble those of HBV infection (Table 11.3).

Hepatitis E Virus (HEV).[5] HEV resembles HAV in all its features and is transmitted mainly by fecal contamination of water supplies. Although rare in the United States and Europe, HEV is fairly common in countries where water supplies may become contaminated by inadequate sewage disposal, particularly during periods of heavy rainfall.

Delta Hepatitis Virus (HDV).[6] HDV is caused by an RNA virus that lacks a protein coat. HDV requires the presence of HBV to become infective; HDV cannot reproduce unless the surface protein (HBsAg) of HBV is available. Apparently, HBV produces HBsAg in an excess amount sufficient for HDV to become infectious. The combined disease, HBV plus HDV, may cause a fulminant hepatitis ending in death. Some patients who recover may be burdened with a chronic hepatitis leading to cirrhosis or hepatic cancer. The presence of antibodies (IgM antiHDV) against HDV can be detected by immunoassay.

With all types of hepatitis, the first antibodies produced are of the IgM class; some months later, the specific IgM antibodies against the hepatitis virus particles are replaced by those of the IgG class.

The only hepatitis vaccine currently available is effective solely against HBV. The vaccine is prepared from the surface antigen (HBsAg), and its use is recommended for high-risk laboratory workers.

Immunochemical tests are available for the detection of IgM antibodies against HAV, HBV, HCV, and HDV, as well as for the detection of hepatitis B surface and core antigens.

Alcoholic Liver Disease

Hepatic injury, induced by the long-time consumption of large amounts of alcoholic beverages, is a common problem. Acetaldehyde, the metabolite formed from ethanol catabolism, becomes a hepatic cell toxin when its concentration becomes too high because of overindulgence. The injury stages progress from fatty infiltration to fibrosis and, if the hepatocellular necrosis and liver insult continue, advance to cirrhosis and even death. The fibrosis and cirrhosis are irreversible.

Hepatic Drug Toxicity

The liver is the metabolic center of the body in the sense that many organic compounds are catabolized or altered there so that they can be solubilized and excreted, inactivated, or broken down into inactive fragments. The portal circulation brings ingested material first to the liver, including all drugs taken orally. Some of the drugs (and their metabolites) are toxic to hepatic cells and induce various injuries that may range from cholestasis to cell injury of particular structures or organelles, and may even cause cell necrosis. The altered hepatic function tests with some drugs resemble those caused by obstruction to biliary flow, whereas others are more like those associated with viral hepatitis. Some of the common drugs notorious for hepatic toxicity in overdosage or in sensitized patients are acetaminophen, isoniazid, tetracyclines, chlorpromazine, thiazides, propylthiouracil, and many others.

Cholestasis

Cholestasis, the obstruction or stagnation of biliary flow, is described in the previous section on disturbances in bilirubin excretion. It is commonly caused either by mechanical obstruction to the flow of bile (stones in bile duct or gallbladder, tumors around the common bile duct or head of the pancreas) or by hepatocellular inflammation or damage, as in viral hepatitis or drug toxicity. Cholesterol is the principal constituent in approximately 80% of gallstones.

Hepatic Encephalopathy (Hepatic Coma)

The liver is the major site for the detoxification of ammonia. Ammonia arises in the gastrointestinal tract from bacterial action; bacteria generate it from amino acids by oxidative deamination and from the hydrolysis of the amide group of glutamine. Portal blood brings the liberated ammonia to the liver, which removes 80% of the load in a single pass-through. The sequestered ammonia is converted into the amide group of glutamine or enters the urea cycle to end up in urea, the final product excreted by the kidney. Ammonia per se is a toxic material that causes an encephalopathy or coma if its concentration in the brain and cerebrospinal fluid becomes sufficiently high. Blood ammonia concentration becomes sufficiently high to affect brain function in the end stages of fulminant hepatitis or in cirrhosis of the liver, where portal blood bypasses the liver because of an extensive collateral circulation or by an operative portal-systemic shunt.

TESTS FOR DISCLOSING HEPATIC DYSFUNCTION

Alterations in hepatic cell activity as a result of disease may affect some of the functions or activities of the liver to a greater or lesser degree and help to indicate liver disorders. Estimation of the presence or absence of hepatic dysfunction is complicated by the large functional reserve of the liver and its power to regenerate rapidly. Under experimental conditions with rats, as much as 80 to 85% of the liver must be removed before certain laboratory tests for liver function become abnormal. Regeneration of new liver tissue is rapid. The amount of hepatic tissue damaged by a disease process is impossible to quantitate; a diffuse minimal involvement of the liver may produce a more grossly abnormal laboratory test than would a focal necrosis.

Because of the many activities engaged in by the liver, numerous tests have been introduced to test a particular hepatic function. Most such tests are nonspecific because diseases other than those involving the liver may produce similar changes. All these tests are difficult to interpret because of the large functional hepatic reserve.

The numerous liver tests proposed is evidence of the complexity of the problem and illustrates the lack of consensus as to which tests are best suited for the purpose. The reasons for requesting liver function tests are threefold; (1) for diagnosis, to see if the hepatobiliary system is impaired; (2) for differentiation, if hepatic impairment exists, to discover whether the problem consists primarily of damage to hepatic cells or is the result of an obstruction to the biliary system; and (3) for prognosis, to estimate the extent of the damage, or the probable outcome with appropriate therapy. The ideal test is specific for hepatic disease, sensitive (capable of detecting early or minimal hepatic disease), and selective (able to detect specific types of liver disorders), but no such test exists. A battery or panel of liver performance (function) tests is usually used, but the composition of the panel may vary according to the medical needs, that is, whether used for diagnosis, differentiation of liver disease, or prognosis. In certain circumstances, special tests may be used.

Several assays (serum bilirubin, urine bilirubin and bile pigments, and serum proteins and enzymes) have been useful in diagnosing and following patients with suspected or known liver disease.

Serum Bilirubin

When serum bilirubin is fractionated by liquid chromatography, the following four separate forms of bilirubin are identifiable[7,8]: (1) nonesterified bilirubin (indirect, unconjugated) that binds strongly but reversibly to albumin and cannot be excreted by the kidney; (2) bilirubin monoglucuronide (conjugated, direct reacting); (3) bilirubin diglucuronide (conjugated, direct reacting); and (4) delta bilirubin, a form that is covalently bound to albumin and hence cannot be excreted by the kidney. Delta bilirubin reacts directly but more slowly with the diazo reagent than do the bilirubin glucuronides, but this reaction varies with the concentration of reagents and the method employed.

The concentration of all four serum bilirubin fractions can be ascertained by the dry film technology[9] of Eastman Kodak. Two different multilayer films are used. The first film is diazo-based and measures total bilirubin. The second film contains a surfactant and caffeine/benzoate to dissociate nonesterified bilirubin (B_u) and the conjugated bilirubin glucuronides (B_c) from their associations with albumin. All compounds, including the delta bilirubin-albumin compound, diffuse into a microporous gelatin layer. The large delta bilirubin complex becomes trapped in the second layer, but B_u and B_c diffuse into a third layer containing a mordant. The mordant shifts the reflectance of B_c to 420 to 425 nm and of B_u to 460 to 465 nm. These can be measured simultaneously from the same slide. The measurements provide directly the concentrations of total bilirubin (B_t), B_u, and B_c. Delta bilirubin $= B_t - (B_u + B_c)$.

Clinical experience with delta bilirubin is short, but the indications are that it is present in low concentrations, if at all, in the blood of normal adults or children, that it is formed from bilirubin glucuronide, and that its serum concentration rises rapidly in situations of cholestasis.[7,10] It cannot be excreted by the kidney, so once formed, it remains in the circulation until the albumin is catabolized (half-life of approximately 17 days). Thus, the total bilirubin of a person with a gallbladder obstruction may not immediately return to normal after surgical relief of the obstruction; it sometimes takes days before the delta bilirubin becomes excretable.

Determination of total and esterified bilirubin is useful for the differential diagnosis of jaundice, particularly in differentiating hemolytic disease from other diseases causing jaundice; the value for the nonesterified fraction is obtained by subtracting the value of the esterified portion from the total.

The classic method for quantitating bilirubin is by converting it into an azo dye and measuring its absorbance at a specific wavelength. The diazo reaction, when carried out by the Jendrassik-Grof method as described by Koch and Doumas[11] splits the bilirubin of all serum fractions (nonesterified, mono- and diglucuronides, and delta bilirubin convalently bound to albumin); each bilirubin molecule splits in half to form two dipyrroles, with loss of the methene group at the scission point. Each dipyrrole forms an azo dye by coupling with the diazonium salt; the resulting azobilirubins have essentially the same spectral absorbance values.

Although many diazonium salts could be used, the one most commonly used is prepared from sulfanilic acid. This salt is diazotized with sodium nitrite in acid solution and then is coupled with bilirubin glucuronide. Because the nonesterified bilirubin is tightly bound to albumin, it requires the presence of a polar solvent (methanol, ethanol) or an accelerator substance like caffeine/benzoate that promotes the displacement of bilirubin from albumin and speeds the diazo coupling.

The azo dye is red in neutral solution and blue in strong alkali. The original Malloy and Evelyn method and its various modifications used methanol as an accelerator; these assays have been used widely for many decades. They are not as accurate as the Jendrassik-Grof method at low concentrations of bilirubin (near the upper limit of normal) and are affected by the protein matrix. Methods based on the Jendrassik-Grof method, with or without modifications, are the most widely used and have been adapted to a great variety of automated instruments. The modified Jendrassik-Grof method described in the following paragraphs is widely accepted as the standard method for total bilirubin.

Principles of Analysis

Total Bilirubin. Serum is mixed with a slightly acidic solution of caffeine and sodium benzoate, compounds that accelerate the diazotization of nonesterified bilirubin. Diazotized sulfanilic acid cleaves both forms of bilirubin (esterified and nonesterified) and forms an azo dye with the dipyrroles. The addition of alkaline tartrate (pH 13) converts azobilirubin to a blue color whose absorbance is measured at 600 nm. The solution appears green because of the reagent's yellow color. The total bilirubin concentration is proportional to the absorbance at 600 nm.

Esterified (Conjugated) Bilirubin. The serum is mixed directly with the diazo reagent in the absence of accelerators. Only esterified bilirubin reacts within 5 min under these conditions. After 5 min, ascorbic acid is added to decompose excess diazo salt and stop the reaction. The solution is made alkaline, and the absorbance is measured at 600 nm. The esterified bilirubin concentration is proportional to the absorbance at 600 nm. Nonesterified bilirubin equals total bilirubin minus esterified.

Precautions

1. When performing bilirubin tests on patients with hemolytic disease of the newborn, a control serum or standard with a concentration close to 20 mg/dL (340 μmol/L) should be used because this is near the critical concentration at which decisions are made concerning exchange transfusions.
2. Bilirubin is altered by ultraviolet light in such a way that it is no longer able to form azobilirubin. Serum samples and standards must be protected from direct sunlight or ultraviolet light.

In SI, μmol/L of bilirubin = mg/dL × 17.1.

Reference Values. Reference values for full-term and premature infants who had no hemolytic problems are presented in Table 11.2. The serum bilirubin concentration of most full-term infants reaches its highest point about 3 to 5 days after birth and begins to fall as the production of the bilirubin-esterifying enzyme (UDPG transferase) matures (Fig. 11.5, Step 4). Adult levels of bilirubin are attained by age 30 days. UDPG transferase takes longer to develop in premature infants, depending on the degree of immaturity; the bilirubin concentration seldom exceeds 16 mg/dL (274 mmol/L) in most uncomplicated cases. All the serum bilirubin is nonesterified until the liver begins to produce UDPG transferase. The serum concentration of esterified bilirubin in adults is only 0 to 0.2 mg/dL (< 3.4 mmol/L) because the ester is water soluble and readily excreted in bile.

Increased Concentration
Total bilirubin. Total bilirubin is increased mildly in chronic hemolytic disease (below 5 mg/dL or 86 μmol/L), moderately to severely in hepatocellular disease, (10 to 30 mg/dL or 170 to 510 μmol/L) and markedly in cholestasis (internal or external obstruction to bile flow, where the concentration could vary from 10 to 60 mg/dL or 170 to 1030 μmol/L).
Esterified bilirubin. This fraction is increased in both hepatocellular disease and cholestasis.
Nonesterified bilirubin. The increase in bilirubin concentration in hemolytic disease, including HDN, is almost entirely in this fraction. It also accounts for about 30 to 50% of the bilirubin rise in hepatocellular disease or cholestasis.

Decreased concentration. Of no clinical significance.

Direct Spectrophotometric Determination

Total bilirubin in the serum of newborns may be estimated directly from the spectral absorbance of bilirubin at 455 nm. This test cannot be done with adult serum because carotene or other pigments that absorb at this wavelength may be present. These pigments are not present in the sera of newborns, but hemolysis gives false elevations in the bilirubin concentration because oxyhemoglobin also absorbs some light at 455 nm. The effect of hemolysis can be avoided by measuring the absorbance at 455 and 575 nm and correcting for hemoglobin as described in a following section.

Principle. The absorbance of a diluted serum is measured at two wavelengths, and a correction is made for the presence of oxyhemoglobin if any should be present. The first wavelength, 455 nm, is approximately the point of maximum absorbance of bilirubin, but is also a measure of the hemoglobin concentration. At 575 nm, the second wavelength, the absorbance of hemoglobin is essentially the same as at 455 nm, whereas bilirubin has zero absorbance. The correction for hemoglobin is made by subtracting A_{575} from A_{455}.

Urine Bilirubin

Bilirubin is not detectable by conventional methods in the urine of a normal, healthy person. When present, urinary bilirubin indicates some pathologic condition of the liver or biliary system. The bile pigment found in the urine in these conditions is bilirubin diglucuronide, the water-soluble ester of bilirubin.

By Tablet

Bilirubin can be detected by the classic procedure of converting it into the dye, azobilirubin. This conversion may be done with commercially available Ictotest tablets (Ames), which contain a diazotized salt in an appropriate buffer.

Five drops of urine are placed on an adsorbent-cellulose mat. If present, bilirubin is adsorbed on the surface of the mat. A tablet containing the diazotized salt and a mixture of $NaHCO_3$, salicylic acid, and boric acid is placed on the wet spot. Two drops of water are flowed down the tablet to the mat. If bilirubin is present, a blue to purple color forms in the mat within 30 sec. A red or orange color is read as a negative test because it means that a urinary compound other than bilirubin has been converted to an azo dye.

By Dipstick

Bilirubin in urine may also be detected by any of the several multitest dipsticks (Ames, Boehringer Mannheim). The diazotized salt in the stick is different from that in the tablet and, consequently, is of a different color. The presence of any bilirubin in the urine is abnormal.

Urine Urobilinogen

Urobilinogen is a colorless compound derived from bilirubin that has been excreted in the bile and partially hydrogenated by bacteria in the intestines. It is partially reabsorbed into the portal vein, almost completely extracted from the blood by normal liver cells, and re-excreted into the bile. The small portion that is not taken up by the hepatocytes is excreted by the kidney as urobilinogen. Normally, about 1 to 4 mg/24 hr are excreted in the urine. The excretion of urinary urobilinogen is elevated in hemolytic disease (excess production from bilirubin), in hepatocellular liver disease (decreased removal by hepatocytes), and in congestive heart failure (impaired circulation to liver).

By Dipstick

A dipstick (Ames, Boehringer Mannheim) impregnated with *p*-dimethylaminobenzaldehyde and an acid buffer turns red in the presence of urobilinogen. Porphobilinogen reacts in a similar fashion. Comparison with a color chart indicates whether the urobilinogen is present in normal amounts or in excess.

Fecal Urobilinogen

Fecal urobilinogen is decreased in hepatocellular disease, as well as in obstructions of the biliary tree. Visual inspection of the feces usually suffices to detect decreased urobilinogen because the stools become pale or clay-colored with decreasing amounts of pigment. The original source of the brownish fecal pigments is the bilirubin that undergoes bacterial action in the gut.

Dye Tests of Excretory Function

In addition to excreting bilirubin, the liver is capable of eliminating various dyes or drugs by the same excretory pathway. Indocyanine green (ICG) is one such dye that is sometimes used as a test of hepatic excretory function when liver disease is suspected in a patient who is *not* jaundiced. Enzyme assays have replaced the ICG test to a large extent. A dye excretion test performed on a jaundiced patient is useless because a liver that cannot excrete bilirubin properly is also unable to excrete the dye in a normal manner.

Tests Based On Abnormalities Of Serum Proteins

The liver is the sole source of synthesis of most of the proteins in plasma except for the gamma globulins, which are produced in lymphocytes and plasma cells. The protein analyses that provide useful information concerning liver disease, however, are those of albumin, the globulin fractions as obtained by electrophoresis or nephelometry, the clotting factors measured by the one-stage prothrombin time, and α-fetoprotein, which is usually elevated in primary hepatic carcinoma. Knowing the concentration of total serum proteins is of little diagnostic help because no characteristic or predictable change is apparent. The proteins synthesized in the liver are usually decreased in hepatocellular disease, but the immunoglobulins are increased in viral hepatitis and in chronic liver infections. Short-term obstruction of the biliary system does not usually affect the protein concentration.

Thus, depending on the type and duration of hepatic disease, the serum total protein concentration may be normal, increased, or decreased.

Albumin

Albumin has a half-life in the body of approximately 17 days; thus, its concentration in plasma would decrease at the rate of about 3% per day if hepatic synthesis of albumin were completely stopped. Because complete stoppage of synthesis is rare, the drop in albumin concentration proceeds somewhat slowly in hepatocellular disease. Albumin concentration is therefore a good indicator for following chronic hepatocellular damage, but is not suited for early detection of hepatic dysfunction. Prealbumin has a much shorter half-life (approximately 2 days) and, thus, may become a better indicator of early hepatocellular damage.

Chapter 9 reviews the method of measuring the concentration of serum albumin. The normal concentration of serum albumin varies from 3.5 to 5.2 g/dL. In parenchymal liver disease (damage to hepatocytes), the decrease in the serum albumin concentration varies directly with the degree and duration of the disease. The prognosis is poor when the hypoalbuminemia is severe and prolonged. A hypoalbuminemia is a common finding in liver cirrhosis. The serum albumin concentration is usually not affected by obstruction of the biliary system unless the condition has been so prolonged that parenchymal damage has also occurred.

Gamma Globulins

Chapter 9 reviews the polyclonal (IgG, IgA, and IgM) rise in the gamma globulins that occurs in viral hepatitis and chronic infections. A typical electrophoretic pattern obtained in this disease is shown in Figure 9.8. Normal gamma globulin concentrations vary from 0.7 to 1.6 g/dL, but these may triple in a severe viral hepatitis. The gamma globulin values are also high in the disease sarcoidosis when it is accompanied by granulomas in the liver and in liver cirrhosis. The protein electrophoresis pattern in cirrhosis usually shows a bridging between the beta and gamma globulin bands.

Alpha Globulins

Serum α_1-globulins are usually decreased along with albumin in parenchymal liver disease, but this decrease is not of great diagnostic value. Changes in α_2- and β-globulins are also not significant.

Clotting Factors

Various proteins that participate in blood coagulation and that are synthesized in the liver are fibrinogen, prothrombin (Factor II), and Factors V, VII, IX, and X. Vitamin K is essential for the hepatic synthesis of prothrombin and Factors VII, IX, and X. The one-stage test of the plasma prothrombin time requires the presence of prothrombin plus Factors V, VII, and X. If any one of these factors is below a critical concentration, the prothrombin time is prolonged. Because the plasma half-life of Factor VII is less than 6 hr, the plasma concentration of this factor is rapidly depleted when damage occurs to hepatocytes.

The measurement of plasma prothrombin time is usually performed in the hematology laboratory or in a coagulation division, if one exists; thus, details of measurement are beyond the scope of this book.

Interpretation. An increased one-stage prothrombin time indicates the failure of hepatic synthesis of one or more of the aforementioned clotting factors. This failure could be caused either by a deficiency of vitamin K, a fat-soluble vitamin that requires the presence of bile salts for its absorption, or by damage to parenchymal cells. To resolve the dilemma, a water-soluble preparation of vitamin K is injected into the patient, and the test for prothrombin time is repeated 24 hr later. The prothrombin time is restored to normal if the fault was the failure of bile salts to reach the intestine (cholestasis or obstruction). In parenchymal damage, the prothrombin time is still prolonged despite the injection of vitamin K. This test helps to differentiate the causes of jaundice. As with all tests of liver function, results overlap when the patient has elements of both parenchymal damage and cholestasis.

Selected Enzyme Tests

A review of the general use of enzyme tests in the diagnosis and management of various diseases appears in Chapter 10. Only those useful in liver disease are considered in this chapter (Table 11.4).

The measurement of the serum activities of several enzymes is a helpful adjunct in the diagnosis and management of patients with suspected liver disease. Numerous enzyme tests have been proposed for this purpose, but

TABLE 11.4
UPPER LIMITS OF NORMAL (ULN) OF SERUM ENZYMES COMMONLY MEASURED IN HEPATOBILIARY DISEASE

ENZYME NAME	CONVENTIONAL ABBREVIATION	UPPER LIMIT OF NORMAL (ULN) IN U/L
Alkaline phosphatase	ALP	Adult 105
Aspartate aminotransferase	AST (GOT)*	25
Alanine aminotransferase	ALT (GPT)*	30
γ-glutamyltransferase	GGT	Male 35 Female 30
Lactate dehydrogenase	LD (LDH)*	290
5′-nucleotidase	5NT	7

*Old abbreviation in parentheses.

many overlap greatly in the information they provide. A rise in the serum activities of some enzymes may be much less specific because of their wide distribution in other tissues. In the interest of utility and economy, one must select the tests that are most suitable for a particular diagnostic or management situation and correlate the results with those obtained from other tests of hepatic function.

The common enzyme tests available in most laboratories for assisting in the diagnosis and management of liver disease are alkaline phosphatase (ALP), aspartate aminotransferase (AST), and alanine aminotransferase (ALT). In special cases, some laboratories may also measure the activities of 5′-nucleotidase (5NT) and gamma glutamyltransferase (GGT).

The procedures for general enzyme tests are described in Chapter 10. Many other enzyme tests exist but are not mentioned because they offer nothing that is not provided by at least one of the mentioned enzyme tests.

Serum Alkaline Phosphatase (ALP)

The determination of serum alkaline phosphatase (ALP) activity was the first enzyme test used in the differential diagnosis of jaundice. It is still a useful procedure when carried out in conjunction with other hepatic tests, despite the proliferation in the number of other enzyme tests available.

Reference Values. The reference range of ALP activity in serum as performed by the method described in Chapter 10 is 20 to 105 U/L for adults. Refer to that chapter for values in different age groups.

Increased Activity. The enzyme ALP is widely distributed in many tissues, including the osteoblasts (the bone-building cells), the cells lining the sinusoids, and bile canaliculi in the liver. Thus, one always must exclude the possibility of osteoblastic activity (for example, in a normally growing child) when using the ALP test in liver disease. The activity of serum ALP rises in both hepatocellular disease and cholestasis, but the rise is usually greater in cholestasis.

The increase in ALP activity is usually great in space-occupying lesions of the liver, such as carcinoma, amebic abscess, amyloidosis, and granulomatous lesions (sarcoidosis, tuberculosis of the liver). Thus, in about 75% of patients with a primary hepatitis, the rise in serum ALP activity is less than 2.5 times the upper limit of normal (ULN). In contrast, the increase in ALP activity exceeds 2.5 × ULN in about 75% of patients with obstruction of the bile duct. The diagnostic value of the test increases when taken in conjunction with a serum protein fractionation and other tests.

Aminotransferases, AST and ALT

The reversible reactions of the two common aminotransferases, or transaminases as they were formerly called, are described in Chapter 10. The two aminotransferases of clinical importance are aspartate aminotransferase (AST) and alanine aminotransferase (ALT). The liver is a rich source of both of these aminotransferases, and their measurements have been of value in the diagnosis of liver disorders.

Reference Values. The normal values for serum AST and ALT as performed by the methods described in Chapter 10 are 6 to 25 U/L for AST and 3 to 30 U/L for ALT (Table 11.4).

Increased Activity. The transaminases are present in normal serum at a low concentration as manifested by a low serum activity, but when tissue cells containing large amounts of these enzymes are injured or killed, the enzymes diffuse into the bloodstream, where a temporary high degree of enzyme activity occurs. The degree of activity depends, of course, on the extent of the tissue damage, the prior concentration of the enzyme in the tissue, and the time course following the tissue injury.

Because the serum levels of both aminotransferases tend to rise or fall more or less together with hepatic cell damage, no real purpose is served in ordering both tests. The rise in serum ALT activity in acute hepatic cell injury is usually a little greater than that in AST, but the determination of either test provides the diagnostic information in most cases.

Normal values for the aspartate aminotransferase and values encountered in various liver diseases are shown in Table 11.5. In general, the serum aminotransferase activity may rise to 100 × ULN in a case of severe viral hepatitis (Fig. 11.7) or of acute toxicity with a drug that severely damages liver tissue (for example, ingestion of chloroform, carbon tetrachloride, or phosphorus compounds). The result depends, of course, on obtaining the blood specimen at the peak of the diffusion of the enzyme from severely damaged hepatic cells into the bloodstream. Because the laboratory and the physician have no control over the time when a patient may seek medical aid, the peak occurrence of the aminotransferase activity in serum may have passed by the time the specimen is obtained, although the activity may still be greatly elevated. Each enzyme has its own particular time course in serum after tissue injury. The rise in serum aminotransferase activity in viral hepatitis begins early in the disease, frequently before jaundice is visible, and reaches a maximum during the acute stage of the disease when destruction of hepatic

TABLE 11.5
TYPICAL RANGE OF AST CONCENTRATION IN VARIOUS LIVER DISORDERS

DISEASE STATE	RANGE OF AST ACTIVITY × UPPER LIMIT OF NORMAL (ULN)
Normal adults	0.3–1.0
Viral hepatitis—nonicteric phase	1.5–8
Viral hepatitis—acute stage	12–100
Acute poisoning ($CHCl_3$ or CCl_4)	12–125
Cholestasis, extrahepatic or intrahepatic	1–8
Primary or metastatic hepatic carcinoma	1–8
Alcoholic cirrhosis	1.3–6

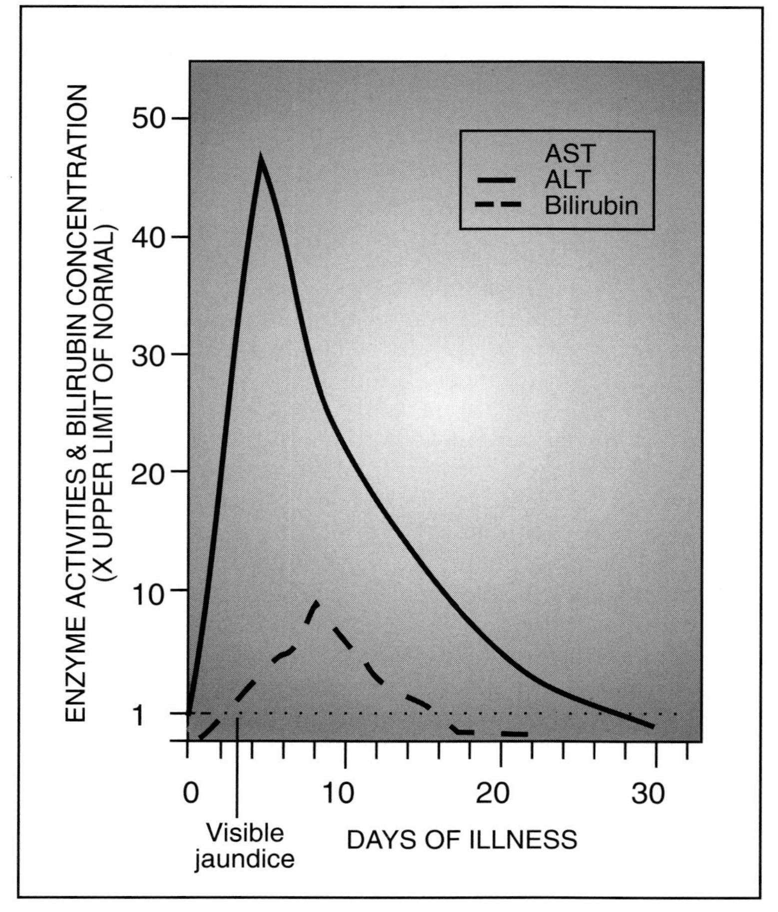

FIG. 11.7. Example of bilirubin concentrations and serum aspartate (AST) and alanine (ALT) aminotransferase activities in serum of patient with acute viral hepatitis. Note that the rise in transaminase activities precedes the jaundice.

cells is at its height. The peak rise in aminotransferase after acute chemical toxicity, such as ingestion of carbon tetrachloride in a suicide attempt may occur within 24 hr and decline rapidly by the second day. The time factor must be taken into consideration in any interpretation.

Moderate elevations (1 to 8 or 9 X ULN) of serum aminotransferase activity usually occur in cholestasis, cirrhosis, hepatic tumors, infections, and infectious mononucleosis. This enzyme test is of value in differentiating these conditions from those causing acute hepatocellular injury. Some of the usual findings are summarized in Table 11.5.

Less Common Enzyme Tests for Special Situations

Most cases of liver disease can be diagnosed by judicious application of the common enzyme tests used in conjunction with the other tests of hepatic function, the patient's history, physical examination, and other studies, such as gallbladder visualization when appropriate, liver biopsy, or a radioactive liver scan. In a few situations, however, the diagnosis may still be in doubt. The special enzyme tests described in this section may be helpful in specific situations.

γ-Glutamyltransferase (GGT)

GGT transfers the γ-glutamyl portion of a peptide to another peptide, amino acid, or water; glutathione is its common substrate in the body. GGT is located primarily on cell membranes and may assist in amino acid transport into cells. The enzyme is present in relatively high concentration in kidney, pancreas, liver, and prostate. GGT is a sensitive indicator of liver disease, particularly hepatobiliary obstruction.[12,13]

Reference Values. Reference values, as obtained by the method of Rosalki and Tarlow, [14] are 3 to 35 U/L for men and 3 to 30 U/L for women.

Increased Activity. Serum GGT activity is usually elevated in both hepatocellular and obstructive liver diseases. The enzyme is not absolutely specific for disorders of the hepatobiliary tree, but its specificity surpasses that of the other enzymes commonly used in the diagnosis of hepatic disorders. Thus, the serum activity of GGT is not elevated in bone disease, as is ALP, or in muscle disease or hemolytic conditions, as is AST. The estimation of serum GGT is most helpful in the following situations:

1. In detecting hepatic injury caused by alcoholism. Serum GGT is considered the enzyme of choice in investigating possible alcoholism.
2. Hepatic metastasis in the anicteric patient. Serum GGT is frequently elevated in hepatic metastasis, but because its activity may also rise in the presence of other malignant conditions, this test is useful in conjunction with tests of ALP or 5NT, which are less sensitive but a little more specific.

3. In the management of patients with infectious hepatitis. If GGT is monitored serially during the course of the disease, a return of serum GGT activity to normal is an excellent prognosis; a later rise in GGT activity indicates an exacerbation of the disease. GGT measurement is of greater value in this situation than is either AST or ALT, although the estimation of 5NT may be equally good.
4. In chronic obstruction of the bile ducts. Usually an elevated serum ALP activity provides this information, but if there is any question of a concomitant bone disease, which also elevates serum ALP, the determination of serum GGT or 5NT resolves the problem because these enzyme tests are more specific for liver disease and are not affected by the proliferation of osteoblasts.

5′-Nucleotidase (5NT)

This enzyme specifically hydrolyzes the phosphate from 5′-adenosine mononucleotide (AMP). It is present as a microsomal enzyme in liver and other tissues, and its activity in serum is increased in hepatobiliary disease.[15]

Reference Values. The reference range varies from 1 to 7 U/L.

Increased Activity. The serum activity of 5NT is increased in various types of hepatic disorders. Its measurement is of value in the following situations:

1. Ascertaining whether an increase in ALP is caused by osteoblastic activity or liver disease. Skeletal disorders usually do not cause an increase in serum 5NT, but liver diseases of various types do. This type of differentiation is frequently necessary in growing children, who usually have an elevated ALP.
2. Diagnosis of hepatic metastases in the anicteric patient. Serum GGT, 5NT, and ALP are usually elevated in this situation. The GGT enzyme test is the most sensitive, but the 5NT test is confirmatory.
3. Management of patients with infectious hepatitis. As mentioned under the GGT test, the activities of both GGT and 5NT rise in infectious hepatitis and fall with recovery; a secondary rise indicates an exacerbation. Tests of GGT and 5NT are sensitive and give the same information.

REFERENCES

1. Sherlock, S.: Diseases of the Liver and Biliary System. 8th Edition. London, Blackwell Scientific Publications, 1989.
2. Isselbacher, K.J., and Podolsky, D.J.: Biological and clinical approaches to liver disease. *In* Harrison's Principles of Internal Medicine. 12th Edition. Edited by J.D. Wilson, et al. New York, McGraw-Hill, 1991.
3. Dienstag, J.L., Wands, J.R., and Isselbacher, K.J.: Acute hepatitis. *In* Harrison's Principles of Internal Medicine. 12th Edition. Edited by J.D. Wilson, et al. New York, McGraw-Hill, 1991.
4. Stevens, C.E., et al.: Epidemiology of hepatitis C virus. JAMA, *263*:49, 1990.
5. Bradley, D.W.: Enterically-transmitted non-A, non-B hepatitis. Br. Med. Bull., *46*:442, 1990.
6. Monjardino, J.P., and Saldanha, J.A.: Delta hepatitis: the disease and the virus. Br. Med. Bull., *46*:423, 1990.

7. Lott, J.A.: New concepts in serum bilirubin measurement. Lab. Management, April, 41, 1987.
8. Lauff, J.J., Kasper, M.E., Wu, T.W., and Ambrose, R.T.: Isolation and preliminary characterization of a fraction of bilirubin in serum that is firmly bound to protein. Clin. Chem., *28*:629, 1982.
9. Dappen, G.M., Sundberg, M.W., Wu, F.W., et al.: A diazobased dry film for determination of total bilirubin in serum. Clin. Chem., *29*:37, 1983.
10. Brett, E.M., Hicks, J.M., Powers, D.M., and Rand, R.N.: Delta bilirubin in serum of pediatrics patients: Correlations with age and disease. Clin. Chem., *30*:1561, 1984.
11. Koch, T.R., and Doumas, B.T.: Bilirubin, total and conjugated, modified Jendrassik-Grof method. *In* Selected Methods of Clinical Chemistry. Vol. 9. Edited by W.R. Faulkner and S. Meites. Washington, D.C., American Association for Clinical Chemistry, 1982.
12. Rej, R.: Aminotransferases in disease. Clin. Lab. Med., *9*:667, 1989.
13. Rosalki, S.B.: Gamma-glutamyl transpeptidase. Adv. Clin. Chem., *17*:53, 1975.
14. Rosalki, S.B., and Tarlow, D.: Optimized determination of γ-glutamyl transferase by reaction-rate analysis. Clin. Chem., *20*:1121, 1974.
15. Bodansky, O., and Schwartz, M.K.: 5′-nucleotidase. *In* Advances in Clinical Chemistry. Vol. 11. Edited by O. Bodansky and C.P. Stewart. New York, Academic Press, 1968.

REVIEW QUESTIONS

1. What are the major functions of the liver?
2. What are the various steps leading to the formation and excretion of bilirubin?
3. What liver tests are based on suppressed protein synthesis in the liver?
4. What role does the liver play in the homeostasis of cholesterol?
5. What role does the liver play in the homeostasis of glucose?
6. What types of viral hepatitis are most commonly spread by drug use, accidental needle puncture, or sexual transmission?
7. In what types of liver disorders should one obtain the results of the enzyme tests ALP and AST?
8. Why is the day that blood was drawn for enzyme tests (in relation to the onset of symptoms) important in liver disease?

Minerals and Trace Elements 12

OUTLINE

OBJECTIVES

After reading this chapter, the student will be able to:

1. Describe the physiologic functions of calcium and how its plasma concentration is regulated so closely.
2. Discuss the relation between calcium and phosphate concentrations and the effect of parathyroid hormone, vitamin D, and calcitonin on plasma phosphate concentration.
3. State the major pathophysiologic conditions associated with disorders of calcium metabolism and know which tests are most commonly used for their diagnosis.
4. Describe the physiologic functions of magnesium.
5. Name the trace elements and describe their physiologic functions.
6. Describe the forms in which iron is distributed within the body.
7. Discuss the principal causes of iron deficiency and know which laboratory tests help to diagnose this disorder.
8. Describe the principles of the following tests in serum: serum iron, total iron binding capacity (TIBC), zinc protoporphyrin/heme ratio, and transferrin.

MINERAL METABOLISM

For proper growth and development, a balanced diet must provide adequate amounts of minerals in addition to sufficient calories, high-quality proteins, and vitamins. Minerals are essential for the adequate function of soft tissues, as well as for the formation of the bony structures and teeth. The minerals fall into two categories, the macrominerals, or major elements, and the microminerals, or trace elements. The major elements are arbitrarily defined as those present in greater than 5-g quantities in the entire body (average adult), with a daily intake requirement for each of 100 mg or more. The total body contents of the 6 major elements (calcium, phosphorus, potassium, sodium, chloride, and magnesium) are shown in Figure 12.1. These quantities range from 1150 g for calcium to 30 g for magnesium, the major mineral that is least plentiful in the body. The recommended daily dietary intake for the major elements, as shown in Table 12.1, ranges from 300 mg for magnesium to several grams for potassium. By contrast, the total quantity in the human body of each essential trace element

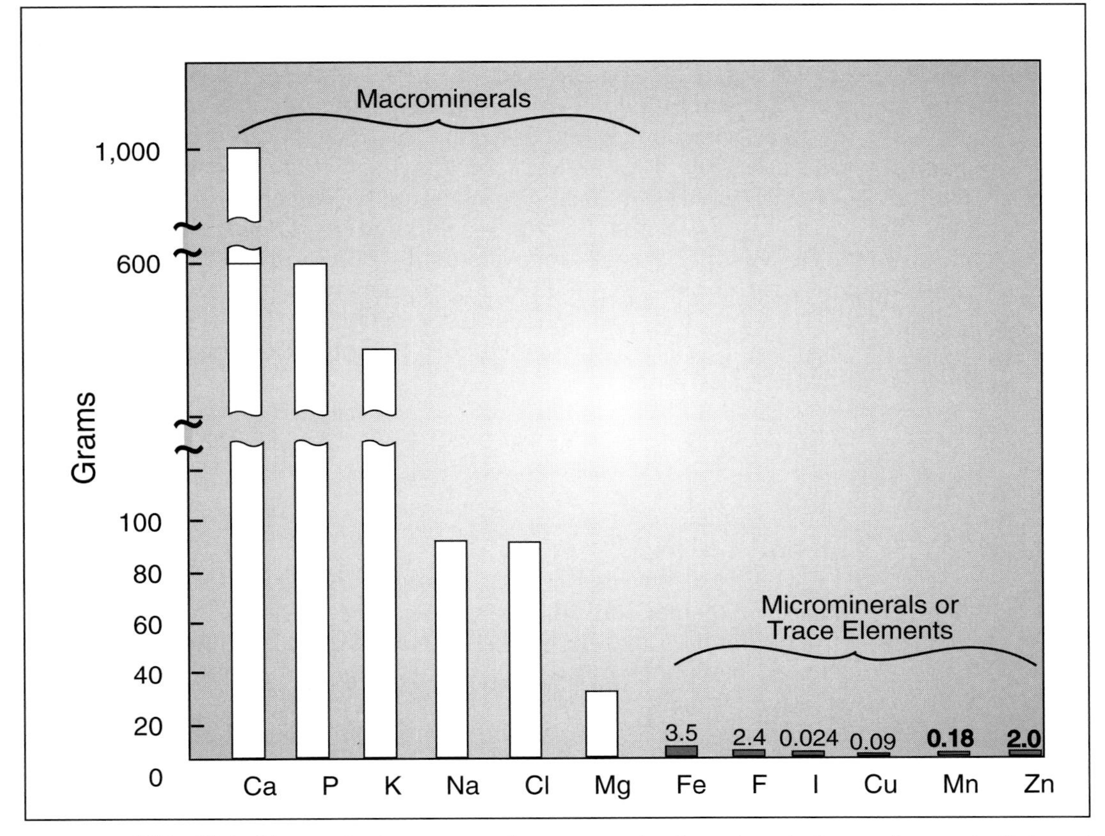

FIG. 12.1. Human body content of macro- and microminerals (trace elements). (Adapted from Hamilton, E.M.N., Whitney, E.N., and Sizer, F.S.: Nutrition, Concepts and Controversies. 5th Edition. New York, West Publishing, 1991.)

TABLE 12.1
MAJOR ELEMENTS REQUIRED BY HUMANS

ELEMENT	RECOMMENDED DAILY DIETARY ALLOWANCE
Calcium	800 mg*
Phosphorous	800 mg
Potassium	2–6 g
Sodium	†
Chloride	†
Magnesium	350 mg (males) 300 mg (females)

*Increased to 1200 mg/day during adolescence (both male and female) and during pregnancy and lactation.
†The average daily intake of sodium and chloride in western societies is about 3 to 4 g/day, but most of this amount is excreted. Because the kidney can conserve sodium, the absolute minimum daily requirement for sodium (and chloride) is probably less than 100 mg/day. This amount varies with the amount of perspiration.

is less than 5 g, as shown in Figure 12.1. These quantities vary from 3.5 g for iron, the most plentiful of the trace elements, to 24 mg for iodide, to even lesser quantities for a few of the trace minerals (chromium, nickel, cobalt, molybdenum) not included in Figure 12.1. Many of the metallic trace elements are necessary components of metalloenzymes in the body; zinc, for example, is a constituent of at least 70 enzymes, including alkaline phosphatase and carbonic anhydrase. The National Research Council of the National Academy of Sciences has established a Recommended Dietary Allowance (RDA) for the four trace elements iron, iodine, zinc, and selenium, and has recommended an RDA for copper, manganese, molybdenum, and fluoride (Table 12.2). No official RDA exists for the remaining trace elements (chromium, cobalt, nickel, silicon, tin, or vanadium) even though they are considered essential.

MAJOR ELEMENTS

Sodium, Potassium, and Chloride[1]

The three major elements, sodium, potassium, and chloride, were described in detail in Chapter 5 under their respective headings. Sodium is unique among the major elements in that the kidney has the ability to restrict its urinary excretion to almost zero in times of insufficient intake. Renal conservation is only partial for the other major elements. Plasma osmotic pressure and water balance depend primarily on the proper intake and distribution of these ions.

TABLE 12.2
TRACE ELEMENTS REQUIRED BY HUMANS

ELEMENT	RECOMMENDED DIETARY ALLOWANCE (RDA)
Established	
Iron	50–200 μg/day
Iodide	150 μg/day
Zinc	Men,15 mg/day; Women, 12 mg/day
Selenium	Men, 70 μg/day; Women, 50 μg/day
Recommended	
Copper	1.5–3.0 mg/day
Fluoride	1.5–4.0 mg/day
Manganese	2.0–5.0 mg/day
Molybdenum	75–250 μg/day
Estimated	
Chromium	50–200 μg/day
Nickel	35–500 μg/day
Cobalt	?
Silicon	?
Tin	?
Vanadium	?

Calcium[2]

Calcium is the mineral present in the largest amount in the body, as much as approximately 1150 g (Fig. 12.1). Approximately 99% of the total body calcium is deposited in the skeleton as hydroxyapatite, $Ca_{10}(PO_4)_6(OH)_2$, and some amorphous Ca salts. The calcium phosphate laid down in bone is by no means an inert component of the skeleton; it is in dynamic equilibrium with the Ca^{2+} and HPO_4^{2-} of body fluids and is constantly turning over by the process of dissolution (resorption) and deposition.

A higher proportion of the nonskeletal calcium is present within cells than in extracellular fluids, but most of the intracellular calcium is bound to proteins in the cell membrane, mitochondria, and nucleus. The concentration of calcium ion in intracellular fluids is reduced considerably by this binding.

Ca^{2+} is physiologically active and functions as an intracellular messenger by binding to or being released from specific intracellular proteins, a process that changes protein conformation and hence its activity or function.[3] All nucleated cells contain a small protein, calmodulin, with four binding sites for calcium. Some enzymes, such as adenylate cyclase, become activated when reversibly bound to a Ca-calmodulin-enzyme complex. The calcium messenger system initiates the contraction of muscle fibers and regulates a host of cellular functions, such as secretion of hormones and fluids, transfer of ions across membranes, cell motility, mitosis, and many others. Ca^{2+} decreases neuromuscular irritability and is essential for blood coagulation.

Most individuals are in a state of calcium balance, ingesting about 500 to 1000 mg daily in the food and excreting a like amount in the urine and feces. Dairy products (milk, cheese) are the best food sources of calcium. A large portion of the dietary calcium is not absorbed because of the formation of insoluble calcium compounds (phosphate, phytate, oxalate, soaps) in the intestines that are excreted in the feces. The calcium balance is maintained by a complex control system that regulates the absorption of calcium from the intestine, its excretion by the kidneys, and the movement of calcium in and out of bone.

Regulation of Plasma Calcium Concentration

Plasma Ca^{2+} concentration is kept within narrow limits by the coordinated action of two principal hormones, parathyroid hormone (PTH) and 1,25-dihydroxy vitamin D_3 [1,25-$(OH)_2D_3$, the active form of vitamin D], with limited assistance by the thyroid hormone calcitonin (CT). The parathyroid secretes PTH in response to a small decrease in plasma Ca^{2+} concentration. PTH raises Ca^{2+} concentration by (1) retrieving calcium and phosphate from bone (promoting bone resorption by osteoclasts), (2) conserving Ca^{2+} by increasing the reabsorption of Ca^{2+} by the renal tubules, (3) facilitating the excretion of phosphate by inhibiting its reabsorption by the renal tubules, and (4) stimulating the kidney to convert 25-$(OH)D_3$, the inactive vitamin D precursor, to active vitamin D, 1,25-$(OH)_2D_3$. The latter hormone increases the intestinal absorption of calcium ingested in the diet (see Chap. 13). An elevated plasma Ca^{2+} turns off the secretion of PTH by negative feedback control and stimulates the secretion of CT, a hormone that inhibits the activity of osteoclasts and thus suppresses bone resorption. The role of CT appears to be minor because surgical removal of the thyroid does not produce hypercalcemia.

Plasma calcium exists in three forms: ionized or free calcium (Ca^{2+}), the physiologically active ion; un-ionized calcium bound primarily as a complex of citrate; and a nondiffusible form bound to plasma proteins. Ca^{2+} and un-ionized calcium pass through the glomerular membrane and appear in the glomerular filtrate, but protein-bound Ca cannot do this. About 43 to 47% of the total plasma calcium is protein-bound, primarily to albumin, but also to some extent to the α-, β - and γ-globulins. The amount of calcium attached to plasma proteins varies with the protein concentration. Ionized calcium constitutes 48 to 52% of the total calcium, and the un-ionized diffusible form constitutes approximately 5%.

Serum is the fluid of choice for measurement of total calcium concentration. The Ca^{2+} level is altered by the following factors: blood pH (alkalosis lowers and acidosis increases Ca^{2+}); diseases of the parathyroid gland or of the calcitonin-producing cells of the thyroid; intestinal disorders of malabsorption; vitamin D deficiency; severe renal disease; and disorders of bone turnover, some of which may be genetic, whereas others may be caused by malignant conditions. A high concentration of Ca^{2+} presents the danger of deposition of calcium salts in soft tissues, including the kidney, and is frequently accompanied by symptoms of fatigue, weakness, and confusion. An elevated Ca^{2+} concentration triggers secretion of the hormone calcitonin, which inhibits the dissolution of bone, but this hormone is insufficient to lower the plasma Ca^{2+} in the face of excess PTH and/or 1,25-$(OH)_2D_3$. The three hormones must operate synchronously to attain a balanced calcium metabolism.

A low concentration of Ca^{2+} increases neuromuscular excitability to a point at which tetany or convulsions may occur. The parathyroid gland is stimulated to secrete PTH, but this secretion alone may not suffice to restore a normal plasma Ca^{2+} concentration if the intake of calcium is inhibited by a malabsorption syndrome or a deficiency of 1,25-$(OH)_2D_3$.

Pregnancy and lactation increase the Ca requirement of women because the body uses Ca for the fetal skeleton and for milk formation. If Ca is not provided in the diet, the necessary Ca and P are obtained by the resorption of maternal bone, with little change in plasma Ca^{2+} concentration.

Serum Total Calcium

Serum total calcium (Ca_T) is determined most frequently by spectrophotometric methods using chelating agents that form highly colored complexes with calcium. The methods are precise, easily automated, and yield results that correlate well with those obtained by atomic absorption spectrophotometry. The calcium chelator *o*-cresolphthalein complexone is widely used for this purpose.

Chelation with *o*-Cresolphthalein Complexone[4]

Principle. The dye o-cresolphthalein complexone binds calcium tightly in alkaline solution to form a highly colored complex with an absorbance measured at 578 nm. The reaction mixture contains the following ingredients to improve performance: 8-hydroxyquinoline to bind magnesium and prevent its interference, urea to decrease the turbidity of a lipemic serum and increase color intensity of the calcium-dye complex, and ethanol to decrease the absorbance of the blank.

Reference Values. The values for Ca_T vary from 8.5 to 10.4 mg/dL (2.13 to 2.60 mmol/L). Multiply mg/dL × 0.25 to convert to mmol/L.

Increased Concentration. The serum calcium concentration is increased in hyperparathyroidism, neoplasms with metastasis to bone, some malignant conditions (lung, kidney, bladder) without bone involvement, hypervitaminosis D, sarcoidosis, multiple myeloma when the plasma proteins are elevated, ingestion of large amounts of milk and alkali as treatment for peptic ulcer, acromegaly (excess growth hormone), some cases of hyperthyroidism, and Paget's disease (a bone disorder). Hypercalcemia may be found in patients who had chronic renal disease and recently received a kidney transplant. The hypercalcemia is ascribed to parathyroid hyperplasia induced by the chronically low serum calcium concentration during the long period before the kidney transplant; the hyperplastic parathyroids secrete too much PTH because the new, normal kidney after the transplant is able to convert 25-(OH)-D_3 to the 1,25-$(OH)_2$ molecule, which improves the intestinal absorption of calcium and helps to elevate the serum calcium concentration. Prolonged elevation of plasma calcium can cause the deposition of insoluble calcium salts in soft tissues and is a factor in the production of renal calculi (stones).

Decreased Concentration. A decrease in the serum ionized calcium concentration below a critical level causes tetany. The physician must be notified immediately if low serum calcium levels are found because tetany can be stopped or prevented by the intravenous injection of calcium gluconate solution. The most common causes of a low serum calcium concentration are hypoparathyroidism, pseudohypoparathyroidism (failure of the target organs to respond to PTH), vitamin D deficiency, gastrointestinal disease that interferes with the absorption of vitamin D and/or calcium (sprue, steatorrhea), nephrosis or other conditions in which the serum protein concentration is low, chronic renal disease (failure of the kidney to hydroxylate 25-hydroxy-D_3), acute pancreatitis, and magnesium deficiency.

The differential diagnosis of hypercalcemia and hypocalcemia is made easier by simultaneous determination of serum calcium, phosphate, alkaline phosphatase, and, in special instances, PTH.

Atomic Absorption Spectrophotometry (AAS)

The determination of Ca_T by AAS is accurate and is considered to be a secondary reference method.[5] AAS methods are so labor intensive, however, that they have been replaced almost completely in routine laboratories by the dye methods for Ca_T.

Principle. The calcium present in a dilute serum sample is converted to free calcium atoms when aspirated into a hot flame.[6] Most of the atoms are in the ground state at the proper flame temperature (see Chap. 3). The atoms in the ground state absorb light at the Ca resonance line, 422.7 nm, when the beam of a calcium hollow cathode lamp is passed through the flame. The absorbance is directly proportional to the calcium concentration.

Ionized Calcium

Only the ionized or free form of calcium (Ca^{2+}) can act as an intracellular messenger.[3] Hence, a knowledge of the concentration of Ca^{2+} provides more valuable information in clinical situations than does that of Ca_T. Bowers and co-workers strongly recommend the measurement of Ca^{2+} instead of Ca_T in most hospital situations even though Ca^{2+} measurement requires more careful sample handling, instrument maintenance, and quality control.[7]

Principle. Ionized calcium is measured in whole blood, plasma, or serum by an ion-selective electrode (ISE) in commercially available instruments. Because the concentration of Ca^{2+} depends on the pH of the specimen at the time of measurement, all the second-generation instruments measure the pH simultaneously with the Ca^{2+}; some instruments also display the Ca^{2+} corrected to pH 7.40.[7]

Reference Values. The Ca^{2+} reference range in plasma, serum, or whole blood is 4.68 to 5.32 mg/dL (1.17 to 1.33 mmol/L). The mean is the same for healthy newborns, but the range is wider; the values rise somewhat after 2 weeks and are maintained above adult levels throughout childhood.

Phosphate

Phosphorus in the body exists only as inorganic phosphate or as organic phosphate esters. Eighty percent of it is laid down in bone matrix as insoluble calcium salts. The organic phosphate esters are primarily confined within cells, associated with nucleoproteins, hexoses (glucose-6-phosphate), deoxygenated hemoglobin in red blood cells (2,3-diphosphoglycerate), and purines (ATP, GTP); phosphate forms high energy bonds in ATP, GTP, and creatine phosphate. The phospholipids are present in cells, particularly as a component of membranes, but they also circulate in plasma in fairly high concentration (see Chap. 8). Inorganic phosphate ions ($H_2PO_4^-$ and HPO_4^{2-}) are mostly confined to the extracellular fluid where they are part of the buffer system. At pH 7.4, 80% of the inorganic phosphate is in the form of HPO_4^{2-}.

Phosphate is ubiquitous in food, and ingestion of sufficient amounts usually is no problem. The organic phosphates in food are hydrolyzed in the gastrointestinal tract, and inorganic phosphate is liberated. A large proportion of the ingested phosphate is precipitated as insoluble salts, however, and is not absorbed. The principal route of excretion for the absorbed phosphate is in the urine.

The control of plasma phosphate concentration is closely linked with that of calcium because they are deposited together in bone as an amorphous or hydrated calcium phosphate; bone resorption increases the plasma concentrations of both calcium and phosphate ions, whereas bone deposition decreases the concentrations of both.

Thus, the hormone regulators of Ca^{2+} concentration (PTH, calcitonin, and $1,25\text{-}(OH)_2D_3$) also affect the phosphate concentration. PTH, however, stimulates the kidney to excrete phosphate while conserving calcium. This action is a protective mechanism because the solubility product of calcium phosphate in body fluids is readily exceeded if the concentrations of both ions are elevated; the concentrations of Ca^{2+} and phosphate tend to vary inversely because of the PTH effect upon the kidney.

The phosphate ion concentration of the plasma is influenced by the calcium ion concentration; by the PTH concentration, which promotes phosphaturia by decreasing the renal tubular reabsorption of phosphate; by calcitonin, which inhibits bone resorption; and by growth hormone, which increases the renal tubular reabsorption of phosphate. In chronic renal disease, phosphate retention occurs because of impaired glomerular filtration.

Serum Phosphate

Serum inorganic phosphate in a protein-free supernate or dialysate is converted into an ammonium or sodium phosphomolybdate complex. The ultraviolet absorbance of the unreduced phosphomolybdate complex (Mo^{VI}) may be measured directly at 340 nm. The phosphate can also be determined colorimetrically after adding a mild reducing agent to convert the complex to the highly colored "molybdenum blue" (Mo^{V}). Although many different reducing agents have been used, we prefer *p*-methylaminophenol because the reagent is stable and easy to prepare. Semidine is more sensitive, however, because it becomes blue when oxidized, in addition to producing molybdenum blue from the complex.

The serum phosphate concentration is customarily reported in terms of mg P per 100 mL rather than mg HPO_4^{2-}. In the SI system, however, mol/L of P are identical with mol/L of HPO_4^{2-}.

Principle. Serum proteins are precipitated by trichloroacetic acid, and the phosphate is converted to a phosphomolybdate (Mo^{VI}) complex by the addition of sodium molybdate. The addition of *p*-methylaminophenol reduces the Mo^{VI} in the complex to yield an intensely blue-colored phosphomolybdate complex (Mo^{V}). The absorbance of the solution at 700 nm is proportional to the serum phosphate concentration.[8]

Reference Values

At birth: 4.2 to 9.5 mg/dL (1.34 to 3.36 mmol/L)
Children: 4.0 to 6.0 mg/dL (1.28 to 1.94 mmol/L)
Adults: 3.0 to 4.5 mg/dL (0.96 to 1.44 mmol/L)

Multiply mg/dL × 0.32 to convert to mmol/L.

Increased Concentration. The serum phosphate concentration is increased in advanced renal insufficiency, true and pseudohypoparathyroidism, hypervitaminosis D, and hypersecretion of growth hormone (GH). The higher phosphate levels in infants and children are associated with an increased concentration of GH.

Decreased Concentration. The serum phosphate concentration is temporarily lowered when carbohydrate is absorbed; glucose and other sugars are phosphorylated when they enter cells. A lowered concentration of serum phosphate typically appears in hyperparathyroidism, rickets (vitamin D deficiency), steatorrhea, and in some renal diseases when tubular absorption of phosphate is impaired (Fanconi syndrome). The prolonged ingestion of antacids containing $Mg(OH)_2$ or $Al(OH)_3$ lowers the serum phosphate because of precipitation of insoluble phosphates in the gastrointestinal tract. Parenteral hyperalimentation causes a movement of phosphate from plasma into muscle and adipose cells, thereby lowering plasma phosphate concentration. A dangerously low hypophosphatemia may be produced unless the fluids contain sufficient phosphate.

Magnesium

Magnesium is the fourth most abundant cation in the body; much of it is present in bone, associated with calcium and phosphate. The magnesium concentration within cells is second only to that of potassium; its level in extracellular fluids is much lower. About 30% of the plasma magnesium is bound to albumin, and 70% exists as ionized magnesium.

The main dietary sources of magnesium are meat and green vegetables. Like calcium, magnesium is absorbed in the upper intestines, but vitamin D is not essential for magnesium absorption. Magnesium compounds have solubility properties similar to those of calcium, and the major portion of the ingested magnesium is not absorbed because of the formation of insoluble phosphates and soaps in the gut. The kidney is the organ responsible for maintaining magnesium balance; it can conserve magnesium when the intake is low and excrete the excess when the intake is high. The exact control mechanism is unknown, but the hormone aldosterone may play a role; aldosterone promotes the excretion of Mg^{2+}, together with K^+, and the retention of Na^+.

Mg^{2+} is an activator for several enzymes, particularly for those involved in phosphorylation, protein synthesis, and DNA metabolism; it also affects neuromuscular excitability. Like calcium, a low serum Mg^{2+} concentration is associated at times with tetany. A high concentration of Mg^{2+} in plasma has a depressing effect on the central nervous system and alters the conducting mechanism of the heart.

Serum Magnesium

Serum magnesium is usually determined spectrophotometrically as a Mg chelate with high molar absorptivity. The most widely used chelates are calmagite and methylthymol blue. Determination by atomic absorption spectrophotometry is accurate but too time-consuming for most clinical laboratories.

Principle. Magnesium forms a colored complex at pH 11.5 with 3-hydroxy-4-[(6-hydroxy-*m*-tolyl)azo]-1-naphthalenesulfonic acid (Calmagite); absorbance is measured at 520 nm. The reagent contains the amphoteric betaine detergent Empigen BB to shift the wavelength of the blank, a strontium chelate to mask the effects of calcium, and triethanolamine to mask the effects of iron.[9]

Reference Values. The concentration of total serum magnesium in normal individuals ranges from 1.3 to 2.1 meq/L or 0.65 to 1.05 mmol/L.

Increased Concentration. Some elevation in the serum magnesium concentration occurs in chronic renal disease, severe dehydration, and adrenal insufficiency with aldosterone deficiency. The oral intake of Epsom salt ($MgSO_4$) for constipation may produce hypermagnesemia in persons with impaired renal function.

Decreased Concentration. Hypomagnesemia is frequently associated with gastrointestinal disorders (malabsorption, prolonged diarrhea, bowel or kidney fistulas, acute pancreatitis), acute alcoholism, prolonged parenteral fluid therapy without magnesium supplementation, and the use of some diuretics. Tetany caused by a low serum magnesium concentration, with a normal serum calcium, has been reported; the parenteral administration of magnesium eliminates the tetany.

TRACE ELEMENTS[10]

The 14 trace elements essential for human nutrition are listed in Table 12.2, along with RDA[10] if known. The elements are listed in 3 groups: those with an established RDA, those without an RDA but for which an adequate and safe daily intake is suggested, and those for which an adequate and safe intake has not been proffered.

The daily requirement for trace elements is small, being μg/day for most and mg/day for the others. Some elements, such as iron, form a high proportion of insoluble compounds while in the digestive tract and are poorly absorbed; the higher RDA of these trace minerals compensates for the inefficient absorption. The RDA is low for most trace elements because (1) the use in the body is restricted to such special molecules as enzymes (for the metallominerals and selenium), the thyroid hormones (for iodide), and heme proteins (for iron) and (2) the kidney excretes little of these trace elements because of the high binding constants of their transport proteins.

Iodide

Iodide is a structural component of T_3 and T_4, the hormones that regulate basal metabolic rate (affect all cells) and are necessary for proper growth and development. Goiters (enlargement of the thyroid gland) usually develop in people living in areas where iodide is lacking in drinking water and food and no program for supplementation exists. The iodine RDA is 150 μg/day (Table 12.2). For a full review of iodide, see Chapter 13.

Assessment of Iodide Nutriture

Determination of serum T_3 is the simplest test to perform. If the T_3 concentration is below the lower limit of the reference range (70 to 200 ng/dL or 1.1 to 3.1 nmol/L), thyroid insufficiency is indicated.

Iron[11]

Iron is an essential component of a group of heme proteins that function in oxygen transport or as enzymes in redox systems. The bulk of the body iron is contained in hemoglobin, the protein of the red blood cell that transports oxygen from the lungs to the tissues, but a portion is also present in myoglobin, a protein in muscle that is capable of binding and releasing oxygen. Iron is also a component of the cytochromes, enzymes that act as electron transfer agents in redox reactions, and of the enzymes catalase and peroxidase, which decompose H_2O_2 or other peroxides. The iron component in all these active compounds is heme (ferroprotoporphyrin, see Fig. 11.4); the iron is covalently bound to the porphyrin. The protein moiety of the different heme proteins confers the properties and specificity peculiar to each. A small amount of iron is contained in several nonheme metalloenzymes.

The total amount of iron in a 75-kg man is approximately 4 g but is less in women during the reproductive years. As shown in Figure 12.2, about 70 to 75% of the iron is incorporated in the following heme compounds: hemoglobin, myoglobin, and heme enzymes. Approximately 25% of body iron stores in men are complexed to a protein, apoferritin, to form soluble ferritin and an insoluble product, hemosiderin. The soluble ferritin comprises the major portion of the iron stores and recycles its iron as needed for heme products; it is present in all tissues but primarily in cells of the reticuloendothelial system (bone marrow, liver, and spleen), whereas hemosiderin is deposited only in the reticuloendothelial cells. The serum concentration of ferritin reflects the amount of iron stored in reticuloendothelial cells. The amount of ferritin and stored iron is greater after puberty in men than in women; children have less of each than adults. Bleeding experiments in humans have revealed that the iron stores in mg approximately equal the serum ferritin concentration (μg/L × 8).

Two to 4 mg of iron circulate in plasma bound to a β-globulin, transferrin; 2 atoms of Fe^{3+} are bound per mole of transferrin. The transferrin binds the iron tightly and makes it unavailable to nonspecific complexing agents in plasma; it releases the iron only to specific receptor sites on cell membranes. The HCO_3^- in plasma is essential for the binding of Fe^{3+} by transferrin. The total amount of iron in plasma is small and does not exceed 1.5 mg/L (150 μg/dL).

Iron, like most trace elements, is recycled so effectively that virtually no body iron is lost through excretion into the urine. The daily loss of iron in urine and sweat in men is less than 1 mg. Because the kidney does not regulate the concentration of plasma iron by excreting excessive intake, the control must

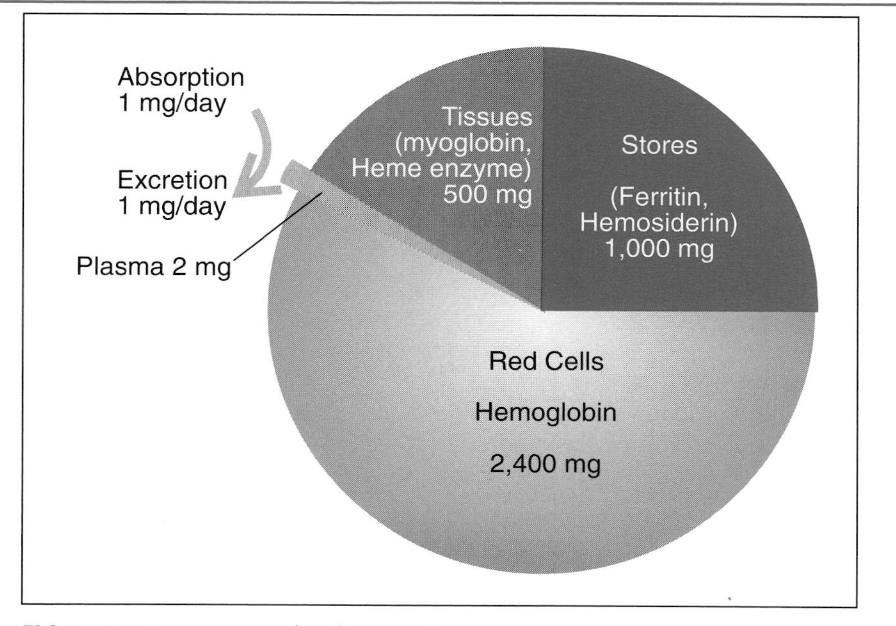

FIG. 12.2. Approximate distribution of iron in an average-sized man. Women have less storage iron and fewer red blood cells.

reside elsewhere. Apparently the intestinal mucosa contains special proteins necessary for the absorption of iron; to some degree, these proteins regulate its absorption.

The RDA for iron (Table 12.2) is 10 mg/day for men and 15 mg/day for women. Only a portion of the iron is absorbed because of the low solubilities of many iron compounds. Some iron is absorbed in the upper bowel as Fe^{2+}, which is oxidized to Fe^{3+} by ceruloplasmin, a copper-containing enzyme, as it traverses the intestinal mucosa and is then stored temporarily in ferritin. On release from the mucosal cell into plasma, the Fe^{3+} becomes tightly bound to transferrin, the iron transport protein. Each molecule of transferrin is capable of binding two atoms of Fe^{3+}, but usually an excess of this compound is in the plasma of normal individuals so that the transferrin is only about 30% saturated with iron.

Formation of Hemoglobin

During the formation of erythrocytes in the bone marrow, storage iron from ferritin is transferred to normoblasts (nucleated red blood cells).[12] The heme moiety of hemoglobin is synthesized in the normoblasts from glycine and succinyl-CoA by a series of enzymatic steps. The enzymic insertion of Fe^{2+} into protoporphyrin IX is the last step in the formation of heme. Heme condenses with a globin protein to form hemoglobin within the nucleated erythrocyte. The iron in hemoglobin is Fe^{2+}; when oxidized to Fe^{3+}, the compound becomes methemoglobin and can no longer transport oxygen. The formation of new erythrocytes keeps pace with the daily destruction of senescent cells with an average lifetime of about 120 days. The iron in the phagocytized old cells is recycled (see Chap. 11).

Iron Deficiency Disorders[13]

Iron deficiency states are widespread and occur most commonly in women and young children. Women have extensive iron losses in pregnancy (because of the need to supply iron to the fetus) and in periodic menstrual bleeding, whereas nursing children have virtually no iron intake. Factors primarily responsible for iron deficiency include an insufficient supply of iron-containing foods in the diet, impaired absorption of iron, and acute or chronic blood loss.

Early diagnosis of iron deficiency is difficult. The serum iron level gives no indication of the size or availability of iron stores. The total iron binding capacity (TIBC) is increased in this state, but the test is not specific. Measurement of serum ferritin concentration is helpful because a low value correlates well with decreased iron stores. The determination of the zinc protoporphyrin/heme ratio (ZPP/H) in a drop of whole blood is the simplest, most sensitive, and least expensive method for detection of iron deficiency.[14] The rationale is that when iron is unavailable for the synthesis of heme, it is replaced by zinc.[15] In lead poisoning, lead interferes with the enzymic transfer of iron into protoporphyrin IX. The results with ZPP/H are comparable to those for free erythrocyte protoporphyrin but require no prior acid extraction.

Serum Iron

Principle. Serum proteins are precipitated in acid solution containing thioglycolic acid that reduces Fe^{3+} to Fe^{2+}, thereby dissociating the iron from its binding to transferrin. A chromogen, ferrozine, is added to the supernate to form a highly colored Fe^{2+} complex with an absorbance measured at 562 nm.[16]

Reference Values. The serum iron concentration is 65 to 165 μg/dL for men (11.6 to 29.5 μmol/L) and 45 to 160 μg/dL (8.1 to 28.6 μmol/L) for women. The concentration is higher at birth but then falls below adult levels, where it remains during infancy and childhood. Diurnal variation reflects higher values in the morning.

Total Iron Binding Capacity (TIBC) and Transferrin Saturation[16]

Principle. A known amount of Fe^{3+}, more than sufficient to fully saturate the serum transferrin with iron, is added to a serum sample. The excess Fe^{3+}, not bound to transferrin, is removed by addition of a small amount of buffered ion exchange resin. The sample is diluted and centrifuged, and an aliquot of the supernate is analyzed for iron content of the fully saturated transferrin; this value is the TIBC.

$$\text{Transferrin saturation (\% saturation)} = \frac{\text{serum Fe}}{\text{TIBC}} \times 100$$

Reference Values. TIBC varies from 260 to 440 μg/dL (46.5 to 78.8 μmol/L). Transferrin saturation ranges from 20 to 50%.

Increased Concentration (Table 12.3)

Serum Iron. The concentration of serum iron is elevated in conditions of (1) increased red blood cell destruction (hemolytic anemia); (2) ineffective or decreased erythrocyte formation (pernicious anemia in relapse, aplastic anemia, marrow damage by toxins); (3) blocks in heme synthesis (lead poisoning, pyridoxine deficiency); (4) increased release of storage iron (acute hepatic cell necrosis); (5) increased intake or impaired control of iron absorption (ingestion of large amounts of iron, hemochromatosis, hemosiderosis); and (6) megaloblastic anemia.

TIBC. An increase in the plasma concentration of transferrin elevates the TIBC. Such elevation may occur during late pregnancy, in iron deficiency anemia, after acute hemorrhage, and after destruction of liver cells.

Transferrin Saturation. Transferrin is increased in iron overload states (hemochromatosis, hemosiderosis), hemolytic anemias, acute hepatitis, and pernicious anemia in relapse.

Decreased Concentration (Table 12.3)

Serum Iron. The concentration of serum iron is decreased when the dietary intake of iron is insufficient (iron deficiency anemia, malabsorption of iron); loss of iron is accelerated (acute or chronic blood loss, late pregnancy); and release of stored iron from reticuloendothelial cells is impaired (infection, neoplasia, rheumatoid arthritis).

TABLE 12.3
CHANGES IN SERUM IRON, TIBC, PERCENT SATURATION, AND FERRITIN CONCENTRATION IN VARIOUS CONDITIONS

CONDITION	SERUM IRON	TIBC	TRANSFERRIN SATURATION	SERUM FERRITIN
Iron deficiency	↓ ↓	↑ ↑	↓ ↓	↓ ↓
Inflammatory states (infection, inflammation, neoplasias)	↓	↓	↓ N	↑ ↑
Reduced transferrin concentration (nephrosis, Kwashiorkor)	↓	↓ ↓	↑	N ↓
Ineffective erythropoiesis (B_6, folate, or B_{12} deficiencies, Pb poisoning, thalassemia)	↑	N	↑	↑
Iron overload (excessive intake, absorption, hemochromatosis)	↑ ↑	↓	↑ ↑	↑ ↑
Liver disease (acute necrosis)	↑	↑	↑	↑ ↑

N signifies no changes from normal, ↑ an elevation, and ↓ a decrease. Double arrows indicate large increases or decreases.

TIBC. The TIBC is decreased when synthesis of transferrin is decreased (infection, neoplasia, uremia) or loss of this protein in the urine is increased (nephrosis).
Transferrin Saturation. The % saturation is decreased in iron deficiency anemia, late pregnancy, infection, neoplasia, and after acute hemorrhage.

Zinc Protoporphyrin/Heme Ratio (ZPP/H)

Measurement of the ZPP/H ratio is an excellent screening test for detecting iron deficiency anemia and for following the results of therapy.[14]

Principle. Hemoglobin in a drop of blood is converted to cyanmethemoglobin by treatment with a cyanide-containing reagent. A portion of the mixture is placed on a coverslip, introduced into a ProtoFluor hematofluorometer (Helena Laboratories), and exposed to a beam of light. The instrument simultaneously measures the light absorbed by the film of cyanmethemoglobin at 424 nm and the fluorescent light emitted at 595 nm by zinc protoporphyrin. The results are displayed as a ratio of μmol ZPP/mol heme.

Reference Values. The ratio in normal individuals is 30 to 80 μmol ZPP/mol heme.

Note: The ratio is within the reference range in patients with thalassemia, a finding that distinguishes this disease from iron deficiency.

Increased Ratio. The ZPP/H ratio is elevated early in (1) all types of iron deficiency syndromes (iron-deficient diets, malabsorption of iron, blockage of iron release from reticuloendothelial cells as in infection or malignant disease) and (2) chronic exposure to lead. The ratio is abnormal before the appearance of anemia.

Serum Ferritin

The protein apoferritin is synthesized in the liver but rapidly scavenges and stores the iron of broken-down red cells to become ferritin; iron may constitute up to 20% of the weight of this protein, which carries out its storage function mostly in the reticuloendothelial system. A low concentration of circulating ferritin reflects decreasing iron stores except when accompanied by infection, liver disease, or chronic inflammatory disease.

Method of Measurement. The concentration of ferritin in serum is so low that it must be measured by immunoassay methods. Commercial kits and instruments are available for the measurement of serum ferritin.

Reference Values. The spread in the serum concentration of ferritin is broad because of variation in iron stores. The mean value for normal women is about 35 μg/L, with a range of 12 to 125 μg/L; the mean value for men is 95 μg/L, with a range from 30 to about 250 μg/L.

Increased Concentration. Elevated concentrations of serum ferritin occur in inflammation produced by various disease states, endotoxins, neoplasms, surgical procedures, and rheumatoid arthritis. Acute and chronic liver disease, iron toxicity, and hemochromatosis also increase the ferritin concentration (see Table 12.3).

Decreased Concentration. A ferritin concentration below 12 or 10 μg/L is indicative of iron deficiency. This finding appears before the anemia.

Serum Transferrin

Although serum transferrin can be easily measured by immunonephelometric methods, the same information may be obtained by measuring the total iron-binding capacity.

Assessment of Iron Nutriture

Measurement of the zinc/protoporphyrin ratio provides information concerning iron stores and the possible need for supplementation. Although the RDA is 50 to 200 μg/day, much more must be given because of the poor absorption of iron salts.

Zinc

Among the trace elements, the abundance of zinc in the body is exceeded only by iron. Zinc is a component or cofactor of various enzymes, including alkaline phosphatase, carbonic anhydrase, carboxypeptidase, alcohol dehydrogenase, DNA and RNA polymerases, and many others. Zinc is involved in the formation of collagen and hence is essential for proper wound healing.

Good dietary sources of zinc are meat, fish, and dairy products. High-fiber diets, however, reduce the availability of dietary zinc because phytates (inositol phosphates) in the fiber bind the zinc tightly and prevent its absorption in the intestines. Most instances of zinc deficiency in humans are produced in one of the three following situations: (1) eating a low-protein, high-fiber diet over a long period of time; (2) receiving total parenteral nutrition or synthetic diets that are not supplemented with zinc; (3) suffering from acrodermatitis enteropathica, a rare genetic skin condition that responds to zinc supplementation. Severe infections, inflammatory bowel disease, and extensive burns can also produce deficiencies that respond to zinc administration.

Assessment of Zinc Nutriture

Determination of the zinc plasma concentration by atomic absorption spectrophotometry (AAS)[17] is an effective way of assessing the nutritional status of zinc. A low plasma concentration of zinc under these circumstances indicates a need for zinc. The reference range for plasma zinc is 70 to 150 μg/dL (10 to 23 μmol/L). Blood samples must be taken during the morning in the fasting state because of diurnal variation in plasma zinc concentration.

Selenium

Selenium, an element related to sulfur, is a component of the enzyme glutathione peroxidase. This enzyme protects tissues from the damaging effects of peroxides (hydrogen peroxide and lipid hydroperoxide) by decomposing these oxidizing compounds; it also helps to reduce the danger of free radical formation. Selenium has no other known function at this time.

Selenium is present in many foods, but its content is related to the selenium content of the soil in which these foods are grown. Some soils in the world happen to be deficient in selenium, and inhabitants of these areas could become selenium deficient if they depended mostly on local sources for food. Most selenium deficiency, however, is associated with malnourishment, chronic bowel disease, or hyperalimentation without selenium supplementation. The RDA is 50 to 70 μg/day (Table 12.2).

Assessment of Selenium Nutriture

The determination of selenium status is best done by the determination of selenium concentration in whole blood by AAS. The reference range for selenium in whole blood is 80 to 320 μg/L (1 to 4 μmol/L.)

Copper[18]

Copper is a constituent of some enzymes (for example, ceruloplasmin, cytochrome oxidase, and uricase) and participates in several different phases of iron metabolism. Copper is involved in the synthesis of hemoglobin, whereas the copper-containing enzyme ceruloplasmin is necessary for the absorption of iron by oxidizing ferrous iron to the ferric state, a prerequisite for binding by transferrin. An anemia develops when a copper deficiency exists. The RDA is 1.5 to 3.0 mg/day (Table 12.2).

Assessment of Copper Nutriture

Serum copper can be determined by AAS, but interpretation must be made with care because of diurnal variation, effect of hormones, and inflammation. The reference range for serum copper is: infants, 20 to 70 μg/dL; children, 80 to 190 μg/dL; men, 70 to 140 μg/dL; nonpregnant women, 80 to 155 μg/dL; pregnant women, 120 to 300 μg/dL.

To convert to SI, multiply μg/dL $\times$ 0.157 = μmol/L.

Cobalt

Vitamin B_{12}, a cobalt-containing member of the vitamin B complex, is an essential nutrient that is abundant in meat proteins; its deficiency results in pernicious anemia. No dietary requirement exists for cobalt per se because no cobalt-containing enzymes have been found.

Fluoride

The only function of fluoride in the human body seems to be its ability to lessen the occurrence of dental caries by making the teeth more resistant to decay. Fluoride replaces some of the hydroxyl groups of hydroxyapatite (see Calcium section) in tooth enamel that apparently inhibit tooth erosion by the acid produced by bacteria in the plaque (decaying food deposit at the site of the incipient cavity). Many communities supplement drinking water with a low concentration of fluoride for this purpose. An excess of fluoride produces mottled teeth and is somewhat toxic when the intake is too high.

Assessment of Fluoride Nutriture

No useful nutrition information is obtained by measuring the serum fluoride concentration. The only occasion for determining serum or urine fluoride concentration is for the study or treatment of fluoride toxicosis; this determination can be done by ion-selective potentiometry.

Chromium, Manganese, Molybdenum, and Nickel

Chromium, manganese, molybdenum, and nickel are grouped together because deficiency states are rare and difficult to investigate. All four elements can be determined by AAS if proper precautions against contamination are taken, but the results provide little useful clinical information. Deficiency states may appear in patients receiving hyperalimentation without supplements, but rarely occur in the general patient population.

Chromium is deemed an essential trace element in humans, because in its absence, insulin in some people seems to be less effective in reducing hyperglycemia, as shown by an abnormal glucose tolerance test curve. Thus, chromium is believed to be a necessary cofactor for insulin in its regulation of glucose metabolism. Chromium is not a constituent of any known enzyme at present.

Manganese is a component of several metalloproteins (concanavalin A, for example), as well as several enzymes, such as pyruvate decarboxylase, superoxide dismutase, and some hydrolases. A manganese deficiency is rare but may be found in some severely malnourished people. An RDA of 2.0 to 5.0 mg/day is recommended (Table 12.2) because manganese appears in so many enzymes and its toxicity is quite low.

Molybdenum is a component of three oxidases (xanthine-, aldehyde-, and sulfite oxidase). No clinical deficiency of molybdenum has ever been reported.

Nickel is linked in some way with iron metabolism because nickel deficiency in animals is accompanied by an iron deficiency. In humans, some conditions affecting iron absorption have been reported to adversely affect nickel absorption. The daily requirement for nickel is listed as 35 to 500 μg/day (Table 12.2). The concentration of nickel in serum can be determined by AAS,[17] although its concentration is low (1 to 4 μg/L or 17 to 70 nmol/L). The demand to assess the nutritional status of nickel in a patient is quite rare; it may possibly be requested when treating a patient for industrial exposure to nickel (usually to nickel carbonyl, a toxic gas).

REFERENCES

1. Luft, F.C.: Sodium, chloride, potassium. *In* Present Knowledge in Nutrition. 6th Edition. Edited by M.L. Brown. Washington, D.C., International Life Sciences Institute, Nutrition Foundation, 1990.
2. Aurbach, G.D., Marx, S.V., and Spiegel, A.M.: Parathyroid hormone, calcitonin and the calciferols. *In* Williams Textbook of Endocrinology, 7th Edition. Edited by J.D. Wilson and D.W. Foster. Philadelphia, Saunders, 1985.
3. Rasmussen, H.: The calcium messenger system. N. Engl. J. Med., *314*:1094, 1164, 1986.
4. Lorenz, K.: Improved determination of serum calcium with *o*-cresolphthalein complexone. Clin. Chim. Acta, *126*:327, 1982.
5. Cali, J.P., Bowers, G.N., Jr., and Young, D.S.: A referee method for the determination of total calcium in serum. Clin. Chem., *19*:1208, 1973.
6. Trudeau, D.L., and Freier, E.F.: Determination of calcium in urine and serum by atomic absorption spectrophotometry (AAS). Clin. Chem., *13*:101, 1967.
7. Bowers, G.N., Jr., Brassard, C., and Sena, S.F.: Measurement of ionized calcium in serum with ion-selective electrodes: a mature technology that can meet the daily service needs. Clin. Chem., *32*:1437, 1986.
8. Power, M.H.: Inorganic phosphate. *In* Standard Methods of Clinical Chemistry. Vol. 1. Edited by M. Reiner. New York, Academic Press, 1953.
9. Abernethy, M.H., and Fowler, R.T.: Micellar improvement of the compleximetric measurement of magnesium in plasma. Clin. Chem., *28*:520, 1982.
10. Hamilton, E.M.N., Whitney, E.N., and Sizer, F.S.: Nutrition, Concepts and Controversies. 5th Edition. New York, West Publishing Co., 1991.
11. Lee, G.R., et al. (Eds.): Wintrobe's Clinical Hematology. 9th Edition. Philadelphia, Lea & Febiger, 1993.
12. Lee, G.R.: Nutritional factors in the production and formation of erythrocytes. *In* Wintrobe's Clinical Hematology. Vol. 1. 9th Edition. Philadelphia, Lea & Febiger, 1993.
13. Lee, G.R.: Iron deficiency and iron-deficiency anemia. *In* Wintrobe's Clinical Hematology. Vol. 1. 9th Edition. Philadelphia, Lea & Febiger, 1993.
14. Labbe, R.F., Rettmer, R.L., Shah, A.G., and Turnlund, J.R.: Zinc protoporphyrin: Past, present and future. Ann. N.Y. Acad. Sci., *514*:7, 1988.
15. Labbe, R.F., and Retmer, R.T.: Zinc protoporphyrin: a product of iron deficient erythropoiesis. Semin. Hematol., *26*:40, 1989.
16. Giovanniello, T.J., and Pecci, J.: Selected Methods of Clinical Chemistry. Vol. 9. Edited by W.R. Faulkner and S. Meites. Washington, D.C., American Association for Clinical Chemistry, 1982.
17. Sunderman, F.W., Jr.: Atomic absorption spectrophotometry of trace elements in clinical pathology. Hum. Pathol., *4*:549, 1973.
18. Mason, K.E.: Copper metabolism and requirements of man. J. Nutr., *109*:1979, 1979.

REVIEW QUESTIONS

1. What are the physiologic functions of Ca^{2+}?
2. How does the body keep the plasma concentration of Ca^{2+} within such close limits?
3. What disease states are associated with a high concentration of calcium, and what laboratory tests are commonly ordered?
4. What are the principles involved in the determinations of serum total calcium and magnesium?

5. Which of the trace elements are essential for hemoglobin formation and why?
6. Why is the RDA of iron listed as 10 to 15 mg/day when the actual daily loss of iron in urine and sweat is only 1 mg/day?
7. What principles are involved in the determination of serum iron?
8. Why are so many trace elements needed in such small amounts?

Endocrinology 13

OUTLINE

I. HORMONES AND THE ENDOCRINE SYSTEM
- A. Functions of hormones
- B. Functional types
- C. Structural classes
- D. Hormonal secretion and regulation
  1. Secretion
  2. Regulation by the central nervous system
  3. Regulation by the hypothalamus
  4. Regulation by the anterior pituitary
  5. Hormonal actions upon target cells

II. ANTERIOR PITUITARY HORMONES
- A. Effector hormones
  1. Growth hormone (GH, somatotropin)
  2. Prolactin (PRL)
- B. Tropic hormones
  1. Thyroid-stimulating hormone (TSH, thyrotropin)
  2. Adrenocorticotropin (ACTH)
  3. Gonadotropins (FSH, LH)
  4. Other anterior pituitary hormones

III. POSTERIOR PITUITARY HORMONES
- A. Antidiuretic hormone (ADH)
- B. Oxytocin
- C. Pineal hormones

IV. THYROID HORMONES, T_4 AND T_3
- A. Evaluation of thyroid function
  1. T_4 and T_3 by immunoassay
     - a. Reference values
     - b. Increased concentration
     - c. Decreased concentration
  2. T_3 uptake (T_3U) and free thyroxine index (FTI)
     - a. Typical reference range
     - b. Increased values
     - c. Decreased values
  3. Free T_4 and free T_3
  4. Thyroid-stimulating hormone (TSH) and thyroid-releasing hormone (TRH)
  5. Guidelines for use of thyroid tests

V. ADRENOCORTICAL HORMONES
- A. Cortisol
  1. Hypersecretion of cortisol
  2. Hyposecretion of cortisol
  3. Evaluation of adrenocortical function
     - a. Serum cortisol determination
       - (1) Reference values
       - (2) Increased concentration
       - (3) Decreased concentration
     - b. Urine free cortisol
- B. Androgens
  1. Androgen excess (hirsutism, virilism)
  2. Steroid metabolites in urine
- C. Aldosterone

VI. FEMALE SEX HORMONES
- A. Ovarian hormones
  1. Estrogens and progesterone
  2. Estriol
- B. Placental hormones
  1. Human chorionic gonadotropin (hCG)
  2. Human placental lactogen (hPL)

VII. MALE SEX HORMONES
- A. Increased concentration
- B. Decreased concentration

OBJECTIVES

After reading this chapter, the student will be able to:

1. Describe the functions of hormones and give examples of five different types of stimuli that result in the secretion of at least one hormone.
2. Give an example of hormonal secretion when the stimulus involves (a) the central nervous system, (b) hypothalamus, (c) anterior pituitary, (d) low serum calcium concentration, and (e) elevated plasma osmolality.
3. Describe the main structural features of the steroid hormones and the characteristic features of each subgroup (androgen, corticoid, estrogen, mineralocorticoid, and progestin).
4. Describe the mode of action of steroid and protein hormones at the cellular level.
5. Describe the different groups of tropic hormones and explain their actions.
6. Describe the role of iodine in the synthesis of thyroid hormones.
7. Describe the difference in primary and secondary thyroid disease.
8. Explain the role of the different thyroid function tests in the diagnosis of thyroid disease.
9. Describe the main pathologic features of hypo- and hypersecretion of each major endocrine gland.
10. Describe the function of human chorionic gonadotropin, its type, and its development during pregnancy.
11. Describe the principle of the test for human chorionic gonadotropin, the purpose of the test, and interpretation of the results.
12. Describe the major neurotransmitters and their mode of action.
13. Describe the adrenomedullary hormones, their immediate precursors, and their physiologic role.
14. Describe the tight hormonal regulation of serum calcium concentration and explain the principle of the assay for parathyroid hormone.

HORMONES AND THE ENDOCRINE SYSTEM[1,2]

The two physiologic regulatory systems are the endocrine system and the nervous system. The endocrine system consists of the ductless glands that produce highly active chemical regulators called hormones. These chemical messengers are secreted into the bloodstream, where they have the opportunity to act upon all organs by interacting with specific receptors on cells. Some hormones, however, flow in the extracellular fluid to stimulate nearby cells. The other regulatory system, the nervous system, acts locally at tissue or cellular sites; the endocrine and nervous systems are, however, intricately related as each system can influence the function of the other.

Functions of Hormones

Hormones regulate metabolism, preserve homeostasis, control the rates of growth and development, and influence behavior by affecting a wide variety of biologic systems. Their functions are the following:

1. To maintain a constant internal environment in the body fluids (homeostasis).
2. To regulate the growth and development of the body as a whole.
3. To promote sexual maturation, maintain sexual rhythms, and facilitate the reproductive process.
4. To regulate energy production and stabilize the metabolic rate.
5. To help the body to adjust to stressful or emergency situations.
6. To promote or inhibit the production and release of certain other hormones.

Hormones usually do not act alone. Any physiologic effect of a particular hormone is the result of it and any counterregulatory hormones that may be circulating; the physiologic process is complex and is the result of actions by both the endocrine and nervous systems.

Individual hormones, however, vary considerably in the extent of their functions. The parathyroid hormone, for example, is concerned with the maintenance of a constant plasma concentration of calcium ion, whereas the thyroid hormones regulate the basal metabolic rate of all cells in the body and, in conjunction with growth hormone, are also essential for proper growth and development.

Functional Types

The four functional types of hormones are (1) releasing hormones from the hypothalamus that promote the secretion of anterior pituitary hormones; (2) inhibitory hormones from the hypothalamus or gastrointestinal tract that suppress the secretion of particular hormones; (3) tropic hormones that stimulate the growth and activity of other endocrine glands; and (4) effector hormones secreted by all the endocrine glands other than the anterior pituitary and hypothalamus. The target cells of the effector hormones are the particular nonendocrine tissue cells upon which these hormones exert their metabolic effects. Thus, gonadal hormones (testosterone in males, estrogens in females) affect a wide variety of tissues during puberty to produce the typical secondary sex characteristics for each sex, thyroid hormones affect the metabolic rates of all tissues, and growth hormone stimulates skeletal growth and affects many metabolic processes.

Structural Classes

From a chemical standpoint, the five general classes of hormones are: (1) proteins (molecular weight > 5,000 daltons) or polypeptides > 20 amino acids); (2) glycoprotein composed of polypeptide chains containing covalently linked carbohydrate; (3) peptides (< 20 amino acids); (4) steroids; and (5) amino acid derivatives. For example, three of the anterior pituitary hormones are glycoproteins (TSH, FSH, LH), and the others are proteins or polypeptides (GH, ACTH, prolactin). All the gonadal and adrenocortical hormones are steroids.

Hormonal Secretion and Regulation

Secretion

The production and secretion of hormones by an endocrine gland may be initiated by one or more of the following signals:

1. Stimulation of the cerebral cortex or neural centers by thoughts; emotions; stress; circadian (periodic) rhythms that may be daily, monthly, or seasonal; and chemical transmitters active at nerve junctions, such as norepinephrine, dopamine, acetylcholine, and serotonin, which stimulate release of hypothalamic or adrenal medullary hormones.
2. A change in the plasma concentration of particular ions or compounds (secretion of parathyroid hormone when the plasma concentration of Ca^{2+} is low or of insulin when the plasma level of glucose is high).
3. Secretion of tropic hormones (for example, ACTH and TSH cause the adrenal and thyroid glands to produce and secrete cortisol and thyroxine, respectively).
4. Variation in blood osmolality (secretion of antidiuretic hormone, ADH, by the posterior pituitary when the plasma osmolality increases).
5. Release of hormones in the gastrointestinal tract in the presence of various foods (food in the stomach stimulates gastrin secretion by the stomach).

Regulation by the Central Nervous System

The secretions of the endocrine glands are carefully regulated by the complex interaction between the nervous and endocrine systems (Fig. 13.1). Neural centers may trigger the release or suppression of particular hormones by means of an action potential or neurotransmitter fired or released within the gland. For example, the feeling of fear or anxiety transmits a message from the brain by

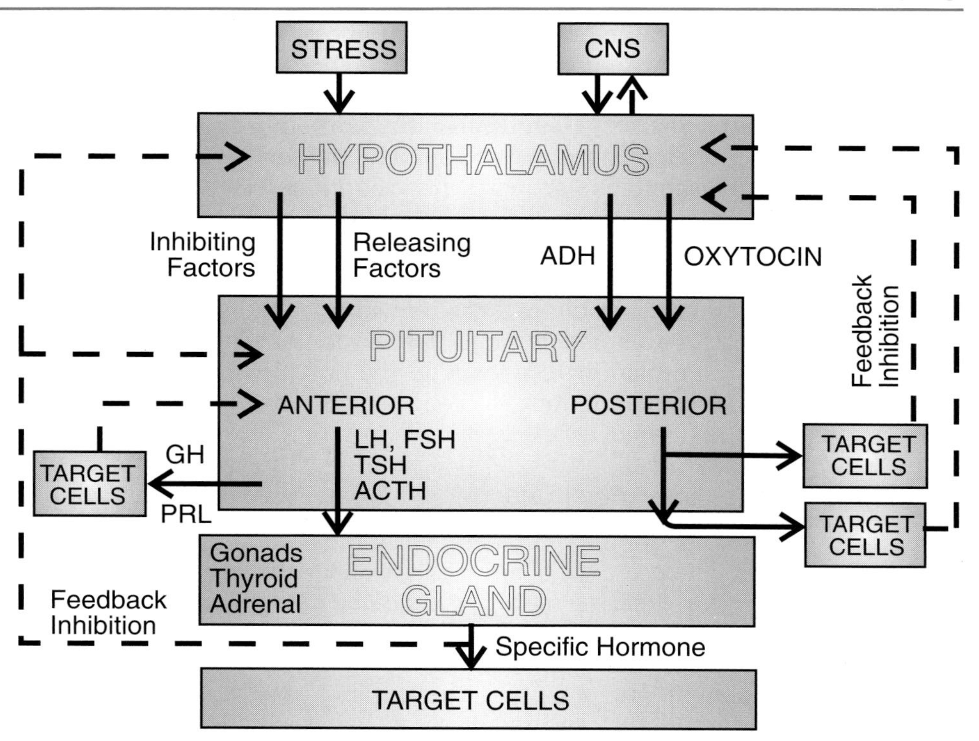

FIG. 13.1. Schematic representation of the control of hormone secretion. The hypothalamus may be stimulated by the central nervous system (CNS) or by stress to secrete releasing factors (GHRF, CRF, TRF, LHRF, or FSHRF), ADH, or oxytocin. GHRF stimulates the anterior pituitary to secrete growth hormone, which acts directly upon target cells. Other releasing factors evoke secretion of tropic hormones that stimulate the adrenal cortex, thyroid, or gonads to secrete their respective hormones, which then act upon target cells. Hormones, ADH, and oxytocin travel down the pituitary stalk into the posterior pituitary, where they are secreted. Broken lines represent feedback control of secretion of hypothalamic or anterior pituitary hormones; some hormones (TSH) may directly inhibit release of tropic factor by the anterior pituitary. GH = growth hormone; PRL = prolactin; TSH = thyroid-stimulating hormone. GHRF, CRF, TRF, LHRF, and FSHRF are the respective releasing factors for growth hormone, adrenocorticotropin (ACTH), thyrotropin (TSH), luteinizing hormone (LH), and follicle-stimulating hormone (FSH).

way of the sympathetic nervous system to the adrenal medulla that results in epinephrine secretion; anxiety or stress can disturb the regularity of the menstrual cycle. On the other hand, hormones can also interact with the brain by affecting mood and emotions. The neurotransmitters, all small molecules, include acetylcholine, norepinephrine, dopamine, serotonin, and gamma-aminobutyric acid.

Regulation by the Hypothalamus[3]

The hypothalamus, located at the base of the brain and connected to the pituitary stalk, contains many neurosecretory cells and neurosecretory nerve fibers. It plays a major role in endocrine regulation because of its ability to release hormones that selectively stimulate or inhibit the secretion of specific anterior pituitary hormones (Table 13.1). These stimulating hormones have the suffix "-liberin" (*somatoliberin,* the somatotropin-releasing hormone); inhibitors end in "-statin" (*somatostatin,* the inhibitor of somatotropin release).

The hypothalamus may be activated by the central nervous system, emotion, or stress to secrete one or more of a group of releasing factors (Fig. 13.1). The releasing factors are peptides or polypeptides that stimulate the anterior pituitary to elaborate and secrete the appropriate tropic hormone. Thus, each endocrine gland in the chain pours out many molecules for each stimulus received, and this produces a cascade effect. The hypothalamus also contains such factors as *somatostatin* and *dopamine,* that inhibit the output of growth hormone and prolactin, respectively, and provide further "fine-tuning" control of hormonal concentrations in blood.

Regulation by the Anterior Pituitary

Four of the six hormones elaborated by the anterior pituitary gland are tropic hormones. The presence of appropriate releasing hormones from the hypothalamus triggers their secretion; a small signal is amplified to give a bigger response. The tropic hormones, adrenocorticotropic hormone (ACTH), thyroid-stimulating hormone (TSH), follicle-stimulating hormone (FSH), and luteinizing hormone (LH), act upon the adrenal cortex, thyroid, and gonads to stimulate production and secretion of their particular hormones (Table 13.2). This process is described further in sections dealing with particular glands.

The effector hormones produced by the endocrine glands exert a feedback inhibition, either on the hypothalamus to retard or block further secretion of releasing factors or on the anterior pituitary to inhibit the secretion of tropins. This mechanism completes the control loop, as indicated schematically in Figure 13.1.

TABLE 13.1
HYPOTHALAMIC HORMONES

HORMONE	TYPE	TARGET TISSUE	PRINCIPAL ACTION
Growth-hormone-releasing hormone (GHRH)	Polypeptide	Anterior pituitary	Release of growth hormone (GH)
Corticotropin-releasing hormone (CRH)	Polypeptide	Anterior pituitary	Release of ACTH
Thyrotropin-releasing hormone (TRH)	Peptide	Anterior pituitary	Release of TSH
Gonadotropin-releasing hormone (GnRH)	Peptide	Anterior pituitary	Release of FSH and LH
Somatostatin	Peptide	Anterior pituitary	Suppression of GH release
Prolactin-releasing factor (PRF)	Peptide	Anterior pituitary	Release of PRL
Prolactin inhibiting factor (PIF)	Tyrosine derivative (dopamine)	Anterior pituitary	Suppression of PRL release

TABLE 13.2
ANTERIOR PITUITARY HORMONES

HORMONE	TYPE	TARGET TISSUE	PRINCIPAL ACTION
Growth hormone (GH)	Protein	Whole body	Growth of bone and muscle
Adrenocorticotropin (ACTH)	Polypeptide	Adrenal cortex	Formation and secretion of cortisol
Thyrotropin (TSH)	Glycoprotein	Thyroid	Formation and secretion of thyroxine (T_4)
Follicle stimulating hormone (FSH)	Glycoprotein	F:* Ovary	F:* Follicular growth, estradiol secretion
		M:* Testis	M:* Seminiferous tubule growth, spermatogenesis
Luteinizing hormone (LH)	Glycoprotein	F:* Ovary	F:* Ovulation (with FSH), corpora lutea, progesterone
		M:* Testis	M:* Interstitial tissue growth, testosterone
Prolactin (PRL)	Protein	Breasts	Mammary gland differentiation, milk secretion

*M = male; F = female.

Hormonal Actions Upon Target Cells

Hormones act in different ways on their target cells. The protein and peptide hormones and the catecholamines (epinephrine and norepinephrine) become attached to target cell membranes and activate adenyl cyclase to produce cyclic adenosine monophosphate (c-AMP). Cyclic AMP activates protein kinases, enzymes that activate phosphorylases; liver phosphorylase, for example, converts glycogen to glucose. Hormonal action on the cell membrane is usually short-lived because of rapid conversion to inactivate metabolites by enzymes.

The chemical nature of the steroid and thyroid hormones enables them to enter their target cells and penetrate to the cell nuclei. Their reaction with nuclear DNA represses some genes and derepresses others (activates genes to make new messenger RNA with instructions for the synthesis of new proteins). Thus, these hormones have relatively long-term effects and indirectly affect many metabolic systems. Special binding proteins in the cytoplasm aid in the transfer of the hormone to the nucleus. The events at the cellular level are beginning to be clarified.

Because the main function of a hormone is to exert a fine regulatory control over metabolic processes or over growth and development, either overproduction or underproduction of various hormones leads to abnormalities. A problem lies, however, in ascertaining whether an endocrine dysfunction is caused by a malfunction of the endocrine gland in question, a deficiency of a tropic hormone (pituitary dysfunction), a deficiency of releasing factors (hypothalamus or feedback control dysfunction), a defect in the target cell, or other factors involving transport proteins or autoimmune disease. Pinpointing the pathologic process responsible for an endocrine abnormality is further complicated by the problem of ectopic hormones. Ectopic hormones are hormones produced elsewhere in the body than in the customary endocrine gland. Many types of malignant disease produce ectopic hormones, which do not respond to the normal regulatory processes of the body. Ectopic production of ACTH is reviewed in the corticosteroid section. Immunologic techniques are commonly used for the assay of most serum hormones and consist primarily of radioimmunoassay, fluorescence immunoassay, fluorescence polarization immunoassay, or enzyme immunoassay. Refer to Chapter 4 for a review of the principles involved. Each endocrine gland and its hormones are described in the following sections.

ANTERIOR PITUITARY HORMONES

The pituitary gland is composed of two distinct parts, the anterior lobe or adenohypophysis and the posterior lobe or neurohypophysis. The anterior pituitary gland controls the hormonal outputs of some of the other endocrine glands by its secretion of tropic hormones, which were described in the preceding section. The secretion of tropic factors is in part regulated by feedback control loops, depending on the concentration of circulating, primary hormone, and, in part, by releasing factors emanating from the hypothalamus. The anterior pituitary also secretes two effector hormones, growth hormone and prolactin.

The concentrations of circulating anterior pituitary hormones are low, except under certain pathologic conditions, and require sensitive methods for measurement. A brief summary of the clinical significance of pituitary hormones appears in the following section. Commercial kits are available for the measurement of all pituitary hormones of clinical interest.

Effector Hormones

Growth Hormone (GH, Somatotropin)

Growth hormone (GH), a protein, is the hormone produced in the largest quantity by the anterior pituitary. It affects many metabolic processes (Table 7.1, Chap. 7) in addition to promoting skeletal growth and protein synthesis in young mammals.

Many of the growth-promoting effects of GH upon cartilage and the long bones are mediated by insulin-like growth factors (IGF), previously called somatomedins, that are secreted by the liver in response to GH secretion. Three IGFs, designated IGF-I, IGF-II, and IGF-III, are structurally related to the glucose-regulating hormone insulin. GH has many metabolic effects that are antagonistic to those of insulin.

The plasma concentration of GH during the day is low in humans (< 5 ng/mL). A sharp increase in concentration occurs during sleep, with the greatest rise occurring at the time of deepest sleep; this level may be eight- or tenfold higher than the basal level during the day. Plasma GH is usually elevated in acromegalic persons, and measurement of it can assist in the diagnosis, especially in the early stages. A deficiency of GH in children causes dwarfism or subnormal growth. This deficiency is difficult to document by measuring the basal concentration of GH, however, because the normal range is low and near the limits of dependable assay. The measurement of serum IGF-I aids in the diagnosis because its concentration is frequently low in GH insufficiency.

Dynamic tests of the capacity of the anterior pituitary to secrete GH are frequently performed because the levels in GH insufficiency are often in the same range as those in normal individuals. The administration of arginine or l-dopa results in an increase in the serum levels of GH to > 10 ng/mL in normal individuals; values < 5ng/mL are highly suggestive of GH insufficiency. Intermediate values require further testing.

Prolonged excess of GH secretion prior to closure of the long bones at puberty causes excessive growth or gigantism. Hypersecretion of GH by a pituitary tumor in adulthood (after closure of the long bones) causes a condition known as acromegaly, in which the person is large and has gross features. Plasma levels of GH and IGF-I are frequently elevated in this condition.

Good commercial radioimmunoassay (RIA) kits are available for measuring the concentration of GH and IGF-I in plasma or serum.[7]

Prolactin

The anterior pituitary secretes prolactin, a hormone similar in structure to GH. GH and prolactin have many amino acid sequences in common. Prolactin participates with gonadal steroids in breast growth during pregnancy. After parturition, prolactin stimulates milk secretory activity, although the initiation of milk flow is induced by one of the posterior pituitary hormones, oxytocin.

Elevated levels of prolactin may result in menstrual irregularity, infertility, and galactorrhea (inappropriate production of breast milk). Certain medications may also cause inappropriate secretion of prolactin. Prolactinomas, prolactin-secreting tumors, are the most common type of secretory pituitary tumor. Elevated levels of prolactin (> 100 ng/mL) are consistent with pituitary tumors.

Reference ranges for prolactin are < 30 ng/mL for women and < 20 ng/mL for men; caution must be exercised in evaluating slightly elevated levels of prolactin because many factors, such as exercise, fasting, stress, or breast examination during a physical examination, may result in elevated levels.

Tropic Hormones

Thyroid-Stimulating Hormone (TSH, Thyrotropin)

TSH, like FSH and LH, is a glycoprotein composed of two chains. The alpha chain is identical in all three glycoproteins, but the beta chain is unique in each. TSH is the major regulator of thyroid secretion and function; its output by the pituitary, however, is under negative feedback control by the level of circulating thyroid hormone, T_4.[4] Serum TSH measurement is of value in the differential diagnosis of thyroid disease.

Serum levels of TSH are elevated in primary hypothyroidism because of the lack of negative feedback by thyroid hormones; levels are depressed in primary hyperthyroidism because of excessive negative feedback by elevated levels of thyroid hormones. The details of TSH measurement and interpretation of results are reviewed in the thyroid section of this chapter.

Adrenocorticotropin (ACTH)

ACTH, a tropic hormone, is a long-chain polypeptide that binds to cells of the adrenal cortex and influences their activities. Its initial action is to stimulate the formation of adrenal steroids by increasing the synthesis of pregnenolone from cholesterol (see Fig. 13.4). The net effect is an increase in the secretion of cortisol and adrenal androgens.

The concentration of ACTH in plasma is highest between 6 and 8 A.M. and lowest in the evening between 6 and 11 P.M.; the plasma concentration in a normal person does not exceed 50 pg/mL at its peak, whereas the basal level may be close to 5 pg/mL. High levels of ACTH are found in 3 pathologic conditions: (1) primary adrenal cortical deficiency; (2) Cushing's disease (hyperactivity of the adrenal cortex caused by excessive pituitary ACTH secretion); and (3) ectopic tumors that produce ACTH. These conditions are reviewed in the section in this chapter dealing with the adrenal cortex.

Gonadotropins (FSH, LH)

FSH and LH are glycoproteins secreted by the anterior pituitary that are necessary for proper maturation and function of the gonads in both men and women. These hormones induce growth of the gonads and secretion of gonadal hormones and are necessary for the reproductive process (development of mature ova in females and of spermatozoa in males). RIA methods are sufficiently sensitive to measure the circulating hormone concentrations. Because FSH, LH, TSH, and human chorionic gonadotropin (hCG) have the same alpha chain, a problem of cross-reactivity with some polyclonal antibodies results; the problem can be eliminated by the use of more selective polyclonal antibodies or of monoclonal antibodies specific for each beta chain.

FSH and LH are present in the plasma of both males and females at all ages. A small rise occurs at puberty in both sexes, but a great increase in the concentration of plasma LH and FSH takes place in women after the menopause and remains elevated for the remainder of their lives. In ovulating females, the concentrations of both LH and FSH rise sharply from the basal level just before ovulation and then fall. FSH and LH measurements are useful in diagnosing menstrual and fertility disorders. Enzyme-linked immunosorbent assay (ELSIA)-based LH kits for home use are now available for ascertaining the day of ovulation.

Other Anterior Pituitary Hormones

The anterior pituitary produces several other peptide hormones that do not have much clinical relevance at this time. A melanocyte-stimulating hormone (α-MSH) is a peptide with an amino acid sequence closely related to a portion of the ACTH chain. Endorphins are polypeptides that raise the pain threshold by exerting an opiate-like action on portions of the brain.

POSTERIOR PITUITARY HORMONES

The posterior lobe of the pituitary gland is connected anatomically to the hypothalamus by a stalk through which a nerve tract and blood vessels pass. The posterior pituitary stores and secretes two closely related peptide hormones, ADH (antidiuretic hormone, vasopressin) and oxytocin (Table 13.3). Each is composed of eight amino acids arranged in a five-membered ring that contains cystine with its disulfide bridge and a three-membered tail. Six of the eight amino acids of ADH and oxytocin are identical in their respective positions in the hormones. ADH and oxytocin are synthesized in the hypothalamus and travel through the nerve tract in the pituitary stalk to the posterior pituitary lobe, where they are stored until secreted.

Antidiuretic Hormone (ADH)

The primary physiologic function of ADH is to increase the reabsorption of water by the renal tubules (see Chap. 6) when the plasma osmolality becomes elevated. The hormone also affects blood pressure. Deficiency of ADH is associated with diabetes insipidus, which is characterized by the passage of large volumes of dilute urine. Because diagnosis is based on clinical evidence, measurement of the hormone concentration in blood is not necessary. The purified hormone may be used therapeutically in those with an ADH deficiency.

Oxytocin

Oxytocin, the other posterior pituitary hormone, is a potent stimulant for the contraction of smooth muscle. Clinically, it is sometimes used to induce labor by promoting uterine contractions. Oxytocin also stimulates the ejection of milk from the mammary glands. No particular medical need is served at present by measuring the concentration of circulating oxytocin.

TABLE 13.3
POSTERIOR PITUITARY HORMONES

HORMONE	TYPE	TARGET TISSUE	PRINCIPAL ACTION
Antidiuretic hormone (ADH, vasopressin)	Peptide	Renal tubules, arterioles	↑ Water reabsorption, ↑ blood pressure
Oxytocin	Peptide	Uterus, breasts	Contracts uterus, ejection of milk

Pineal Hormones

The pineal gland, a small gland weighing less than 200 mg and attached to the midbrain, secretes 2 neurotransmitters, melatonin and serotonin, that presumably modulate some circadian rhythms of the endocrine system (Table 13.4). The pineal secretions are controlled by certain nerve stimuli and are enhanced by light and inhibited by darkness. At present, the measurement of pineal hormones is not clinically indicated.

THYROID HORMONES, T_4 AND T_3

The thyroid gland is a small tissue situated in the neck just below the larynx (voice box). Its hormones increase the basal metabolic rate and are necessary for proper growth and development (Table 13.5). The thyroid also secretes calcitonin, a hormone that participates in the regulation of plasma Ca^{2+} concentration by inhibiting bone resorption (see Parathyroid section in this chapter).

The circulating thyroid hormones, thyroxine (3,5,3′,5′-tetraiodothyronine, T_4) and 3,5,3′-triiodothyronine (T_3), the structures of which are shown in Figure 13.2, are iodinated derivatives of the amino acid tyrosine. Their immediate precursors are monoiodotyrosine (MIT) and diiodotyrosine (DIT), but these compounds do not appear to any significant extent in plasma except under rare pathologic conditions. Neither MIT nor DIT has any hormonal activity.

The various events occurring in the synthesis and release of thyroid hormones are depicted in Figure 13.3 In step 1, the entry of inorganic iodide ion into the thyroid cell is facilitated by an active transport system that can be inhibited by such anions as ClO_4^- and SCN^-, or by high concentrations of I^-. The I^- is trapped once it enters the thyroid gland and does not diffuse out readily. In step 2, the I^- is oxidized with the help of a peroxidase enzyme system to an active state, presumably I_2, which prepares it for step 3. The activated iodine reacts in step 3 with tyrosyl residues in the protein, thyroglobulin, a large glycoprotein present in the thyroid cells. The products are MIT and DIT residues attached to thyroglobulin, with the iodine in the 3-position in MIT and in the 3,5-positions in DIT. The MIT and DIT residues are oxidatively coupled in step 4 by an enzyme system to form T_3 and T_4 residues attached to thyroglobulin; two DIT residues are condensed to form T_4, whereas the coupling of MIT and DIT yields T_3. All the steps in thyroid

TABLE 13.4
PINEAL HORMONES

HORMONE	TYPE	TARGET TISSUE	PRINCIPAL ACTION
Melatonin	Tryptophan derivative	Hypothalamus and anterior pituitary	Suppresses GH and GnRH release
Serotonin	Tryptophan derivative	Brain, smooth muscle	Neurotransmitter

TABLE 13.5
THYROID HORMONES

HORMONE	TYPE	TARGET TISSUE	PRINCIPAL ACTION
Thyroxine (T_4) and triiodothyronine (T_3)	Iodo-derivatives of tyrosine	All tissues	↑ Metabolic rate (O_2 consumption)
Calcitonin (CT)	Polypeptide	Bone osteoclasts	Inhibits bone resorption

hormone synthesis (iodide trapping, iodination of tyrosine, condensation, and the proteolytic release of the hormones) are under regulatory control by TSH.

T_4 is the principal iodinated hormone secreted by the thyroid; peripheral tissues convert T_4 to T_3, the active hormonal entity that enters tissue cells. About 70% of the plasma T_3 is derived from the tissue deiodination of T_4, whereas the rest of the plasma T_3 is of thyroidal origin. A reverse T_3 (rT_3), in which the iodine molecules are in the 3,3′,5′ position instead of the 3,5,3′ position of T_3, is formed to a minor extent during the deiodination process. Reverse T_3 has no hormonal activity, and its determination is of limited clinical usefulness.

The thyroglobulin with its mixture of iodinated tyrosyl and thyronine residues is stored in the thyroid follicles until the signals for release are given. In step 5, some of the stored iodinated thyroglobulin is degraded by a cellular protease to yield a mixture of free MIT, DIT, T_3, and T_4. MIT and DIT are enzymatically deiodinated, and the I^- is recycled within the gland (step 6). T_3 and T_4 are poorly soluble in plasma and are transported primarily by a thyroxine-binding globulin (TBG), a plasma protein that carries 70 to 75% of circulating T_4 and T_3. The remaining 25 to 30% of T_4 is transported by albumin and prealbumin, but T_3 has no affinity for prealbumin and circulates only with TBG and albumin. TBG has a single binding site for T_4 or T_3, but in normal individuals, only about 30% of the binding sites are occupied. It has the potential for binding up to 20 μg/dL (258 nmol/L) of T_4.

A low concentration of free (unbound) T_4 and T_3 exists in plasma in equilibrium with the protein-bound forms. *Only the free hormone (FT_4 and*

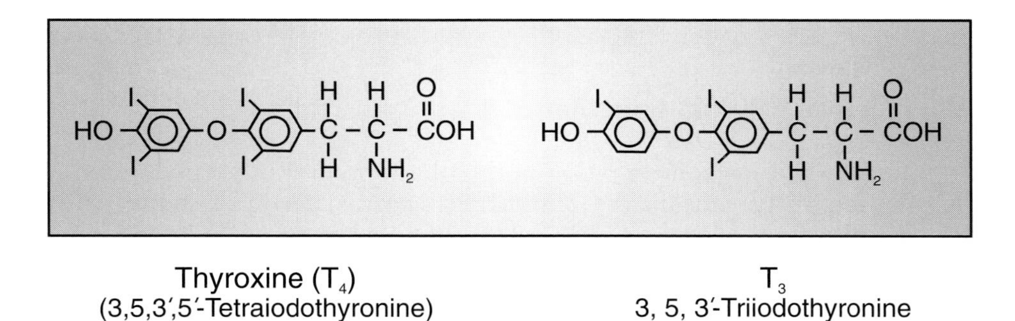

Thyroxine (T_4)
(3,5,3′,5′-Tetraiodothyronine)

T_3
3, 5, 3′-Triiodothyronine

FIG. 13.2. Structures of the thyroid hormones, T_4 and T_3. The parent molecule with no iodine atoms is thyronine. T_4 and T_3 are tetraiodo- and triiodothyronine, respectively. Reverse T_3 (rT_3) is not shown, but its iodine molecules are in the 3,3′,5′ position of triiodothyronine.

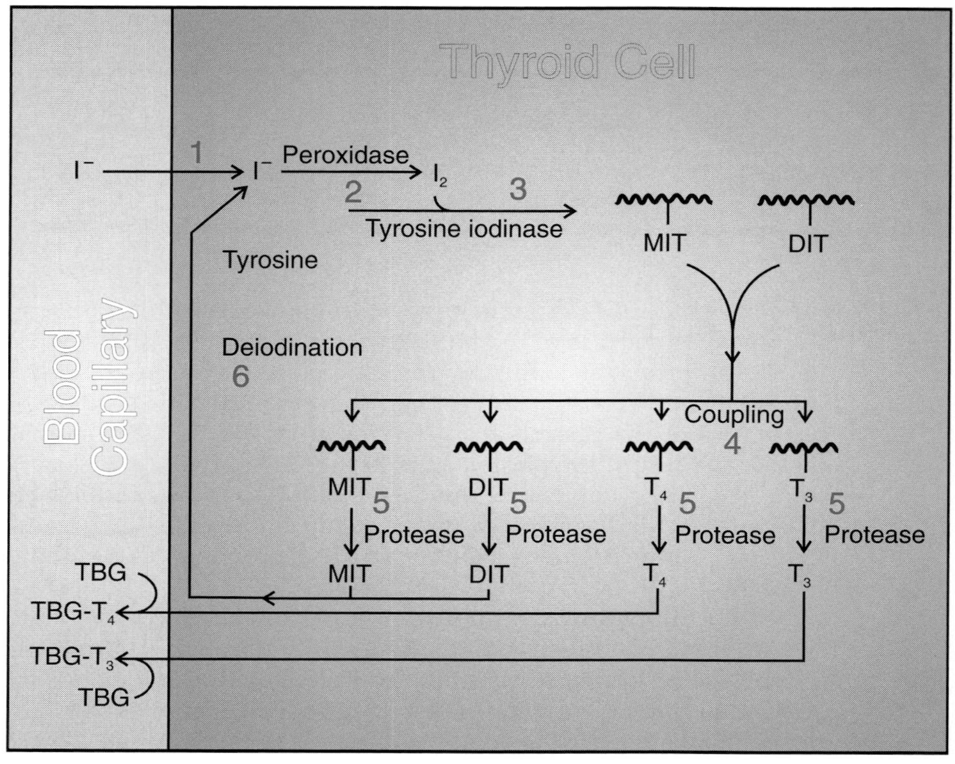

FIG. 13.3. Schematic representation of the production of thyroid hormones. MIT = monoiodotyrosine; DIT = diiodotyrosine; T_3 = 3,5,3′-triiodothyronine; T_4 = thyroxine (3,5,3′,5′-tetraiodothyronine); TBP = T_3- and T_4-binding proteins; wavy line represents the thyroglobulin molecule. Defects may occur at any of the numbered steps; these are reviewed in the text.

FT_3) can bind to receptors on cell surfaces. The cell converts FT_4 to FT_3, which then delivers its hormonal message by reacting with DNA. The ratio of FT_4 to total T_4 is about 1:3000, with a corresponding figure of 1:300 for T_3. The plasma distribution of thyroid hormones ($>$ 99.9% bound) is approximately 97% T_4, 2% T_3, and $<$ 1% rT_3.

Evaluation of Thyroid Function

The plasma concentration of thyroid hormones in normal individuals is kept relatively constant by a sensitive negative feedback control by free T_4 on pituitary release of TSH. The hypothalamus with its releasing hormone (thyrotropin-releasing hormone, TRH); the anterior pituitary with its tropic hormone (TSH); and the thyroid gland with its secretion of T_4 interact in a dynamic fashion to regulate T_4 (and T_3) levels. Thus the evaluation of thyroid status is not a simple procedure because it does not depend solely on the measurement of circulating thyroid hormones. One or more of the following factors may be abnormal and must be sorted out and evaluated: the concentration of TBG and its degree of saturation with thyroid hormone, the actual concentrations of free (unbound) T_4 and T_3, the status of the hypothalamus and the anterior pituitary with their respective outputs of TRH and TSH, the response of the pituitary to TRH, and the response of the thyroid to TSH. Different strategies for diagnosing thyroid disease have been proposed in an attempt to obtain the greatest diagnostic information for the least expenditure of effort and money. Sensitive immunoassays for measurement of these factors are reviewed in the following section. Tests for diagnosis of thyroid dysfunction make up a large percentage of all endocrine tests.

T_4 and T_3 by Immunoassay

The serum T_4 concentration is a better indicator of the thyroid secretory rate than is T_3 because T_4 is the thyroid's principal secretory product.[5] Most of the circulating T_3 comes from the peripheral deiodination of T_4, a process that is depressed by severe illness or stress. Although the serum concentrations of T_4 and T_3 frequently rise and fall in a parallel fashion, a rise in the concentration of serum T_3 may be the first abnormality in some cases of hyperthyroidism and, in rare cases, the only abnormality.

Serum concentrations of T_4 and T_3 are frequently measured in most clinical laboratories. All the methods used for determination of T_4 and T_3 employ immunoassay of one type or another. Details of immunoassay are described in Chapter 4. In brief, labeled T_4 or T_3 compete with unlabeled hormone (either in patient's samples or in calibrators) for binding sites on antibodies specific for either T_4 or T_3. The amount of labeled hormone bound to the antibody is inversely proportional to the amount of unlabeled hormone present. After separation of the bound from the free fraction of labeled hormone, the bound fraction is counted in an appropriate counter and a standard curve is drawn where the amount of bound hormone is plotted as a function of the concentration of hormone in the calibrators. The concentration of hormone in the patient's sample is then determined from the standard curve.

Reference Values (Table 13.6). Our reference values for T_4 are 5.1 to 11.0 μg/dL (66 to 142 nmol/L); those for T_3 are 70 to 200 ng/dL (1.08 to 3.08 nmol/L).

TABLE 13.6
TYPICAL REFERENCE VALUES FOR TOTAL AND FREE T_2 AND T_3

	T_4		T_3	
THYROID STATUS	TOTAL	FREE	TOTAL	FREE
Euthyroid	5.1–11.0 μg/dL 66–142 nmol/L	1.5–3.5 ng/dL 19–45 pmol/L	70–200 ng/dL 1.1–3.1 nmol/L	200–750 pg/dL 3–11.6 pmol/L
Hyperthyroid	13–31 μg/dL 168–400 nmol/L	> 4.0 ng/dL > 52 pmol/L	210–2000 ng/dL 3.2–31 pmol/L	> 850 pg/dL > 13 pmol/L
Hypothyroid	< 6 μg/dL < 77 nmol/L	< 2.0 ng/dL < 26 pmol/L	< 75 ng/dL < 1.2 nmol/L	< 210 pg/dL < 3.2 pmol/L

Increased Concentration. The serum concentrations of both T_4 and T_3 are usually elevated in hyperthyroidism, with the few exceptions noted earlier in which only the T_3 may be abnormal. Both T_4 and T_3 levels are increased in conditions that raise the plasma concentration of TBG (pregnancy, taking of birth control pills, or estrogens). No increase occurs in the basal metabolic rate, however, because the extra T_4 and T_3 are bound to the newly synthesized TBG as equilibrium between the free and bound hormones is maintained. The person is euthyroid (normal thyroid function) because the level of *free* hormone, the biologically active entity, remains essentially unchanged.

Decreased Concentration. The serum concentrations of T_4 and T_3 are usually decreased in hypothyroidism, but the level may overlap with the reference range in mild cases. Concentrations of T_4 and T_3 are also lowered when the TBG level has been decreased by disease (nephrosis, in which there is loss of TBG and the other binding proteins into the urine, and in liver disease, in which there is decreased synthesis of TBG), by medications that reduce the synthesis of TBG (androgens, anabolic steroids), or by medications that compete with T_4 and T_3 for binding sites (aspirin, phenytoin, tolbutamide, and others). The lowering of plasma T_4 and T_3 concentrations accompanying a decrease in TBG concentration has no effect on the basal metabolic rate.

T_3 Uptake (T_3U) and Free Thyroxine Index (FTI)

Physiologic variation in concentrations of TBG and the other thyroid-binding proteins may occur in a fashion unrelated to thyroid disease. Thus, TBG synthesis and plasma concentration are increased by estrogens (in birth control pills) and pregnancy; they are decreased by depressed synthesis (severe liver disease and administration of anabolic steroids) or loss through the kidney (advanced renal disease). In these instances, the individual is usually euthyroid because the thyroid gland adjusts its secretory rate so that the plasma concentration of free hormone remains within normal limits after equilibration with bound hormone is reached.

The T_3 uptake (T_3U) test provides an indirect estimate of the binding capacity of the plasma thyroid-binding proteins (either TBG or TBG + albumin + prealbumin, depending on the conditions of the assay). Serum is incubated with $^{125}I\text{-}T_3$; the labeled hormone becomes firmly attached to all *unoccupied* binding sites on the TBG. The excess, unattached $^{125}I\text{-}T_3$ is removed by the addition of a suitable solid adsorbent (frequently a resin, but occasionally other materials, including antibody bound to a solid matrix) and is counted. T_3U is the percent of labeled hormone *taken up by the adsorbent and is inversely related to the unoccupied binding sites on TBG.* A serum of known T_3U is used for calibration.

T_3U values are usually calculated as follows:

$$\%T_3U = cpm_p/cpm_R \times \%T_3U_R,$$

where cpm_p and cpm_R are counts/min (on adsorbent) in patient's and in normal reference serum, respectively; $\%T_3U_R$ is the stated value of the reference serum.

Typical Reference Range. Each laboratory should establish its own range because of variability in methods and techniques. Typical values for %T_3U are 25 to 35%.

Increased Values. The T_3U is increased in hyperthyroidism because more of the TBG binding sites are occupied by the patients' T_3 and T_4, thus leaving more of the labeled T_3 available for adsorption by the secondary binder (adsorbent). In like manner, the uptake is elevated in all situations of a decrease in TBG concentration, whether through loss of binding proteins by the kidney or through decreased synthesis by the liver. Drugs that lower the serum T_3 and T_4 concentration by competitive binding to TBG increase the T_3U because the labeled T_3 cannot displace the drugs from their binding to TBG; more of the labeled T_3 is taken up by the adsorbent.

Decreased Values. The T_3U is decreased in hypothyroidism because hormonal output by the thyroid is lowered and, consequently, TBG is less saturated. The TBG binds more of the added label and less is taken up by the adsorbent. Estrogens (increased in pregnancy, present in birth control pills) increase the plasma concentration of TBG. More binding sites are available for accepting the $^{125}I\text{-}T_3$, so less of the label appears in the adsorbent.

Because the T_3U test varies inversely with the T_3 and T_4 concentrations in a wide variety of conditions in which thyroid function is normal, and varies directly with the T_3 and T_4 concentration in hypothyroidism and hyperthyroidism, the T_3U test is much more useful in conjunction with the T_4 measurement. The free thyroxine index (FTI) has become a better way of clarifying this relationship.

$$\text{FTI} = T_4 \times \frac{T_3U_P}{T_3U_C},$$

where T_3U_P refers to the resin uptake in patient's sample and T_3U_C refers to resin uptake in a control or normal sample. When the FTI is calculated in this manner, the reference range is essentially the same as that for T_4. The main relationships are the following:

1. *Normal FTI:* In euthyroid subjects, pregnancy, women taking estrogens, patients with nephrosis or hepatitis, and persons taking drugs that elevate T_3U.
2. *Increased FTI:* In hyperthyroidism.
3. *Decreased FTI:* In hypothyroidism.

Free T_4 and Free T_3

The effective thyroid hormones are free T_4 and T_3, because these hormones act upon tissue cells; the T_4 and T_3 bound to plasma proteins function as a reservoir and are in equilibrium with the free forms. Free T_4 and free T_3 in serum can be separated from the protein-bound hormones by dialysis or other separation techniques and then determined by the same methods as for total T_4 or T_3. The levels of free T_4 and T_3 correlate much better with the thyroid status than does total T_4 or T_3 and are not affected by an abnormal TBG concentration. Some common reference values are shown in Table 13.6. Free T_4 comprises only 0.03% of the total T_4, and its concentration is 1.5 to 3.5 ng/dL (19 to 45 pmol/L). The concentration of free T_3 is extremely low, approximately 200 to 750 pg/dL (3 to 11.6 pmol/L).

Thyroid-Stimulating Hormone (TSH) and Thyroid-Releasing Hormone (TRH)

Serum TSH levels are elevated in hypothyroidism primary to the thyroid gland because of the absence of the negative feedback control. Hypothyroidism in newborns can be detected by the presence of an elevated TSH. The reference values for TSH in euthyroid persons vary with the different methodologies employed but appear to be in the range of 0.4 to 6 μU/mL for adults and 3 to 15 μU/mL for newborns; in newborns, the TSH concentration decreases to adult levels in about 3 weeks. Values as high as 500 μU/mL may be found in primary hypothyroidism, but most values are above 30 μU/mL.

TSH levels in primary hyperthyroid patients (< 0.4 μU/mL) are considerably lower than the levels in euthyroid patients (0.4 to 6 μU/mL). The advent of "sensitive" TSH assays, utilizing the immunoradiometric assay (IRMA) or sandwich technique, has revolutionized the diagnosis of hyperthyroidism because of the ability of these assays to differentiate euthyroid from hyperthyroid patients; such differentiation was not possible with the less sensitive RIA for TSH (see Chap. 4).

Thyrotropin-Releasing Hormone (TRH) Stimulation Test
Injection of TRH and measurement of the output of TSH have some value as a test for indicating combined pituitary and thyroid function or for separating hypothalamic from pituitary disease.

Guidelines for Use of Thyroid Tests

When clinical evidence of hyperthyroidism exists, measure T_4:

1. Elevated serum concentration of T_4 is found in most cases, thus confirming the diagnosis.
2. If T_4 is normal, measure serum T_3 and free T_4 index (FTI). Increased T_3 indicates T_3 hyperthyroidism, whereas increased FTI signifies a decreased plasma concentration of TBG. A free T_4 assay may be substituted for the FTI. Use of sensitive TSH tests usually reveals a decreased TSH in primary hyperthyroidism.

When clinical evidence of hypothyroidism exists:

1. Decreased T_4 is found in many cases.
2. Increased TSH confirms the diagnosis.
3. If T_4 is normal and TSH elevated, an FTI is indicated for evaluation of the TBG concentration.
4. In a few cases, measurement of free T_4 or T_3 may be necessary; sometimes an rT_3 assay may help to clarify an ambiguous situation.

Many patients in an intensive care unit have a low serum T_4 level but no specific symptoms of hypothyroidism; thyroid function returns to normal upon recovery. This situation is known as the "sick low T_4 euthyroid syndrome."

Apparently, binding of T_4 to its carrier proteins is decreased during severe illness because most of these patients have a normal FT_4 concentration. Measurement of serum TSH differentiates those with primary hypothyroidism (TSH $>$ 20 μU/mL) from those with "sick low T_4 euthyroid syndrome" (TSH $<$ 10 μU/mL).

Screening of newborns for hypothyroidism: In most states, measurement of serum T_4 and checking of all low or questionable values with a TSH test are customary; an increased TSH confirms a diagnosis of hypothyroidism. In other states, the TSH assay may be performed first, and elevated values confirmed by a decreased T_4 concentration. The diagnosis must be made early so that treatment with T_3 or T_4 can be started before arrested development occurs. A serum $T_4 < 6.0$ μg/dL should be confirmed by finding a serum TSH $>$ 30 μU/mL and vice versa.

ADRENOCORTICAL HORMONES

An adrenal gland is situated above each kidney. It is composed of an outer cortex and an inner core or medulla. The cortex produces many steroid hormones derived from cholesterol, a waxlike lipid found in all cells (Figs. 13.4 and 13.5). The synthesis of the cortical hormones proceeds by a series of enzymatic steps involving cleavage of part of the side chain of cholesterol, followed by dehydrogenation (conversion of the hydroxyl group to a keto group or formation of a double bond in one of the rings), hydroxylation in a few favored locations, and other reactions. The adrenal cortex is the only gland containing the enzymes with the ability to hydroxylate steroid molecules in the C-21 (21-hydroxylase) and C-11 (11β-hydroxylase) positions (Fig. 13.4, Reactions A and B, respectively).

An outline of the steps involved in the conversion of cholesterol to an adrenal steroid appears in most biochemistry and endocrinology textbooks and is therefore not presented here.

The different types of steroid hormones synthesized by the adrenal cortex (Table 13.7 and Fig. 13.4) are glucocorticoids (cortisol primarily), which affect protein, carbohydrate, and lipid metabolism, mineralocorticoids (aldosterone is the most potent), which affect the concentrations of plasma Na^+ and K^+, androgens (masculinizing hormones), and minute amounts of estrogens (female sex hormones). An abbreviated schema (Fig. 13.4) illustrates the multiple and interrelated pathways for production of the various adrenal hormones.

The main hormonal product of the adrenal cortex in both males and females is the glucocorticoid cortisol. The physiologically important mineralocorticoid aldosterone is present in plasma at approximately one-thousandth the concentration of cortisol. The most potent male sex hormone, testosterone, is secreted primarily by the testis and is mainly responsible for the production of male secondary sex characteristics. The adrenal cortex does produce in small amount some weaker androgens (androstenedione and dehydroepiandrosterone) and their sulfates, which peripheral tissues may convert into testosterone. An abnormally high production of androgens may cause hirsutism (growth of facial hair) and virilism in females and (rarely) precocious sexual development

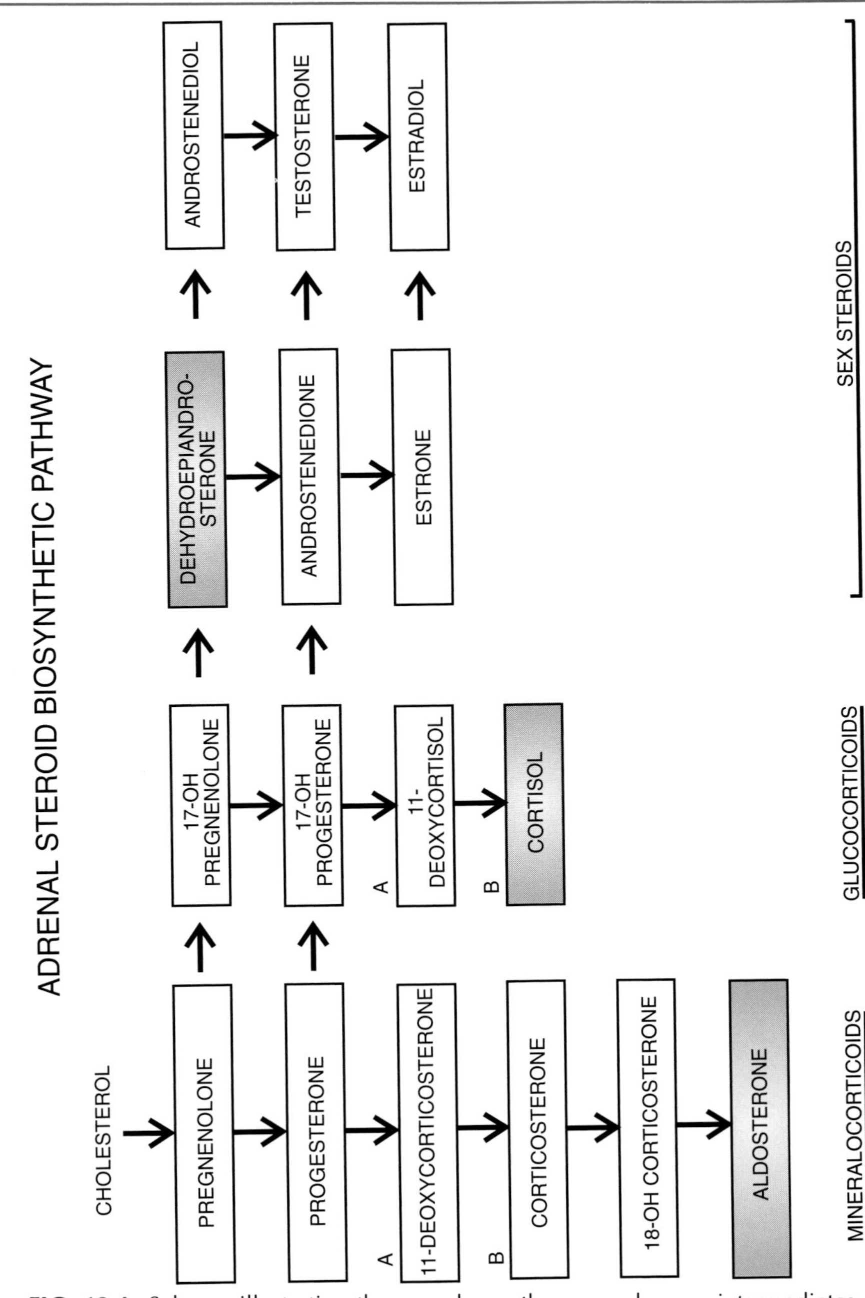

FIG. 13.4. Schema illustrating the complex pathways and many intermediates formed during the synthesis of adrenal cortex hormones. The principal corticosteroids are cortisol (glucocorticoid) and aldosterone (mineralocorticoid). The main androgen, dehydroepiandrosterone (DHEA), is produced in much lesser quantity than the corticosteroids. The estrogens (estradiol and estrone) are produced in the least amount. Each rectangular block in the diagram represents a steroid intermediate that may or may not have some hormonal activity; names have been omitted from some of the transitory intermediates. Reactions A and B refer to 21-hydroxylase- and 11-hydroxylase-catalyzed reactions, respectively, which are lacking in patients with congenital adrenal hyperplasia (CAH).

A. Cholesterol

B. Steroid Skeleton

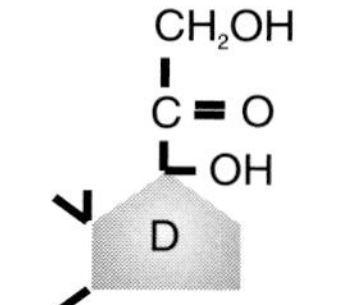

C. 17-Hydroxycortico-steroids. C-21 (17-OHCS)
17, 21 hydroxy, 20-keto

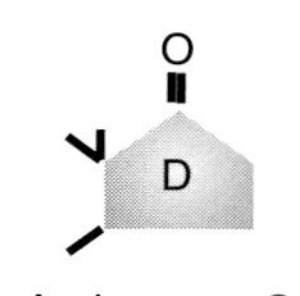

D. Androgens. C-19 (17-KS)
All 17-keto except testosterone (17-hydroxy)

E. Estrogens. C-18

Ring A reduced to phenol, C_{19} removed

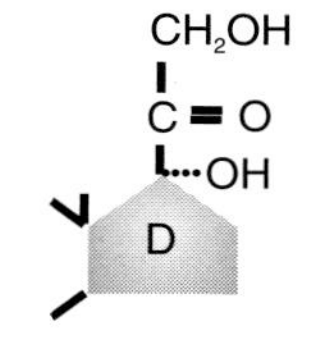

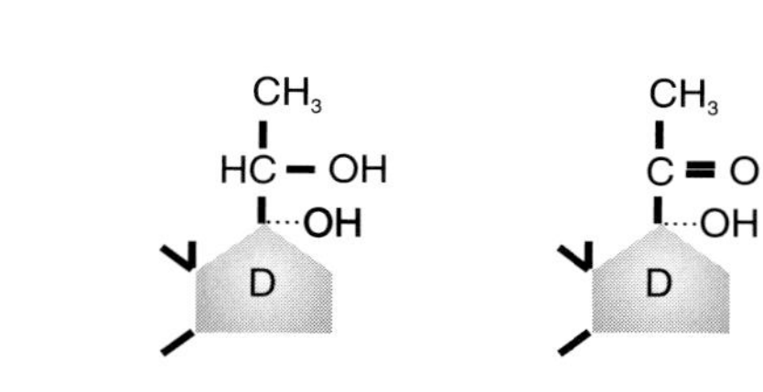

F. 17-Ketogenic Steroids (17-KGS)
Like 17-OHCS, with different side chains

FIG. 13.5. Configuration of cholesterol and some steroid hormones.

in immature males. Estradiol, the most potent female sex hormone, is secreted primarily by the ovary, but small amounts may be derived from adrenal androgens.

A change in the proportions or activities of some of the key enzymes in the adrenal cortex, however, can lead to a serious hormonal imbalance (overproduction of some hormones and underproduction of others). This

TABLE 13.7
ADRENAL CORTICAL HORMONES

HORMONE	TYPE	TARGET TISSUE	PRINCIPAL ACTION
Cortisol	C_{21} steroid	All tissues	Affects carbohydrate, protein and fat metabolism
Aldosterone	C_{21} steroid	Kidney	↓ Na^+ excretion, ↑ K^+ excretion

imbalance may be caused by disease processes, radiation damage, genetic defects, or drugs that inhibit some of the key enzymes. The most commonly encountered defect is a congenital deficiency or absence of the 21-hydroxylase enzyme (see Chap. 15).

Cortisol

Cortisol, the principal glucocorticoid, acts on target cells by penetration and transport to the cell nucleus, binding to DNA, and altering the transcription of RNA. Accordingly, cortisol alters various metabolic processes. It accelerates the enzymatic breakdown of muscle proteins and conversion of their amino acids into glucose by the liver; it mobilizes fat in adipose tissue for energy purposes; it inhibits the uptake of glucose by muscle, and thus cortisol acts as an insulin antagonist (see Chap. 7, Table 7.1); it reduces cellular reaction to inflammatory agents; and it lessens the immune response by inhibiting antibody formation.

ACTH stimulates the adrenocortical production and secretion of cortisol, and to a lesser extent, androgen, but has no such effect on aldosterone. Cortisol is the only adrenal hormone to inhibit the anterior pituitary secretion of ACTH by negative feedback. The secretion of ACTH responds in turn to neurogenic control in its circadian rhythm and in its episodic stimulation by stress, and also by the corticotropin-releasing hormone (CRH) produced by the hypothalamus.

The daily cortisol secretion of the adrenal gland is approximately 25 mg, but the usual concentration of the hormone in plasma varies from 5.0 to 22.0 μg/dL. The plasma concentration of cortisol closely follows the diurnal variation pattern of ACTH and is highest in the early morning hours and lowest at night. Cortisol is transported in plasma in 3 ways: 75% is transported by corticosteroid-binding globulin (CBG, transcortin), 15% is bound to albumin, and approximately 10% is free (not bound to proteins). Cortisol and its oxidation product, cortisone, are inactivated in the liver by two separate reduction processes (hydrogenation) to tetrahydro-derivatives, which are rapidly conjugated with glucuronic acid and excreted in the urine.

Not more than 1% of the total cortisol synthesized daily is excreted as such in the urine. About 30 to 50% appears in the form of glucuronide conjugates of tetrahydro-derivatives of cortisol and cortisone. All these compounds contain the dihydroxyacetone group in the side chain (Fig. 13.5) and are known as 17-hydroxycorticosteroids (17-OHCS).

Hypersecretion of Cortisol

Overproduction of cortisol over a period of time produces a constellation of symptoms known as Cushing's syndrome. The cause can be primary to the adrenal cortex (adrenal hyperplasia or, rarely, carcinoma) or secondary to overproduction of ACTH (pituitary adenoma or an ectopic carcinoma elsewhere that produces ACTH, for example, oat cell carcinoma of the lung). Plasma ACTH levels are low in primary adrenal disease and high with uncontrolled production of ACTH by a neoplasm. The diagnosis can be confirmed by means of a dexamethasone suppression test; administration of the drug suppresses normal secretion of ACTH, as evidenced by a fall in plasma cortisol concentration and excretion of free cortisol in the urine, but has no effect on ACTH secretion by tumors.

Hyposecretion of Cortisol

Adrenocortical insufficiency, evident by a low plasma cortisol concentration, may be primary to the adrenal cortex because of destruction of cortical tissue by autoimmune disease or infection (Addison's disease) or secondary to ACTH deficiency. Plasma ACTH levels are high in Addison's disease and low in ACTH deficiency. The diagnosis is confirmed by finding a subnormal cortisol response to the administration of exogenous ACTH (rapid ACTH stimulation test).

Evaluation of Adrenocortical Function[10]

Serum Cortisol Determination. Blood should be collected at approximately 8 A.M. (near peak time for cortisol concentration) and allowed to clot. Serum cortisol is most frequently measured by RIA or fluorescence polarization immunoassay (FPIA) after displacing the cortisol from CBG (by lowering the pH, by heat treatment, or by displacing the cortisol by chemical means). The principle of these assays is the same as that for the measurement of T_4 and T_3.

Reference Values. The morning plasma cortisol concentration in normal adults varies from 5 to 22 μg/dL (138 to 607 nmol/L), with a mean of 11 μg/dL (300 nmol/L). Cortisol levels in the afternoon and evening are considerably lower.

Increased Concentration. Serum cortisol is elevated in Cushing's syndrome, whether primary to the adrenal cortex (adrenal adenoma or carcinoma) or secondary to overproduction of ACTH.

Decreased Concentration. Serum cortisol is low in Addison's disease and in ACTH deficiency. Prolonged therapy with anti-inflammatory steroids (for example, prednisone) can suppress ACTH secretion sufficiently to cause adrenal insufficiency.

Urine Free Cortisol. The small amount of cortisol in urine reflects the plasma level of free cortisol (not bound to CBG) because only the unbound hormone can be excreted. The method for urine free cortisol is similar to that described for serum. Elevated urine free cortisol is the hallmark for the diagnosis of Cushing's syndrome.

Androgens

The adrenal androgens are dehydroepiandrosterone (DHEA), DHEA sulfate, androstenedione, and testosterone (Figs. 13.4 and 13.5D). All but testosterone are weak androgens, but peripheral tissues convert a portion of them into testosterone. All the androgens are C_{19} compounds; all are 17-ketosteroids (17-KS) except for testosterone, which has a hydroxyl instead of a ketone group at C_{17} and is a 17-ketogenic steroid (17-KGS).

Androgen Excess (Hirsutism, Virilism)

Androgen excess in children is usually caused by congenital adrenal hyperplasia or adrenal carcinoma; in boys, it is characterized by precocious puberty. The most common manifestation in women is hirsutism, the growth of body hair in a male-like pattern. Virilism is rarely caused by androgen excess alone, but usually results from a combination of adrenal and ovarian disorders.

Measurement of individual plasma androgens greatly assists in the diagnosis of androgen excess; the use of urinary 17-KS excretion is of limited help because normal values are found in 80% of women with hirsutism.

Steroid Metabolites in Urine

The metabolites of all the androgens appear in the urine as 17-KS or as their glucuronide or sulfate conjugates (Fig. 13.5D). The measurement of urinary excretion of 17-KS was widely used before the development of immunoassays for individual hormones. After a laborious extraction process, the 17-KS were quantitated colorimetrically after reaction with *m*-dinitrobenzene (Zimmerman reaction). Reference values are given in Appendix 1.

Two different chemical methods were formerly used to provide an estimate of the urinary excretion of cortisol and its derivatives. The 17-KGS compounds illustrated in Figure 13.5F were converted to 17-KS by reduction with sodium borohydride, followed by oxidation with sodium metaperiodate; they were then quantitated in the same manner as 17-KS. The method for 17-hydroxycorticosteroids (17-OHCS) is more restrictive and measures fluorometrically only cortisol-like compounds with the structure shown in Figure 13.5C. The Porter-Silber fluorometric method was used for quantitation of 17-OHCS.

Aldosterone

Aldosterone is by far the most potent mineralocorticoid secreted by the adrenal cortex, even though several corticosteroids, including the glucocorticoids, have some mineralocorticoid activity. Like cortisol, aldosterone is a C_{21} compound and has a 11-hydroxy group, but lacks a hydroxyl at C_{17}. Aldosterone differs from all the other steroids by the presence of an aldehyde group at C_{18}, which is in equilibrium with the hemiacetal form. Aldosterone increases the plasma concentration of Na^+ by increasing Na^+ reabsorption in the renal tubules; the plasma K^+ concentration falls because of the concomitant exchange of K^+ for Na^+ (see Chap. 6 and Fig. 6.2).

Aldosterone is present in plasma at low concentrations and is difficult to measure accurately. A normal person who has been in the upright position for several hours has a plasma aldosterone concentration between 5 and 20 ng/dL (0.14 to 0.55 nmol/L). If the subject is recumbent for several hours, the plasma aldosterone level falls to 10 to 40% of that concentration. The liver converts aldosterone into a glucuronide or a tetrahydroglucuronide, which passes into the urine; the kidney also inactivates aldosterone by transformation into water-soluble glucuronides.

Hyperaldosteronism has been found in patients with some adrenal tumors. These patients usually have elevated serum Na^+ concentration, lowered K^+, and hypertension. Measurement reveals an increased plasma aldosterone concentration and an increased excretion of this hormone in the urine.

FEMALE SEX HORMONES

Ovarian Hormones[8]

The female gonad, or ovary, has a double function; it not only produces and secretes the female sex hormones, but it also is the site of production and maturation of the ova. One mature ovum is released approximately every 4 weeks by a nonpregnant woman during the years between the onset of menstruation and the menopause.

The reproductive system of females is far more complicated than that of males because of the cyclical events that take place during the menstrual cycle and the even greater changes that occur during pregnancy. Two different chemical types of steroid hormones are produced and secreted by the ovary in nonpregnant women. During pregnancy, the same hormones are produced by the ovary, but in different proportions. The placenta also makes the hormones that are necessary for the maintenance of pregnancy.

The first group of female sex hormones, the estrogens, (Table 13.8) originate in the ovarian follicles (and also in the placenta during pregnancy). The estrogens participate in the menstrual cycle and are essential for the development and maintenance of the reproductive organs and secondary sex characteristics.

TABLE 13.8
OVARIAN HORMONES

HORMONE	TYPE	TARGET TISSUE	PRINCIPAL ACTION
Estrogens (E_1, E_2)	C_{18} steroids	Female accessory sex organs	Promotes secondary sex characteristics
Progesterone	C_{21} steroid	Uterus	Prepares for ovum implantation, maintains pregnancy

The second group includes progesterone and its metabolites (progestational hormones), which are formed in the corpus luteum, the body that develops from the ruptured ovarian follicle (Table 13.8). Progesterone stimulates the uterus to undergo changes that prepare it for implantation of the fertilized ovum and suppresses ovulation and secretion of pituitary LH. If pregnancy occurs, the secretion of progesterone by the corpus luteum and also by the placenta suppresses menstruation for the duration of the pregnancy. The hormonal events that occur in a typical 28-day menstrual cycle are shown in Figure 13.6.

Estrogens and Progesterone

The estrogens comprise a number of C_{18} steroids. They differ structurally from the androgens in that one methyl group has been lost and ring A of the estrogen steroid is phenolic (has three double bonds, as shown in Fig. 13.7). Progesterone, the other important ovarian hormone, is a C_{21} compound and chemically more closely related to the adrenal steroids; as a matter of fact, progesterone is an intermediate in the production of adrenal steroids (Figs. 13.4 and 13.7).

The three clinically important estrogens are estradiol (E_2), estrone (E_1), and estriol (E_3), although several other metabolites exist. The principal and most potent estrogen, however, is E_2. It exists in a reversible state with E_1, a hormone with weaker biologic action, but it must be converted into E_1 before it is degraded. The final degradation product is E_3, a steroid without biologic activity.

RIA kit methods are available for the measurement of serum estrogens; the principles for determination are similar to those for cortisol and the thyroid hormones. Plasma E_2 levels are valuable for the investigation of women with menstrual difficulties, because E_2 is an indicator of ovarian function. More information can be gained if the pituitary tropic hormones, FSH and LH, are measured at the same time to ascertain whether a problem is of pituitary or ovarian origin. The concentration of estrogens in the plasma of normal nonpregnant women is in the pg/mL range and varies with the stage of the menstrual cycle.

Estriol (E_3)

E_3 has no hormonal activity. It is produced in relatively large quantities during the last trimester of pregnancy by the placental conversion of fetal adrenal steroids. Thus, its concentration in the urine or plasma of pregnant women provides some indication of fetal well-being (fetoplacental viability). Serial measurement of serum (or urine) estriol during late pregnancy is an important test in obstetrics because a sudden drop in estriol concentration or output is a danger signal of fetoplacental dysfunction; the serum test is preferred.

Placental Hormones

The placenta in the pregnant woman serves as an endocrine organ in addition to its physiologic functions of providing nutrients to the developing embryo and removing its waste products. In humans, the placental hormones (Table

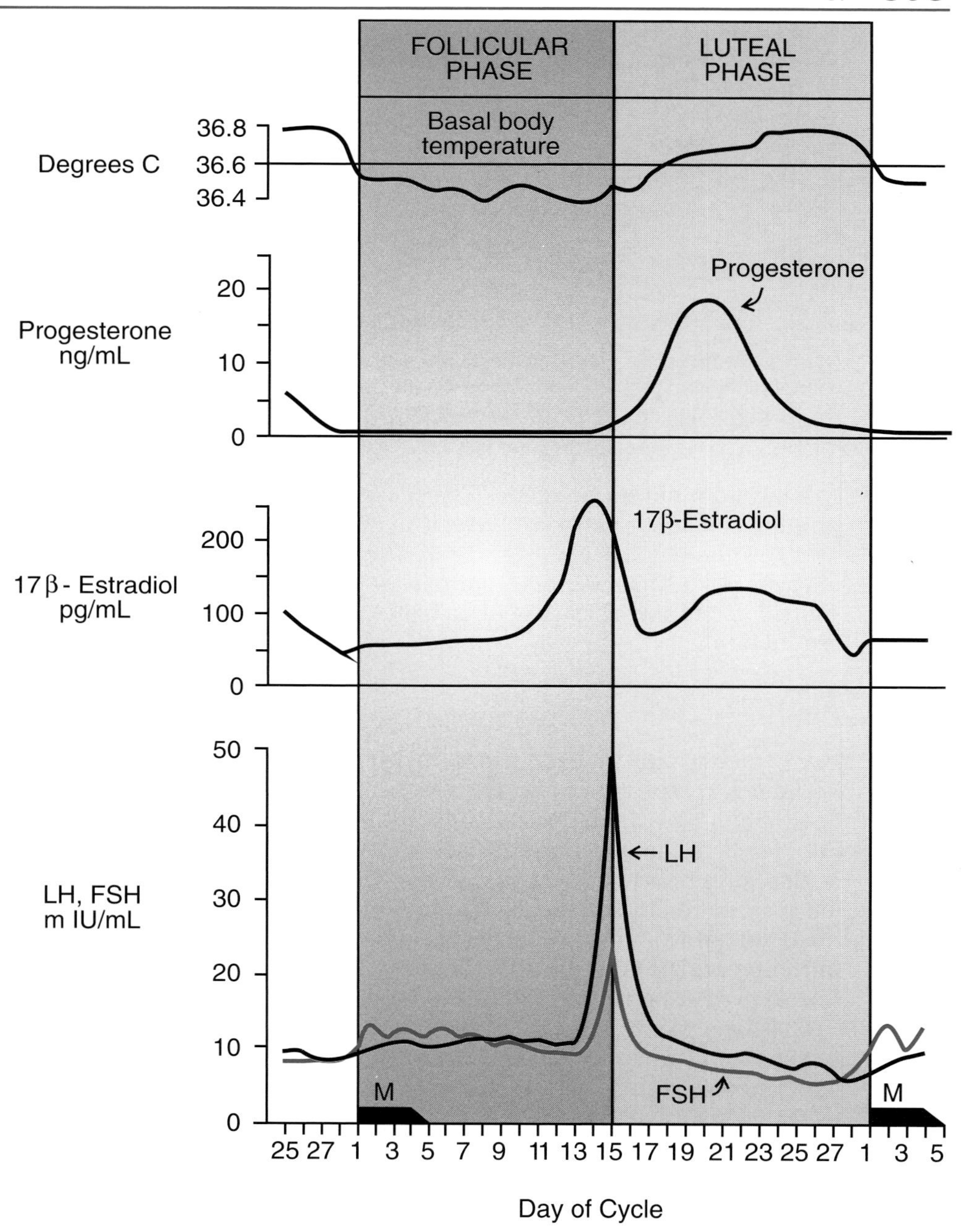

FIG. 13.6. Hormonal changes during normal menstrual cycle.

13.9) are identical to those produced by the ovarian follicle and corpus luteum except for chorionic gonadotropin (hCG) and lactogen (hPL, similar to pituitary prolactin and growth hormone). The placenta, unlike the gonads and the adrenal cortex, cannot synthesize steroids from acetyl CoA units (see Chap. 7, Fig. 7.5), but requires sterols or steroids as precursors, depending on the

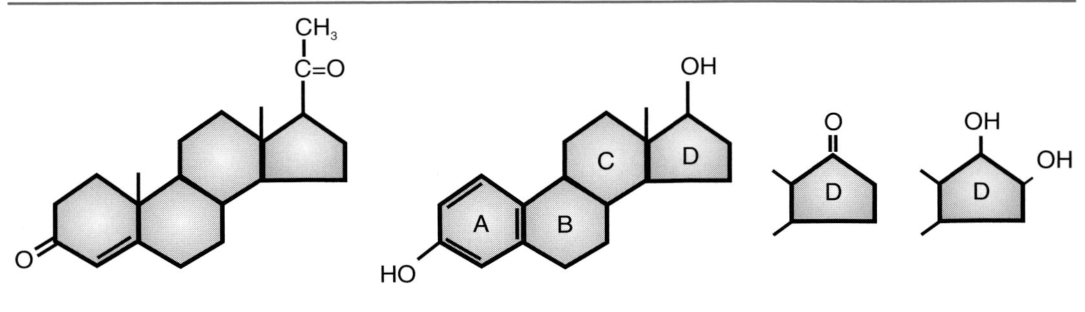

Progesterone Estradiol, E_2 Estrone, E_1 Estriol, E_3

FIG. 13.7. Steroids produced by the ovary (and also by the placenta). Ring A of estradiol is a phenolic ring that makes the hormone slightly acidic. Rings A, B, and C of E_1, E_2, and E_3 are identical; the differences are confined to ring D.

specific hormone to be made. The placenta produces progesterone primarily from plasma cholesterol, whereas it requires a C_{19} steroid, dehydroepiandrosterone sulfate (DHEAS, Fig. 13.5) for synthesis of estriol and the other estrogens. The fetal adrenal cortex, as well as the maternal adrenal cortex, contributes to the plasma pool of DHEAS that the placenta uses for its estriol production.

Human Chorionic Gonadotropin (hCG)

Human chorionic gonadotropin (hCG) is one of a group of glycoprotein hormones composed of two chains that includes the pituitary hormones, LH, FSH, and TSH. The alpha chain is essentially the same in all four hormones, but each beta chain is unique. The action of hCG is similar to that of LH because it also stimulates the corpus luteum to produce progesterone. Progesterone helps to maintain the pregnancy by preventing menstruation. The chorion of the developing placenta begins to secrete hCG shortly after implantation of the fertilized egg. Its concentration in plasma and urine rises steadily from the first few days after conception until the tenth or twelfth week of pregnancy.

The detection of hCG in urine or serum is biochemical confirmation that a patient is pregnant. The most sensitive tests, IRMA assays for hCG, may confirm pregnancy 7 to 10 days post conception (4 to 7 days before the next expected menstrual period); most tests for hCG should be positive in 90% of patients 14 to 21 days after the next expected menstrual period. Many assays

TABLE 13.9
PLACENTAL HORMONES

HORMONE	TYPE	TARGET TISSUE	PRINCIPAL ACTION
Estrogens and progesterone	See ovary	See ovary	See ovary
Chorionic gonadotropin (hCG)	Glycoprotein	Same as LH	Same as LH
Placental lactogen (hPL)	Protein	Same as PRL	Same as PRL

for hCG employ antibodies specific for the beta subunit (so-called beta hCG assays) to eliminate cross-reactivity with LH. The most sensitive and specific assays for hCG are the IRMA assays, which utilize two antibodies to recognize the hormone (one specific for the alpha subunit and another for the beta subunit). The urine test is usually sufficient to confirm pregnancy in a woman who has missed a menstrual period. The serum test is usually performed to confirm an equivocal urine test. A quantitative hCG assay, performed on serum, may also be used to confirm an equivocal urine test, to confirm an ectopic pregnancy (hCG levels are considerably lower than those in a normal intrauterine pregnancy), or to diagnose a hydatidiform mole* in pregnancy.

Women with choriocarcinoma, a malignant condition resulting from retained products of conception, have greatly elevated concentrations of hCG (usually > 150,000 mIU/mL). Levels of hCG may also be elevated in approximately 40 to 60% of men with testicular carcinoma. The levels of hCG are useful for monitoring the course of therapy or the reoccurrence of tumor.

Human Placental Lactogen (hPL)

The placenta also elaborates a protein hormone, placental lactogen, that shares some properties with both growth hormone and prolactin. Acting in concert with prolactin, hPL helps to prepare the mammary gland for lactation. It also has some metabolic activities similar to those of GH.

MALE SEX HORMONES

The male gonads are the testes. Like the ovary, they have a double function: to produce and secrete the male hormone testosterone and to produce the spermatozoa, which are essential for fertilization of the ovum in the reproductive process (Table 13.10). The pituitary gonadotropin, LH, stimulates interstitial cells (Leydig's cells) in the testis to produce testosterone, and FSH promotes spermatogenesis by the germinal cells in the seminiferous tubules.

A specific plasma globulin, sex hormone binding globulin (SHBG) transports about 80% of the circulating testosterone, with albumin binding about 17 or 18%; the remainder is the unbound, active hormone. Practically all the testosterone in males is derived from the testis; the contribution of the adrenal cortex is negligible. Hence, measurement of plasma testosterone concentration is a good way of studying hypogonadism and hypergonadism. Of course, the role of the pituitary still must be assessed to determine whether an abnormality is primary to the testis or secondary to an LH deficiency or excess. Plasma total testosterone levels in normal adult men range from 350 to 850 ng/dL (12.1 to 29.5 nmol/L) and from 20 to 80 ng/dL (0.8 to 2.8 nmol/L) in women.

Plasma testosterone levels are much lower in women, usually amounting to only 5% of those found in men, but they can still be measured by sensitive RIA methods; the principles are the same as those for the T_4 or T_3 assay. Kits are available. Testosterone in women arises from the tissue conversion of androgens.

*A cyst-like structure in the uterus that develops from an aberration in a normal pregnancy. It may become locally invasive or even malignant.

TABLE 13.10
TESTICULAR HORMONE

HORMONE	TYPE	TARGET TISSUE	PRINCIPAL ACTION
Testosterone	C_{19} steroid	Male accessory sex organs	Promotes secondary sex characteristics

Increased Concentration. Elevated plasma testosterone concentration in men may be caused by testicular carcinomas or by some abnormalities of pituitary gonadotropin. In women, plasma testosterone may be elevated in some cases of virilism or hirsutism as a result of adrenal or ovarian tumor.

Decreased Concentration. Plasma testosterone concentration in men may be decreased in a variety of conditions directly affecting the testes, by pituitary failure, and in certain chromosomal abnormalities involving the sex chromosomes.

ADRENOMEDULLARY HORMONES

The adrenal medulla originates in the developing fetus from cells of the sympathetic nervous system, and its hormones reflect this close relationship. One hormone, norepinephrine (noradrenaline), is also a neurotransmitter produced in the brain and at synapses with peripheral nerves (Table 13.11). Epinephrine (adrenaline), the principal medullary hormone, produces effects on tissues and organs similar to those following stimulation of the sympathetic nervous system by fear, anger, or aggression (for example, increases in heart rate and blood pressure). Both hormones have some metabolic effects that facilitate the generation of energy by the organism; they raise the plasma concentration of glucose by inducing liver glycogenolysis and of free fatty acids by promoting lipolysis in adipose tissue (see Chap. 7, Table 7.1).

Epinephrine and norepinephrine are derived from tyrosine via dopamine as illustrated in Figure 13.8. The three amines are known collectively as catecholamines because they have the dihydroxybenzene structure of catechol. Epinephrine is formed from norepinephrine by methylation of the amino group. Dopamine is an intermediary product in the adrenal medulla, but when

TABLE 13.11
ADRENAL MEDULLARY HORMONES

HORMONE	TYPE	TARGET TISSUE	PRINCIPAL ACTION
Epinephrine	Tyrosine derivative	Sympathetic receptors, liver, muscle, adipose cells	Stimulates sympathetic nerves, ↑ glycogenolysis (liver and muscle), ↑ lipolysis (adipose cells)
Norepinephrine	Tyrosine derivative	Sympathetic receptors	Stimulates sympathetic nerves

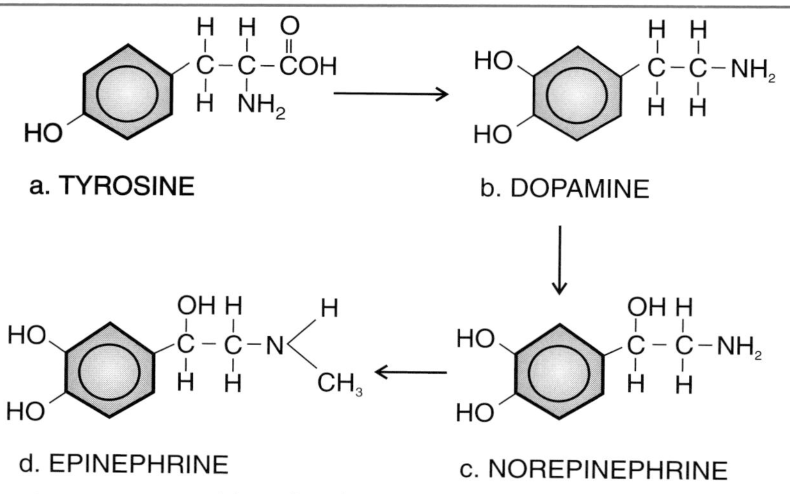

FIG. 13.8. Structural formulas of tyrosine (A), the precursor of the adrenal medulla hormones, dopamine (B), norepinephrine (C), epinephrine (D).

synthesized in brain cells, it serves as a neurotransmitter. Epinephrine comprises 80 to 90% of the medullary hormones.

The catecholamines exert their rapid action on tissues by binding to the cell membranes and activating adenyl cyclase, which produces cyclic AMP within the cell. The second messenger, cyclic AMP, promotes the various metabolic and physiologic effects.

The hormones are so potent physiologically that only small amounts are needed to obtain their effects. Their action is transitory because of rapid inactivation. The catecholamines are catabolized or inactivated by several different pathways, which involve O-methylation of the hydroxyl group on C_3 and/or oxidative deamination. The principal metabolites are vanillylmandelic acid (VMA, 3-methoxy-4-hydroxymandelic acid), metanephrine, normetanephrine, and sometimes homovanillic acid. The metanephrines are formed by O-methylation of the hydroxyl group of epinephrine and norepinephrine. Unchanged catecholamines in a 24-hr urine usually account for about 1% of the medullary hormones that have been turned over, whereas VMA accounts for 75%, the metanephrines for 10%, and closely related metabolites for the remainder. The metabolites in urine are a mixture of the free form and sulfate or glucuronate conjugates.

The measurement of urinary catecholamines or their metabolites is of value in two clinical conditions: (1) in unexplained high blood pressure (hypertension) to rule out or to uncover the possible presence of a pheochromocytoma, a tumor of the adrenal medulla that is a cause of hypertension; and (2) to detect a neuroblastoma, an often fatal malignant condition in children in which cancer of the nervous system causes excess production of norepinephrine.

Screening for Pheochromocytoma

The incidence of pheochromocytoma *in a population of hypertensive patients* is about 1 or 2 per 1000. A good screening test is essential because 90% of pheochromocytomas are benign and can be removed completely by surgery; a missed diagnosis is a lost opportunity for correction and may lead to an early death. The tumor may secrete catecholamines intermittently in spurts, so the plasma level of these hormones may be elevated for only a short time; a randomly drawn blood sample may miss the rise. For general screening purposes, collection of a 24-hr urine may bracket periods of tumor activity and reveal an abnormality. Because a reliable 24-hr urine collection is difficult to achieve, especially in young children, urine results are more meaningful if expressed as a ratio of catecholamine or metabolite excreted per mg of creatinine. A positive test should be confirmed by direct measurement of plasma catecholamines or by analysis of a different catecholamine or metabolite in urine. Principles of the methods are reviewed in the following sections.

Screening for Neuroblastoma

A neuroblastoma may secrete metabolites of all three catecholamines, but in some cases, dopamine and its metabolite, homovanillic acid (HVA), predominate. A high excretion of HVA or VMA (or both) occurs in most cases of neuroblastoma. Both VMA and HVA may be measured by use of gas chromatography or high-performance liquid chromatography (HPLC).

Urinary Metanephrines (Total)[9]

Principle. The metanephrine conjugates in an acidified urine aliquot are hydrolyzed by heating. The free and liberated metanephrines are adsorbed onto a weak cation exchange resin, eluted with ammonium hydroxide, and extracted into an ethyl acetate/acetone mixture. After evaporation of the solvent, the residue is dissolved in dilute acid. The metanephrines may then be determined (1) after injection onto an HPLC column with an electrochemical detector or (2) spectrophotometrically after oxidation with sodium periodate to vanillin. The HPLC method is preferred because it allows for the quantitation of the individual metanephrines—normetanephrine, metanephrine, and 3-methoxytyramine, a dopamine metabolite.

Reference Values. 0.3 to 0.9 mg/day (1.5 to 1.6 μmol/day) for adults.

Plasma Catecholamines by Radioenzymatic Assay*

Principle. Catecholamines in an aliquot of plasma are labeled with a tritiated methyl group (^{3}H-methyl) by incubation with the enzyme catechol-O-methyltransferase, and tritiated S-adenosyl-methionine (^{3}H-SAM). This process converts the plasma catecholamines into their respective ^{3}H-metanephrines (metanephrine, normetanephrine, and 3-methoxytyramine). The radioactivity

*Available from Upjohn Co., Kalamazoo, MI 49001

in the metanephrines is counted and is directly proportional to the catecholamines present in a patient's sample.

Plasma Catecholamines by HPLC

Catecholamines in plasma samples are absorbed onto alumina, the alumina is washed to eliminate other compounds, and then the catecholamines are eluted from the alumina with acid. The acid extract is then injected into an HPLC with a weak cationic exchange column, and electrochemical detection is used to determine the concentration of epinephrine, norepinephrine, and dopamine.

PARATHYROID HORMONE (PARATHORMONE, PARATHYRIN, PTH)[5]

The parathyroid glands consist of four small glands that are found at the poles of the thyroid. The secretory product of the parathyroid glands is a single-chain polypeptide hormone consisting of 84 amino acids, the major role of which is the maintenance of calcium homeostasis. PTH is synthesized as a larger 116 amino acid precursor (prepro-PTH), which undergoes successive degradation to pro-PTH and finally to PTH, which is the secretory product (Table 13.12). PTH in the circulation undergoes rapid cleaveage into an N-terminal fragment, which has biologic activity, and a C-terminal fragment, which is biologically inactive. The half-life of PTH is approximately 12 min. Historically, PTH determinations were performed by measuring either the N-terminal or the C-terminal fragments, both of which were fraught with difficulty. The current approach is to measure intact PTH with IRMA assays (see Chap. 4), which utilize antibodies to both the C- and N-terminal portions.

The parathyroid maintains the plasma Ca^{2+} concentration within narrow limits by secreting PTH in response to a small decrease in Ca^{2+} concentration. A negative feedback control inhibits the secretion of PTH when Ca^{2+} is elevated. PTH raises Ca^{2+} directly by (1) mobilizing calcium from bone and (2) decreasing renal excretion of Ca^{2+} by stimulating tubular reabsorption. It further increases Ca^{2+} indirectly by enhancing the renal formation of 1,25-$(OH)_2D_3$, the biologically active form of vitamin D that increases the intestinal absorption of calcium (see Chap. 12). PTH simultaneously reduces the phosphate load arising from the bone reabsorption by promoting its excretion by the kidney (inhibiting tubular reabsorption).

TABLE 13.12
PARATHYROID HORMONE

HORMONE	TYPE	TARGET TISSUE	PRINCIPAL ACTION
Parathyroid hormone (PTH)	Polypeptide	Bone, kidney	↑ Plasma Ca^{2+}, ↑ bone resorption ↓ Ca^{2+} excretion, ↑ phosphate excretion

The PTH assay is used for investigating primary hyperparathyroidism, for assessing the hypercalcemia of malignant growths, and for checking for possible hypoparathyroidism. PTH is elevated in primary hyperparathyroidism and is normal or undetectable in hypercalcemia and hypoparathyroidism.

PTH assays should always be performed in conjunction with either a total or an ionized Ca measurement.[6] A PTH-Ca nomogram is commonly used to report PTH determinations (Fig. 13.9).

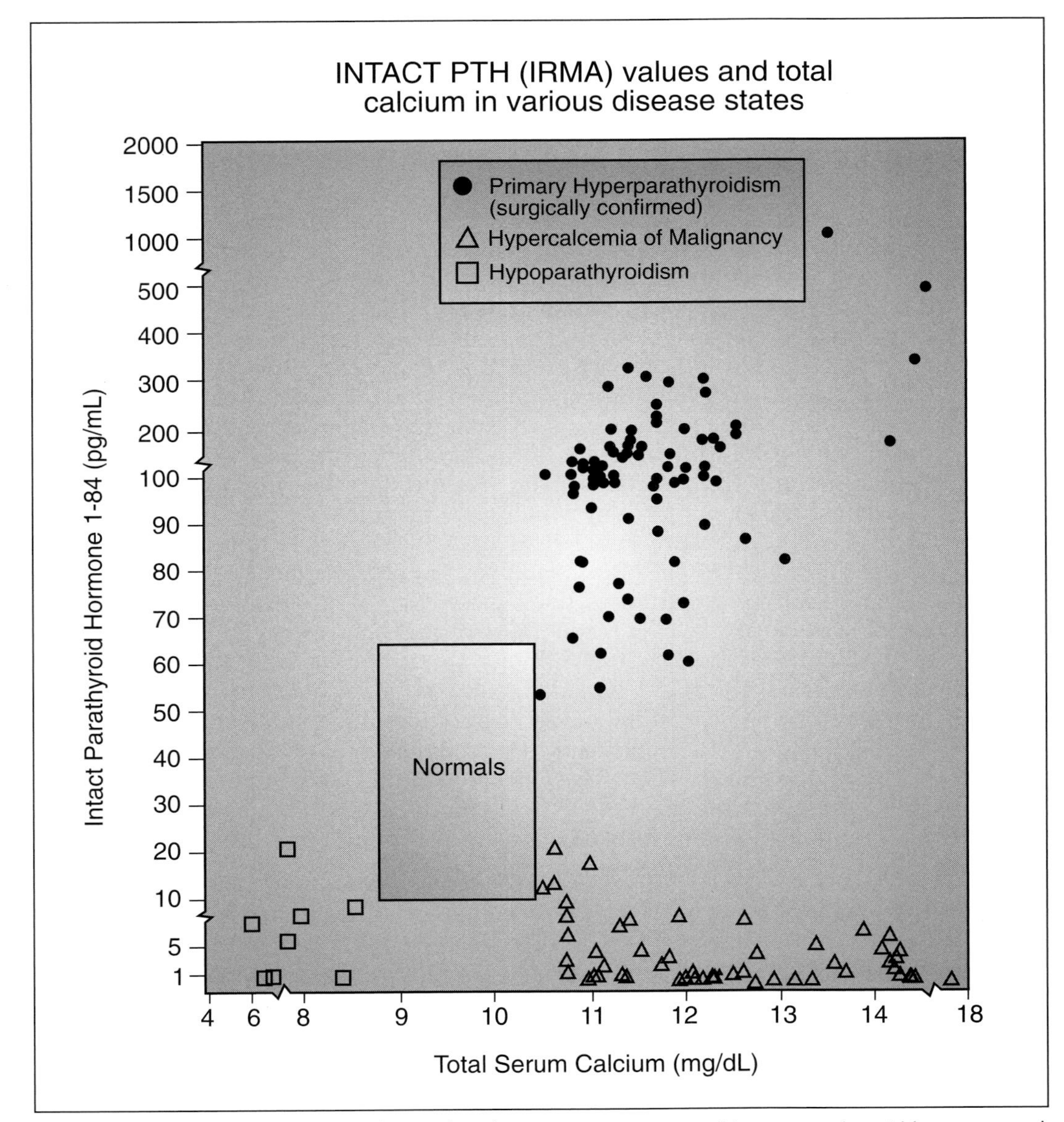

FIG. 13.9. Nomogram relating the plasma concentrations of intact parathyroid hormone and calcium with various parathyroid diseases. (Courtesy of Nichols Institute Diagnostics.)

CALCITONIN

Calcitonin (CT), a hormone synthesized in the thyroid gland, participates to a limited extent in calcium homeostasis by responding to a hypercalcemia. CT depresses the release of calcium from bone by inhibiting the bone-dissolving activity of osteoclasts. The importance of CT in the regulation of plasma Ca^{2+} concentration is not clear because individuals without CT do not develop hypercalcemia (for example, after complete removal of the thyroid). CT can be measured by RIA, but its diagnostic usefulness has not been established in Ca disorders; however, CT can be used as a marker for medullary thyroid carcinoma.

PANCREATIC HORMONES

The pancreas is one of the few glands with both an exocrine and an endocrine function. It plays an important role in the digestive process by secreting into the intestine by way of the common bile duct a juice that is rich in bicarbonate and a multiplicity of digestive enzymes; it also secretes somatostatin, a hormone that helps to regulate the secretion of various gastrointestinal fluids and enzymes. Moreover, the pancreas is the source of insulin and glucagon, the two major hormones for regulating plasma glucose concentration (Table 13.13).

Insulin

Insulin is originally synthesized as part of a larger inactive protein, proinsulin, that is stored as granules in the islet beta cells. A rise in plasma glucose concentration induces an enzyme to cleave the stored protein into the active hormone, insulin, and an inactive fragment called C-peptide; the insulin is secreted into plasma. Normally, insulin and small amounts of proinsulin and C-peptide can be found in plasma by appropriate RIA.

Insulin is the only hormone with the ability to lower the plasma glucose concentration (see Table 7.1, Chap. 7). It does so by (1) promoting glucose uptake by muscle and fat cells, (2) inducing glycogen storage in the liver, (3) inhibiting lipolysis and promoting triglyceride synthesis in adipose tissue, (4) increasing protein synthesis from amino acids, and (5) enhancing glucose

TABLE 13.13
PANCREATIC HORMONES

HORMONE	TYPE	TARGET TISSUE	PRINCIPAL ACTION
Insulin	Protein	All tissues	↓ plasma glucose, ↑ glycogenesis, ↑ lipogenesis
Glucagon	Polypeptide	Liver	↑ glycogenolysis, ↑ gluconeogenesis
Somatostatin	Peptide	Pancreas, alpha and beta cells	Inhibits release of insulin and glucagon

utilization. An absolute deficiency of insulin causes a severe type of diabetes mellitus (see Chap. 7).

Measurement of the plasma insulin concentration is of value in the diagnosis of adenoma of the pancreas (insulinoma), in which the insulin concentration is high. Insulin analysis is performed by RIA. The plasma insulin concentration of normal fasting individuals does not exceed 860 pg/mL (20 mU/L). It is elevated in patients with an insulinoma and in those with excessive GH secretion (gigantism in children, acromegaly in adults). Quantitation of plasma insulin is of no value in the diagnosis of diabetes mellitus.

Glucagon

Glucagon, a polypeptide made by islet alpha cells, is the primary counterregulatory hormone to insulin action (see Chap. 7); the pancreas secretes glucagon in response to a fall in plasma glucose level. It increases the glucose concentration by inducing rapid breakdown of stored liver glycogen and by promoting the formation of glucose from amino acids.

Measurement of plasma glucagon levels has no clinical value at this time. Glucagon is sometimes injected intravenously as a provocative test in certain diagnostic situations. In patients with insulinoma, the injection of glucagon produces hyperglycemia, followed by hypoglycemia; the hypoglycemia does not occur to a significant extent in individuals without an insulinoma. If glucagon is injected into patients with von Gierke's disease (Type 1 glycogen storage disease), the plasma glucose concentration does not rise because the hepatic enzyme, glucose-6-phosphatase, is missing; this outcome helps to confirm the diagnosis.

Somatostatin

Somatostatin is a polypeptide hormone made in the delta cells of the pancreatic islets, in the hypothalamus, and elsewhere. In addition to regulating gastrointestinal secretions, somatostatin inhibits the secretion of insulin and glucagon by nearby islet cells. No clinical demand exists at present for the measurement of plasma somatostatin concentration.

HORMONES OF THE GASTROINTESTINAL TRACT

The gastrointestinal tract secretes large amounts of a variety of hormones that regulate the entire digestive process, from the secretion of gastrointestinal juices to the absorption of the final products of digestion (Table 13.14). These hormones may exert their actions through one or more of the following ways: (1) act as "true" hormones on distant target cells after transport in plasma (for example, secretin); (2) act upon neighboring target cells after secretion into extracellular fluid or a lumen, as does somatostatin; and (3) mediate their effects through stimulation of nearby nerve cells (vasoactive intestinal peptide, for example). Food in the gut is the usual stimulus for gastrointestinal hormonal release, but some degree of neural control plays a part also. The

TABLE 13.14
GASTROINTESTINAL TRACT HORMONES

HORMONE	TYPE	TARGET TISSUE	PRINCIPAL ACTION
Gastrin	Peptide	Stomach	↑ Secretion of gastric HCl
Secretin	Polypeptide	Pancreas	↑ Secretion of pancreatic fluid and HCO_3^-
Cholecystokinin-pancreozymin	Polypeptide	Gallbladder, pancreas	Gallbladder contraction, ↑ pancreatic secretion
Somatostatin*	Peptide	Gastrointestinal tract	↓ Secretions and motility
Vasointestinal peptide (VIP)	Polypeptide	Gastrointestinal tract	↑ Secretions, relaxes gut muscles

*Somatostatin is also secreted by the hypothalamus; it is carried to the anterior pituitary, where it inhibits the release of growth hormone.

following review is limited to the few gastrointestinal hormones with measurements that may aid clinical diagnosis.

Gastrin

Gastrins are hormones, secreted primarily by the antrum of the stomach, that are powerful inducers of gastric HCl secretion. Three biologically active gastrins differ only in the chain length of the N-terminal end: (1) "big gastrin," composed of 34 amino acids; (2) "little gastrin," the most potent of the group, with 17 amino acids; and (3) "mini gastrin," with 14 amino acids and the least activity. Total gastrins are usually measured by RIA methods because polyclonal antibodies react with all three compounds.

Overproduction of gastrin causes ulcers in the upper gastrointestinal tract. The most important clinical use of the gastrin test is to detect a condition known as the Zollinger-Ellison syndrome. In this disease, non-beta cell tumors of the pancreas produce large amounts of gastrin, and the patient suffers severely from ulcers; the plasma gastrin level may be grossly elevated over the reference range of 46 to 140 pg/mL (46 to 140 ng/L).

Vasoactive Intestinal Peptide (VIP)

VIP is an intestinal hormone that stimulates nerve cells in the gastrointestinal tract. Its overproduction by tumor cells produces an intractable diarrhea, sometimes called pancreatic cholera. Plasma levels of VIP in normal individuals are not sufficiently high to assay, but are measurable by RIA methods in patients with tumors that cause the pancreatic cholera syndrome. The test is used for screening patients with severe, persistent diarrhea.

Most of the other hormones secreted by the gastrointestinal tract either are too difficult to measure accurately in plasma or have no urgent medical reason for analysis. Secretin and cholecystokinin are sometimes injected to study pancreatic response, but no clinical demand exists as yet for measurement of the plasma concentration of these hormones.

TABLE 13.15
RENAL HORMONES

HORMONE	TYPE	TARGET TISSUE	PRINCIPAL ACTION
1,25-dihydroxyvitamin D_3	Sterol	Intestine, bone, kidney	↑ Ca^{2+} absorption in gut, synergistic with PTH
Erythropoietin	Glycoprotein	Bone marrow	↑ Erythrocyte formation

RENAL HORMONES

The kidneys are usually not considered endocrine glands, but they secrete several hormones that are important in regulating calcium metabolism, erythropoeisis, and salt metabolism (Table 13.15). Renin, an enzyme secreted by the kidney, is important in the regulation of sodium homeostasis because of its role in the angiotensin-aldosterone system described in Chapter 6.

1,25-Dihydroxyvitamin D_3

Vitamin D *per se* is an inert substance until a liver enzyme inserts a hydroxyl group in the 25-position of the molecule, followed by hydroxylation at the 1-position in the kidney. This product, 1,25-dihydroxyvitamin D_3 (also called 1,25-dihydroxycholecalciferol) is the active form of vitamin D that has hormonal action in the homeostatsis of calcium metabolism (see Chap. 12).

Erythropoietin

Erythropoietin is a 166 amino acid glycoprotein that is an important factor in the regulation of hematopoiesis. Erythropoietin is secreted in response to anoxia resulting from anemia, exposure to high altitude, or hypoventilation; it results in increased red blood cell production. Patients with end-stage renal failure are usually anemic because of the inability of their kidneys to secrete erythropoietin. The availability of recombinant erythropoietin has become an important therapeutic aid in the treatment of anemia in patients with end-stage renal failure.

REFERENCES

1. Wilson, J.D., and Foster, D.W. (Eds.): Williams Textbook of Endocrinology. 8th Edition. Philadelphia, W.B. Saunders, 1992.
2. Greenspan, F.S. (Ed.): Basic and Clinical Endocrinology. 3rd Edition. Norwalk, Appleton and Lange, 1991.
3. Hershman, J.M. (Ed.): Endocrine Pathophysiology: A Patient Oriented Approach. Philadelphia, Lea & Febiger, 1988.
4. Barkan, A.L., Beitins, I.Z., and Kelch, R.P.: Plasma insulin-like growth factor-I/somatomedin-C

in acromegaly: correlation with the degree of growth hormone hypersecretion. J. Clin. Endocrinol. Metab., *67*:69, 1988.

5. Braverman, L.E., and Utiger, R.D. (Eds.): Werner and Ingbar's The Thyroid: A Fundamental and Clinical Text. 6th Edition. Philadelphia, J.B. Lippincott, 1991.
6. Mallette, L.E.: Regulation of blood calcium in humans in hypercalcemia. Endocrinology and Metabolism Clinics of North America, *18*:3, 1989.
7. Snow, K., Jiang, N.-S., Kao, P.C., and Scheithauer, B.W.: Biochemical evaluation of adrenal dysfunction: the laboratory perspective. Mayo Clin. Proc., *67*:1055, 1992.
8. Adashi, E.Y.: The ovarian cycle. *In* Reproductive Endocrinology. 3rd Edition. Edited by S.S.C. Yen and R.B. Jaffe. Philadelphia, W.B. Saunders, 1991.
9. Foti, A., Kimura, S., DeQuattro, V., and Lee, D.: Liquid-chromatographic measurement of catecholamines and metabolites in plasma and urine. Clin. Chem., *33*:2209, 1987.
10. Nussbaum, S.R., Zahradnik, R.J., Lavigne, J.R., et al: Highly sensitive two-site immunoradiometric assay of parathyrin and its clinical utility in evaluating patients with hypercalcemia. Clin. Chem., *33*:1364, 1987.

REVIEW QUESTIONS

1. What are the different types of stimuli that may cause secretion of hormones?
2. What are the different mechanisms for controlling the plasma level of a hormone?
3. What is the difference between primary and secondary endocrine disease?
4. How do protein or polypeptide hormones (growth hormone or parathyrin, for example) transmit their message to target cells?
5. How do steroid hormones transmit their message to target cells?
6. How are poorly soluble hormones, such as cortisol or thyroxine, transported in plasma?
7. What hormones are involved in the regulation of plasma glucose concentrations? What effect does each of these hormones have on glucose levels?
8. What laboratory tests are commonly used for the diagnosis of thyroid disorders?
9. A patient's total thyroxine level is 3.0 μg/dL and the radioactive $^{125}I\text{-}T_3$ resin uptake is 20% (normal is 40%). What is this patient's free thyroxine index (FTI)? Is this patient euthyroid, hypothyroid, or hyperthyroid? What test could be performed to confirm this diagnosis?
10. The presence of what hormone in either urine or plasma is confirmation of pregnancy? What is the source of this hormone? What is the function of this hormone?
11. Which three hormones significantly alter the plasma levels of calcium? What is the mechanism of action of each of the hormones?

14 Clinical Toxicology

OUTLINE

I. **FUNCTIONS OF A CLINICAL TOXICOLOGY LABORATORY**

II. **PHARMACOKINETICS**

III. **THERAPEUTIC DRUG MONITORING (TDM)**
- A. Optimal time for drawing blood specimen
- B. Free-drug concentrations

IV. **TECHNIQUES OF DRUG ANALYSIS**
- A. Immunologic techniques
  - 1. Enzyme immunoassay systems
    - a. Enzyme multiplied immunoassay technique (EMIT)
    - b. Enzyme-linked immunosorbent assay (ELISA)
    - c. Enzyme immunochromatography
  - 2. Fluorescence immunoassay (FIA)
  - 3. Radioimmunoassay (RIA)
- B. Chromatography
  - 1. High-performance liquid chromatography (HPLC)
  - 2. Gas liquid chromatography (GLC)
  - 3. Thin-layer chromatography (TLC) for drug screening
- C. Spectrophotometry
  - 1. Atomic absorption spectrophotometry (AAS)

V. **TECHNIQUES FOR TDM**
- A. Chloramphenicol in serum by HPLC
  - 1. Reagents and materials
  - 2. Procedure
- B. Lithium by flame emission
  - 1. Serum lithium
    - a. Principle
    - b. Reference values
    - c. Increased concentration

VI. **TECHNIQUES FOR DRUG OVERDOSE TESTING**
- A. Ethanol in serum
  - 1. Principle
  - 2. Interpretation
- B. Salicylate in serum
  - 1. Principle
  - 2. Reagents
  - 3. Procedure
  - 4. Interpretation
- C. Acetaminophen in serum
  - 1. Principle
  - 2. Reagents
  - 3. Procedure
  - 4. Interpretation
- D. Barbiturates in serum
  - 1. Principle
- E. Drug screening by TLC

VII. **HEAVY METALS**

OBJECTIVES

After reading this chapter, the student will be able to:

1. Describe the important physiologic and nonphysiologic factors that can affect the blood level of an administered drug.
2. Understand the rationale for therapeutic drug monitoring and the circumstances in which such monitoring is advisable.
3. Define the term "$t_{1/2}$ of a drug" and describe how this concept is useful in managing drug therapy.
4. Understand what is meant by "free drug" concentration and know in what situations its measurement is most useful clinically.
5. Describe the most commonly used techniques of drug analysis and the principles upon which they rest.
6. Understand the analytical principles involved in the determination of some of the drugs commonly associated with overdosage (accidental or planned), such as ethanol, salicylate, acetaminophen, and barbiturate.
7. Describe a technique of drug screening (identification) by thin-layer chromatography.

We are surrounded in our daily lives by a many chemicals that may provide relief from disease and suffering but may produce toxic effects if more than certain limited amounts enter our bodies. The chemicals may be in the medicine chest (prescription drugs or over-the-counter remedies), kitchen (detergents and cleaning compounds), garden area (insecticides), or workplace (industrial chemicals and solvents). When we are sick, physicians prescribe and administer many different medications that could have adverse effects under certain conditions; many people take over-the-counter drugs without medical supervision. Even in our social lives, we may consume large amounts of alcohol and some persons experiment with or become addicted to various drugs of abuse. Accidental or deliberate exposure to toxic levels of drugs or chemicals occurs frequently.

FUNCTIONS OF A CLINICAL TOXICOLOGY LABORATORY

The main efforts of the modern clinical toxicology laboratory are devoted to therapeutic drug monitoring, which is the measurement of the serum concentration of a wide spectrum of medications to achieve optimum concentrations to benefit patients. Advances in analytical technology and the commercial availability of reliable and convenient kits for rapid assay of many therapeutic agents have made such monitoring possible. The list of monitored drugs includes such groups as sedatives, tranquilizers, anticonvulsants, antihypertensives, antibiotics, antiarrhythmics, cardiac stimulants, antidepressants, antiasthmatics, and others (Table 14.1); the list is growing steadily. In addition, clinical laboratories may provide screening assays for patients arriving in the emergency room with possible drug overdose, and more recently, some laboratories offer screening assays for drugs of abuse in urine and for exposure to environmental toxins. The clinical toxicology laboratory may have four functions:

1. *Therapeutic drug monitoring (TDM):* The laboratory measurement of serum drug concentrations enables the physician to adjust and optimize the dosage on an individual basis. (Refer to the book by Baer and Dito[1] for a review of TDM. Consult toxicology books by Moffat[2] and Baselt and Cravey[3] for methodologies and the books by Gibaldi[4] or Winter[5] for pharmacokinetics.)
2. *Identification of drugs in acute intoxication.*[6,7] When poisoning is suspected, the laboratory attempts to identify the offending drug or drugs to help to establish the diagnosis, to assess the level of intoxication, and, in a few instances when antidotes or specific treatments are available, to suggest the course of therapy. In general, effective antidotes are few, and clinical treatment relies heavily on supportive measures.

   Drug identification in the comatose patient can be a difficult problem without clues; the number of possible poisons is enormous and identification may be tenuous. The laboratory can "play the odds," however, by setting up analyses for the 8 or 10 drugs involved in 80 to 90% of the overdosage cases. These drugs include ethanol, tricyclic antidepressants, acetaminophen, salicylates, barbiturates, benzodiazepines, phenothiazines, opiates, and cocaine.[8] Nonisotopic immunoassays, such as fluorescence polarization immunoassay (FPIA) or enzyme multiplied immunoassay technique (EMIT) (see Chap. 4), are widely used for rapid identification of intoxicants (Table 14.2).
3. *Urine testing for drugs of abuse:*[9] During the last several years, some hospital laboratories have begun to provide urine screening for drugs of abuse (Table 14.2). Demand for such testing is increasing because of a growing concern over the use of illicit or banned drugs in the workplace. Drug testing by urinalysis has been implemented in many industries as a pre-employment condition for new applicants, as well as for current employees under specified conditions, for the military, in methadone or other drug rehabilitation programs, for parolees, and for some athletes (for example, in the Olympic games and the National Collegiate Athletic Association). Immunoassays and thin-layer chromatography are most frequently used for screening urine samples (Table 14.2). All drugs detected with these screening techniques must be treated as "presumptive" positive results. Because of the potential legal consequences, all such "presumptive" results *must be confirmed by a more specific method;* at present, gas chromatography/mass spectrometry (GC/MS) is considered the most conclusive method of confirming the presence of drugs in urine. Only when the GC/MS confirmation agrees with the initial screening test should a result of "positive" for that drug be issued.

   Most toxicology laboratories in hospitals are concerned with the care of patients and not with medicolegal aspects. This purpose greatly simplifies

TABLE 14.1
DRUGS FREQUENTLY MONITORED IN SERUM (CONCENTRATIONS ARE EXPRESSED IN µg/mL [mg/L] EXCEPT WHERE OTHERWISE INDICATED)

DRUG NAME		SERUM CONCENTRATION	
GENERIC	PROPRIETARY	THERAPEUTIC	POTENTIALLY TOXIC
Analgesics, anti-inflammatory			
Acetylsalicylic acid (salicylate)	Aspirin	150–300 (anti-inflammatory)	>350
Acetaminophen	Tylenol	10–20	>300 or $t_{1/2}$ > 4h
Anti-asthmatic, anti-apnea			
Theophylline	Aminophylline*	10–20 (asthma) 5–15 (newborn, apnea)†	>25
Caffeine		5–15 (newborn, apnea)	
Anticonvulsants			
Carbamazepine	Tegretol	4–10	>10
Free-carbamazepine		(25% of total carbamazepine level)	
Ethosuximide	Zarontin	40–100	>150
Phenobarbital	Luminal	15–40	>40
Phenytoin	Dilantin	10–20	>20
Free-phenytoin		1–2 µg/mL (10% of total phenytoin level)	
Primidone‡	Mysoline	5–15	>15
Valproic acid	Depekane	50–100	
Free-valproic acid		(variable; 10–30% of total valproic acid level)	
Antineoplastic			
Methotrexate		Depends on treatment protocol	
Antibiotics			
Aminoglycosides			
Amikacin	Amikin	15–25, peak; <5, trough	>30, peak
Gentamicin	Garamycin	5–12, peak; <2, trough	>12, peak; >2, trough
Tobramycin	Nebcin	5–12, peak; <2, trough	>12, peak; >2, trough
Chloramphenicol	Chlormycetin	10–20, peak; 5–15, trough	>25
Vancomycin	Vancocin	20–40, peak; 5–10, trough	>80, peak; >20, trough
Cardioactive			
Antiarrhythmics			
Disopyramide	Norpace	2–5	>7
Lidocaine	Xylocaine	1.5–5	>9
Procainamide/	Pronestyl	4–10	>12
N-acetylprocainamide§		6–20	
Quinidine	Kinidin	2–5	>6
Digitoxin	Digitaline	13–25 ng/mL	
Digoxin	Lanoxin	1–2 ng/mL	>2 ng/mL
Psychoactive			
Antidepressants			
Amitriptyline/ Nortriptyline§	Elavil Aventyl	120–250 ng/mL, amitriptyline + nortriptyline	>1000 ng/mL, amitriptyline + nortriptyline
Imipramine/ Desipramine§	Tofranil Pertofrane	150–250 ng/mL, imipramine + desipramine	>1000 ng/mL, imipramine + desipramine
Doxepin/ Desmethyldoxepin§	Sinequan	150–300 ng/mL, doxepin + desmethyldoxepin	>1000 ng/mL, doxepin + desmethyldoxepin
Lithium		0.3–1.3 mmol/L	>2 mmol/L
Immunosuppressants			
Cyclosporine	Immuran	(by HPLC)‖ 100–200 ng/mL, trough	>600 ng/mL

*Aminophylline is a mixture of theophylline and ethylenediamine; only theophylline is measured.
†Theophylline is metabolized to caffeine in a newborn (but not in a child or adult); both drugs should be monitored in the newborn.
‡Primidone is metabolized to phenobarbital; both drugs should be monitored.
§This compound is an active metabolite of the drug above it.
‖Serum concentrations of cyclosporine measured by RIA may be up to two times higher than those measured by HPLC because of cross-reactivity with cyclosporine metabolites in the RIA procedure.
**These assays are performed on an ultrafiltrate of serum.

TABLE 14.1
(continued)

SERUM-HALF-LIFE $t_{1/2}$ (ADULT)	METHODS FOR ASSAY							SI CONVERSION FACTORS µg/mL TO µmol/L (OR * ng/mL TO nmol/L)
	FPIA	EMIT	aca	RIA	HPLC	GLC	OTHER	
3hr for single dose	X		X				Colorimetric	7.24
<4hr for single dose	X	X	X				Colorimetric	6.62
4–12 hr 30 hr (newborn)	X	X	X		X			5.55
100 hr (newborn)		X			X			5.15
10–25 hr	X	X	X		X			4.23
	X**	X**			X**			
40–60 hr	X	X	X			X		7.08
49–120 hr	X	X	X		X			4.31
Variable; > 22 hr	X	X	X		X			3.96
	X**	X**			X**			
6–8 hr	X	X	X		X			4.58
8–15 hr	X	X	X			X		6.93
	X**	X**			X**			
2–15 hr	X	X					Enzymatic	2.20
0.5–15 hr	X	X	X	X				1.72
0.5–15 hr	X	X	X	X				2.09
0.5–15 hr	X	X	X	X				2.14
2–7 hr		X			X			3.09
4–10 hr	X	X						0.69
4.5–9 hr	X	X			X			2.95
1–3hr	X	X			X			4.27
3–5 hr	X	X			X			4.25
6–10 hr	X	X			X			3.60
6–7 hr	X	X			X			3.08
6–8 days								1.31
40 hr	X	X	X	X				*1.28
17–40 hr		X			X			*3.61
18–93hr		X			X			*3.80
9–24 hr		X			X			*3.57
12–24 hr		X			X			*3.75
8–36 hr					X			
14–33 hr						X	Flame photometry; atomic absorption; ion selective electrode	—
11 hr, renal transplant 7 hr, heart transplant				X	X			*0.83

TABLE 14.2
DRUGS AND TOXINS FREQUENTLY ENCOUNTERED IN ACUTE INTOXICATION AND DRUG ABUSE SCREENING

DRUG OR DRUG CLASS	PROPRIETARY OR STREET NAME	METHODS FOR ASSAY						
		FPIA	EMIT	RIA	TLC	GLC	GC/MS	OTHER
Acetaminophen	Tylenol	S	S		U	S,U		Colorimetric
Alcohols								
Ethanol	Ethyl alcohol	S	S,U			S,U		Enzymatic
Ethylene glycol						S		
Isopropanol	Rubbing alcohol					S		
Methanol	Wood alcohol					S		
Amphetamine/ methamphetamine	Speed	U	U	U	U	U	U	
Anabolic steroids	e.g., Testosterone and related compounds					U	U	
Barbiturates		S,U	S,U	U	S,U	S,U		Spectrophotometric
Amobarbital	Amytal				S,U	S,U	U	
Pentobarbital	Nembutal				S,U	S,U	U	
Phenobarbital	Luminal	S	S		S,U	S,U	U	
Secobarbital	Seconal				S,U	S,U	U	
Benzodiazepines		S,U	S,U	U	S,U	S,U	S,U	
Chlordiazepoxide	Librium				S,U	S,U	S,U	
Diazepam	Valium				S,U	S,U	S,U	
Carbon monoxide								Co-oximetry
Cocaine (metabolites)	Crack, Coke	U	U	U	U	U	U	
Iron		S						Colorimetric
Lysergic acid diethylamide	LSD			U				
Methadone	Dolophine	U	U	U	U	U	U	
Methaqualone	Quaalude		U		U	U	U	
Nicotine	Tobacco				U	U	U	
Cotinine (nicotine metabolite)					U	U	U	
Opiates		U	U	U				
Diacetylmorphine	Heroin					U	U	
Morphine						U	U	
Codeine						U	U	
Organophosphates	e.g., Malathion, Parathion							Cholinesterase inhibition
Phencyclidine	PCP, angel dust	U	U		U	U	U	
Phenothiazines	e.g., chlorpromazine				U	U	U	
Propoxyphene	Darvon	U	U			U	U	
Salicylates	Aspirin	S			U	U	U	Colorimetric (Trinder)
Tetrahydrocannabinol metabolites	THC, marijuana, pot	U	U	U	U		U	
Tricyclic antidepressants	See Table 14.1	S	S				S	

S = Serum; U = Urine

the problem of specimen handling because, in cases that come to court, the laws of evidence are strict concerning the custody of the specimen from the moment it is obtained until the analysis is made (that is, a documented chain-of-custody). An analysis is not admitted into evidence when an opposing counsel can successfully challenge the identity of the sample. Textbooks on forensic (used in court) toxicology or an experienced toxicologist should be consulted before engaging in toxicology work that has legal implications.

To attain the highest quality and accuracy in drug abuse testing, the National Institute for Drug Abuse (NIDA) has initiated a certification program for laboratories performing work for federal agencies and the military. To be certified by NIDA, a laboratory must pass a rigorous inspection and participate in an on-going, blinded proficiency testing program. Poor performance during an inspection or in proficiency testing can result in loss of certification.

In addition to testing for illicit drugs, certain groups (for example, insurance companies) may request that their applicants be screened for the use of certain therapeutic drugs (such as those used to treat hypertension and diabetes) and for such substances as cotinine (a nicotine metabolite, Table 14.2), which indicates tobacco use. Although insurance companies desire such information to assess actuarial risk, the potential for invasion of an individual's privacy is obvious.

4. *Testing for toxins from environmental or occupational exposure:* The use of myriad toxic chemicals in modern industry and agriculture has resulted in increased unintentional exposure to humans. Examples of environmental and occupational contamination from pesticides, heavy metals, and organic solvents (including oil spills) are reported all too frequently in the news. Some clinical laboratories are now offering their services to detect and monitor such mishaps. Of particular concern is the recent report concerning the effects of low blood lead levels in children and the effect lead has on their development. The concern has prompted the Centers for Disease Control (CDC) to issue new guidelines lowering the permissible blood lead level in children from 25 μg/dL to 10 μg/dL. The clinical significance of this new lower limit and the means by which it can be monitored are currently being examined (see Heavy Metal section).

PHARMACOKINETICS[4,5]

The aim of drug administration is for the patient to rapidly attain the therapeutic effect without deleterious side reactions. In general, the therapeutic effects of a drug, as well as its toxicity, correlate better with the serum drug concentration than with the drug dosage. Hence, serum drug levels must be monitored to accommodate individual variation and achieve the optimum effect.

The important physiologic factors that affect the blood level of an administered drug are *absorption* of the drug into the blood; *distribution* or equilibration of the drug with body tissues; *metabolism* (biotransformation) of the circulating drug, primarily by the liver; and *excretion* of the drug, usually by the kidney. Genetic factors are often responsible for individual differences, but for any person, the processes change with time as a result of maturation or aging, organ disease, diurnal variation, and drug interactions. In addition, nonphysiologic factors, such as differences in dosage regimen, differences in drug formulation, medication errors, and patient compliance, also influence the processes. Many patients do not follow directions and may take the drug erratically, or take too much or too little. The immature organ systems of the pediatric population also must be borne in mind. Patients with liver or renal disease present problems because of impairment of drug metabolism and excretion; the usual drug dose may reach toxic levels when malfunction of the liver or kidney exists.

THERAPEUTIC DRUG MONITORING (TDM)

The administration of all drugs has a potential for hazards and side reactions; optimal drug concentration in blood can be assured only by TDM. Furthermore, the margin between therapeutic and toxic levels is small with some medications (see digoxin, Table 14.1) and requires monitoring on a regular basis. TDM helps to provide an estimate of patient compliance in taking the medication as directed. Therapeutic and toxic concentrations of several drugs are listed in Table 14.1.

TDM is necessary for drugs with a low therapeutic index (the ratio between the minimum toxic and maximum therapeutic serum concentrations) that cannot be monitored by clinical observation or other laboratory tests. For these drugs, the therapeutic range should be well defined and a reliable assay must be available. TDM should be considered under the following circumstances:

1. The serum concentration should be determined after initiation of therapy or when the dosage regimen or drug formulation has been changed. This procedure is usually done after the drug has reached a steady-state concentration (see the following).
2. In neonates and infants, more frequent monitoring is necessary because renal and hepatic functions change rapidly during development.
3. Drug levels should be monitored during illness, especially when the illness involves impairment of renal, hepatic, gastrointestinal, or cardiovascular function, because significant changes in serum concentration may occur.
4. Monitoring is indicated whenever the medication is unexpectedly ineffective or toxic.
5. Dosage adjustment may be required when a new drug is added to the patient's regimen because of possible drug interactions.

A new class of therapeutic agents, the antiretroviral drugs, has been created during the last decade. These drugs (such as zidovudine, also known as azidothymidine or AZT) were developed to treat HIV infection by specifically preventing the causal retroviruses, HIV-1 and HIV-2, from replicating. Besides AZT, two other similar drugs, dideoxycytocine (ddC) and dideoxyinosine (ddI), are now approved by the Food and Drug Administration (FDA) for treating HIV infection. Extensive pharmacokinetic studies with these drugs in humans have not shown any significant correlation of blood levels and corresponding therapeutic effects. Hence, at present, therapeutic monitoring of these drugs is not needed, except for research.

Optimal Time for Drawing Blood Specimens

The concept of the plasma half-life ($t_{1/2}$, the time necessary for the serum concentration of a drug to decrease by ½) helps to explain the peaks and troughs obtained in the serum concentration of a drug during a steady-state period (see Table 14.1 for the $t_{1/2}$ of specific drugs). The trough occurs just before the administration of a dose, whereas the peak occurs during the period of maximal absorption. The steady state is achieved when the drug intake balances the drug inactivation and excretion so that an average constant level between peaks and troughs is maintained.

Drug monitoring is of maximum benefit when blood samples are drawn from a patient who is at the steady-state concentration, that is, one who has received regular maintenance doses of the drug for about five half-lives ($5 \times t_{1/2}$) of the drug. For most orally administered drugs, the most consistent results are obtained if the sample is taken during a trough or predose period (that is, just before the next dose). The trough level is the lowest drug concentration during the dosing interval. Blood may be drawn at the peak or postdose period (the highest drug concentration during the dosing interval) in patients who exhibit toxic symptoms shortly after receiving the drug. The time for orally administered drugs to reach peak concentrations may range from 1 to 5 hr after administration and depends on many variables, including drug type, formulation (tablet, capsule, or solution), and the patient's state of fasting and gastrointestinal motility. When patients receive their drug by intravenous infusion or intramuscular injection, generally the best times to draw peak blood samples are 30 min after the infusion has stopped and about 60 min after the intramuscular injection.

Hundreds of different drugs are administered to patients daily, but the bulk of the monitoring (80 to 90%) is confined to fewer than 20 drugs. Table 14.1 lists the drugs most commonly assayed in 1993 by most hospitals. The number will increase, of course, as new methods become available to perform the assays rapidly. The most recent drug to require monitoring is cyclosporine, which is given to organ transplant recipients to suppress their immune systems and prevent rejection.

A particular drug usually can be assayed by several of the techniques described in the following section, but time is a factor in drug monitoring. The physician wants the results on hospitalized patients sometimes within a few hours to modify the regimen; in outpatient clinics, the results may be desired within 30 min, while the patient is still present. Hence, the most widely used technique is nonisotopic immunoassay (for example, FPIA or EMIT), which can be performed within several minutes. Commercially available kits exist for all of the drugs listed in Table 14.1 except lithium and several of the tricyclic antidepressants. Fluorescent immunoassay can compete with enzyme immunoassay for rapidity of performance. The principles for both of these techniques are outlined in a following section. No details are given because the manufacturers' directions must be followed for each different drug.

Free-Drug Concentrations

Many drugs circulating in the blood bind reversibly to serum proteins, particularly albumin. For some drugs, the degree of binding is significant, and only 5% of the drug may exist in the free-drug form. Only the free-drug form can diffuse out of the bloodstream (the drug-protein complex is too large) to produce therapeutic effects in other tissues. This situation is similar to that for the thyroid hormone thyroxine, or T_4. Only the free T_4 (T_4 not bound to serum proteins) can produce the biologic effects.

Changes in the concentration of albumin (for example, in renal or hepatic failure or in pregnancy) or the presence of competitive ligands for drug binding sites on albumin (for example, in chronic renal failure) can cause significant changes in free-drug level, whereas the *total drug* may still be in the therapeutic range. Drug assays listed in Table 14.1 measure the total-drug concentration in the blood; measurement of the free-drug concentration requires separation of the free drug from protein-bound drug using a process such as ultrafiltration. Currently, free-drug levels are measured clinically for only three drugs—phenytoin, carbamazepine, and valproic acid (Table 14.1). Although free-drug concentrations may be preferable to total-drug concentrations (just as use of free T_4 is preferable to use of total T_4), free-drug measurements are still technically difficult and more time consuming and expensive.

TECHNIQUES OF DRUG ANALYSIS

The techniques for identifying and measuring the concentration of small amounts of drugs in such a complex protein matrix as serum are varied and sometimes intricate. No single instrument or technique can be used for all the different types of drugs to be measured. A large medical center laboratory may use many, if not all, of the following chromatographic, immunoassay, and spectrophotometric techniques to perform many various drug analyses within a reasonable period.

Immunologic Techniques (see Chapter 4)

The immunologic techniques include enzyme immunoassay (EIA), fluorescence polarization immunoassay (FPIA), fluorescence immunoassay (FIA), and radioimmunoassay (RIA).

Enzyme Immunoassay Systems

Enzyme Multiplied Immunoassay Technique (EMIT, Syva). This homogeneous system is widely used in the United States (Table 14.1). The drug to be measured is the hapten part of an antigen to which antibody has been prepared.

The pure drug is labeled by covalently linking it to an enzyme. This technique, more rapid than RIA, consists of first incubating the serum with a buffered mixture containing limiting amounts of antibody, a small amount of enzyme-labeled drug, substrate, and cofactors for the enzyme. Enzyme activity is measured kinetically in a spectrophotometer for less than 1 min; the drug concentration is obtained from a standard curve in which enzyme activity is plotted against the drug concentration. Many of the EMIT tests have been repackaged for use on a variety of laboratory instruments.

Enzyme-Linked Immunosorbent Assay (ELISA). The ELISA system has been used extensively for the identification of microbiologic agents but is now applied to drug analyses. The ELISA system is a heterogeneous assay, for the antibody is bound to a solid-state carrier (membrane, beads, or some other surface) instead of being free in solution. The drug to be measured is also the hapten of the antigen against which the antibodies have been prepared. The principle is the same as that described for the EMIT system, but one extra step is necessary—the separation of the bound drug from the unbound. In the tests for digoxin and digitoxin (Stratus, American Dade), the antibody is bound to a membrane; thus, the separation step of bound from unbound merely consists of washing the membrane by applying substrate solution.

Enzyme Immunochromatography[10] AccuLevel, a noninstrumented method for therapeutic drug monitoring in the physician's office, was introduced by the Syntex Corporation. The assay is based on the principle of immunochromatography (Fig. 14.1). For the assay of the anticonvulsant drug phenytoin (Dilantin), a small amount of the patient's blood (10 μL) is added to a vial containing a solution of the enzyme glucose oxidase and a trace amount of phenytoin attached to the enzyme horseradish peroxidase (phenytoin-HRP). A small stick with antibody against phenytoin immobilized on it is inserted into the vial. The phenytoin from the patient's blood and the attached phenytoin-HRP migrate up the stick; the greater the amount of phenytoin in the patient's sample, the farther up the stick the phenytoin and the phenytoin-HRP will migrate. When the solvent front has reached the top of the stick, the stick is removed from the vial and immersed in a tube containing a solution of glucose and a chromogenic substrate (4-chloro-1-naphthol) for the phenytoin-HRP. A blue color is produced on the stick as high as the phenytoin-HRP has migrated. A numeric scale etched on the side of the stick converts the height of the color into phenytoin concentration units.

Fluorescence Immunoassay (FIA)

Approaches to the application of FIA to the measurement of serum analytes were described in Chapter 4. To recapitulate briefly, all the systems use antibodies to react with the antigen (drugs or other substances). The systems differ in their use of fluorophors, the molecules that fluoresce when irradiated with light of the proper wavelength.

At present, the most widely used procedure for monitoring therapeutic drugs is FPIA (TDx, Abbott Laboratories, Table 14.1). The principle of this procedure

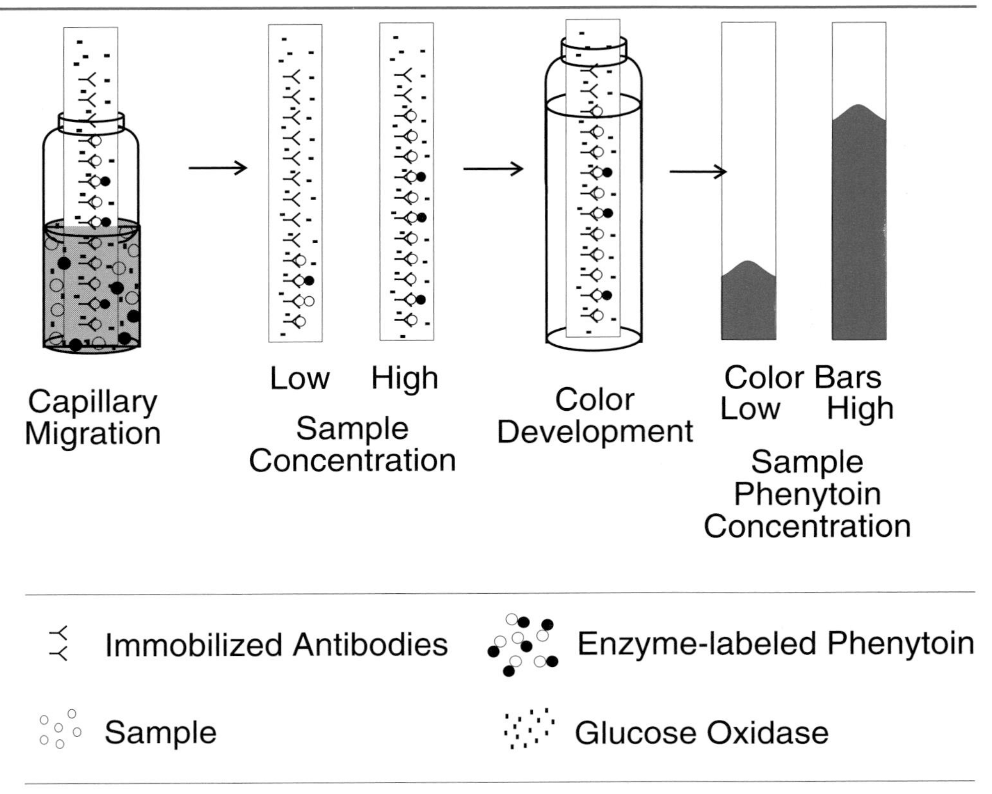

FIG. 14.1. Schematic representation of enzyme immunochromatography. (From Opheim, K.E., Statland, B.E., Tillson, S.A., and Litman, D.J.: Ther. Drug Monit., 9:190, 1987.)

has been described in Chapter 4. In another system (Ames), the drug is covalently bound to an enzyme substrate (umbelliferyl-β-D-galactoside) that is nonfluorescent. The fluorogenic drug reagent (FDR) plus antibody plus β-galactosidase are incubated with a serum sample. The drug and FDR compete for binding; the amount of unbound FDR varies directly with the serum drug concentration. The enzyme splits the unbound FDR in the homogeneous reaction mixture and the fluorescence is measured. The system is capable of assaying several antibiotics and theophylline in serum.

The EMIT system can also be used in a fluorescent mode by measuring the fluorescence of NADH instead of its absorbance at 340 nm. This approach greatly increases the sensitivity of the assay and represents an extension of their EIA system.

Radioimmunoassay (RIA)

This technique is not frequently used in toxicology. The drug to be measured is covalently bound as a hapten to a protein (for example, bovine serum albumin) and antibodies to it are produced by repeated injections into an animal. The antibody is able to recognize and react with the hapten (drug), even in a complex mixture of proteins. The principle of a RIA test for a drug is as described in Chapter 4, namely the following: (1) incubating serum with a limiting amount of antibody plus radioactively labeled drug; the serum and labeled drug compete for antibody binding sites; (2) separating the free from the antibody-bound drug; (3) counting the radioactivity in either the free or the antibody-bound fraction; and (4) calculating the amount of drug in serum from a standard curve of % antibody-bound drug versus drug concentration. The method is cumbersome compared to automated enzyme immunoassay methods.

Chromatography (see Chapter 3)

Three types of chromatography, high-performance liquid chromatography (HPLC), gas liquid chromatography (GLC), and thin-layer chromatography (TLC), are commonly used in a toxicology laboratory. HPLC and GLC are used primarily for quantitating serum drug levels in TDM but may also be used for confirming drug identification. TLC is primarily a screening technique for drug identification. All three techniques are based on the adsorption of drug to a solid support and elution by means of a mobile, liquid phase.

High-Performance Liquid Chromatography (HPLC)[11]

A sample or extract is injected onto a chromatographic column, eluted from the column at high pressure (up to 6000 lb/in^2) with an appropriate mobile phase, detected, recorded, and quantitated by peak height or by curve area integration. This procedure has become a versatile, useful technique for drug analysis. An example of an HPLC assay for measuring serum concentrations of the toxic antibiotic chloramphenicol is given at the end of this section.

Gas Liquid Chromatography (GLC)

GLC usually involves solvent extraction, concentration of extract, conversion to a volatile derivative (derivatization), injection into a gas chromatograph, elution from the column, and detection and quantitation. Positive identification of peaks may require the coupling of the gas chromatograph with a mass spectrometer (MS).

Thin-Layer Chromatography (TLC) for Drug Screening

This screening procedure is efficient and useful for drug identification. The process requires a solvent extraction, concentration of the extract, application of concentrate to a TLC plate, separation of the components by means of a mobile phase, visualization of the separated drugs by chemical reactions, light absorbance, or fluorescence, and identification by comparison of R_f values and colors with known standards. A drug screening procedure is given at the end of this chapter.

Spectrophotometry

Tentative identification of some drugs may be made by means of a spectral scan. When the identity of the drug is known, quantitative analysis may be performed by spectrophotometry after protein precipitation or extraction. The measurement may be made by the following:

1. *In the visible spectrum:* Measurement is made after the conversion of the drug into a colored compound. The determination of salicylate concentration is an example.
2. *In the ultraviolet spectrum:* Serum barbiturates may be quantitated by measuring the differential ultraviolet absorption curve at pH 10.3 and 13.5 from 320 nm to 220 nm.
3. *By fluorescence:* Quinidine is an example of a drug with a concentration in a protein-free filtrate of serum that can be measured readily by fluorescence.

Atomic Absorption Spectrophotometry (AAS)

The serum concentration of many elements can be determined readily by AAS. The AAS determination most frequently requested is for the TDM of serum lithium, the carbonate salt of which is used in the treatment of manic-depressive psychosis. Flame emission photometry could also be used for lithium analysis, but the precision is better with AAS. AAS is well suited for the determination of some heavy elements in serum, particularly copper and zinc, but the determination of the concentration of lead in blood is best done with a graphite furnace or rod attachment to the atomic absorption spectrophotometer.

TECHNIQUES FOR TDM

Chloramphenicol in Serum by HPLC[12]

Chloramphenicol (Fig. 14.2) is a potent and toxic broad-spectrum antibiotic. Its only valid medical indication is for the treatment of life-threatening bacterial or

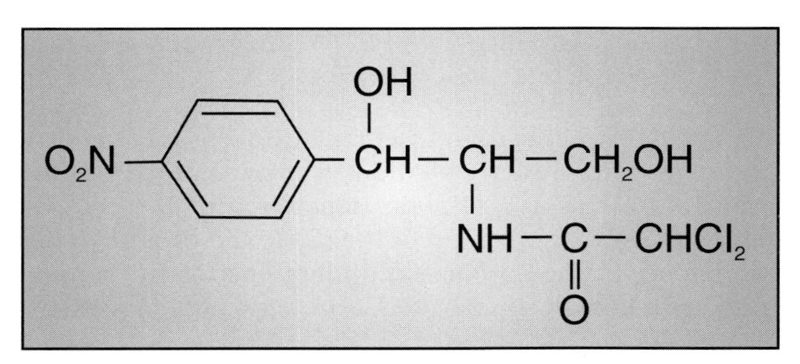

FIG. 14.2. Structural formula of chloramphenicol.

rickettsial infections for which no other less toxic antibiotic is available. It is frequently used in pediatrics to treat meningitis caused by *Hemophilus influenzae* that is resistant to ampicillin, an increasingly common problem in many hospitals. Toxicity from chloramphenicol is revealed as aplastic anemia, "gray-baby syndrome" (caused by inhibition of the mitochondria's ribosomes), and reversible bone marrow suppression. These adverse reactions are related to the serum concentration of chloramphenicol and may be minimized by maintaining a peak serum concentration between 10 and 20 μg/mL and a trough level between 5 and 15 μg/mL.

A widely used method to measure serum concentrations of chloramphenicol is by HPLC, and an example of this procedure is given in the following section. An EMIT assay for this drug has been introduced, offering more laboratories a means to assay this drug.

Reagents and Materials

1. High-performance liquid chromatograph (Waters Associates), with sample injector, solvent delivery system, ultraviolet absorbance detector (variable wavelength), C_{18} column, dual pen recorder, and integrator. Flow rate: 2.5 mL/min.
2. Millipore filtration system with 0.22-μm filters.
3. High-speed centrifuge with Eppendorf polypropylene centrifuge tubes.
4. SMI pipets, 25 and 50 μL.
5. Methanol, HPLC grade (Burdick and Jackson).
6. Stock sodium acetate buffer, 2.0 mol/L. Dissolve 16.4 g sodium acetate (anhydrous; reagent grade) in water and make up to 100-mL volume.
7. Working acetate buffer, 0.02 mol/L, pH 3.5. Dilute 20 mL of stock buffer to 2.0 L with water. Adjust to pH 3.5 with glacial acetic acid.
8. Mobile phase. Combine 450 mL of working buffer with 350 mL methanol. Filter through Millipore system. Discard after 1 week.
9. Stock standards:
   a. Chloramphenicol (Sigma Chemical Co.). Prepare a 1.00 mg/mL solution in methanol.
   b. *p*-Nitropropionanilide.

      Prepare by combining 10 mg of *p*-nitroaniline (Aldrich, N985-3), 50 μL of pyridine, and 100 μL of propionic anhydride (Aldrich, P5147-8) in a small vial with a Teflon-lined cap. The mixture is heated at 37° C overnight. The vial is cooled and 1 mL of methanol is added. This solution is evaporated to dryness at 37° C under a stream of air. The residue, consisting of about 14 mg of *p*-nitropropionanilide, is dissolved in 280 mL of methanol to give a final *p*-nitropropionanilide concentration of 50 μg/mL. (*Note:* All work with pyridine and propionic anhydride must be done in a well-ventilated hood.)

      Both of these solutions are stable for 1 year at −20° C.
10. Working standard, chloramphenicol, 20 μg/mL.

    Prepare by adding 200 μL of stock standard chloramphenicol and 10 mg sodium azide to calf serum and making up to a final volume of 10.0 mL with calf serum. Store in 1-mL aliquots at −20° C. Stable for 1 year frozen or 1 month if stored at 0° C after each use.

11. Internal standard, *p*-nitropropionanilide, 50 μg/mL in methanol, as prepared in step No. 9b.

Procedure

1. Pipet 25 μL of working standard, controls, and patients' samples into 500-μL Eppendorf tubes with an SMI pipet.
2. Pipet 50 μL of internal standard/methanol solution into each tube with an SMI pipet.
3. Vortex tubes for about 5 sec.
4. Centrifuge at 7000 × g for 2 min.
5. Inject 25 μL of supernate into the chromatograph. The retention times for chloramphenicol and internal standard are approximately 5.0 and 9.0 min, respectively (Fig. 14.3).

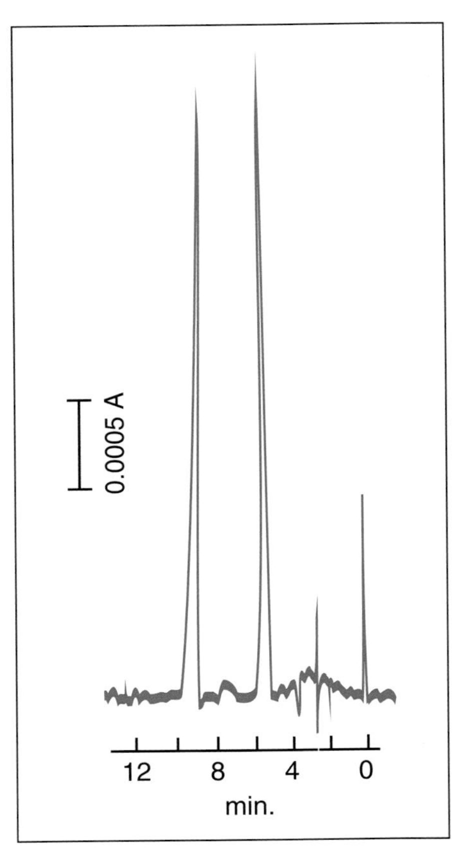

FIG. 14.3. HPLC chromatogram, measured at 280 nm, of an aqueous solution of chloramphenicol (peak at 5 min.) and internal standard (peak at 9 min.). Chloramphenicol concentration is 20 μg/mL; retention times in min are measured on the x-axis and absorbance units are indicated on the y-axis.

6. Calculation (from absorbance at 280 nm):

Chloramphenicol, μg/mL

$$= \frac{\text{peak height sample}}{\text{peak height IS}} \times \mu\text{g/mL S} \times \frac{\text{peak height IS}}{\text{peak height S}}$$

where IS = internal standard and S = standard chloramphenicol solution.

Note: Variations in the extraction or chromatographic processes are automatically corrected by the use of these two ratios. The first ratio of peak heights of sample to internal standard adjusts for variations in the run. The second ratio of peak heights of internal standard to standard chloramphenicol corrects for variation from run to run or day to day.

Lithium by Flame Emission

Lithium is an element in the same periodic series as sodium and potassium. Its salts are widely distributed on earth but in small quantities; even though lithium salts are ubiquitous, their concentrations in river and well water are usually low. Consequently, the concentration of lithium compounds in plant or animal tissue is also small.

Lithium carbonate (Li_2CO_3) is of value in the treatment of manic-depressive psychosis; it has a calming effect on the manic stage and is frequently given as a therapeutic measure to forestall possible attacks. Lithium seems to decrease the concentration of neurotransmitters at synapses by promoting their reabsorption.

Because the concentration of Li^+ is toxic at a plasma concentration of about 2 mmol/L or greater, the serum Li^+ level of patients receiving this drug is usually monitored weekly during the early phase of treatment. A serum concentration of 5 mmol/L can be lethal. Therapeutic levels are usually considered to be from 0.3 to 1.3 meq/L or mmol/L.

Serum Lithium

The concentration of serum Li^+ is readily measured by flame emission photometry, ion selective electrode, or by AAS.[13] The flame emission method is described below.

Principle. Serum is diluted 1:50 with a diluent containing cesium or potassium as an internal standard. When the serum is aspirated into a hot flame, some of the lithium atoms are forced into an activated state by thermal collisions; on returning to the ground state, these atoms emit light at the characteristic wavelength for lithium, 671 nm. The emitted light of lithium is electronically compared to that given off by the internal standard.

Reference Values
Therapeutic levels: 0.3 to 1.3 mmol/L

Li^+ is not measurable in the serum of individuals who are not taking any lithium medication.

Increased Concentration. Toxic levels have been reported as low as 2.0 mmol/L. The physician should be notified if serum levels exceed 2.5 mmol/L.

TECHNIQUES FOR DRUG OVERDOSE TESTING[6,7]

Procedures for the analysis of ethanol, salicylate, acetaminophen, and barbiturate are detailed because accidental or deliberate overdose with these drugs is relatively common. The screening for various classes of drugs by TLC is also given.

Ethanol in Serum

Ethanol can be measured easily by gas chromatography, by diffusion into potassium dichromate in acid solution and oxidation to acetic acid, and by action of the enzyme alcohol dehydrogenase. Analysis by gas chromatography is a good method if one has a chromatograph dedicated to this purpose because the presence of other volatile alcohols (methanol, isopropanol) can be determined at the same time. The dichromate method is not specific for ethanol because it reacts with other volatile alcohols, if present. The enzymatic method is described because it is simple, rapid, and reasonably specific.[14]

Principle. Ethanol is selectively oxidized to acetaldehyde by yeast alcohol dehydrogenase while NAD^+ is reduced to NADH in the process. The reaction is forced to completion by chemically trapping the acetaldehyde by means of amino-oxyacetic acid. The alcohol concentration is proportional to the absorbance of NADH at 340 nm.

Interpretation. If no alcohol has been ingested, the serum alcohol levels should be 0 to 10 mg/dL. For legal purposes, some states in the United States and many foreign countries consider a blood level of 80 mg/dL or greater as evidence of intoxication, that is, slower reaction time and impairment of visual acuity. Other states have set the legal definition of intoxication as 100 mg/dL.

Concentrations between 100 and 150 mg/dL correlate with greater uncoordination, slower reaction time, slurring speech, perhaps difficulty in balance, and general central nervous system depression. Concentrations over 400 mg/dL may be lethal.

Salicylate in Serum[15]

Salicylates are present in numerous over-the-counter medicines, particularly in cold remedies and analgesics; the most common salicylate is aspirin. Despite the mandatory use of child-proof caps for aspirin bottles, salicylates are still a major cause of accidental poisoning in children. Overdose with salicylate

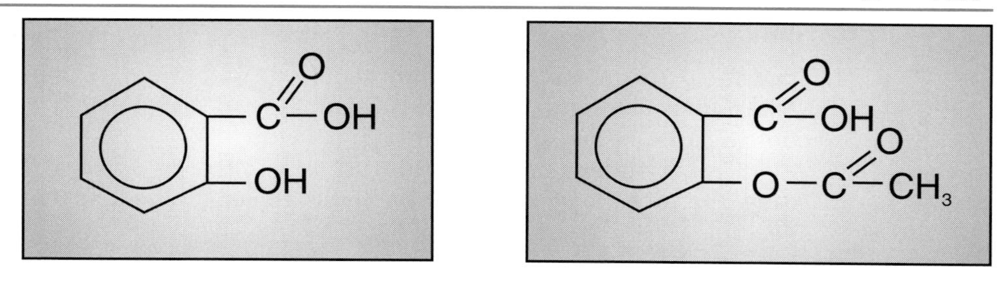

FIG. 14.4. Structural formulas of salicylic acid and aspirin. Aspirin is converted into salicylic acid by body esterases.

causes hyperventilation at first as the respiratory center in the brain is stimulated; the hyperventilation produces a respiratory alkalosis that is characterized by a lowered serum total CO_2 concentration and a low pCO_2 (see Chap. 5). This later (2 to 5 hr) changes to a metabolic acidosis as salicylate anion begins to accumulate; the serum CO_2 content is still low, but the concentration of pCO_2 rises to nearly normal levels.

Aspirin is acetylsalicylic acid, the compound that is produced by acetylating the hydroxyl group of salicylic acid (Fig. 14.4). Because the acetyl group of aspirin is hydrolyzed by esterases shortly after the drug is ingested, partially in the gastrointestinal tract and the rest in tissues, the active form of aspirin in plasma and tissues is salicylate. Hence, the analysis is made for salicylate and not acetylsalicylate (aspirin).

The medical uses of aspirin are antipyretic (reduce fever), analgesic (reduce the feeling of pain), and anti-inflammatory (reduce the pain and inflammation of rheumatoid arthritis).

Principle. A serum sample is treated with an acid solution of $HgCl_2$ and $Fe(NO_3)_3$. The mercuric salt precipitates the serum proteins while the Fe^{3+} reacts with the phenolic group of salicylic acid to form a violet-colored complex. After centrifugation, the absorbance of the supernatant fluid is measured at 540 nm.

Reagents

1. Trinder's reagent. Dissolve 40 g $HgCl_2$ in about 800 mL water. Add 120 mL 1 mol/L HCl and 40 g $Fe(NO_3)_3 \cdot 9H_2O$, and when completely dissolved, make up to 1 L volume with water. The solution is stable indefinitely at room temperature.
2. Standard salicylate, 400 μg/mL. Dissolve 46.4 mg sodium salicylate (equivalent to 40.0 mg salicylic acid) in about 80 mL water and dilute to 100 mL volume with water. Add a few drops of chloroform as a preservative. The solution is stable for 6 months at 4° C. A low standard of 100 μg/mL is prepared by diluting 1 volume of above standard with 3 volumes of water.

Procedure

1. Transfer 0.20 mL each of serum, standard, and water to tubes labeled "Unknown," "Standard," and "Blank," respectively. Always run a control serum.
2. Add 1 mL Trinder's reagent to all tubes and mix.
3. Allow the tubes to stand for 5 min and then centrifuge for 10 min at high speed.
4. Transfer the supernatant fluid to a microcuvet and read the absorbance against the blank at 540 nm.
5. Calculation:

$$\frac{A_u}{A_s} \times 400 = \mu\text{g/mL of salicylate}$$

In SI, μg/mL × 7.24 = μmol/L.

Interpretation

1. No salicylate should be in the serum of individuals who are not receiving the drug. Trinder's reagent does react slightly with some nonsalicylate compounds in serum to give a serum blank that varies from about 5 to 12 μg/mL salicylate equivalent.
2. *Therapeutic levels:* A serum concentration of up to 300 μg/mL (2170 μmol/L) may be reached in treating patients with rheumatoid arthritis. The concentration is usually lower in other conditions.
3. *Toxic levels:* A ringing sensation in the ears and hyperventilation may occur at serum concentrations of 300 to 400 μg/mL (2170 to 2900 μmol/L). In accidental poisoning or suicide attempts, the serum concentration may vary from 500 to 1200 μg/mL, depending on the amount of overdose and the elapsed time since ingestion.

Acetaminophen in Serum[16]

Acetaminophen (Tylenol, Fig. 14.5) is a common over-the-counter drug that is present in many medications to relieve pain or to reduce the extent of fever. It is of low toxicity when taken in the usual therapeutic dose, but an overdose

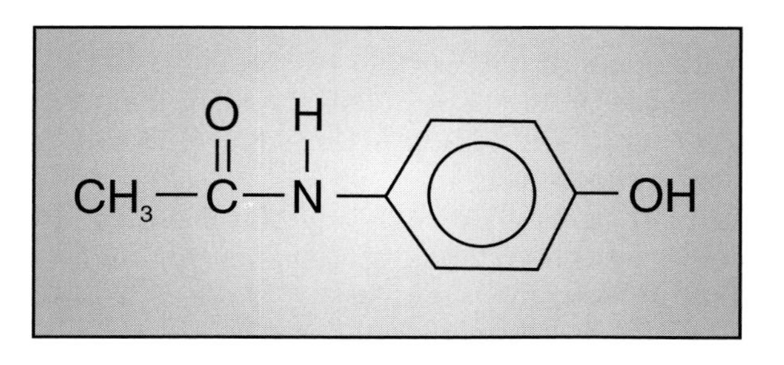

ACETAMINOPHEN

FIG. 14.5. Structure of acetaminophen.

produces severe liver damage. The common causes of an overdose are accidental ingestion of large quantities, particularly by children, and suicide attempts. The assay is of value for determining whether or not acetaminophen has been ingested, for assessing the level of overdose, and for deciding whether or not an antidote is required.

Principle. Acetaminophen in protein-free serum is converted into 2-nitro-5-acetaminophenol by mild nitration with nitrous acid ($NaNO_2$ + HCl). The nitroderivative turns yellow in alkali, and its absorbance at 430 nm is measured.

Reagents

1. Stock standard acetaminophen, 1.00 mg/mL. Transfer 100.0 mg acetaminophen (Aldrich reagent grade) to a 100-mL volumetric flask, dissolve in water, and make up to volume.
2. High working standard, 150 μg/mL. Dilute 15.0 mL of stock standard to 100 mL volume with serum or plasma devoid of acetaminophen.
   Low working standard, 50 μg/mL. Dilute 5.0 mL of stock standard to 100 mL volume with negative serum or plasma.
3. Sodium nitrite, $NaNO_2$, 100 g/L. Place 10.0 g $NaNO_2$ in a 100-mL volumetric flask, dissolve in water, and make up to volume.
4. Ammonium sulfamate, $NH_4NH_2SO_3$, 150 g/L. Place 15.0 g of ammonium sulfamate in a 100-mL volumetric flask, dissolve in water, and make up to volume.
5. NaOH, 500 g/L. Place 50 g NaOH in a 250-mL beaker. Dissolve in 50 mL water while stirring. (Caution: solution becomes hot.) When cool, transfer to a 100-mL volumetric cylinder and make up to volume with water.
6. HCl, 6 mol/L. In a hood, add 255 mL concentrated HCl to 245 mL water in a liter bottle. Mix and stopper.
7. Trichloroacetic acid (TCA), 100 g/L. Place 10 g TCA in a beaker containing 50 mL water. When dissolved, make up to 100 mL volume with water.

Procedure

1. Transfer 1 mL of high and low working standard, patient serum, and water, respectively, into 125 × 25 mm test tubes, appropriately labeled.
2. Add 2 mL TCA solution to each.
3. Vortex and let stand 5 min. Centrifuge at high speed.
4. Quantitatively transfer the supernatant to clean, labeled test tubes.
5. Add 0.5 mL 6 mol/L HCl to each tube.
6. Add 0.5 mL $NaNO_2$ to each tube. Agitate by hand and wait 2 min. Tubes containing acetaminophen turn a faint yellow color.
7. Slowly add 0.5 mL ammonium sulfamate solution to each tube. The mixture fizzes as excess HNO_2 is decomposed.
8. Add 0.5 mL NaOH solution and mix. The yellow color intensifies.
9. Centrifuge at high speed to remove bubbles.
10. Zero the spectrophotometer with the reagent blank at 430 nm. Read absorbance of samples. The cuvets must be bubble-free and read immediately after pouring. Do not use a flow-through cuvet.

11. Calculation:

$$\text{Acetaminophen, } \mu g/mL = \frac{A_u}{A_s} \times 150$$

where A_u and A_s are absorbances of unknown and high standard, respectively.

Notes:

1. Report values of 10 μg/mL or less as negative for acetaminophen.
2. If the patient value is greater than 200 μg/mL, dilute the colored specimen with an equal volume of reagent blank, read the absorbance, and repeat the calculation after multiplying absorbance × 2.

Interpretation

Toxic Level: The following serum concentrations are toxic and certain to produce severe liver damage, which may not be evident for several days:

1. If the serum level exceeds 300 μg/mL 4 hr after drug ingestion.
2. If the serum half-life exceeds 4 hr.
3. If the half-life exceeds 10 hr, hepatic coma should be expected.

Antidote: The administration within 12 hr after drug exposure of N-acetyl-cysteine, a sulfhydryl compound that replaces glutathione for detoxifying acetaminophen metabolites. The antidote causes gastrointestinal irritation, and neurologic disturbances in some people, however, so it is not given unless clearly warranted.

Barbiturates in Serum

A barbiturate is a derivative of barbituric acid that has been developed for its sedative properties. Barbiturates are general depressants taken most frequently for preventing seizures (phenobarbital) or for inducing sleep (short- and intermediate-acting barbiturates). Barbiturates account for a large percentage of the adult patients treated in a general hospital for toxicity because of accidental or intentional (suicidal) overdose; these drugs are responsible for many deaths each year.

The barbiturates are divided into four groups based on their pharmacologic action: long-acting (phenobarbital, barbital, and others), intermediate-acting (amobarbital, butabarbital), short-acting (secobarbital, pentobarbital), and ultrashort-acting (thiopental and others). The toxicity (serum concentration at which coma is induced) of the different barbiturates varies with the type; the fastest-acting barbiturates cause coma at the lowest serum level, and the critical serum concentration is highest for the longest-acting compounds (Table 14.3). The ultrashort-acting barbiturates are not prescribed for patients, because they are used almost exclusively for anesthesia; they are rarely involved in problems of overdose. The serum concentrations causing coma are about 10, 30, and 55 μg/mL (10, 30, and 55 mg/L), respectively, for the short-, intermediate-, and long-acting barbiturates.

TABLE 14.3
RELATION BETWEEN A PATIENT'S CLINICAL CONDITION AND THE BARBITURATE CONCENTRATION* IN HIS/HER BLOOD

	CLINICAL STAGES†				
BARBITURATE	1	2	3	4	5
Amobarbital	7.0	15.0	30.0	52.0	66.0
Pentobarbital	4.0	6.0	15.0	20.0	30.0
Phenobarbital	10.0	34.0	55.0	80.0	150.0
Secobarbital	3.0	5.0	10.0	15.0	20.0

From Manual of Analytical Toxicology. Edited by I. Sunshine. Copyright The Chemical Rubber Co., Cleveland, 1971.
*Concentration is expressed in μg per mL of blood, which is identical with mg/L. Each figure represents the concentration of barbiturate at which the average nontolerant patient enters into a given clinical stage.
†The clinical conditions of the patients were grouped into five arbitrary categories on the basis of the following criteria:
Stage 1: awake, competent, and mildly sedated.
Stage 2: sedated, reflexes present, prefers sleep, answers questions when roused, does not cerebrate.
Stage 3: comatose, reflexes present.
Stage 4: comatose, reflexes absent.
Stage 5: comatose, circulatory difficulty, and/or depression of the respiratory center in the medulla.

The barbiturates are inactivated in the liver by a series of enzymatic steps that may include oxidation of substituted radicals, removal of N-alkyl groups if present, removal of sulfur from thiobarbiturates, and opening of the barbital ring. The drug metabolites and a portion of the unchanged barbiturate are excreted in the urine.

Barbiturate concentrations in serum may be measured by FPIA or EMIT, and by GLC. GLC has an advantage in that it can identify the type of barbiturate from its retention time, and quantitate at the same time; it is, however, more labor-intensive than are the immunoassays. In addition, a spectrophotometric method that depends on the differential ultraviolet absorption at two different pH levels is presented.[17] (See Figures 14.6 and 14.7.)

Principle. Barbiturates are quantitated by automatic differential spectrophotometry. Barbiturates have spectra in 0.45 mol/L NaOH (approximately pH 13.5) different from those at pH 10.3. Typical spectral curves, together with a differential spectral curve (automatically subtracting the absorbance at pH 10.3 from that at 13.5) are shown (Figs. 14.6 and 14.7). The differential spectrum of most barbiturates has a maximum absorbance at 260 ± 5 nm and a minimum absorbance at 240 ± 5 nm; the difference between the two peaks is used for quantitation by comparing the differences between the unknown and a standard.

Drug Screening by TLC

Cases of suspected drug overdose in comatose patients or of possible drug abuse are the most frequent situations that require the identification of drugs in body fluids. Drug screening for overdose or abuse can be accomplished using

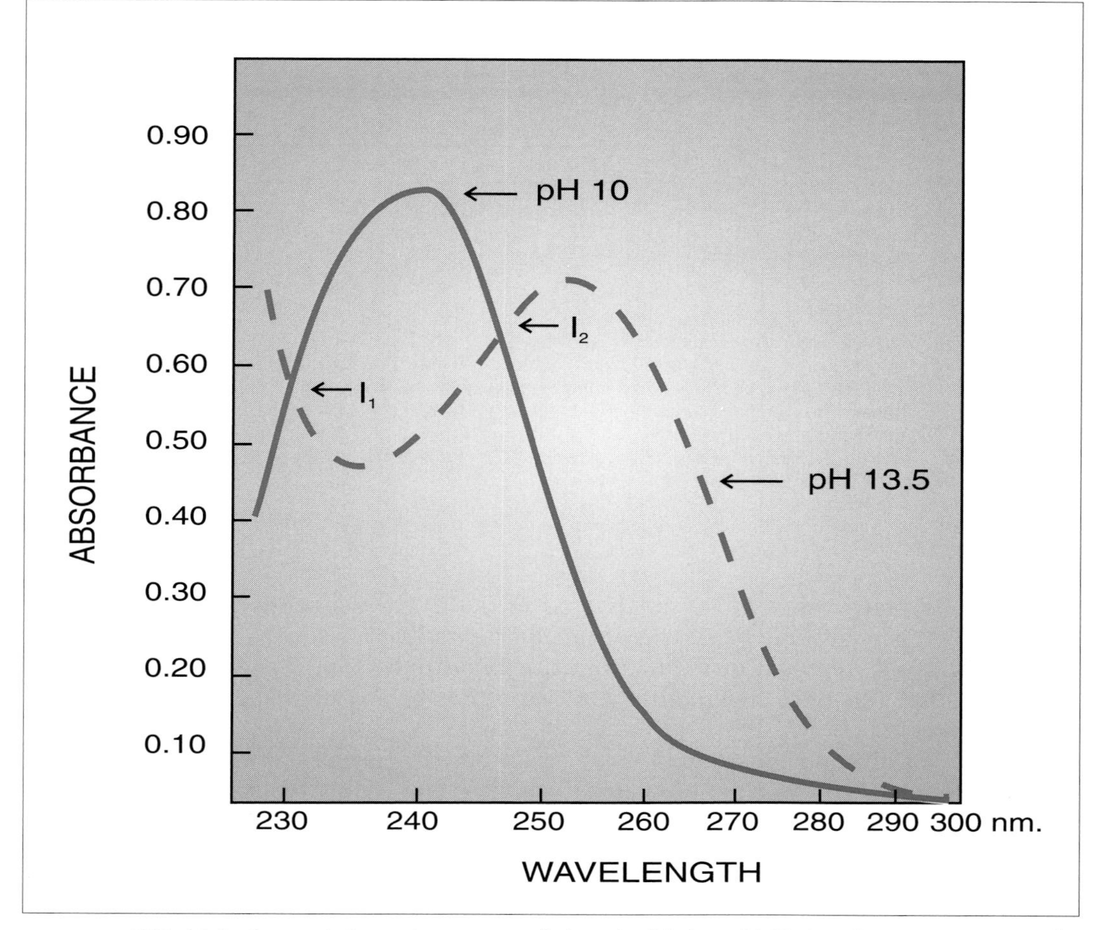

FIG. 14.6. Spectral absorption curves of phenobarbital at pH 13.5 and at pH 10.3. I_1 and I_2 are isobestic points, the places where the two curves intersect. (Modified from Manual of Analytical Toxicology, edited by I. Sunshine. Copyright The Chemical Rubber Co., Cleveland, 1971.)

GC, GC/MS, EMIT, FPIA, or TLC. Screening by TLC is an efficient, rapid, and economical way of identifying the offending drug(s) in such situations. Urine is the best specimen for analysis; gastric fluid is good for drugs taken orally but should be diluted before analysis.

The many different TLC separation systems are all based on the same general principles. The drugs are separated from urine by first adjusting the urine to a desired pH and then extracting the urine with an appropriate organic solvent. Next, the extracted drugs are separated by TLC in two different solvent systems. The separated drugs on the chromatogram are visualized by several techniques: by their fluorescence when examined under ultraviolet light, by their colored derivatives when sprayed with or dipped into particular chemical solutions, and by their absorbance of ultraviolet light.

In the past, many laboratories prepared their own TLC systems by making or purchasing the individual components, such as thin-layer plates, migrating

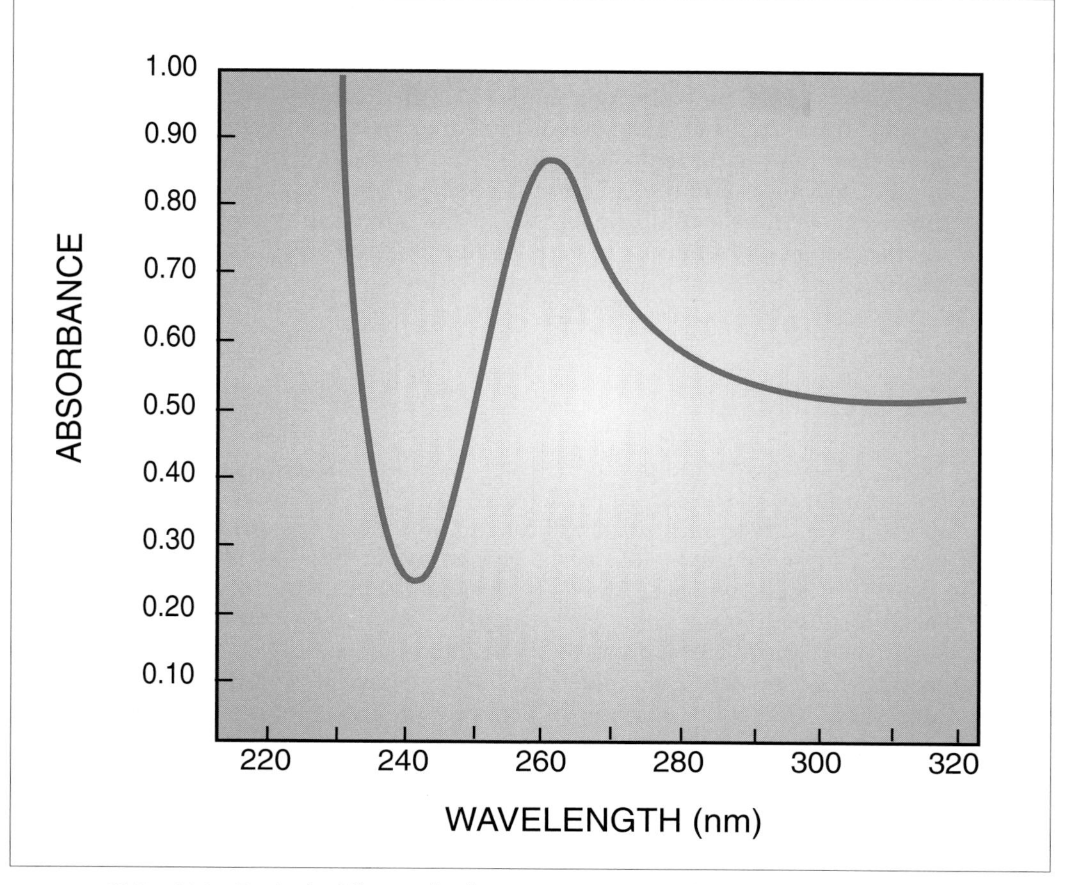

FIG. 14.7. Typical differential absorption spectrum between pH 10.3 and 13.5 for secobarbital.

solvents, drug standards, and developing solutions. Although success could be achieved by experienced laboratories, little consistency existed between laboratories. Currently the most widely used TLC system is a commercially available kit called Toxi-Lab (TOXI-LAB, Inc.). The Toxi-Lab system has standardized the critical steps of the TLC procedure, thereby resulting in more accurate and consistent results than those that were achieved with the previous "home-made" TLC systems.

The procedure starts by performing extraction of drugs from the urine using two types of prefilled (with appropriate buffers and organic extraction solvents) extraction tubes. Tube "A" is optimized to extract basic (opiates, amphetamines, antidepressants) and neutral (benzodiazepines) compounds; tube "B" is optimized to extract acidic drugs (barbiturates). After the tubes containing samples of the urine are thoroughly mixed, the organic solvent layers (containing extracted drugs) are transferred to a small, warmed evaporating plate, and the drug residues are evaporated onto a small fiberglass disk. The tiny disk is placed in a precut hole in the standardized thin-layer plate (fiberglass backing, impregnated with silica gel; Fig. 14.8). There are two types of thin-layer plates, one for the "A" extract (Fig. 14.9) and one for the "B" extract (Fig. 14.10). After the plates have developed, they are dipped in a series of color-developing solutions and examined under ultraviolet light. The standardized Toxi-Lab plates have control drug standards placed in the outside lanes flanking the patient's sample (Figs. 14.9 and 14.10). Drugs are identified by their R_f values and the characteristic series of color changes they go through during the color-developing process. The Toxi-Lab system has been documented to detect hundreds of drugs. The sensitivity of the method is about 0.5 to 3.0 μg/mL for most drugs present in urine.

HEAVY METALS

Toxic metals suspected in a poisoning case (for example, Cu, Zn, Pb, and Hg in body fluids or tissue) are now detected by AAS (see Chap. 3). The venerable Reinsch's test was used to screen for the presence of heavy metals (As, Sb, Hg, and Bi),[18] but the test is not satisfactory when the concentration of metal is low.

Chronic lead exposure may also be determined by its effect on the zinc protoporphyrin/heme ratio (see Chap. 12), however, this method has poor sensitivity below corresponding blood lead levels < 25 μg/dL; this procedure would be unsatisfactory to meet the CDC's new guidelines for measuring blood levels of 10 μg/dL. The most accurate method to measure lead at low concentration is still AAS.

FIG. 14.8. Placement of concentrated drugs on the fiberglass disc into the precut hole of a prestandardized Toxi-Lab chromatographic plate.

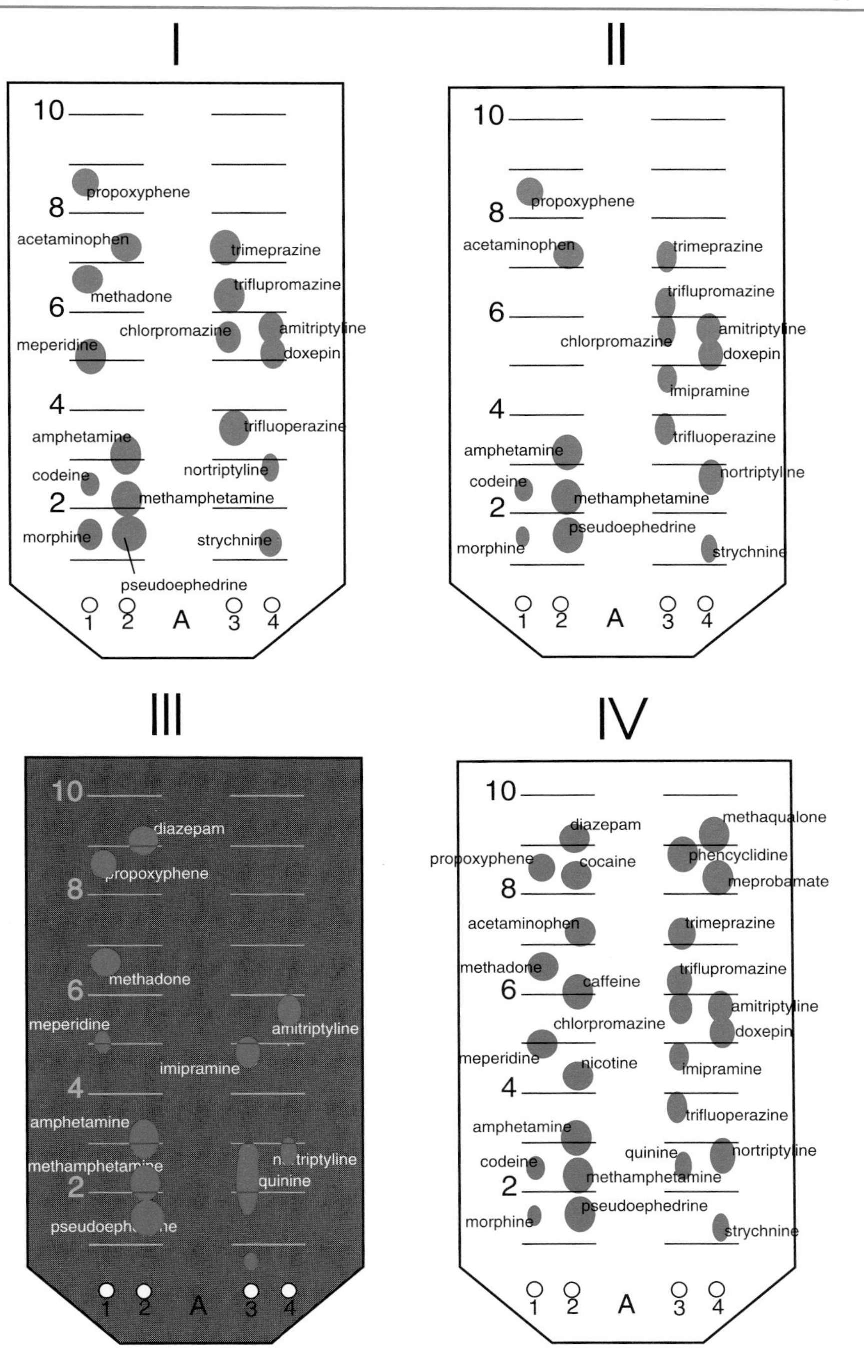

FIG. 14.9. Four stages of color development of a Toxi-Lab "A" chromatogram for detection of neutral and basic drugs. Stages I, II, and IV are after chemical treatment, and Stage III is examination under ultraviolet light.

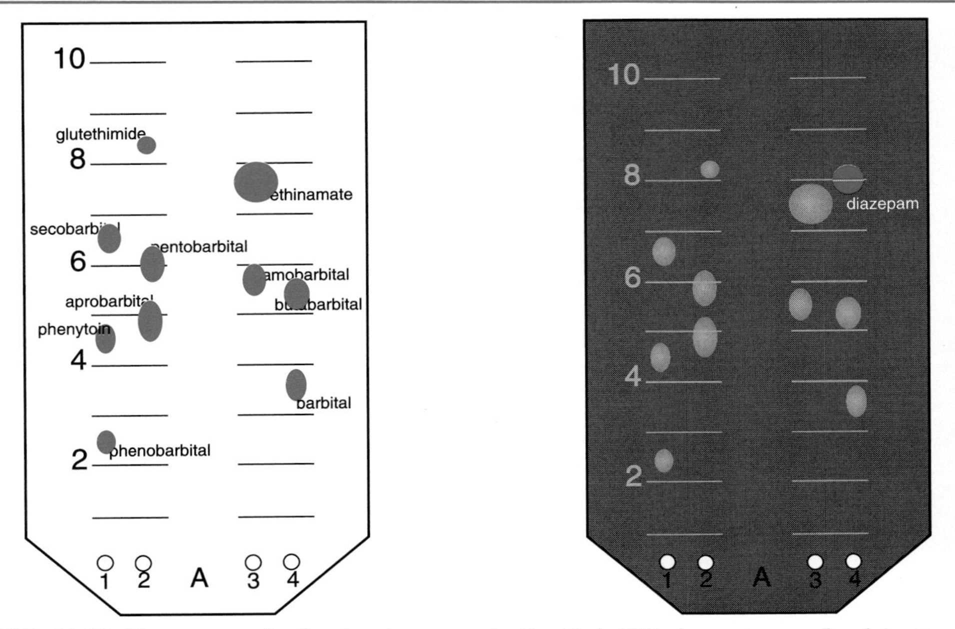

FIG. 14.10. Two stages of color development of a Toxi-Lab "B" chromatogram for detection of acidic drugs. Stage I is after chemical treatment, and Stage II is examination under ultraviolet light.

REFERENCES

1. Baer, D.M., and Dito, W.R. (Eds.): Interpretations in Therapeutic Drug Monitoring. Chicago, American Society of Clinical Pathologists, 1981.
2. Moffat, A.C. (Ed.): Clarke's Isolation and Identification of Drugs. 2nd Edition. London, The Pharmaceutical Press, 1986.
3. Baselt, R.C., and Cravey, B.S.: Disposition of Toxic Drugs and Chemicals in Man. 3rd Edition. Chicago, Year Book Medical Publishers, 1989.
4. Gibaldi, M.: Biopharmaceutics and Clinical Pharmacokinetics. 4th Edition. Philadelphia, Lea & Febiger, 1991.
5. Winter, M.E.: Basic Clinical Pharmacokinetics. 2nd Edition. San Francisco, Applied Therapeutics, 1988.
6. Goldfrank, L.R. (Ed.): Toxicologic Emergencies, A Comprehensive Handbook in Problem Solving. 4th Edition. New York, Appleton-Century-Crofts, 1990.
7. Ellenhorn, M.J., and Barceloux, D.G.: Medical Toxicology, Diagnosis and Treatment of Human Poisoning. New York, Elsevier, 1987.
8. Litovitz, T.L., et al.: 1985 Annual Report of the American Association of Poison Control Centers National Data Collection System. Am. J. Emerg. Med., *4*:427, 1986.
9. Hawks, R.L., and Chiang, C.N.: Urine Testing for Drugs of Abuse. Washington, D.C., National Institute on Drug Abuse, Research Monograph #73, Department of Health and Human Services, 1986.
10. Zuk, R.F., et al.: Enzyme immunochromatography—a quantitative immunoassay requiring no instrumentation. Clin. Chem., *31*:1144, 1985.
11. Gerson, B., and Anhalt, J.P.: High-pressure Liquid Chromatography and Therapeutic Drug Monitoring. Chicago, American Society of Clinical Pathologists, 1980.

12. Petersdorf, S.H., Raisys, V.A., and Opheim K.E.: Micro-scale method for liquid chromatographic determination of chloramphenicol in serum. Clin. Chem., *25*:1300, 1979.
13. Levy, A.H., and Katz, E.M.: Comparison of serum lithium determinations by flame photometry and atomic absorption spectrophotometry. Clin. Chem., *16*:840, 1970.
14. Jones, D., Gerber, L.P., and Drell, W.A.: A rapid enzymatic method for estimating ethanol in body fluids. Clin. Chem., *16*:402, 1970.
15. MacDonald, R.P.: Salicylates. *In* Standard Methods of Clinical Chemistry, Vol. 5. Edited by S. Meites. New York, Academic Press, 1965.
16. Walberg, C.B.: Determination of acetaminophen in serum. J. Anal. Toxicol., *1*:79, 1977.
17. Williams, L.A., and Zak, B.: Determination of barbiturates by automatic differential spectrophotometry. Clin. Chim. Acta, *4*:170, 1959.
18 Gettler, A.O., and Kaye, S.: A simple and rapid method for Hg, Bi, Sb, and As in biological material. J. Lab. Clin. Med., *35*:146, 1950.

REVIEW QUESTIONS

1. What factors affect the blood concentration of an administered drug?
2. When and under what circumstances is therapeutic drug monitoring necessary or advisable?
3. What causes peaks and troughs in the serum concentration of an administered drug?
4. How is the $t_{1/2}$ concept used in therapeutic drug monitoring?
5. What is meant by "free"-drug concentration, and in what clinical situations would its determination be most valuable?
6. What are some of the common immunologic techniques used in therapeutic drug monitoring and the principles upon which they are based?
7. How does high-performance liquid chromatography differ in principle from thin-layer chromatography?
8. How can a drug be identified using thin-layer chromatography?
9. Why is serum the specimen of choice for therapeutic drug monitoring assays?
10. Why is urine the specimen of choice for emergency toxicology and drugs of abuse assays?

15 Prenatal and Perinatal Testing

OUTLINE

I. PRENATAL TESTING
- A. Amniotic fluid
- B. Neural tube defects
- C. Down syndrome
- D. Isoimmune hemolytic disease (IHD)
  1. Materials
  2. Principle and procedure

II. ASSESSMENT OF FETAL MATURITY
- A. Lecithin/sphingomyelin (L/S) ratio
- B. Individual phospholipids
- C. Foam stability test and foam stability index (FSI)
- D. Lamellar body phospholipid
- E. Fluorescence polarization

III. GENETIC COUNSELING
- A. Examination or culture of cells in amniotic fluid

IV. POSTNATAL TESTING
- A. Congenital hypothyroidism
- B. Congenital adrenal hyperplasia
- C. Phenylketonuria
- D. Sickle cell disease

OBJECTIVES

After reading this chapter, the student will be able to:

1. Define amniotic fluid and describe its role in pregnancy.
2. Describe two conditions in pregnancy that place the fetus at risk and for which prenatal testing of amniotic fluid assists the obstetrician in assessing the degree of risk.
3. Describe the use and importance of bilirubin measurements in assessing the severity of isoimmune hemolytic disease.
4. Describe the physiologic basis for the tests of fetal maturity.
5. Describe the origin and physiologic role of α-fetoprotein in the fetus and explain how its measurement in maternal serum is of value in assessing the risk of a neural tube defect.
6. Explain why confirmatory tests of α-fetoprotein are essential.

PRENATAL TESTING

Profound hormonal and physiologic changes occur during the course of a normal pregnancy. (Some of the hormonal changes were reviewed in the section on Female Sex Hormones in Chapter 13.) Although most pregnancies in an average population are uneventful, vigilance and monitoring are required to detect abnormalities should they occur. The monitoring may vary from measuring the maternal concentration of specific hormones to analyzing certain constituents in amniotic fluid. (The concentration of maternal serum estriol (E_3) as an indicator of fetoplacental viability in late pregnancy was reviewed in the Estriol Section of Chapter 13.)

Amniotic Fluid

Amniotic fluid is the fluid contained in the amniotic sac that envelops the fetus. A portion of the fluid arises from the fetal respiratory tract, urine, the amniotic membrane, and the umbilical cord. The volume of the amniotic fluid at term usually ranges from 500 to 1100 mL. Significant water transfer occurs between the three intrauterine compartments: the fetus, the placenta, and the amniotic fluid.

The amniotic fluid has become increasingly important in the clinical chemistry laboratory as a fluid for analysis. It is obtained by amniocentesis (aspiration by hypodermic needle and syringe) and is used for screening for such congenital disorders as neural tube defects and Down syndrome; as a guide to action in isoimmune hemolytic disease; for assessing fetal maturity; and in genetic counseling. Amniocentesis is illustrated in Figure 15.1.

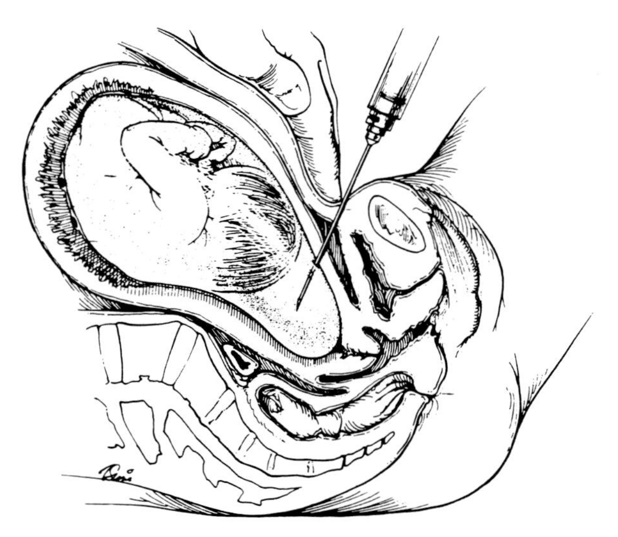

FIG. 15.1. Illustration of amniocentesis, the technique used for obtaining amniotic fluid. (Adapted from Pritchard, J.A., MacDonald, P.C., and Gant, N.F., (Eds.): Williams Obstetrics. 17th Edition. New York, Appleton and Lange, 1985.

Neural Tube Defects[1,2]

The development of the central nervous system (brain and spinal cord) in the human embryo begins early in pregnancy with the conversion of a flat plate of precursor cells into a neural tube. This conversion is initiated by the rolling up and fusion of the ends of the plate to form the neural tube, which then develops into the spinal column, its enclosed spinal cord, and the brain. If, for some reason, the normal process of development does not occur, serious malformations called neural tube defects occur. Neural tube defects are of two types, anencephaly and spina bifida. In anencephaly, the child is born with a large portion of the brain missing, and death ensues within a few days; in spina bifida, the spinal column fails to fuse completely and is left open to the environment. Children with open spina bifida have significant mortality and morbidity; approximately 60% of children so affected die within 5 years. Survivors with open spina bifida may be moderately to severely handicapped with lower limb paralysis, sensory deprivation, chronic urinary tract infections, bowel and bladder incontinence, and mental retardation. The causes of neural tube defects are unknown. The incidence of neural tube defects is approximately 1 to 2 per 1000 live births and are found in approximately 3 to 5% of spontaneously aborted fetuses.

Fortunately, laboratory tests are available that can detect the presence of neural tube defects early in pregnancy. Maternal serum screening for the presence of α-fetoprotein (AFP) is now an accepted practice among obstetricians.[3] AFP is produced by the fetal liver and appears to function as a fetal equivalent of albumin in the adult. Fetal serum AFP is excreted into fetal urine, and subsequently into the amniotic fluid, and crosses the placental barrier into the maternal circulation. Amniotic fluid AFP reaches peak concentration at approximately 13 weeks' gestation; maternal serum AFP levels are first detectable at 10 to 12 weeks' gestation and continue to increase during the course of the pregnancy (Fig. 15.2). In the presence of a neural tube defect (open spina bifida or anencephaly), AFP leaks into the amniotic fluid at an increased rate, crosses the placental-maternal barrier and is detectable in the maternal circulation at an elevated concentration. Maternal AFP testing is usually performed between 15 to 20 weeks' gestation (16 to 18 weeks being optimal). The American College of Obstetrics and Gynecology in 1985 issued a policy statement that recommended that every pregnant woman be offered maternal serum AFP testing as standard of care in obstetrics practice.[4]

The results of the AFP test are expressed in concentration and also in multiples of the median (MoM).* Results between 0.5 MoM and 2.5 MoM are usually considered to be normal (Fig. 15.3). Results less than 0.5 MoM are suggestive of Down syndrome, and results greater than 2.5 MoM are suggestive of neural tube defects and should be investigated further.

*The MoM is derived by calculating the median concentration of AFP from a reference population. The MoM is then calculated by dividing the patient's concentration by the median concentration. A MoM of 2.5 means that the patient's concentration is 2.5 times the median concentration of the reference group. The MoM is used because of the tremendous variability in results obtained with the different AFP assays that are available, thus making intra-assay comparisons difficult. If the assumption is valid that the distribution of AFP concentrations obtained with different assay methods is similar, then the comparison of MoM between assays is warranted.

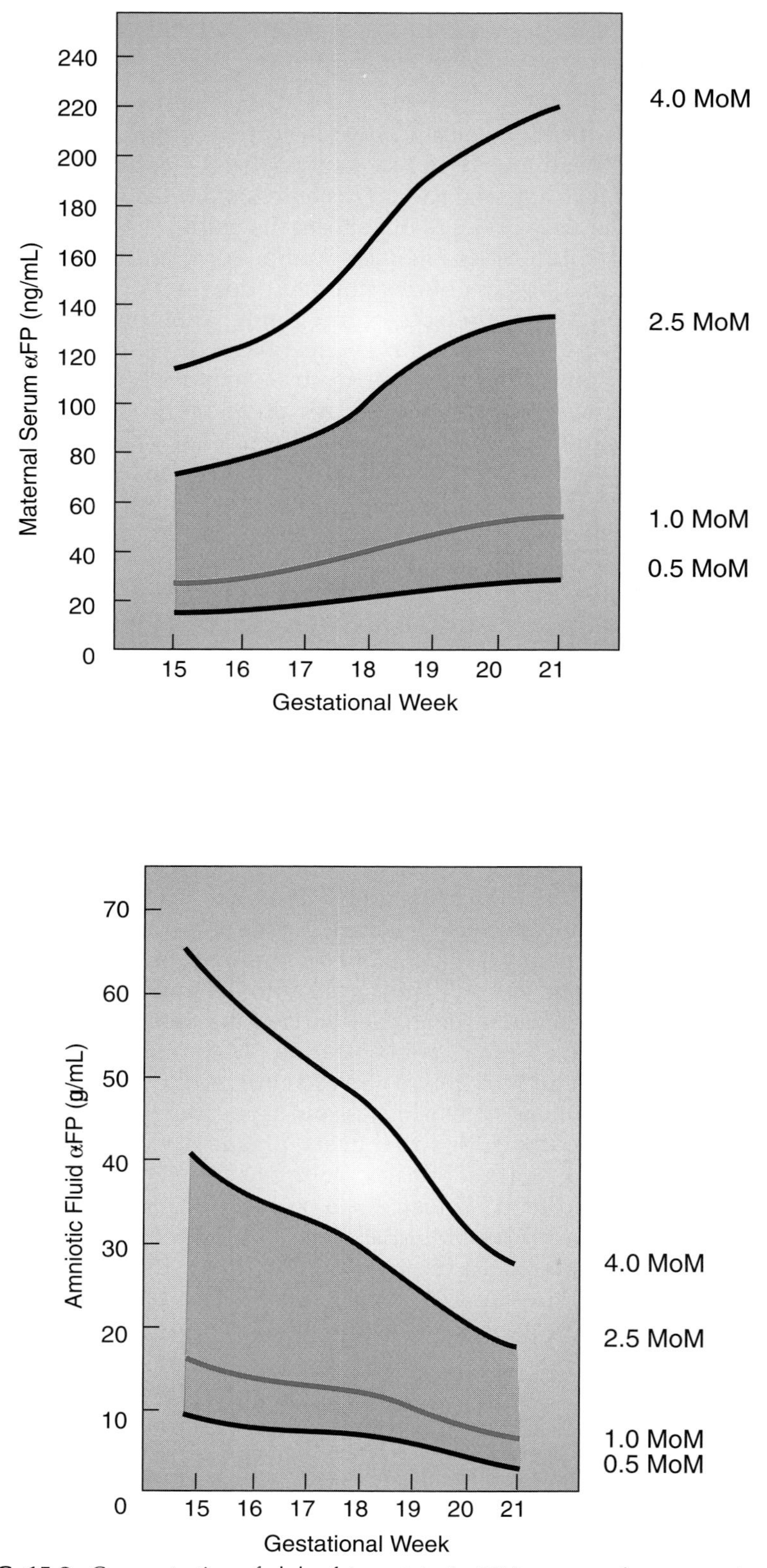

FIG. 15.2. Concentration of alpha-fetoprotein (α-FP) in maternal serum and amniotic fluid expressed as both concentration and MoM during weeks 15 to 21 of gestation. Note the difference in concentration units between maternal serum (ng/mL) and amniotic fluid (μg/mL). The shaded area between 0.5 MoM and 2.5 MoM is considered low risk for neural tube defects and Down syndrome.

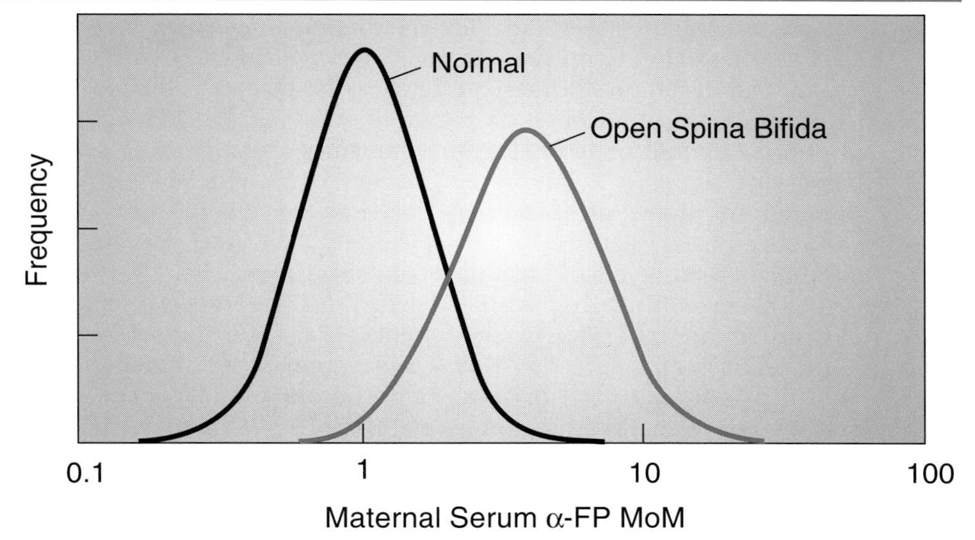

FIG. 15.3. Frequency distribution of maternal serum alpha-fetoprotein (α-FP) concentrations in unaffected and open spina bifida pregnancies, expressed in log MoM units.

Numerous conditions other than neural tube defects may result in elevated maternal AFP levels, including multiple gestation pregnancies, incorrect dates of gestation, and race (Blacks have higher levels than Caucasians); a partial list is shown in Table 15.1. Any unexplained elevation of AFP level should always be followed with additional testing. A repeat AFP test should be performed, and if still elevated, amniocentesis should be performed and an amniotic fluid AFP level determined. If the amniotic fluid AFP is elevated, the risk of a neural tube defect is high. The activity of the enzyme acetylcholinesterase (AChE) should be measured. A high AChE activity (> 5 SD above the mean) or the presence of a fast-moving isoenzyme confirms the diagnosis.[5] High-resolution ultrasound imaging then may be performed to confirm further the laboratory studies.

Down Syndrome[6]

Down syndrome is the most common fetal chromosomal abnormality. The incidence rises markedly with advancing maternal age, reaching at age 35 a risk of approximately 1:350 and at age 40 approximately 1:40. Down syndrome is

TABLE 15.1
CONDITIONS ASSOCIATED WITH INCREASES IN MATERNAL SERUM AFP

Gestational age underestimated	Other congenital neoplasms
Multiple pregnancy	Open neural tube defects
Growth retardation	Intrauterine death
Fetal distress	Maternal AFP-producing neoplasms
Gastroschisis	Hydrocephaly
Omphalocele	Turner's syndrome
Congenital nephropathies	Placental defects
Fetomaternal hemorrhage	Microcephaly

characterized by severe mental retardation and other associated physical deformities. The definitive test for Down syndrome is karyotyping (chromosomal analysis) performed on fetal cells obtained during amniocentesis. Unfortunately, amniocentesis has an approximate 1:100 risk of complications to either mother or fetus. Perhaps someday a noninvasive screening test for assessing the risk of having an affected fetus will be discovered to enable parents to make an informed decision on whether or not to undergo amniocentesis. Decreased levels of AFP (< 0.5 MoM), elevated levels of hCG, and decreased levels of unconjugated estriol (see Chap. 13) in maternal serum have been empirically associated with an increased risk of having a Down syndrome fetus, although the biologic basis for the association of these biochemical changes with Down syndrome is not understood. The test is usually performed early in the second trimester of pregnancy. The triple screen consisting of maternal serum levels of AFP, hCG, and unconjugated estriol, along with maternal age, may be used to calculate a "risk factor" for an affected Down syndrome fetus (Fig. 15.4). If the calculated risk factor from the triple screen is significantly lower than the risk based on maternal age alone, the risk of performing amniocentesis may not be warranted. On the other hand, if a significantly greater risk exists, the risks associated with amniocentesis are warranted. This strategy prevents the performance of unnecessary amniocentesis with its attendant risks, but still provides a high detection rate of affected pregnancies.[7]

Isoimmune Hemolytic Disease (IHD)

Prenatal testing of amniotic fluid is also employed to estimate the degree of fetal danger in isoimmune hemolytic disease (IHD), also known as erythroblastosis fetalis or hemolytic disease of the newborn. IHD is an isoimmune disease resulting from a blood type incompatibility between mother and fetus (see Chap. 11). Maternal blood group antibodies (IgG) cross the placenta and are capable of destroying the "foreign" fetal red cells. The risk is great for an RhD-positive fetus in an Rh-negative mother who is already sensitized to the D antigen. Destruction of fetal red cells can begin in utero and continue for several weeks after delivery because its red cell surfaces usually become coated with maternal antibodies. IHD may result from other surface antigen incompatibilities (ABO, Duffy, Lutheran, C and E of the Rh system, and others). Every type O (ABO system) individual has antibodies against type A and B surface antigens, the risk always exists that a blood type A or B fetus in a type O mother may develop IHD. Fortunately, an ABO incompatibility engenders a less severe hemolytic reaction than that caused by the D antigen in an Rh-negative sensitized mother; ABO incompatibility seldom requires an exchange transfusion.

The fetal reticuloendothelial cells phagocytize the injured red blood cells and convert the heme portion of hemoglobin to bilirubin (see Fig. 11.4); a portion of the bilirubin is excreted into the amniotic fluid as its concentration in fetal plasma rises. The fetus becomes progressively more anemic as the hemolysis continues, may develop edema, and could have respiratory problems after delivery. Fetal death in utero can occur in severe cases.

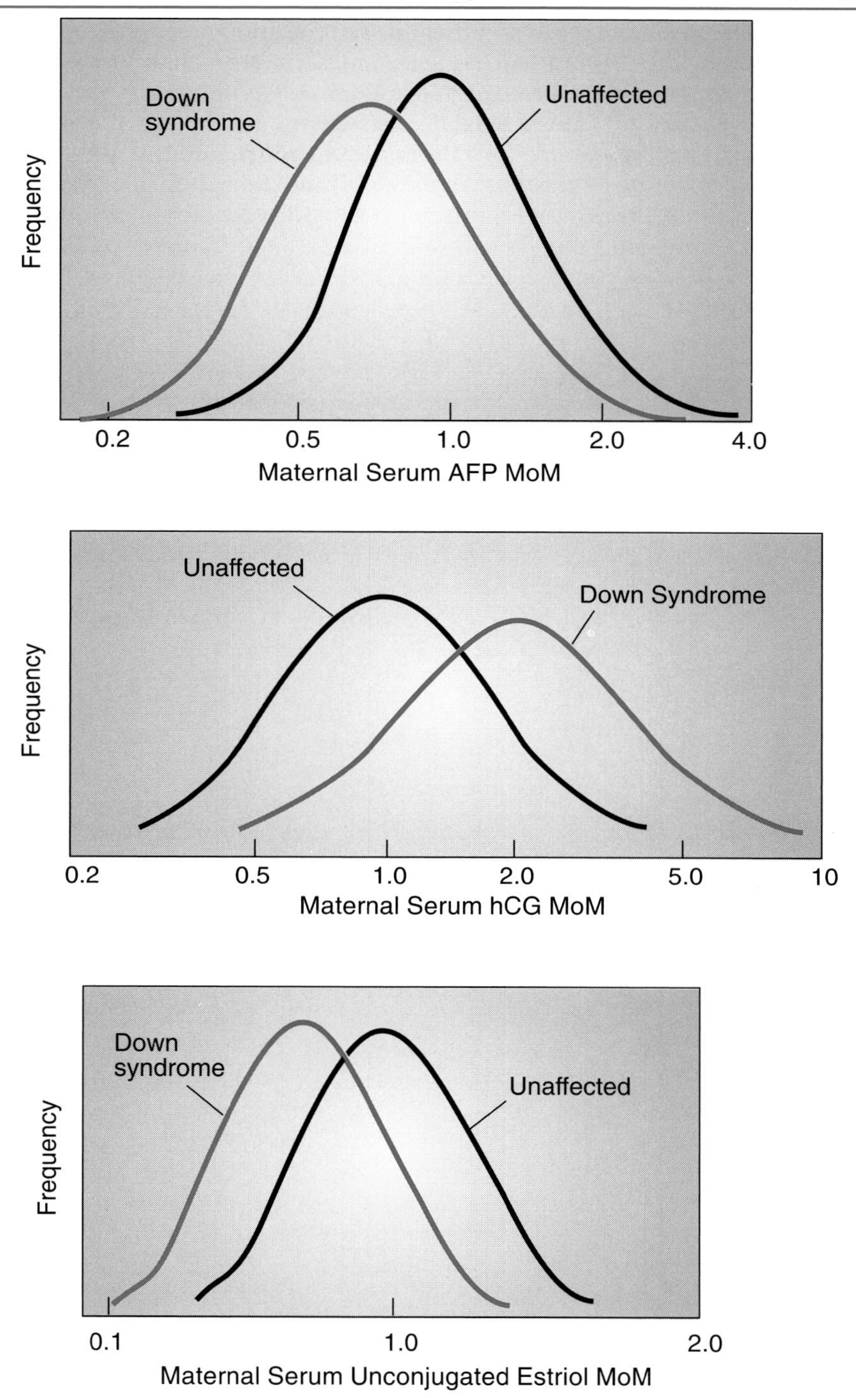

FIG. 15.4. Frequency distribution of maternal serum AFP, hCG, and unconjugated estriol in unaffected and in Down syndrome pregnancies expressed as log MoM units. The distribution of these three analytes, along with the maternal age, constitute the "triple screen" that is used to calculate the risk of a Down syndrome pregnancy.

The measurement of bilirubin in amniotic fluid is used to estimate fetal danger.[8,9] Liley plotted the spectrophotometric absorbance of bilirubin at 450 nm against gestational age in a large series of newborns, with and without IHD, and established three zones of increasing fetal risk;[9] the higher the bilirubin concentration for any gestational age, the greater was the danger of IHD (Fig. 15.5). The presence of oxyhemoglobin in the sample elevates the absorbance at 450 nm somewhat. Figure 15.6 illustrates normal and abnormal bilirubin absorbance curves.

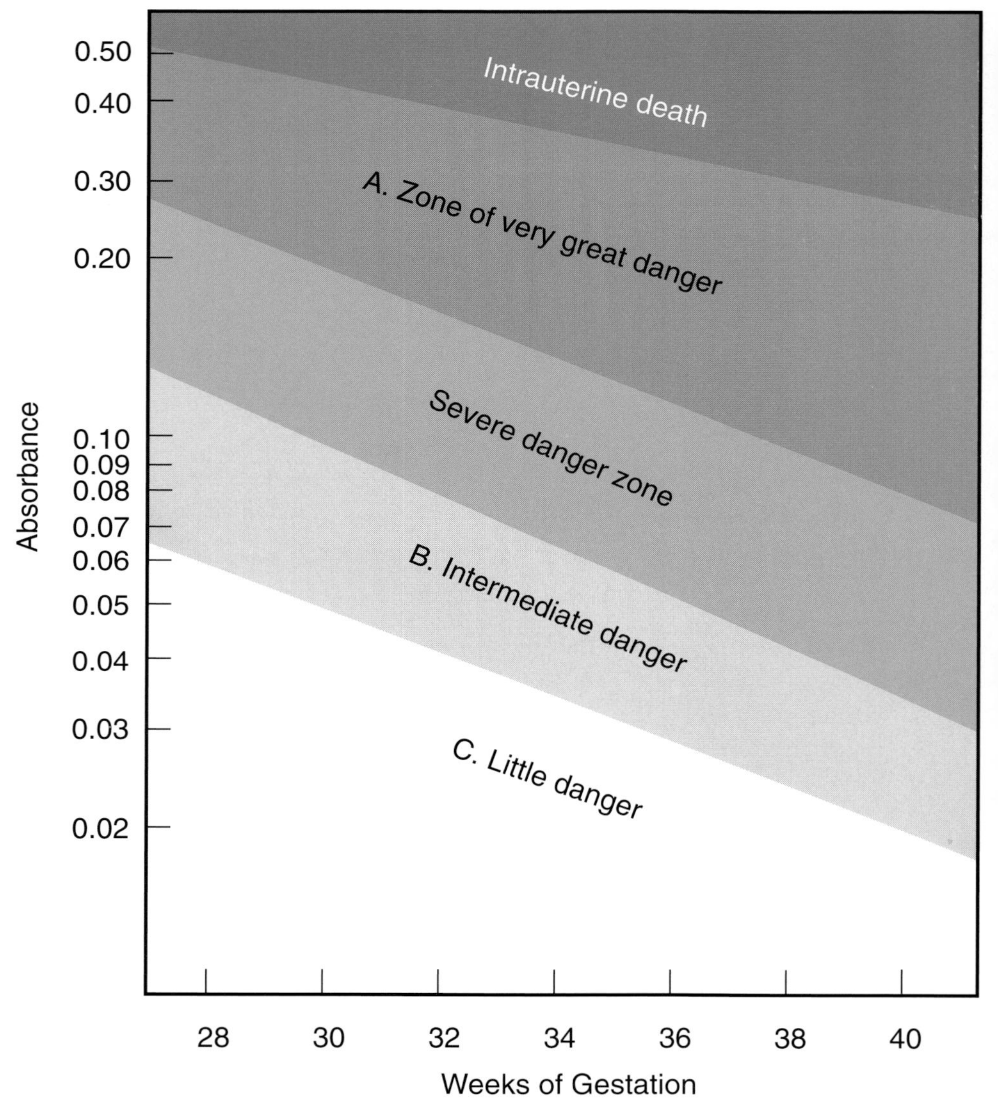

FIG. 15.5. Chart for estimating the danger of hemolytic disease of the newborn from the bile pigment concentration in amniotic fluid. (Modified from Liley, A.W.: Liquor amnii analysis in the management of the pregnancy complicated by rhesus sensitization. Am. J. Obstet. Gynecol., *82*:1359, 1961.)

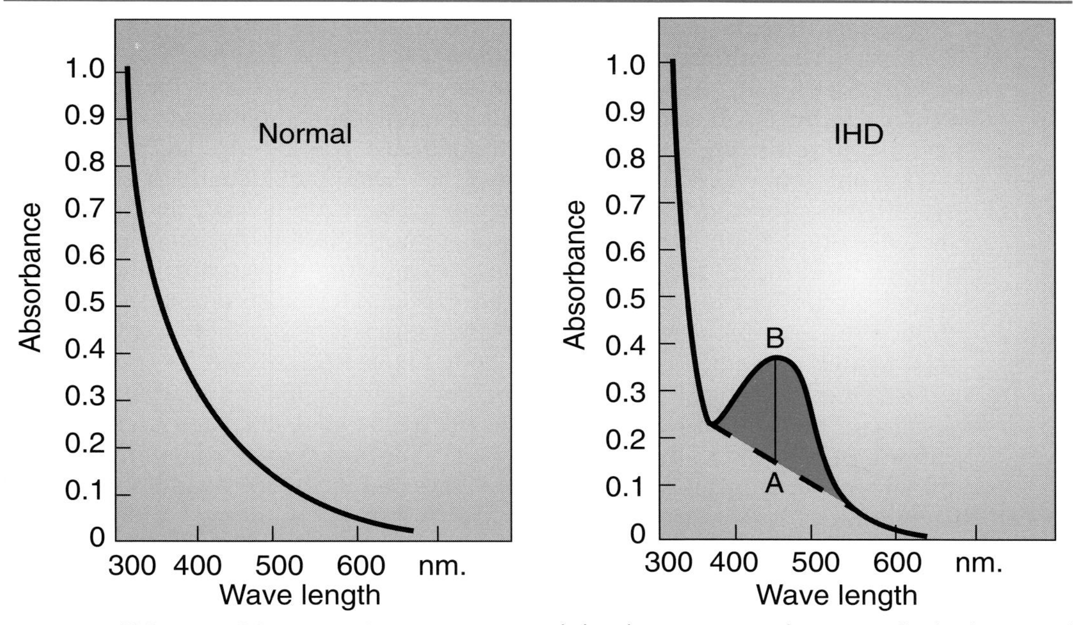

FIG. 15.6. Diagrammatic representation of absorbance curves of amniotic fluid. The normal curve shows the absence of bilirubin. The IHD curve shows a distinct bilirubin peak (the stippled area) at 450 nm and indicates that the fetus has a hemolytic problem. The line AB is the absorbance at 450 nm (A_{450}) ascribed to bilirubin. (Modified from Liley, A.W.: Liquor amnii analysis in the management of the pregnancy complicated by rhesus sensitization. Am. J. Obstet. Gynecol., *82*:1359, 1961.)

Materials. A good recording spectrophotometer with a narrow bandpass.

Principle and Procedure. A centrifuged specimen of amniotic fluid is scanned in a recording spectrophotometer from 350 to 700 nm. The absorbance is plotted on semilog paper as the ordinate (y-axis) versus wavelength as the abscissa (x-axis). A straight line (baseline) is drawn on the curve from about 365 nm to 550 or 600 nm. The difference in absorbance, A, is obtained by measuring the difference between the baseline and the peak at 450 nm (Fig. 15.6). The fetal risk is estimated by plotting the A_{450} versus weeks of gestation (Fig. 15.5) and noting into which zone the point falls.

Note: The wavelength accuracy of the spectrophotometer should be checked in the vicinity of 450 nm, and adjustments should be made, if necessary.

ASSESSMENT OF FETAL MATURITY[1]

Early indication of a high-risk pregnancy may be advantageous (IHD, diabetes mellitus, pre-eclampsia, premature rupture of membrane, premature labor), provided that the fetus has attained a sufficient degree of lung maturity to cope with life in the outside world. If the fetal lungs have not matured sufficiently, the infant may have serious respiratory difficulties (respiratory distress syndrome, RDS). RDS is the most frequent complication in premature infants.

An important factor in fetal lung maturity is the ability of the lung to synthesize in sufficient quantity the phospholipid lecithin, which coats alveolar sac linings. Lecithin is a surfactant that is essential for respiration because it lowers alveolar surface tension and permits the lungs to retain some residual air on expiration, instead of collapsing.

The pulmonary synthesis and release of surfactant usually become well established after 32 to 36 weeks of gestation; maturity is indicated by a sharp increase in the lecithin concentration of amniotic fluid as some of the alveolar surfactant is regurgitated during normal respiratory movements of the fetus. Direct or indirect measurement of the surfactant concentration in amniotic fluid is the best method for assessing lung maturity in utero. Most infants born after evidence of sufficient surfactant do not have RDS, except for some infants from diabetic mothers. Infants born before a critical amount of amniotic fluid surfactant is reached have a high risk for RDS. In RDS, many alveolar sacs cannot remain open for gaseous exchange. The sacs may become filled with a proteinaceous fluid in severe cases, the first step in the formation of a hyaline membrane.

Lecithin/Sphingomyelin (L/S) Ratio

The lecithin/sphingomyelin (L/S) ratio in amniotic fluid was the first laboratory test for estimating fetal lung maturity.[10] The rationale is based on the large increase in amniotic fluid lecithin between the 32nd week of gestation and term, the slight decrease in amniotic fluid sphingomyelin during the same period, and the assumption that changes in lecithin in amniotic fluid reflect those taking place in the lungs. Sphingomyelin is not a surfactant and has no effect on lung maturity; it serves as an internal standard that corrects for volume variability in the amniotic fluid and variability in extraction of the lipids, because both lecithin and sphingomyelin are affected to the same extent. Determination of the L/S ratio is a long, cumbersome procedure that involves extraction of the amniotic fluid lipids, partial purification of the phospholipids by acetone precipitation, separation by thin-layer chromatography (TLC), color production, and quantitation of the lecithin and sphingomyelin spots by densitometry. Also, most TLC systems do not clearly separate lecithin and sphingomyelin from other phospholipids in amniotic fluid, particularly from phosphatidyl inositol and phosphatidyl serine.

An L/S ratio of 2 is the critical value usually accepted as the indicator for lung maturity. Few newborns develop RDS when the L/S ratio before birth is > 2. In contrast, the predictive value for *pulmonary immaturity* is not high; nearly half of the newborns with an L/S ratio of < 2 do *not* develop RDS. Many variations in the technique for obtaining the L/S ratio have resulted, as well as the introduction of other measurements in an attempt to improve the discriminant capabilities of a surfactant test. Some are described in the following sections.

Individual Phospholipids

Some laboratories have improved extraction and TLC techniques to quantitate particular phospholipid constituents, such as phosphatidyl glycerol, phosphatidyl inositol, and/or disaturated lecithin. Each variant has its proponents, but no general consensus has been reached concerning which variant is best. The concentration of phosphatidyl glycerol is usually too low to measure until the 35th week of gestation; it then rises rapidly until term. Phosphatidyl inositol, however, is detectable in amniotic fluid before 28 weeks of gestational age, increases slowly to a peak around the 36th week (when phosphatidyl glycerol rises sharply), and then slowly decreases. The measurement of these two substances provides additional evidence for cross-checking the results of an L/S ratio or for making a decision regarding fetal maturity.

Foam Stability Test and Foam Stability Index (FSI)

Clements and colleagues introduced the foam stability (shake) test as a rapid method for evaluating fetal lung maturity.[11] The presence of surfactant stabilizes the foam when amniotic fluid is shaken with a nonfoaming agent, such as ethanol, and indicates lung maturity. The reading of the test is subjective, however, and quality assurance is a problem because slight contamination of the tubes or stoppers with detergent causes false results. The shake test has been modified, standardized, and converted into a foam stability index (FSI) test.[12] Varying amounts of ethanol and amniotic fluid are added to a series of tubes, shaken in a prescribed manner, allowed to settle for 30 sec, and read. The FSI is the highest ethanol concentration that permits an uninterrupted ring of bubbles to persist. A commercial kit, Lumadex-FSI Test (Beckman Instruments), is now available that avoids some of the problems with the shake test. The tubes are detergent-free and sealed, the alcohol concentration is correct, and a control containing dipalmitoyl lecithin provides a model for matching the endpoint. Several laboratories have reported the results of their clinical studies.[13,14]

Lamellar Body Phospholipid

Alveolar epithelial cells produce lamellar bodies that store the pulmonary surfactant. These organelles appear in amniotic fluid after secretion by the cells. Duck-Chong and associates were among the first to isolate the lamellar bodies and use their content of phospholipids as an index of fetal maturity;[15] a lamellar body phospholipid value of $\geqq$ 4.0 mg/L was taken as the index of maturity. Ivie and co-workers achieved better prediction of maturity and immaturity by using 3.5 mg/L as the cutoff point.[16]

A definitive test for predicting fetal lung maturity probably will not emerge until obstetric centers cooperate in a large-scale study conducted under an agreed-upon protocol for methods to be tested, details of procedures to be used; and criteria for diagnosis of respiratory conditions.[1]

Fluorescence Polarization

The addition of a lipophilic fluorescent compound to amniotic fluid may be used to predict the microviscosity of amniotic fluid and, by inference, the amount of lung surfactant that is present. The initial description by Shinitzky[17] used diphenylhexatriene as the fluorescent probe; this technique was subsequently modified by Tait[18,19] and colleagues employing 1-palmitoyl-2{6-[(7-nitro-2,1,3-benzoxadiazol-4-yl)amino]caproyl}phosphatidylcholine (NBD-PC) as the probe and an Abbott TDX as the analytical instrument. An automated commercial version of this assay is also available from Abbott on the TDX analyzer.[20] In practice, an amount of NBD-PC is added to the patient's amniotic fluid and the fluorescence polarization intensity is measured. The degree of fluorescence polarization is inversely proportional to the quantity of pulmonary surfactant present. An amniotic fluorescence polarization value of < 260 mPol (polarization units $\times$ 1000) is consistent with fetal lung maturity ($< 1\%$ of these infants develop RDS); values of > 290 mPol are consistent with fetal lung immaturity (approximately 80% of such infants develop RDS), and values between 260 and 290 mPol are in a transitory zone (approximately 25% of these infants develop RDS). The diagnostic sensitivity and specificity of the fluorescence polarization assay are comparable to those of the L/S ratio assay, but the enhanced precision, ease of analysis, rapid turnaround time, and wide availability of the assay make it the test of choice for assessment of fetal lung maturity.[21]

GENETIC COUNSELING

Various types of amniotic fluid examination have proved useful in ascertaining in utero whether a fetus is afflicted with, or is a carrier of, certain genetic diseases. This information is particularly useful in women whose previous pregnancies have produced at least one infant with a genetic defect or who are older than 35 years of age. Some information may be obtained by direct examination and analysis of the cells in the amniotic fluid, but most of the techniques require cell culture. This specialized area of investigation is performed principally in large medical centers.

Examination or Culture of Cells in Amniotic Fluid

The information obtained through study of the cells from the amniotic fluid (either directly or after cell culture) falls into three general classes:

1. *Determination of the sex of the fetus:* Because some genetic diseases are sex-linked, the sex and, hence, the degree of risk, if any, must be known.
2. *Identification of fetal karyotypes:* Many chromosome abnormalities can be detected by chromosomal analysis of amniotic fluid cells.[22] Chromosomal abnormalities appear with relatively greater frequency when the mother is older than 35; the abnormality rate rises to 1 in 40 births for mothers older than 40. Down syndrome, an extra chromosome No. 21 (Trisomy 21), is perhaps the most commonly detected abnormality.
3. *Specific enzyme defects in cultured cells:* Galjaard[22] and Milunsky[23] list and describe more than 50 genetic diseases that have been predicted successfully in utero by the lack of a specific fetal enzyme in cultured amniotic fluid cells. The list includes lipidoses (for example, Tay-Sachs

disease, with its absence of hexosaminidase A); mucopolysaccharidoses (for example, Hurler's syndrome, with abnormal deposits of mucopolysaccharides); amino acid disorders (for example, maple syrup urine disease, with a lack of branched chain, keto acid decarboxylase); disorders of carbohydrate metabolism (for example, galactosemia, with a deficiency of galactose-1-phosphate-uridyltransferase); and a large miscellaneous group. The list of defects amenable to prenatal diagnosis will increase as our knowledge of biochemical lesions in genetic diseases grows.

POSTNATAL TESTING

Numerous tests have been mandated by state health authorities to be performed on all newborns by a prescribed age; the testing is usually performed by state laboratories on blood samples collected before the infants leave the hospital. Many inborn errors of metabolism have been described; testing is available for many of them, but only a few conditions warrant the performance of screening tests on all newborns. Postnatal screening is warranted for only conditions that have a relatively high rate of incidence, are fatal, or have a high degree of morbidity if not treated; for which effective treatment is readily available; and for which an inexpensive and valid screening test is available. Some of the common mandated tests follow.

Congenital Hypothyroidism

Congenital hypothyroidism has an incidence of approximately 1/10,000 live births. Mental retardation and growth abnormalities, called cretinism, result from neonatal hypothyroidism. The condition is irreversible if not corrected shortly after birth. Treatment is simple (administration of exogenous thyroxine), inexpensive, and effective. Blood samples are usually collected by heel stick; blood is absorbed onto an absorbent cardboard card (Guthrie card) and submitted for analysis. A disk is punched out of the Guthrie card, the blood sample is eluted by addition of buffer, and a radioimmunoassay (RIA) is performed to analyze for thyroxine (T_4). The patient's result is compared to appropriate age-adjusted reference ranges and repeat testing is performed on all specimens below the established cutoff values. Cutoff values are carefully chosen to eliminate the possibility of reporting false-negative results. Confirmatory thyroxine and thyroid-stimulating hormone (TSH) tests are performed on a freshly collected specimen from all infants in whom the screening tests were positive. In one state recently, 23 confirmed cases of congenital hypothyroidism were detected out of approximately 75,000 neonates who were screened.

Congenital Adrenal Hyperplasia

Congenital adrenal hyperplasia (CAH) has an incidence of approximately 1 in 5000, but in certain ethnic groups, the incidence may approach 1 in 250. The defect in CAH is total or partial absence of adrenal cortical enzymes that are necessary for the synthesis of adrenal steroids. Numerous variants have been described, several of which result in fetal demise; the most common (90%) is an absence of the 21-hydroxylase enzyme (see Fig. 13.4). Absence of the 11-hydroxylase enzyme accounts for 5% of all cases. All the variants result in diminished production of cortisol and consequently an increase in adrenocorticotropic hormone (ACTH) resulting from lack of negative feedback on the pituitary by cortisol. Adrenal androgen production is not affected, and precursors of cortisol synthesis are pushed toward excessive androgen synthesis by the chronic elevation of ACTH. Mineralocorticoid (aldosterone) deficiency also occurs and results in salt-wasting in approximately 70% of cases. In girls, the overproduction of fetal androgens may result in virilization in utero, occasionally to the extent that genital ambiguity is so pronounced that inappropriate sex assignment results. In boys, CAH may result in precocious puberty and permanent short stature; however, normal sexual development occurs in many cases. The screening test for CAH is a RIA test for 17-OH progesterone performed on a blood disk punched out from the Guthrie card as described for hypothyroidism. Repeat testing is performed on all samples with elevated levels of 17-OH progesterone. Treatment is life long and consists of the administration of cortisol or a synthetic glucocorticoid. Adequacy of therapy may be assessed by monitoring androstenedione levels in serum. In one state recently, 3 patients (2 males, 1 female) with 21-hydroxylase deficiency (all salt-wasting) were detected by screening tests performed on approximately 75,000 neonates.

Phenylketonuria

Phenylketonuria (PKU, see Chap. 16) is another disease that meets the criteria for newborn screening. The disease is relatively prevalent, between 1 in 10,000 and 1 in 20,000 births in the United States. The incidence is 1 in 30,000 worldwide, meaning that the gene is present in many ethnic groups. One state reports detection of 6 newborns affected with PKU in a year totaling 78,000 births.

Phenylalanine can be measured in blood spots to detect PKU by the Guthrie microbiologic assay. This test uses the principle of bacterial inhibition: phenylalanine in the test sample overcomes the metabolic inhibition of β-2-thienylalanine in a strain of *Bacillus subtilis.* Phenylalanine can be quantitated by the amount of bacterial growth after severals days of incubation with test sample, *B. subtilis,* and β-2-thienylalanine. The more bacterial growth, the higher the phenylalanine concentration.

PKU also fulfills the "available treatment" criteria for newborn screening. Dietary restriction of phenylalanine prevents mental retardation and the occurrence of an otherwise debilitating disease.

Sickle Cell Disease

Many states also screen for sickle cell anemia, which has an incidence of 1 in 400 in the Black population. The incidence in the Caucasian population is 1 in 1,000,000. Many states have chosen to screen only the newborn Black population for the disease, whereas others have chosen to test all ethnic groups. The methodology used for screening is isoelectric focusing, followed by high-performance liquid chromatology (HPLC) to confirm the presence of any abnormal hemoglobins. Although the testing is directed toward detection of sickle cell disease (homozygous Hb-S), it detects any aberrant hemoglobin. Any hemoglobinopathy (such as β-thallasemia) is detected by this method. One state reports the detection of 11 sickle cell births in 100,000 babies born, plus the detection of more than 30 additional hemoglobinopathies. Carriers for these diseases are also discovered by the screening process.

REFERENCES

1. Ashwood, E.R.: Evaluating health and maturation of the unborn: The role of the clinical laboratory. Clin. Chem., *38*:1523, 1992.
2. Kirkpatrick, A.M., and Nakamura, R.M.: Alpha-fetoprotein: Laboratory Proceedings and Clinical Application. New York, Masson Publishing, 1981.
3. Simpson, J.L., and Nadler, H.L.: Maternal serum alpha-fetoprotein screening in 1987. Obstet. Gynecol., *69*:134, 1987.
4. American College of Obstetricians and Gynecologists, Department of Professional Liability: Professional liability implications of AFP testing. American College of Obstetricians and Gynecologists, 1985.
5. Hodgson, A.J., Pilowsky, P.M., Robertson, E.F., et al.: Combined analysis of acetylcholinesterase and alpha-fetoprotein improves the accuracy of antenatal diagnosis of neural tube defects. Med. J. Aust., *1*:457, 1981.
6. Haddow, J.E., Palomaki, G.E., Knight, G.J., et al.: Prenatal screening for Down's syndrome with use of maternal serum markers. N. Engl. J. Med., *327*:588, 1992.
7. Cheng, E.Y., Luthy, D.A., Zebelman, A.M., et al.: A prospective evaluation of a second-trimester screening test for fetal Down syndrome using maternal serum alpha-fetoprotein, hCG and unconjugated estriol. Obstet. Gynecol., *81*:72, 1993.
8. Foerster, J.: Alloimmune hemolytic anemias. *In* Wintrobe's Clinical Hematology. 9th Edition. Edited by G.R. Lee, et al. Philadelphia, Lea & Febiger, 1993.
9. Liley, A.W.: Liquor amnii analysis in the management of the pregnancy complicated by rhesus sensitization. Am. J. Obstet. Gynecol., *82*:1359, 1961.
10. Gluck, L., and Kulovich, M.: Lecithin/sphingomyelin ratio in normal and abnormal pregnancy. Am. J. Obstet. Gynecol., *115*:539, 1973.
11. Clements, J.A., Platzker, A.C.G., Tierney, D.F., et al.: Assessment of the risk of the respiratory distress syndrome by a rapid test for surfactant in amniotic fluid. N. Engl. J. Med., *286*: 1077, 1972.
12. Sher, G., Statland, B.E., Freer, D., et al.: Assessing fetal lung maturity by the foam stability index test. Obstet. Gynecol., *52*:673, 1978.
13. Sher, G., and Statland, B.E.: Assessment of fetal pulmonary maturity by the Lumadex foam stability index test. Obstet. Gynecol., *61*:444, 1983.
14. Lockitch, G., Wittmann, B.K., Snow, B.E., and Campbell, D.J.: Prediction of fetal lung maturity by use of the Lumadex-FSI test. Clin. Chem., *32*:361, 1986.
15. Duck-Chong, C.G.: Lamellar body content of amniotic fluid; a possible index of fetal maturity. Ann. Clin. Biochem., *16*:191, 1979.
16. Ivie, W.M., Novy, M.J., Reynolds, J.W., et al.: Modified lamellar body phospholipid assay compared with L/S ratio and phosphatidyl glycerol assay for assessment of pulmonary status. Clin. Chem., *33*:24, 1987.
17. Shinitzky, M., Goldfisher, A., Bruck, A., et al.: A new method for assessment of fetal lung maturity. Br. J. Obstet. Gynaecol., *83*:838, 1976.
18. Tait, J.F., Franklin, R.W., Simpson, J.B., and Ashwood, E.R.: Improved fluorescence polarization assay for use in evaluating fetal lung maturity. I. Development of the assay procedure. Clin. Chem., *32*:248, 1986.
19. Ashwood, E.R., Tait, J.F., Foerder, C.A., et al.: Improved fluorescence polarization assay for use

in evaluating fetal lung maturity. III. Retrospective clinical evaluation and comparison with the lecithin/sphingomyelin ratio. Clin. Chem., *32*:260, 1986.

20. Russell, J.C.: A calibrated fluorescence polarization assay for assessment of fetal lung maturity. Clin. Chem., *33*:1177, 1987.
21. Chen, C., Roby, P.V., Weiss, N.S., et al.: Clinical evaluation of the NBDF-PC fluorescence polarization assay for prediction of fetal lung maturity. Obstet. Gynecol., *80*:688, 1992.
22. Galjaard, H.: Genetic Metabolic Diseases: Early Diagnosis and Prenatal Analysis. New York, Elsevier, North Holland Medical Press, 1980.
23. Milunsky, A. (Ed.): Genetic Disorders and the Fetus. 3rd Edition. Baltimore, Johns Hopkins, 1992.

REVIEW QUESTIONS

1. What is amniotic fluid and for what purpose is it generally used in the laboratory?
2. How does analysis of amniotic fluid assist in estimating fetal maturity? In what situations is this important?
3. How does the serum concentration of bilirubin affect the severity of isoimmune hemolytic disease (hemolytic disease of the newborn)?
4. What is analyzed when estimating fetal maturity by fluorescence polarization? By the L/S ratio?
5. What is the source of α-fetoprotein in maternal serum during the 12th to 16th week of gestation, and why is an elevated value taken as an indication of a neural tube defect? Why is a confirmatory test essential?
6. What is the triple screen for Down syndrome and why is it used?
7. What are some congenital defects amenable to adequate replacement therapy that can be detected in the newborn's postnatal blood (within the first 2 or 3 days of life)?

Genetic Disorders and Their Diagnosis 16

OUTLINE

OBJECTIVES

After reading this chapter, the student will be able to:

1. Describe the three-dimensional structure of DNA and the process by which genetic information is transmitted.
2. Describe the steps in a Southern blot transfer of DNA.
3. Describe how DNA probes may be used in a clinical assay.
4. Give clinical indications for a metabolic workup in a newborn.
5. List three metabolic diseases that are diagnosed through plasma amino acid quantitation and three that would require organic acid analysis.
6. Understand the importance of Tay-Sachs screening in a family at risk.

Genetic disease and congenital malformations occur in 2 to 5% of all live births and account for as many as 30% of pediatric admissions to the hospital. About 10% of the adult population has a chronic disease with a significant genetic component.[1]

DNA: STRUCTURE AND FUNCTION

Until recently, genetic disorders were investigated by examining gene products (enzymes or proteins). Now, defects can be discovered in the genetic material itself. Deoxyribonucleic acid (DNA) is composed of two linear strands intertwined to form a double helix (Fig 16.1). The DNA strands contain the code for the production of the actual amino acid sequence of each protein and therefore each enzyme. The structure and function of these protein products are determined by their amino acid sequence. Each strand consists of alternating sugar (deoxyribose) and phosphate groups (Fig. 16.2). The phosphate group connects the 5′ carbon atom of one deoxyribose residue to the 3′ carbon atom of the deoxyribose in the adjacent nucleotide. Every nucleic acid chain has a 5′ phosphate group at the beginning and a 3′ hydroxyl group at the end. The two strands of DNA are complementary, with one strand starting with the 5′ end and terminating with the 3′ end and its partner starting with the 3′ end and terminating with the 5′ end. A strand of DNA is linked to the opposite strand by hydrogen bonding between pairs of nucleotide bases: adenine, guanine, thymine, and cytosine (Fig. 16.2). Adenine pairs to thymine, and guanine pairs to cytosine (Fig. 16.3).

The human genome contains more than 3 billion base pairs, or 3 million kilobases (kb) of DNA sequence. The genetic code is really a 4-letter alphabet used to write out information in a linear form. The genetic information contained in DNA can be copied onto messenger ribonucleic acid (mRNA) by a process called transcription (Fig. 16.4). The strands of DNA separate, and one strand acts as a template for RNA. RNA retains all the information of the DNA

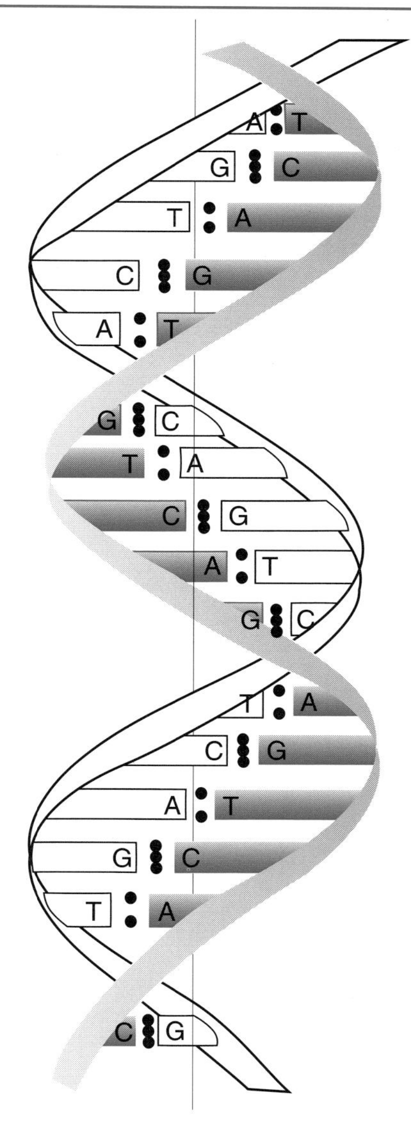

FIG. 16.1. The structure of DNA. The two helical strands represent the backbone of the DNA, with the alternating sugar and phosphate groups. The base pairs are shown with linking hydrogen bonds.

sequence from which it was copied, but is single stranded, contains the sugar ribose instead of deoxyribose, and substitutes the base uracil for thymine (Fig. 16.3). After transcription, the mRNA moves from the nucleus of the cell to the cytoplasm, where it acts as a template for protein synthesis. Translation is the process of decoding mRNA into a corresponding sequence of amino acids with the help of transfer RNA (tRNA) (Fig. 16.4). Each sequence of three bases of

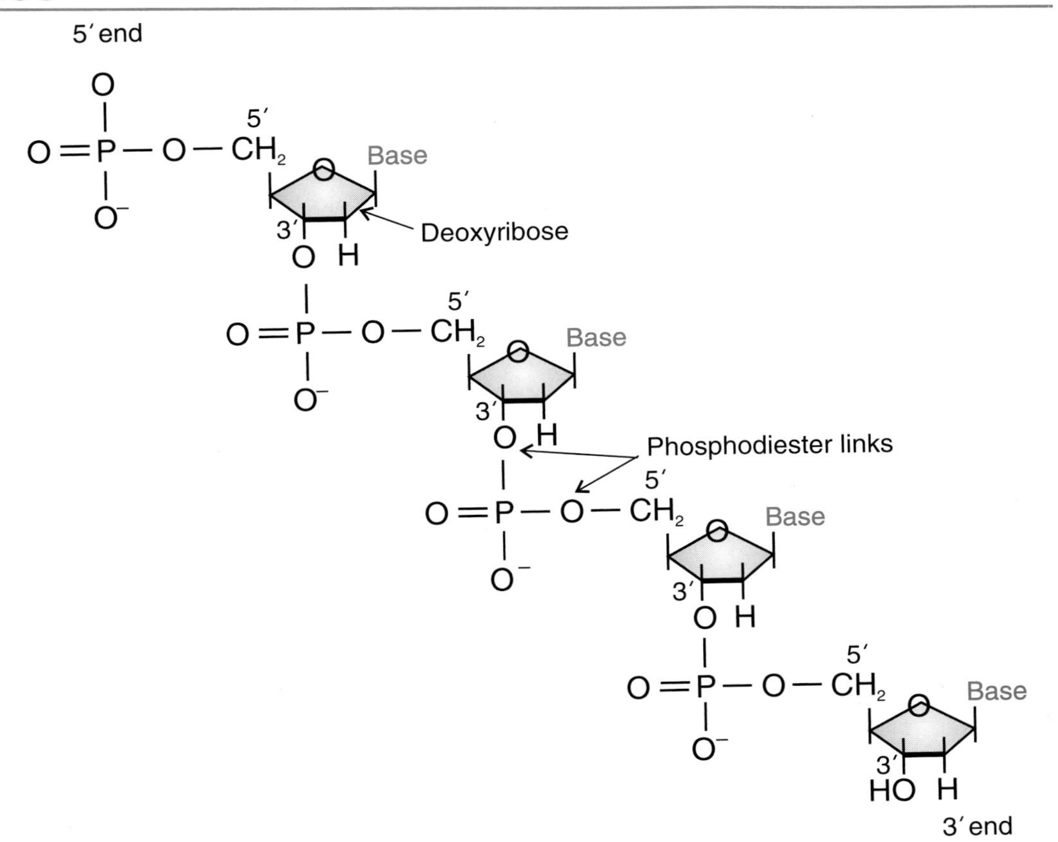

FIG. 16.2. A portion of a DNA chain shows the deoxyribose and phosphate residues that form the backbone.

DNA (or codon) codes for one amino acid. The result is a flow of genetic information from DNA to RNA to protein (Fig. 16.5).

DIAGNOSIS OF DISEASE BY MOLECULAR BIOLOGY TECHNIQUES

Revolutionary developments in molecular biology over the past decade have facilitated the understanding of diseases at the molecular level. The discovery of restriction endonucleases, bacterial enzymes that recognize and cleave specific nucleotide sequences in DNA, has facilitated these developments. Restriction enzymes have made possible the use of DNA probes that can identify mutations in the cleaved nucleotide sequence.

DNA Extraction

Many of the following techniques require purification of the DNA as a preliminary step. Although DNA can be obtained from most tissues, the most common source is whole blood, or white cells isolated from whole blood

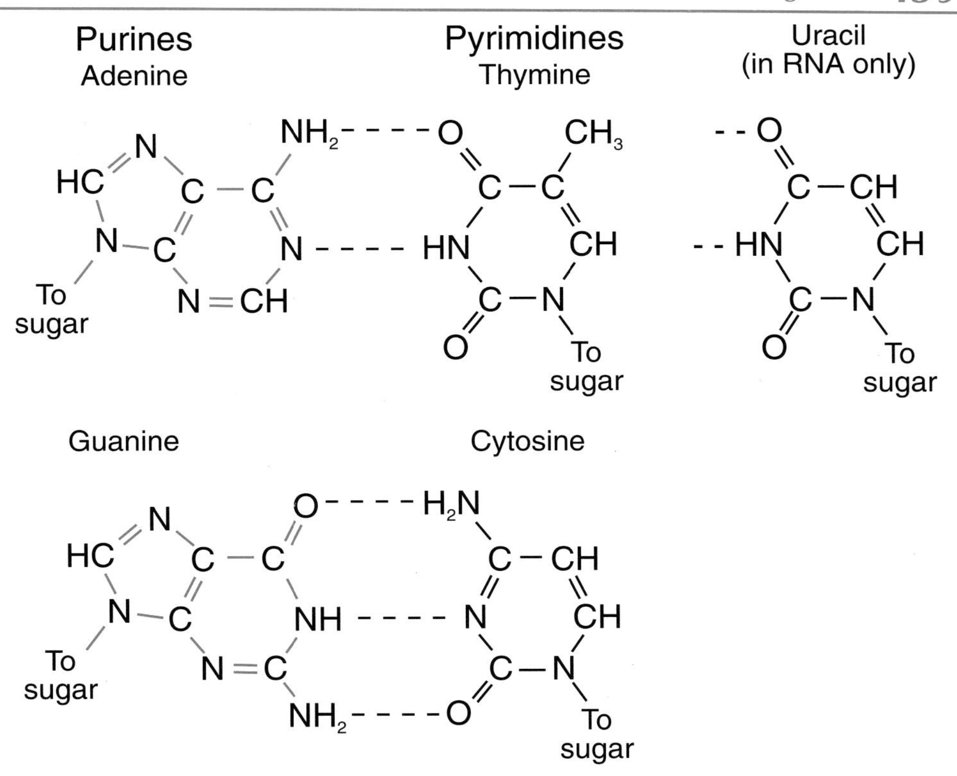

FIG. 16.3. The chemical structure of the nucleotide bases—adenine, guanine, thymine, cytosine, and uracil (which takes the place of thymine in RNA). Hydrogen bonds that hold the two strands of DNA together are indicated by dotted lines.

Before DNA purification, white cells or tissue homogenates are treated with proteinase K, an enzyme that digests cell membranes and liberates the DNA.

Many techniques allow DNA extraction, but the standard method is a phenol/chloroform extraction.[2] The phenol denatures cellular proteins, and the DNA remains in the aqueous fraction. Subsequent chloroform extraction is helpful in removing any traces of phenol, which would interfere with restriction enzymes. DNA is precipitated from the aqueous fraction by means of a sodium acetate/ethanol mixture, centrifuged, washed, and stored in buffer.

DNA extracted directly from cells (genomic DNA) consists of large molecules ($>$ 100 kb). Precautions should be taken to avoid shearing of DNA, which results in smaller fragments. Ideally, for extraction of high-molecular-weight DNA, vortexing must be gentle and kept to a minimum; mixing with phenol should be performed by slow rocking or rotation (not shaking), and large-bore pipets should be used to transfer the aqueous phase.

An automatic DNA extraction system manufactured by Applied Biosystems is now available (Fig. 16.6). This instrument processes as many as 8 DNA samples simultaneously in 4.5 hr. Faster, nonorganic solvent extraction techniques are also available for automated or manual use. These are acceptable for use in the polymerase chain reaction (PCR, described later) but are not recommended for Southern blot analysis or long-term storage of DNA.

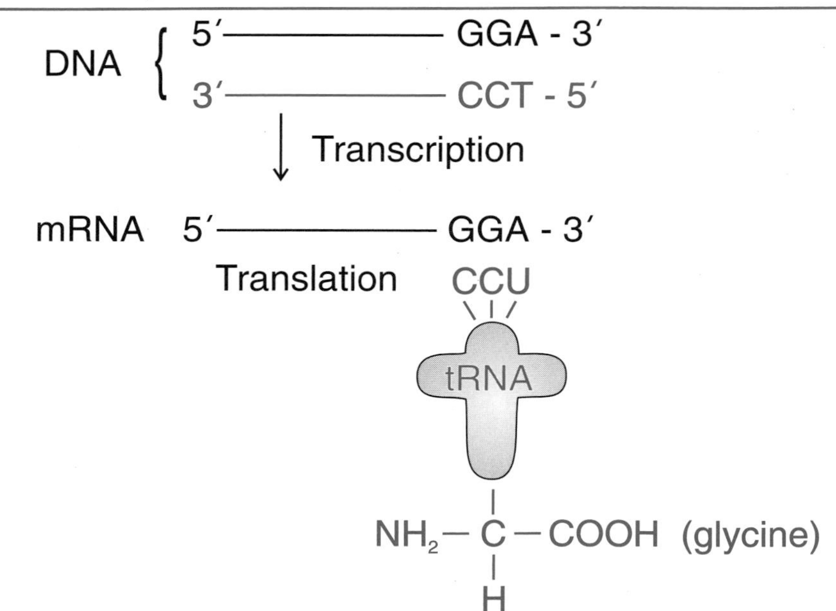

FIG. 16.4. The process of transcribing DNA onto RNA and translating information encoded on RNA into protein. Each sequence of three base pairs of DNA, or codon, codes for one amino acid. The amino acid glycine is formed.

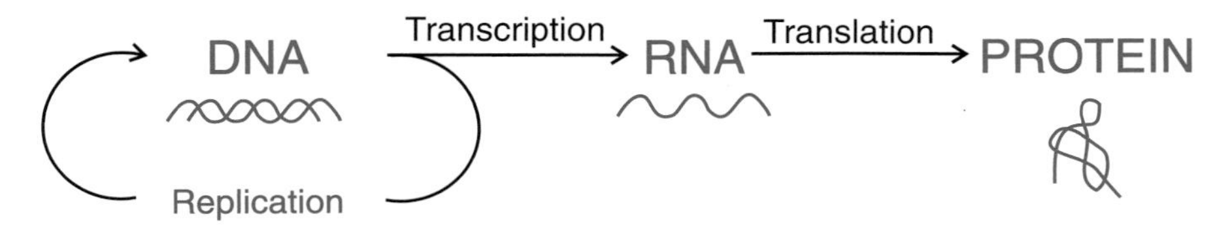

FIG. 16.5. The flow of genetic information.

Several companies also market columns for DNA extraction that use anion exchange to separate DNA from RNA and other proteins. These columns can produce pure, high-molecular-weight DNA and have the added benefit of avoiding the use and disposal of hazardous organic solvents.

Care also must be taken to prevent bacterial contamination of DNA. Gloves should always be worn because DNAases (enzymes that can cleave DNA) in skin bacteria can break down DNA. All reagents, buffers, pipets, pipet tips, and glassware used in DNA preparation and analysis should be autoclaved. Bacterial contamination of commonly used reagents can wreak havoc in a DNA laboratory, because DNA exposed to bacterial contamination is unsuitable for further analysis.

The possibility of cross contamination of DNA samples must be eliminated by the use of careful technique. Contamination could cause ambiguous or erroneous test results.

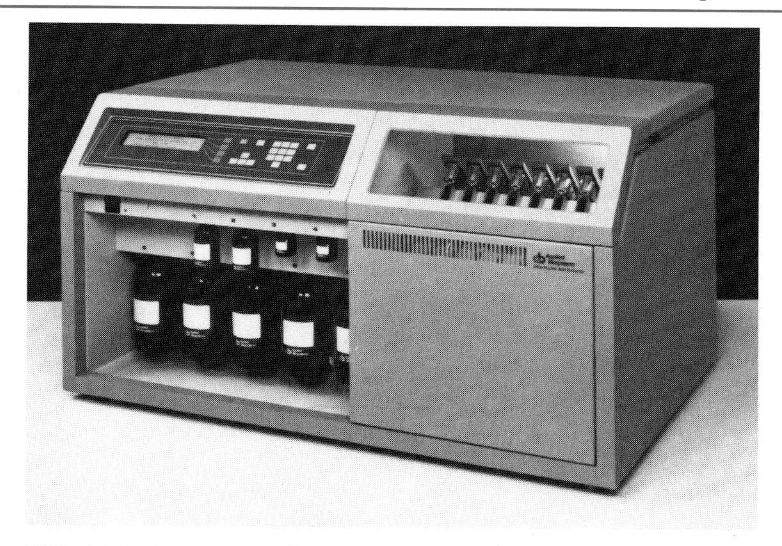

FIG. 16.6. An automated DNA extractor. (Photo courtesy of Applied Biosystems.)

The purity and concentration of extracted DNA can be assessed by measuring its absorption at 260 nm and 280 nm. Pure preparations of DNA have A_{260}/A_{280} ratios of 1.8 to 2.0. A ratio lower than 1.8 reflects some contamination with protein or phenol. The concentration of DNA can be measured by the absorbance at 260 nm as follows:

$$\text{concentration in } \mu\text{g/mL} = \frac{A_{260} \times \text{dilution}}{0.02 \times 1.0}$$

where 0.02 = extinction coefficient of DNA at 260 nm and 1.0 = light path in cm.

One can usually obtain 30 μg of DNA from 1 mL of whole blood.

Separation of DNA by Electrophoresis

Agarose gel electrophoresis is used extensively in DNA analysis to separate DNA fragments according to their size after digestion with a restriction enzyme. Agarose, a polysaccharide derivative from seaweed, is cast into a gel by melting it in a buffer and pouring the solution into a mold. The concentration of agarose used to make the gel varies, depending on the size of DNA fragments one wants to separate most efficiently.[2] The higher the percentage of agarose in the gel, the more efficient separation of low-molecular-weight DNA. Standard agarose gels are about 0.6 to 1.0% (w/v) agarose in buffer. Polyacrylamide gels are used for separation of smaller fragments of DNA (3 to 2000 base pairs).

As an electrical field is applied across the gel, the negatively charged DNA migrates toward the anode. The rate of migration depends mainly on the agarose concentration, the buffer composition and ionic strength, and the voltage (see Chap. 3).

If undenatured DNA undergoes electrophoresis without treatment with a restriction enzyme, the DNA should appear as a compact region near the sample application point. DNA that has been cut with a restriction enzyme or is partially degraded appears as a streak along the lane in which it was applied (Fig. 16.7).

DNA Probes

A schematic description for the preparation of DNA probes is outlined in Figure 16.8. Restriction endonucleases allow for the production of specific fragments of DNA, which then can be enzymatically spliced into bacterial plasmids. Plasmids are circular molecules of DNA that are separate from the main chromosome of the bacteria. The foreign DNA then can be propagated in bacteria (*Escherichia coli*, for example). The *E. coli* host produces large amounts of the recombinant plasmid. Restriction enzymes are used again to liberate the cloned (amplified) fragment from the plasmid DNA.[3,4]

These DNA fragments now can be used as probes to seek out and hybridize (bind) to complementary sequences in clinical samples of a patient's DNA. Because DNA is double stranded, it must be dissociated into single strands (usually by heat treatment) before use in hybridization.

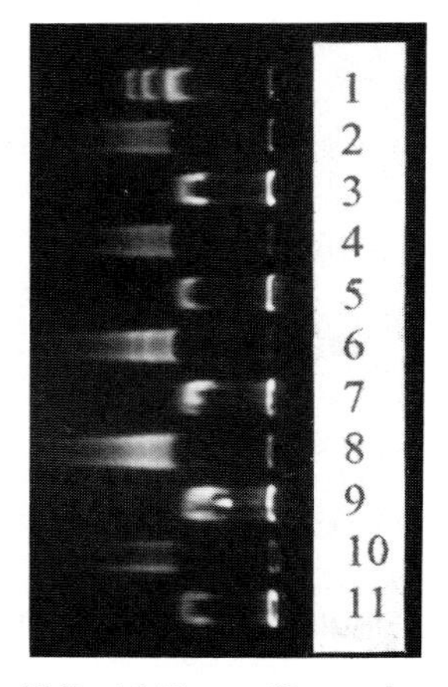

FIG. 16.7. Genomic DNA and DNA digested with a restriction enzyme on agarose gel electrophoresis. Lane 1 is a size marker; lanes 2, 4, 6, 8, and 10 are digested DNA; and lanes 3, 5, 7, 9, and 11 are high-molecular-weight undigested DNA.

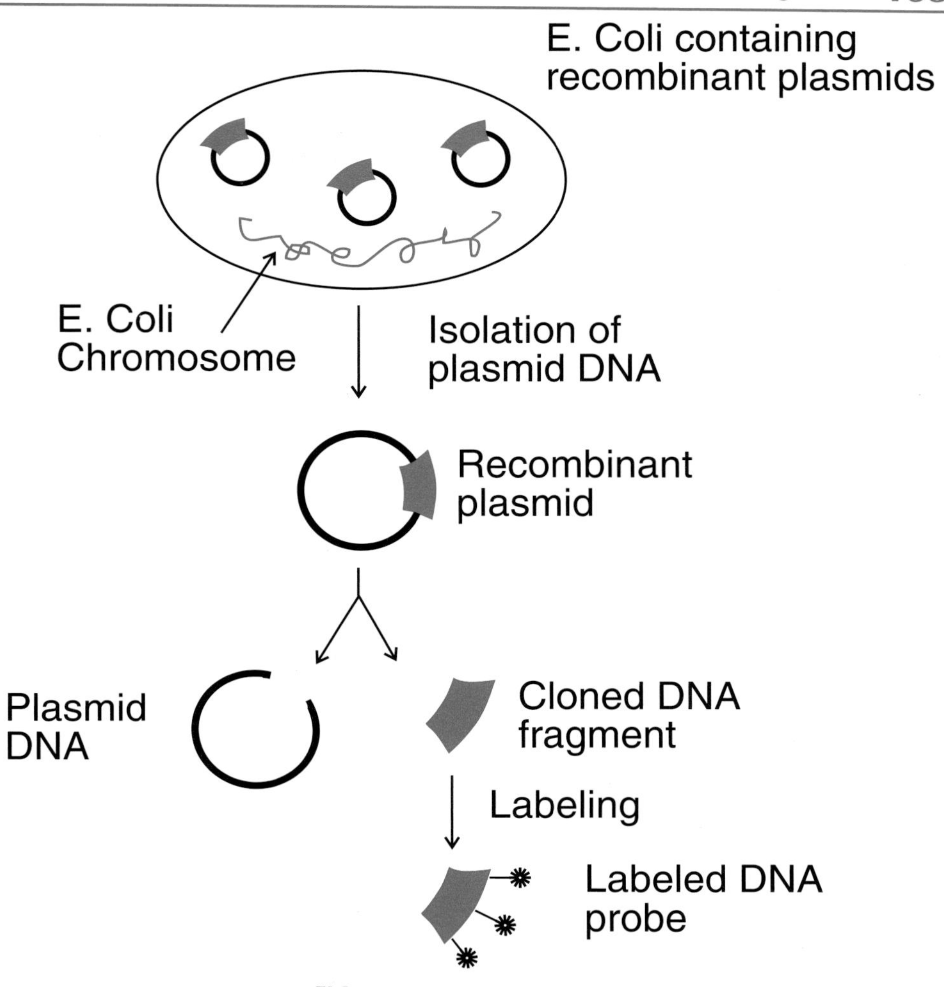

FIG. 16.8. Probe preparation.

Probes must be labeled in some way to be detected. One of the most common markers is a radioactive isotope, either ^{125}I or ^{32}P. Labeled nucleotides can be incorporated enzymatically into DNA, and the radioactive DNA probe can be detected by autoradiography or by counting in a gamma counter.

Although radioactive probes still offer the most sensitivity, the use of radioactivity always should be avoided if possible. An alternative is a biotin-labeled probe. After hybridization, either fluorescein or enzyme-labeled avidin is added. Because avidin has a high affinity for biotin, detection of probe hybridization takes place through fluorescence or a colorimetric change produced by the enzyme (Fig. 16.9).

RNA, as well as synthetic DNA, also can be used for probes. The nucleotide sequence of a cloned DNA fragment can be determined by DNA sequencing techniques. This sequence then can be reproduced as a single strand of synthetic DNA. Synthetic DNA probes can be produced in vast quantities in a

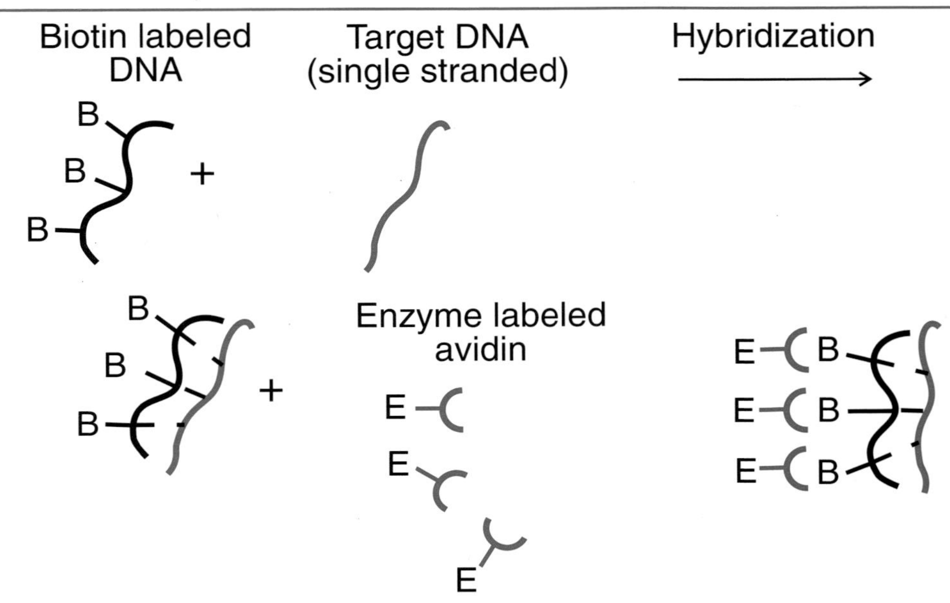

FIG. 16.9. Signal production with biotin-labeled DNA. Enzyme-labeled avidin is added to bind to biotin and produce a color wherever probe is bound.

matter of hours, in contrast to the 1 to 2 weeks required for preparation of cloned DNA probes.

Dot Blot Hybridization of DNA

Dot blot hybridization is so named because the sample DNA is usually applied as a spot on a membrane. The assay steps are as illustrated in Figure 16.10.

1. DNA is extracted from the sample (white blood cells, tissue).
2. Sample DNA is denatured into single strands and bound to a membrane.
3. Hybridization solution containing a labeled single-stranded probe is added. The probe seeks out and hybridizes selectively to complementary sequences in the immobilized sample DNA.
4. The membrane is washed to remove any unbound probe.
5. The appropriate detection system (for example, fluorescence) is used to reveal the presence of hybridized probe molecules.

In Situ Hybridization of DNA

In situ hybridization refers to hybridization of probe to DNA within intact cells. This technique is advantageous because DNA purification is not a prerequisite, and testing can be performed directly on cytologic smears or histologic sections

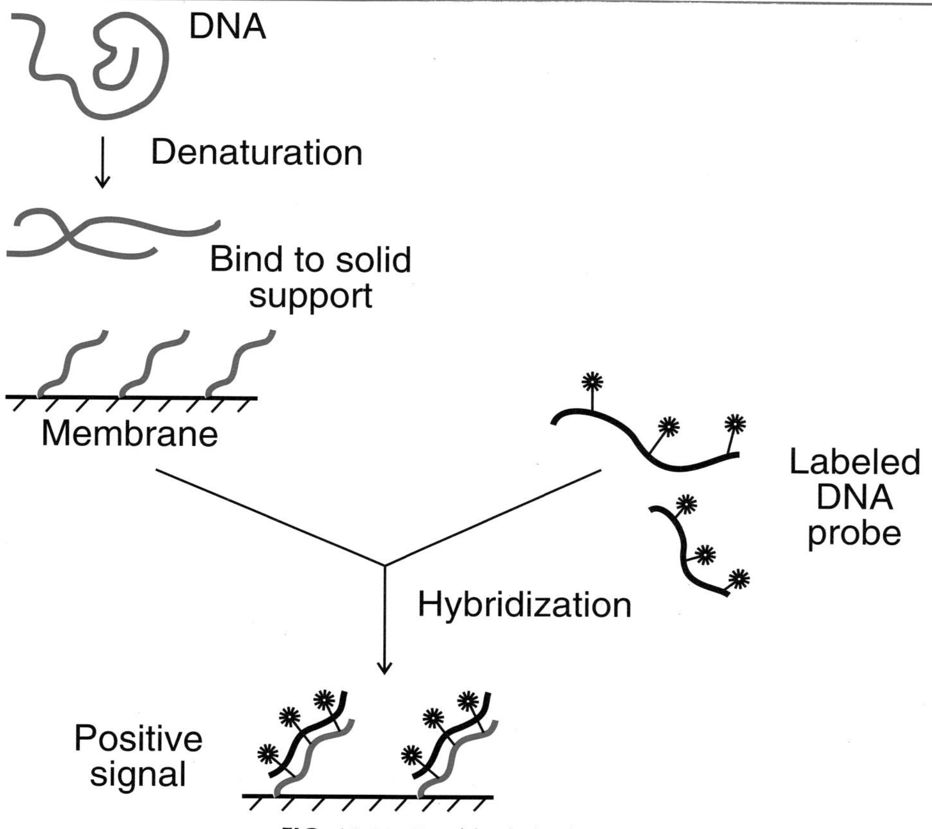

FIG. 16.10. Dot blot hybridization.

on slides. Commercial chromosome probes are available for detection of chromosomal abnormalities.

Both dot blot and in situ hybridization can be useful in diagnosing infectious diseases.[4,5] One can tailor the specificity of the assay by designing a probe to pinpoint a unique DNA sequence in a particular strain within a microbial species. DNA testing is also more rapid than traditional culture methods. Commercially available kits have been developed for *E. coli*, *Neisseria gonorrhoeae*, *Chlamydia*, *Campylobacter jejuni*, *Mycoplasma pneumonia*, *Legionella*, and *Mycobacterium tuberculosis*. Viruses for which DNA probe kits are available include hepatitis B, cytomegalovirus, Epstein-Barr, herpes simplex, adenovirus, and human immunodeficiency virus (HIV).

Southern Blot Analysis of DNA

The diagnosis of inherited disease is somewhat more complicated than the diagnosis of infectious disease. Much of the need in this area is for prenatal diagnosis and carrier detection. The diagnosis of genetic disease requires the detection of subtle qualitative changes in the DNA sequence that can be as small as one base pair. The Southern blot (named for E.M. Southern) is utilized for such structural analysis of DNA.

Figure 16.11 illustrates the steps in a Southern blot analysis.

1. The DNA is cut with restriction endonucleases to generate a mixture of fragments.
2. The fragments are separated according to size by agarose gel electrophoresis.
3. The DNA is denatured and transferred from a gel to a solid support, such as nitrocellulose or nylon.
4. The immobilized complementary DNA fragments are hybridized with labeled probe.
5. An autoradiograph is made of the hybridized blot with x-ray film. (A nonisotopic visualization method such as fluorescence may also be used.)

Sickle cell anemia can be diagnosed this way. The abnormality in sickle cell disease is caused by a single base pair substitution that gives rise to one incorrect amino acid (substitution of valine for glutamic acid) and an abnormal hemoglobin molecule.[1] Coincidentally, the mutation destroys a restriction enzyme recognition site. Normal DNA yields 2 fragments, a 1.15-kb and a

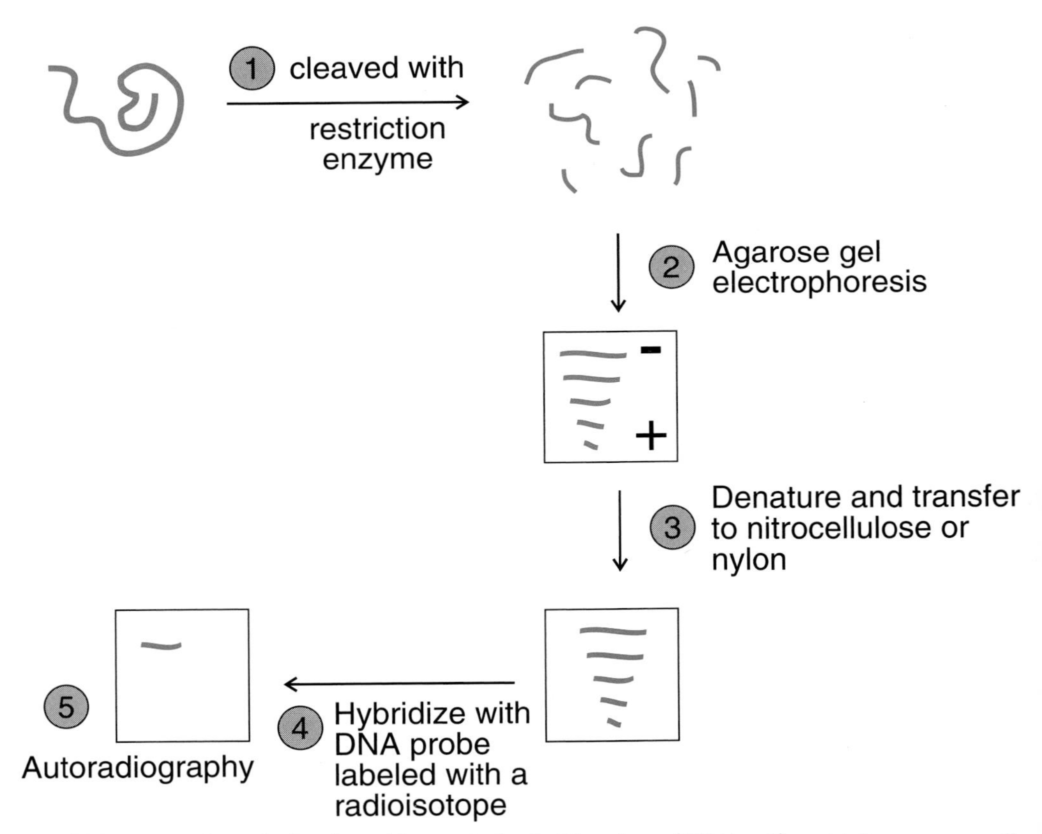

FIG. 16.11. Steps in Southern blot analysis. 1, Digestion of DNA with restriction enzyme. 2, Agarose gel electrophoresis. 3, Transfer. 4, Hybridization with labeled probe. 5, Detection.

0.2-kb fragment. Sickle cell DNA yields a single 1.35-kb fragment (Fig. 16.12). The difference in size of the restriction fragments is detected by Southern blot analysis, using a labeled probe specific for a DNA sequence in the hemoglobin gene.

DNA probe methodology is sensitive, and can be performed on small amounts of blood or cells from amniotic fluid. Probes are becoming available for more genetic diseases, including Lesch-Nyhan syndrome, medium-chain acyl CoA dehydrogenase deficiency, fragile X syndrome, myotonic dystrophy, muscular dystrophy, cystic fibrosis, phenylketonuria, hemophilia A, sickle cell anemia, the thalassemias, Huntington's disease, and many genes associated with cancer (oncogenes).[6–8]

Although mutations in DNA sometimes can be detected directly, as in sickle cell anemia, most inherited disorders are caused by mutations that do not fall within a restriction enzyme sequence. Restriction fragment length polymorphism (RFLP) analysis has been devised to circumvent this problem. RFLP relies on the presence of marker genes (polymorphisms), which are relatively common in the population. The marker genes are close to and are coinherited with the disease-causing mutation, but are functionally independent. The marker gene serves as an indirect indicator for the mutation by the way it affects the restriction fragment pattern.

Sometimes the RFLP is linked to an inherited disease only within a particular family; thus, the study of DNA from family members is necessary to make a diagnosis. In other instances, an RFLP frequently may be coinherited with the disease in the general population. Theoretically, the diagnosis of any genetic disease can be accomplished by establishing a linkage relationship between a useful marker and the disease-causing mutation, even when the primary genetic defect is unknown. RFLP analysis is used in many inherited disorders when common mutations cannot be found.[8,9]

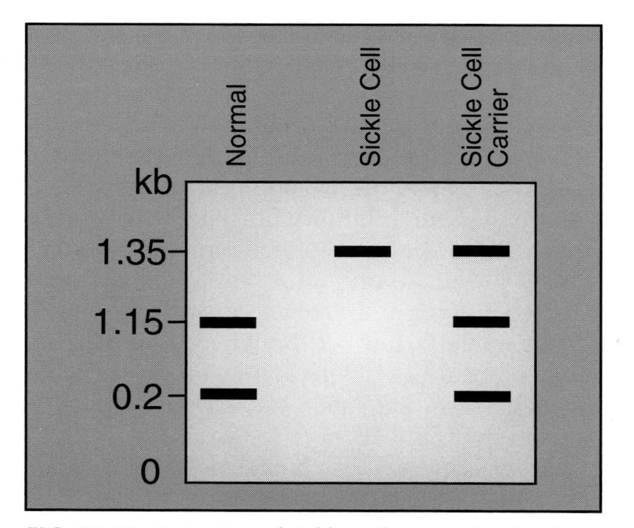

FIG. 16.12. Detection of sickle cell anemia by Southern blot analysis.

Amplification of DNA by the Polymerase Chain Reaction (PCR)

The polymerase chain reaction (PCR) is a technique used to amplify or make many copies of a sequence of DNA.[10] The strength of this system is sensitivity. The amplification makes possible the study of small amounts of biologic material, even the DNA from dried blood spots.

To utilize PCR, one first must determine the DNA sequence to be amplified, and then obtain chemically synthesized oligonucleotide primers (short segments of DNA that bracket the DNA sequence to be amplified and are complementary to that sequence). The DNA to be amplified is denatured by heating to 95° C, and the oligonucleotide primers are then allowed to anneal to the DNA, usually at a temperature approximately 10° C. DNA synthesis is then accomplished enzymatically by a heat-stable DNA polymerase, known as *Taq* polymerase, at 73° C. The reaction takes place in a thermal cycler, which has the ability to heat and cool rapidly and also to maintain a constant temperature. The result is a two-fold increase in amplification products for every PCR cycle[11] (Fig. 16.13). Amplification of a target DNA sequence by over 1 million fold can be done easily in a few hours.

PCR has proved useful in detecting genetic disorders (cystic fibrosis, hemophilia A and B, β-thalassemia, muscular dystrophy, Huntington's disease, α_1-antitrypsin deficiency, and phenylketonuria to name a few), infectious agents (HIV, herpes, hepatitis), and chromosomal rearrangements associated with cancer.

BIOCHEMICAL DIAGNOSIS OF DISEASE

Although the diagnosis of hereditary disease at the molecular level has become more feasible, it usually is not the first line of testing in a clinical setting. The symptoms exhibited by a newborn with a genetic metabolic disease are diverse and often nonspecific. The use of screening procedures that look at the biochemical changes manifested by the disease is far more cost-effective than is the initial use of DNA testing. Because the potential gene defects could be numerous, the use of DNA analysis as an initial investigative step would be time-consuming, costly, and probably unrewarding.

Infants who have a metabolic acidosis associated with vomiting, ketosis, hypoglycemia, or seizures are potential candidates for metabolic screening in the newborn period. Infants who feed poorly or have a crisis initiated by feeding also warrant a metabolic laboratory workup. A plasma ammonia level should be required in any child who deteriorates after feeding, as hyperammonemia is common in urea cycle disorders and disorders of branched-chain amino acid metabolism. Although symptoms are somewhat inscrutable, distinctive odors are associated with certain metabolic diseases, such as maple syrup urine disease. The laboratory should take notice of the odor of the urine when screening (Table 16.1).

Disorders of intermediary metabolism of amino acids, carbohydrates, or lipids are usually a result of an enzyme or transport defect. The biochemical result of such a defect is either the accumulation of an intermediate to a toxic level or the deficiency of an essential compound. Excellent reference texts are recommended.[12–15]

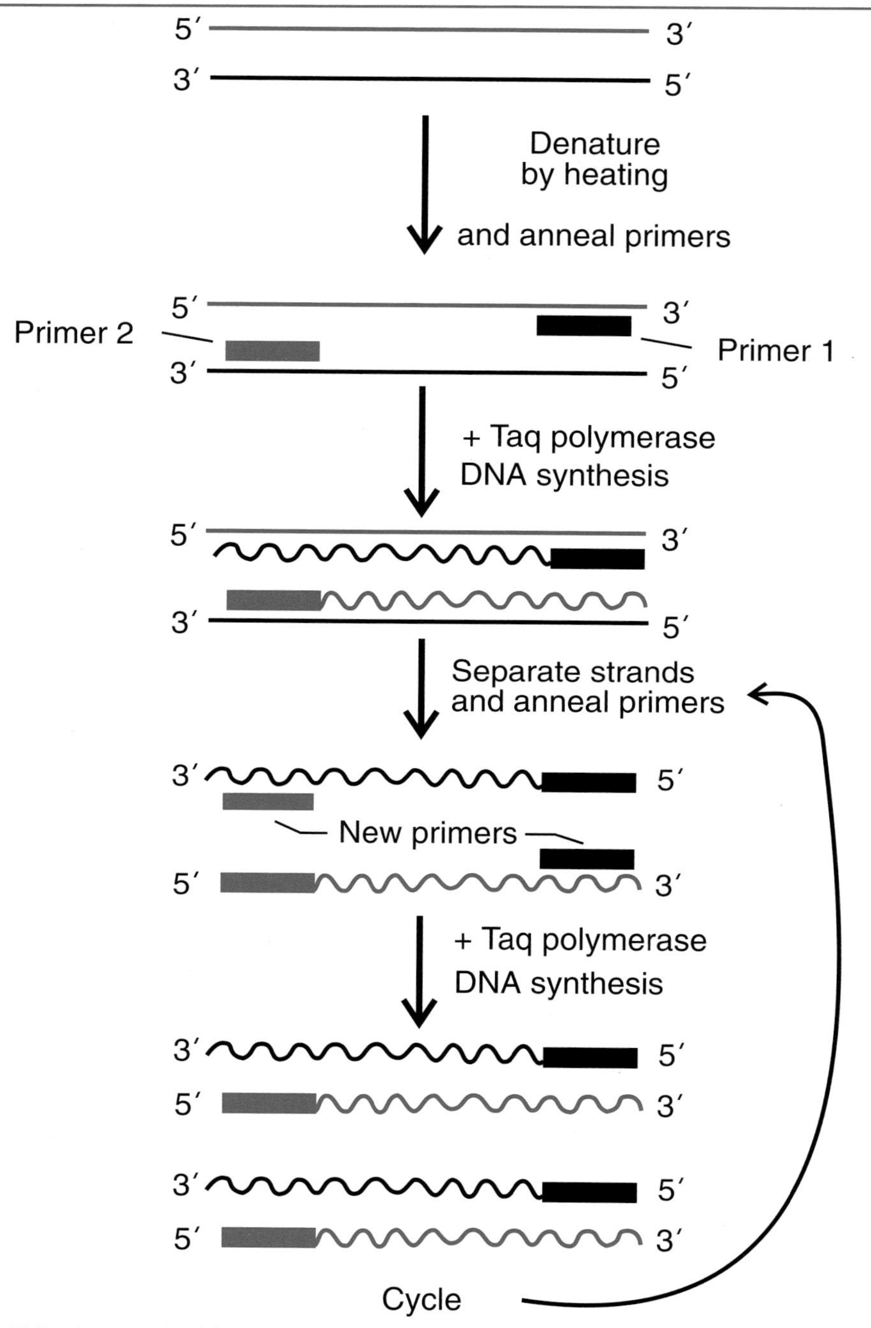

FIG. 16.13. Amplification of DNA by the polymerase chain reaction (PCR). Double-stranded DNA is separated by heating. The reaction is then cooled to allow primers that flank the target region to anneal. *Taq* polymerase is added to synthesize DNA complementary to the template. The process is repeated for 20 to 40 cycles to amplify the sequence between the primer $> 10^6$ fold.

TABLE 16.1
ODORS ASSOCIATED WITH METABOLIC DISEASE

DISORDER	ODOR
Maple syrup urine disease	Maple syrup
Isovaleric aciduria	Old sweat socks
Glutaric aciduria type II	Old sweat socks
Phenylketonuria	Musty
Tyrosinemia	Cabbage-like
Methionine malabsorption (Oasthouse urine disease)	Malt or hops
β-methylcrotonic aciduria	Tomcat urine
β-OH-β-methylglutaryl CoA lyase deficiency	Tomcat urine
Diabetic ketoacidosis	Acetone
Methylmalonic aciduria	Ammonia
Propionic aciduria	Ammonia
Urea cycle defects	Ammonia
Hawkinsinuria	Chlorine
Trimethylamine	Fish

Metabolic Screening Tests

Metabolic screening tests of urine are used to identify abnormal compounds or elevated quantities of metabolites that indicate genetic metabolic disease. Urine is commonly used in screening procedures because the abnormal metabolites are concentrated in the urine and excreted (not reabsorbed by the kidney tubule). The first line of screening includes tube tests and spot tests. These tests are simple to perform and are rapid. Alone, they are not considered diagnostic because of so many false-positive results (many therapeutic drugs react positively in these tests), but they do alert the laboratory to a possible abnormality and indicate the need for further testing. Useful screening tests are described in the following sections.

Ferric Chloride Test[16]

The ferric chloride test is a rather general nonspecific screening tool because a color change is evident in numerous different situations (Table 16.2). One should always use a dipstick for bilirubin if a ferric chloride test shows a positive green color, because bilirubin in the urine, as well as the metabolic disease tyrosinemia, produces a green ferric chloride test.

Reagent. 10% (w/v) ferric chloride. Dissolve 10 g ferric chloride in 2M HCl and dilute to 100 mL.

Procedure

1. Place 0.5 mL well-mixed urine in a tube.
2. Add 2 to 5 drops ferric chloride 1 drop at a time and observe color change (see Table 16.2 for typical color changes).

TABLE 16.2
INTERPRETATION OF COLOR CHANGES IN THE FERRIC CHLORIDE TEST

DISORDER	COLOR
Normal urine (occasionally)	brown ppt caused by phosphates
Phenylketonuria	green
Tyrosinemia	green
Bilirubinuria	green
Homogentistic acid	transient blue-green
Melanoma	gray turning black
Ketones	purple
Salicylate	purple
Phenothiazines	purple or green
Maple syrup urine disease	greenish gray

Reducing Sugars[16]

A test for reducing substances is important in a newborn to rule out galactosemia. The Clinitest tablets (Ames) yield a positive result with all reducing sugars (Table 16.3). The principle of the test is the reduction of cupric sulfate by reducing substances (sugars) in urine (see Chap. 7). A range of colors (greenish to yellow) correlating with the concentration of reducing substances develops. The manufacturer supplies directions for testing and a color chart for comparison, which should be used. A positive test for reducing substances should be followed up with a urine dipstick for glucose (Clinistix, Ames) to determine if glucose is the reducing sugar present. Glucose, galactose, lactose, fructose, and xylulose cause a positive test for reducing substances. False-positive results can occur from many compounds, including amino acids, certain antibiotics (ampicillin and similar penicillin derivatives), and ascorbic acid (vitamin C).

Dinitrophenylhydrazine Test for Ketoacids[16]

Reagents. 2,4-dinitrophenylhydrazine solution: Dissolve 150 mg of 2,4-dinitrophenylhydrazine in 40 mL of absolute methanol. Add 10 mL of 6 M HCl. Store at 4° C.

TABLE 16.3
POSITIVE REDUCING SUBSTANCES

SUGAR	CLINICAL IMPORTANCE
Glucose	Diabetes mellitus
Galactose	Galactosemia
Fructose	Hereditary fructose intolerance
Lactose	Lactose intolerance
Xylulose	Essential pentosuria (harmless)

Procedure. Add 0.5 mL of reagent 1 drop at a time to 1.0 mL clear urine.

The formation of a dense yellow precipitate indicates the presence of excess ketoacids. The precipitate is insoluble hydrazones formed by the reaction of carbonyl groups with dinitrophenylhydrazine. Ketones (acetoacetate and acetone) cause a positive reaction in this test. Other α-ketoacids, which cause a positive test, might result from any number of organic acidurias, maple syrup urine disease, or valproic acid treatment (for seizures). One should distinguish ketones normally present in a fasting sample from other ketoacids by using the Ames Acetest, which is specific for acetoacetate and acetone and does not react with β-hydroxybutyrate. To perform, place one drop of sample on the Acetest tablet. Color changes are compared to the color guide in the manufacturer's package insert to determine the presence of ketones.

Cyanide Nitroprusside Reaction for Sulfhydryl Groups[16]

Reagents

1. 5% (w/v) aqueous sodium cyanide solution (caution: poisonous reagent).
2. 5% (w/v) aqueous nitroprusside reagent (sodium nitroferricyanide).

Procedure

1. To 1 mL of patient's urine and positive control, add 0.4 mL 5% sodium cyanide.
2. Mix and let stand 10 min at room temperature.
3. Add 1 to 2 drops sodium nitroprusside reagent and observe color change.

A positive reaction is an immediate beet red or magenta color, which fades with time. The test is positive with homocystine or cystine. Weak positives are often seen in concentrated urine because of normal amounts of cystine. Differentiation must be made between cystine, which would indicate cystinuria, and homocystine, which is present in homocystinuria. Cystinuria is a renal transport defect of cystine and the dibasic amino acids lysine, ornithine, and arginine. It is the most common cause of renal stones in children. Homocystinuria is a deficiency of cystathionine β-synthetase, an enzyme in the pathway of methionine metabolism. The following silver nitroprusside test reacts only with homocystine and, thus, differentiates between the two diseases.

Silver Nitroprusside Test for Homocystine[16]

Reagents

1. Sodium chloride.
2. 3% (v/v) ammonia in water.
3. Silver nitrate reagent: Dissolve 100 mg of silver nitrate in 3% ammonia and dilute to 10 mL.
4. 1% (w/v) aqueous sodium nitroprusside reagent.
5. 0.7% (w/v) aqueous sodium cyanide reagent (caution: poisonous reagent).

Procedure

1. Saturate 2 mL of urine with sodium chloride.
2. To 0.5 mL sodium chloride-saturated sample add 0.05 mL 1% silver nitrate reagent.
3. Mix and let stand 1 min.
4. Add 0.05 mL sodium nitroprusside reagent.
5. Add 0.05 mL sodium cyanide reagent.
6. Observe sample for immediate pink or purple color change, which indicates the presence of homocystine.

Mucopolysaccharide Screening

Mucopolysaccharide storage diseases, such as Hurler's or Hunter's syndrome, are inborn errors of glycosaminoglycan metabolism resulting from the deficiency of one of several lysosomal enzymes involved in the breakdown of these compounds (Table 16.4). The glycosaminoglycans are a group of hexose-containing polysaccharide derivatives, such as keratan sulfate, heparan sulfate, and dermatin sulfate. Diagnosis is often made by detection of excess excretion of mucopolysaccharides in the urine. Several tests are available. Perry's test is a turbidity test. Cetyltrimethylammonium bromide forms a turbid solution or precipitate with urinary mucopolysaccharides.[17] All samples give a false-positive result if they are not performed with reagents and urine samples at room temperature. The toluidine blue test,[18] also called Berry's test, is a dye-binding method performed on urine-saturated filter paper. Another dye, 1,9-dimethylmethylene blue (DMB), can be used to quantitate the mucopolysaccharides in urine.[19] Newborns excrete high levels of mucopolysaccharides, and often produce false-positive results in this period. The quantitative dye-binding test is reliable only in samples with creatinine concentrations > 10 mg/dL (880 μmol/L). Dilute urines with lower creatinine levels do not give reliable results.

TABLE 16.4
MUCOPOLYSACCHARIDE DISORDERS

TYPE	SYNDROME	ENZYME DEFICIENCY
I H	Hurler's	α-L-Iduronidase
I H/I S	Hurler's/Scheie's	α-L-Iduronidase
I S	Scheie's	α-L-Iduronidase
II	Hunter's	Iduronate sulfatase
III type A	Sanfilippo's type A	Heparan sulfatase
type B	Sanfilippo's type B	N-acetylglucosamidase
type C	Sanfilippo's type C	Acetyl CoA: α-glucosaminide-N-acetyltransferase
type D	Sanfilippo's type D	N-acetylglucosamine-6-sulfatase
IV A	Morquio's	Galactose-6-sulfatase
IV B	Morquio's	β-galactosidase
VI	Maroteaux-Lamy	Arylsulfatase B (acetylgalactosamine sulfatase)
VII	Sly syndrome	β-glucuronidase

Although mucopolysaccharidoses are a diverse group of storage diseases, clinical findings compatible with these conditions include mental retardation, dysostosis multiplex (bony changes), hepatosplenomegaly, coarse facial features, cardiac involvement, deafness, and eye changes, such as corneal clouding. A diagnosis is confirmed by demonstrating a lack of specific lysosomal enzyme activity associated with the disease. Enzyme activities are readily measured on cultured fibroblasts from skin biopsy or, in some cases, isolated white cells. All mucopolysaccharidoses are inherited in an autosomal recessive manner, except for Hunter's disease, which is X-linked.

Other lysosomal enzyme deficiencies do not show excess excretion of mucopolysaccharides in the urine. These conditions are reviewed in Table 16.5. All except Fabry's disease (which is X-linked) are inherited as an autosomal recessive trait. Again, they are clinically diverse,[12] with central nervous system degeneration, hepatosplenomegaly, eye changes, and bone changes as features common to many.

Tay-Sachs Disease

Tay-Sachs disease is one of the more common lipid storage diseases. The disease is caused by a deficiency of the enzyme hexosaminidase A, without which lipid accumulates and damages the central nervous system. The gene is quite frequent in the Ashkenazi (Eastern European) Jewish population (1 in 30 to 1 in 40). The gene frequency in the general population is about 1 in 400. Individuals affected with Tay-Sachs appear normal at birth, but undergo progressive mental and motor deterioration, with onset in childhood (1 to 2 years of age). The outcome is fatal, usually by age 4 or 5 years. Mass screening programs among the at-risk population have been carried out worldwide to

TABLE 16.5
OTHER STORAGE DISEASES

DISEASE	ENZYME DEFECT
Sphingolipidoses	
Fabry's disease	α-galactosidase
Metachromatic leukodystrophy	Arylsulfatase A
GM-1-gangliosidosis	β-galactosidase
Gaucher's disease	β-glucosidase
Krabbe's disease	Cerebroside β-galactosidase
Niemann-Pick disease	Sphingomyelinase
Sandhoff's disease (GM-2-gangliosidosis)	Total hexosaminidase
Tay-Sachs disease (GM-2-gangliosidosis)	Hexosaminidase A
Glycoprotein Storage	
Mannosidosis	α-mannosidase
Fucosidosis	α-fucosidase
Sialidosis	α-neuraminidase
Glycogen storage disease type II (Pompe's)	α-glucosidase
Mucolipidosis	
Type I	Neuraminidase
I cell disease (type II, type III)	Incorporation of hydrolases into lysosome

identify carriers of this disease. Hundreds of at-risk carrier couples have been identified by less than normal hexosaminidase A activity, and prenatal diagnosis has led to the identification of Tay-Sachs disease in utero. Tay-Sachs disease is the first example of a genetic disease in which the birth of an affected child has been prevented by mass screening for heterozygotes in at-risk populations.[12]

Disorders of Amino Acid Metabolism

Amino acids are normal constituents of plasma. When blood flows through the kidneys, amino acids pass through the glomerular membrane and appear in the glomerular filtrate along with a host of other small molecules. Most of the amino acids are reabsorbed in the tubules, however, by special transport mechanisms of limited reabsorptive capacity. Amino acids may be excreted in excess in the urine (aminoaciduria) as the result of a genetic defect (lack of a crucial enzyme) in a metabolic pathway or of a deficient reabsorption transport system. These conditions are known as primary aminoacidurias. Secondary aminoacidurias are nongenetic in origin and may occur during periods of massive protein breakdown, severe damage to the liver, or severe renal tubular disease that interferes with renal tubular reabsorption.

A screening test for aminoaciduria is a good first step in screening for genetic metabolic disorders; differentiation between primary and secondary aminoacidurias is relatively easy on the basis of history and clinical findings. Genetic defects are usually characterized by the presence in urine of specific amino acids and their metabolites, depending on the disorder, and not by the generalized aminoaciduria that is common in the secondary aminoacidurias.

Thin-Layer Chromatography

Two-dimensional thin-layer chromatography is often performed on urine to look for excess excretion of normal amino acids or for the excretion of unusual or abnormally occurring amino acids. This procedure is an integral part of a good metabolic screen that discloses most of the potential abnormalities.

Procedure. Urine is spotted on a silica plate and eluted with a mobile phase containing 40 mL chloroform; 40 mL methanol, 12 mL NH_4OH, and 8 mL H_2O. At the end of the migration period, the plates are dried, rotated 90°, and eluted with a second mobile phase containing 60 mL butanol, 20 mL glacial acetic acid, and 20 mL H_2O. Plates are then sprayed with the following to visualize the amino acids: triketohydrindene hydrate (Ninhydrin), 150 mg; isatin, 5 mg; butanol, 50 mL; and collidine, 1.5 mL. A normal pattern can be seen in Figure 16.14. An abnormal pattern should be confirmed in most cases by a quantitative plasma amino acid analysis.

Quantitation of Plasma Amino Acids

The several methods for quantitating amino acids all use high-pressure liquid chromatography to separate the amino acids. A chemical derivative is made of the amino acids so that they absorb light at a particular wavelength and can be

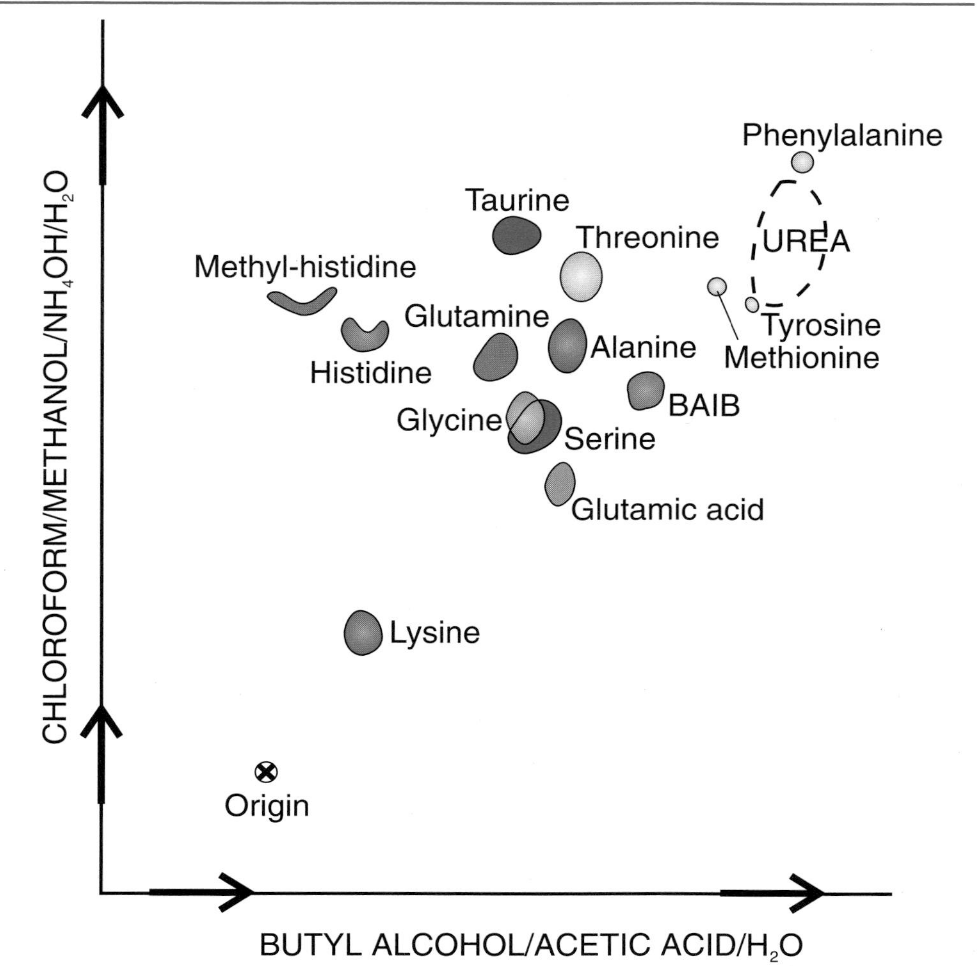

FIG. 16.14. Two-dimensional thin-layer chromatography of amino acids in urine. Normal pattern.

detected when they elute from the analysis column. Choices for precolumn derivatization include phenylisothiocyanate (PITC)[20] or dansyl chloride.[21] In these techniques, a derivatizing agent is added in sample preparation before amino acid separation takes place on the column. Choices for postcolumn derivatization include *o*-phthalaldehyde[22] or ninhydrin.[23] Many metabolic genetics laboratories use an automated instrument, such as the Beckman amino acid analyzer, which uses ion exchange chromatography with postcolumn ninhydrin derivatization and detection at 2 wavelengths, 570 nm and 440 nm.

Major Amino Acid Disorders

Phenylketonuria (PKU)

Many states mandate the screening of all newborns for phenylketonuria (PKU) at birth. PKU is a primary disorder of amino acid metabolism that results from the deficiency of an enzyme, phenylalanine hydroxylase (Fig. 16.15), that converts phenylalanine to tyrosine by hydroxylation. PKU is inherited as an autosomal recessive disorder, so both parents must be carriers of the disease; statistical considerations predict that 1 of 4 children should have the disease if both parents are carriers. If untreated, phenylalanine builds up to toxic levels in the blood and central nervous system and causes mental retardation. The treatment is a low phenylalanine diet, with tyrosine supplementation. Each patient must be monitored for phenylalanine and tyrosine throughout childhood and must stay on a restricted diet as long as possible. If a woman with PKU decides to pursue a pregnancy, her levels must be carefully controlled. Mental retardation of the fetus could occur because of elevated phenylalanine in utero even though the fetus may not carry the two recessive genes that cause PKU.

Maple Syrup Urine Disease

Another classic example of a primary amino acid disorder is maple syrup urine disease (MSUD). The disease is named for the distinctive maple syrup odor of the urine from patients with the disorder. The defect is a deficiency of the enzyme that catalyzes the decarboxylation of the three essential branched-chain amino acids: valine, leucine, and isoleucine (Fig. 16.16). Diagnosis is

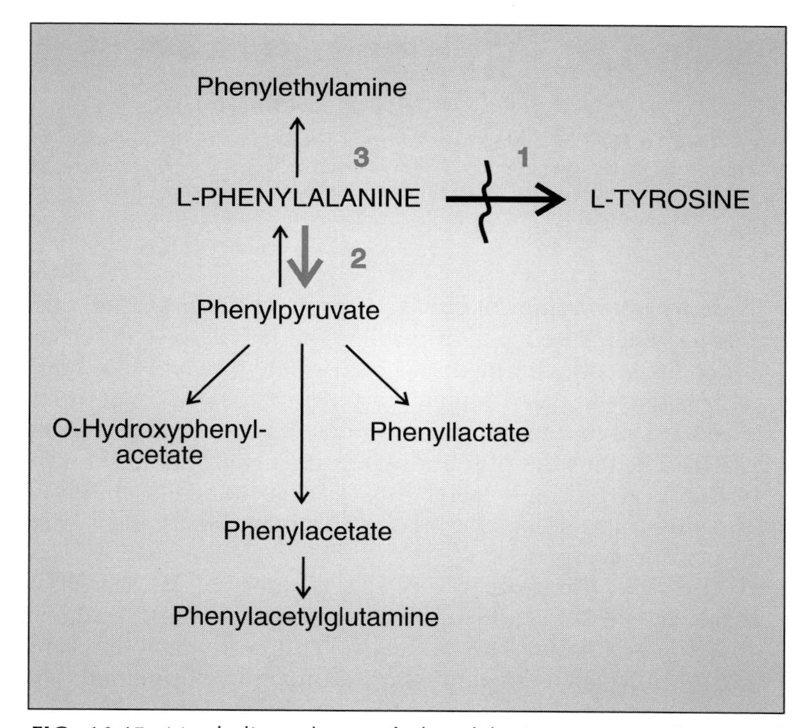

FIG. 16.15. Metabolic pathway of phenylalanine. In PKU, the normal hydroxylation pathway at (1) is blocked because of absence of the enzyme phenylalanine hydroxylase. Consequently, pathway (2) predominates, with accumulation of phenylalanine in blood and other metabolites in urine.

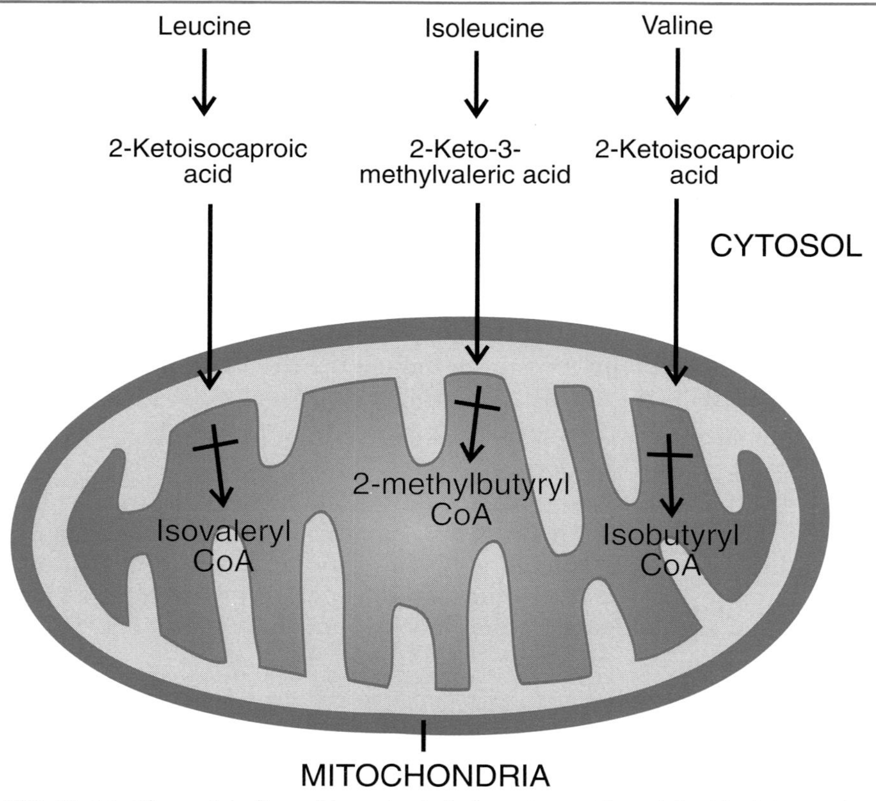

FIG. 16.16. The metabolism of branched-chain amino acids and the block in maple syrup urine disease. The keto-derivatives of the branched-chain amino acids (leucine, isoleucine, and valine) are not metabolized further by the mitochondrial enzyme complex.

made by observation of elevated levels of the branched-chain amino acids in plasma. The presence of alloisoleucine (an abnormal derivative of isoleucine) is also diagnostic for the disease. The renal threshold for branched-chain amino acids is quite high. Although the amino acids spill over into the urine if plasma levels are high enough, urine is not the most reliable screening material for MSUD. The most helpful findings in the urine are the abnormal branched-chain ketoacids, which are metabolites of the branched-chain amino acids. These ketoacids can be detected by gas chromatography/mass spectrometry of urine for organic acids.

Clinically, the more severe classic form of MSUD has an acute neonatal onset. Severe ketoacidosis ensues shortly after birth with seizures, hypoglycemia, coma, and sometimes death. With early detection and dietary intervention, children can develop with minimal developmental delays. The availability of commercial MSUD formulas low in leucine, isoleucine, and valine (Mead Johnson) has greatly benefited the care and outcome of these children. Milder and intermittent forms of the disease also exist. Children with this form may exhibit symptoms only when stressed (bacterial or viral infection). Some may respond favorably to thiamine treatment.

Tyrosinemia

Elevation of the amino acid tyrosine is a common finding in premature infants because of the immaturity of liver enzymes (transient neonatal tyrosinemia). Tyrosine is a hydroxyl derivative of phenylalanine and is metabolized by the liver, as seen in Figure 16.17. Hereditary tyrosinemia is caused by a deficiency of the enzyme fumarylacetoacetic acid hydrolase. Early in life (usually before age 6 months) children develop hepatomegaly, jaundice, failure to thrive, vomiting, diarrhea, and a cabbage-like odor. They often progress to liver failure, and many die of liver tumors. Liver transplant has been attempted in several patients to prevent the formation of tumors. Experimental trials with drug therapy to block the formation of succinylacetone (which is associated with toxicity) and with tyrosine-restricted diets are also now underway.

A diagnostic workup for a child suspected of tyrosinemia should include quantitative plasma amino acid analysis (which reveals elevated tyrosine and methionine) and urine organic acid analysis to document the presence of succinylacetone. An elevated plasma tyrosine may be quite nonspecific and be a result of a liver impairment. The finding of succinylacetone in urine is diagnostic for hereditary tyrosinemia (type I).

Nonketotic Hyperglycinemia

An isolated elevation of plasma and cerebrospinal fluid (CSF) glycine in a newborn without ketone formation is diagnostic for nonketotic hyperglycinemia. Elevations of glycine are usually quite extreme, about 10 times the upper limit of normal. Because plasma glycine may be elevated for other reasons, glycine in CSF should be quantitated to determine a CSF/plasma glycine ratio. A normal ratio is approximately 0.02. Affected individuals have a CSF/plasma glycine ratio of approximately 0.2.[13] Excessive excretion of glycine in the urine should also be a clue to examine plasma and CSF for this particular diagnosis.

The classic presentation is an overwhelming illness in the newborn period. Because glycine acts as a neurotoxin, patients have intractable seizures and severe mental retardation. Quite often they are unable to ventilate independently. Although no successful treatment exists, nonketotic hyperglycinemia must be differentiated from other disorders or conditions with elevations in plasma glycine for which dietary intervention may be helpful. Plasma glycine may also be elevated in disorders of organic acid metabolism, in patients receiving hyperalimentation fluid, and in patients receiving the drug valproic acid for seizure control.

Urea Cycle Defects

Urea is the nitrogenous end product of protein catabolism. After protein hydrolysis, the amino groups of the amino acids are converted to ammonia by oxidative deamination. The ammonia enters the urea cycle by enzymatic conversion to carbamyl phosphate in the liver and leaves the cycle as urea (Fig. 16.18).

Urea is synthesized only in the liver and is excreted by the kidneys. Ammonia is also produced in the gastrointestinal tract by bacterial action on

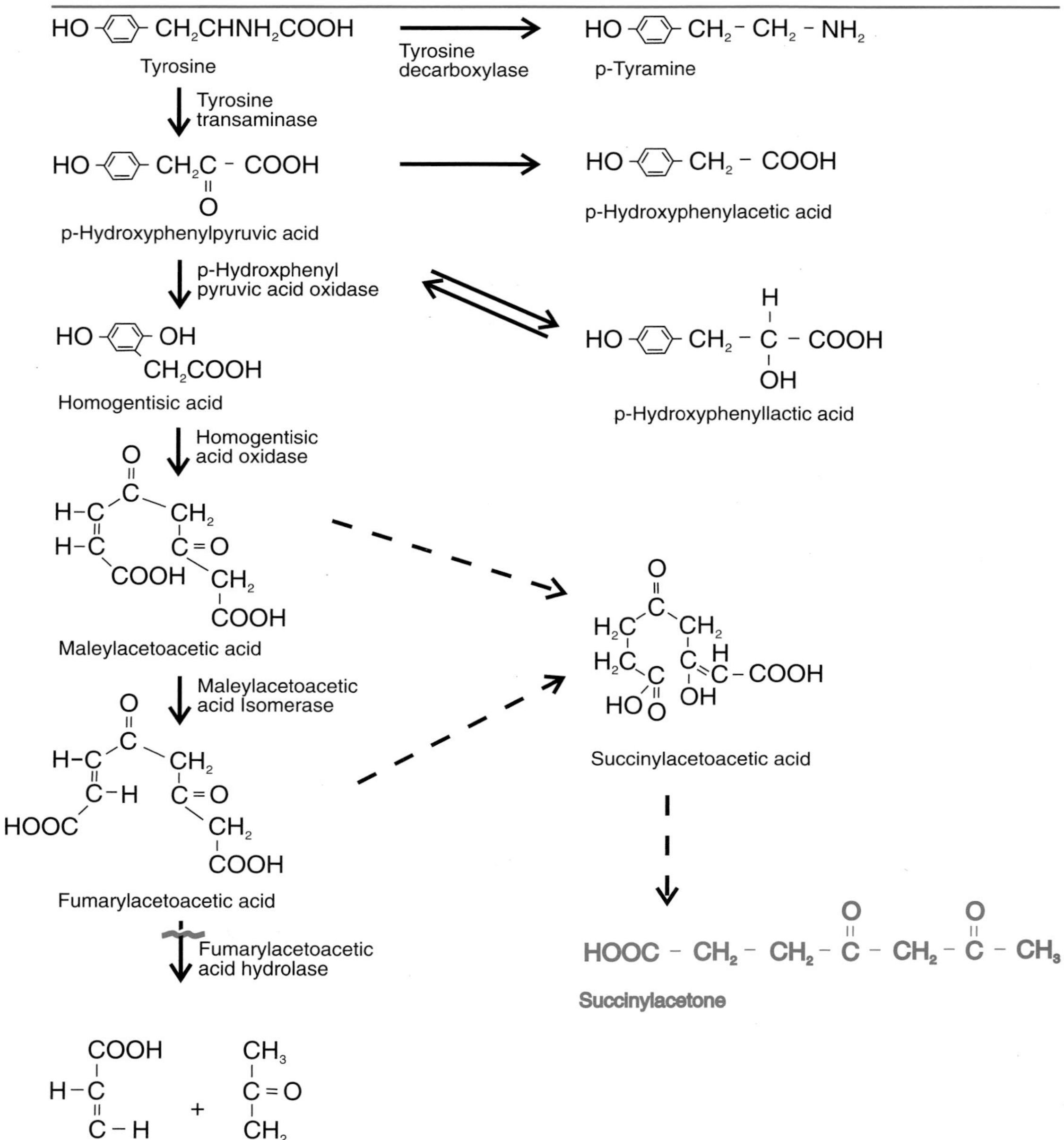

FIG. 16.17. Metabolic pathway of tyrosine. The enzyme fumarylacetoacetic acid hydrolase is blocked in type I tyrosinemia.

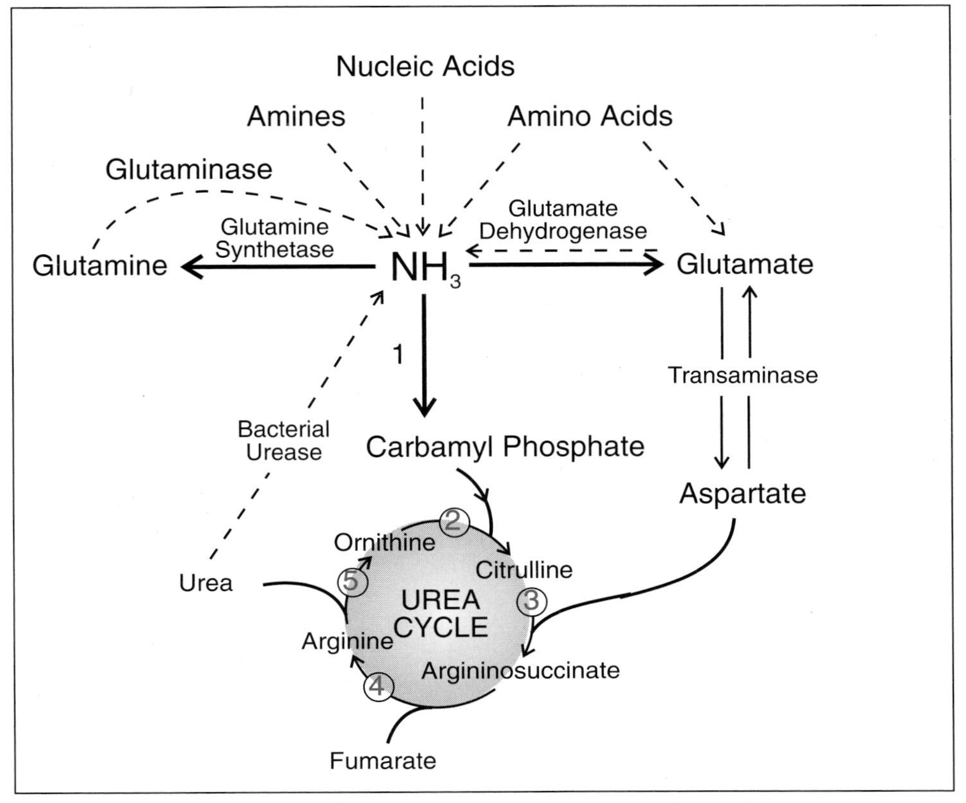

FIG. 16.18. Ammonia metabolism in the liver. Enzymes of the cycle are (1) carbamyl phosphate synthetase (mitochondrial), (2) ornithine transcarbamylase (mitochondrial), (3) argininosuccinate synthetase (cytoplasmic), (4) argininosuccinase (cytoplasmic), and (5) arginase (cytoplasmic). All enzymes are sites for potential defects.

amino acids and is carried by the portal vein to the liver, where it is converted to urea. Adequate conversion of ammonia to urea is extremely important because ammonia is toxic to the central nervous system, whereas urea is an innocuous compound that is readily excreted.

Several potential defects in the urea cycle may arise from one of six enzyme deficiencies (Fig. 16.18). Hyperammonemia is the characteristic biochemical manifestation common to all the urea cycle defects. Severely affected patients may appear clinically with overwhelming illness in the newborn period with extremely high blood ammonia levels. Less severely affected individuals may appear with recurrent vomiting, protein aversion, and elevated ammonia levels only with the stress of other illness. Treatment, mainly protein restriction, may be possible in some cases. Because ammonia is believed to be a neurotoxin that effects carbohydrate metabolism and nerve transmission, successful treatment must maintain a reasonably low ammonia level. Table 16.6 outlines the different types of urea cycle defects. Other findings that help to differentiate the disorders from one another are noted.

TABLE 16.6
UREA CYCLE DEFECTS

DISEASE	ENZYME DEFICIENCY	PLASMA AMINO ACIDS	URINE METABOLITES
Carbamyl phosphate synthetase deficiency	Carbamyl phosphate synthetase	Glutamine, alanine, lysine elevated	No orotic acid
Ornithine transcarbamylase deficiency (X-linked)	Ornithine transcarbamylase	Glutamine, alanine, lysine elevated, citrulline absent	Elevated orotic acid
Citrullinemia	Argininosuccinic acid synthetase	Citrulline elevated	Elevated citrulline; +/− orotic acid
Argininosuccinic aciduria	Argininosuccinase	Argininosuccinic acid elevated	Argininosuccinic acid elevated
Hyperargininemia	Arginase	Elevated arginine	Elevated arginine, cystine, lysine, and ornithine; +/− orotic acid
Lysinuric protein intolerance	Defective transport of dibasic amino acids	Elevated ornithine	Elevated lysine, cystine, citrulline, and arginine

Organic Acidurias

The organic acidurias are a broad group of diseases that are screened for by gas chromatography/mass spectrometry. The technique requires solvent extraction of organic acids from urine and subsequent derivatization, usually with a trimethylsilyl group. Organic acid analysis may be helpful in diagnosis for any child with metabolic acidosis, hypoglycemia, vomiting, seizures, hyperglycinemia, hyperammonemia, or α-ketoacids in the urine. Many organic acidurias are associated with peculiar odors; urine from a child with maple syrup urine disease smells as the name implies, and urine from a child with isovaleric or propionic aciduria smells like "old sweat socks." An unusual odor emanating from the child may suggest an organic acid analysis. Table 16.7 outlines the major disorders of organic acid metabolism and biochemical findings.

Beta-Oxidation Defects

Fat is utilized for energy by the hydrolysis of triglycerides and the subsequent breakdown of fatty acids, two carbon units at a time, to form acetyl-CoA. Acetyl-CoA is oxidized in mitochondria to CO_2 and water in the tricarboxylic acid (TCA) cycle. The entire process is referred to as β-oxidation (see Chap. 8). During fasting, some of the acetyl-CoA is diverted to the production of ketones that can also be utilized as a source of energy by muscle cells (see Chap. 7). The production of ketones during fasting is more pronounced in children than in adults. β-oxidation of fat is an essential pathway for energy production after depletion of glycogen stores.

TABLE 16.7
ORGANIC ACIDURIAS

DISORDER	CHARACTERISTIC COMPOUNDS
Phenylketonuria	Phenylpyruvic acid Phenylacetic acid 2-Hydroxyphenylacetic acid
Tyrosinemia (type I)	4-Hydroxyphenylpyruvic acid 4-Hydroxyphenylactic acid Succinylacetone
Alcaptonuria	Homogentisic acid
Maple syrup urine disease	2-Ketoisocaproic acid 2-Keto-3-methylvaleric acid 2-Ketoisovaleric acid 2-Hydroxyisovaleric acid 2-Hydroxyisocaproic acid 2-Hydroxy-3-methylvaleric acid
Isovaleric aciduria	Isovalerylglycine 3-Hydroxyisovaleric acid 4-Hydroxyisovaleric acid
Multiple carboxylase deficiency	3-Hydroxyisovaleric acid 3-Methylcrotonylglycine Methylcitric acid Lactic acid 3-Hydroxypropionic acid
Propionic aciduria	Methylcitric acid Propionylglycine 3-Hydroxypropionic acid 3-Hydroxyvaleric acid
Methylmalonic aciduria	Methylmalonic acid, plus same metabolites as propionic aciduria
Glutaric aciduria type I	Glutaric acid Glutaconic acid 3-Hydroxyglutaric acid

Disorders in the β-oxidation of fatty acids are usually caused by a deficiency in one or more of the acyl-CoA dehydrogenases necessary for the degradation of short-, medium-, and long-chain fatty acids (Fig 16.19). The most common of these defects is medium-chain acyl-CoA dehydrogenase (MCAD) deficiency. MCAD deficiency may go unnoticed until a child is stressed by fasting. Clinically these individuals can present with sudden death following a viral or bacterial illness associated with vomiting. They can also present with a Reye's-like illness characterized by coma, hypoglycemia, and hyperammonemia. If these individuals can be detected early enough, treatment is available in the form of frequent feeding (or glucose supplementation when ill) and carnitine therapy (see the following section). Diagnosis can be made through analysis of the urine for organic acids (Table 16.8). Many of the other defects in β-oxidation of fat are not responsive to treatment.

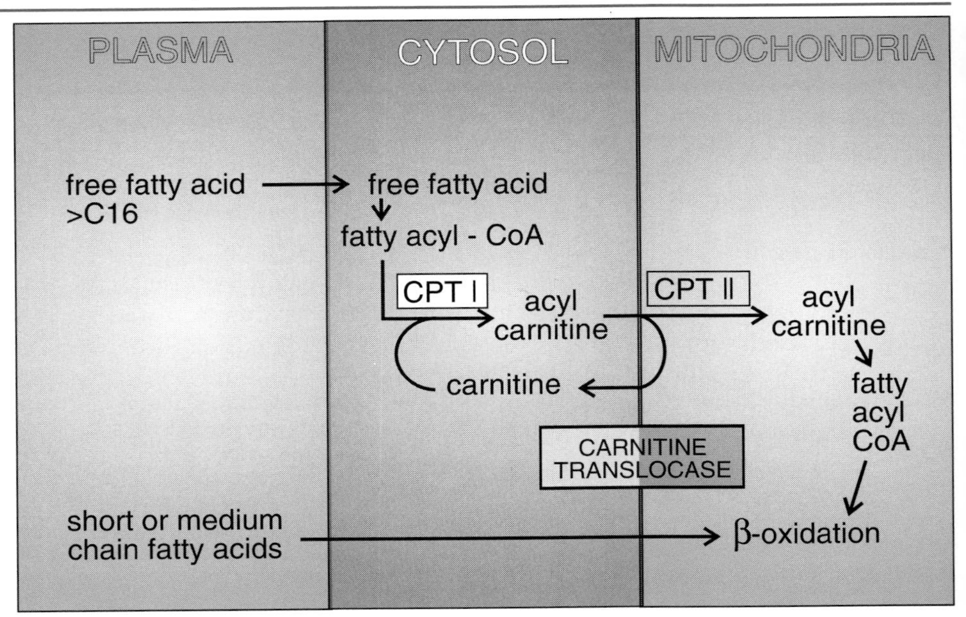

FIG. 16.19. Fatty acid transport across the mitochondrial membrane with carnitine serving as the carrier.

TABLE 16.8
DISORDERS OF FATTY ACID OXIDATION

Long-chain acyl CoA dehydrogenase deficiency	Suberic acid, sebacic acid
Medium-chain acyl CoA dehydrogenase deficiency	Octanoic acid, adipic acid, suberic acid, sebacic acid, 5-hydroxyhexanoic acid, 7-hydroxyoctanoic acid, octenedioic acid, decenedioic acid, hexanoylglycine, phenylpropionylglycine, suberylglycine
Short-chain acyl CoA dehydrogenase deficiency	Ethylmalonic acid, adipic acid, suberic acid, sebacic acid, methylsuccinic acid
Multiple acyl CoA dehydrogenase deficiency	Glutaric acid, ethylmalonic acid, adipic acid, suberic acid, 2-hydroxyglutaric acid, isovalerylglycine, isobutyrylglycine, 2-methylbutyrylglycine

Carnitine Assessment

Carnitine is an amino acid derivative that is produced endogenously from methionine and lysine (Fig. 16.20). Many foods, mainly meat and dairy products, are rich dietary sources of carnitine.

Carnitine is essential for the mitochondrial β-oxidation of fat. During periods of fasting and prolonged exercise, liver glycogen stores are depleted. Fatty acids, stored as triglycerides, are then called upon as an energy source. Long-chain fatty acids require carnitine for transport into the mitochondria as

$$(CH_3)_3-\overset{\oplus}{N}-CH_2-CHOH-CH_2-COOH$$

FIG. 16.20. Structure of carnitine, transmembrane carrier for long-chain fatty acids.

acyl-carnitine moieties (Fig. 16.19). Two primary enzymes facilitate the transport across the outer mitochondrial membrane (carnitine palmityltransferase I) and the inner mitochondrial membrane (carnitine palmityltransferase II). Inside the mitochondria, they enter the β-oxidation pathway to produce acetyl-CoA. The acetyl-CoA enters the TCA cycle for energy production or is converted to ketones in the liver. This pathway is especially important in a sick infant who is not eating and needs fatty acids and ketones as an energy source.

Carnitine exists as a free fraction and as an acyl-bound fraction; the two together make up the total carnitine. Carnitine binds to long-chain fatty acids, organic acids, drugs (primarily valproic acid), and other fatty acids. Free carnitine is reabsorbed in the kidney, whereas bound carnitine is lost in the urine.

Lack of free carnitine in the cell causes problems with cellular energy production, even if total carnitine in plasma is adequate. For this reason, carnitine assessment must measure free short-chain acyl-carnitine and long-chain acyl-carnitine to be valid. The ratio of acyl/free carnitine is important in assessing available carnitine.

Few, if any, reasons are known for a primary deficiency in carnitine. Secondary causes for carnitine deficiency abound, however, because most organic acidurias and β-oxidation defects have characteristically low free carnitines. This low amount is the result of increased esterification of carnitine (and an increase in short- or long-chain acyl-carnitine), and subsequent renal loss. The same phenomenon is seen in patients on long-term valproate therapy because valproate also binds to carnitine.

Another potential cause for carnitine deficiency is long-term hyperalimentation feeding. Infant formulas and total parenteral nutrition solutions are usually not supplemented with carnitine; thus, patients fed in these manners may eventually develop a dietary insufficiency.

Symptoms of carnitine deficiency are those of altered fat metabolism: skeletal muscle weakness, cardiomyopathy, liver dysfunction, and hypoglycemia.[24]

The reference method for carnitine analysis is a radioenzymatic assay[25] that uses ^{14}C-acetyl-CoA in the following reaction:

$$\text{L-carnitine} + {}^{14}\text{C-acetyl-CoA} \xrightarrow{\text{carnitine acyltransferase}} {}^{14}\text{C-acetyl-carnitine} + \text{CoASH}$$

The labeled acetylcarnitine formed is separated from unreacted ^{14}C-acetyl-CoA by passing the mixture through a column of anion exchange resin and determining the isotope content of the effluent fluid. Plasma carnitine is determined in three fractions: total, short-chain, and long-chain. The free carnitine is calculated.

Reference ranges	
Free carnitine	18 to 50 μmol/L
Short-chain acylcarnitine	3 to 10 μmol/L
Long-chain acylcarnitine	3 to 5 μmol/L
Acyl/free ratio	< 0.4

REFERENCES

1. Weatherall, D.J.: The New Genetics and Clinical Practice. New York, Oxford University Press, 1985.
2. Sambrook, J., Fritsch, E.F., and Maniatis, T.: Molecular Cloning: A Laboratory Manual. 2nd Edition. Plainview, New York, Cold Spring Harbor Laboratory Press, 1989.
3. Donis-Keller, H., and Botstein, D.: Recombinant DNA methods: applications to human genetics. *In* Molecular Genetics in Medicine. Edited by Childs, B., Holtzman, N.A., Kazazian, H.H., and Valle, D.L. New York, Elsevier, 1988.
4. Fenoglio-Preiser, C.M., and Willman, C.L.: Molecular biology and the pathologist: general principles and applications. Arch. Pathol. Lab. Med., *111*:601, 1987.
5. Lowe, J.B.: Critical review: clinical applications of gene probes in human genetic disease, malignancy, and infectious disease. Clin. Chim. Acta, *157*:1, 1988.
6. Sklar, J.: Current topics: DNA hybridization in diagnostic pathology. Hum. Pathol., *16*;654, 1985.
7. Caskey, C.T.: Diseases diagnosis by recombinant DNA methods. Science, *236*:1223, 1987.
8. Connor, J.M., Ferguson-Smith, M.A.: Essential Medical Genetics. 3rd Edition. Oxford, Blackwell Scientific Publications, 1991.
9. Thompson, M.W., McInnes, R.R., and Willard, H.F.: Genetics in Medicine. 5th Edition. Philadelphia, W.B. Saunders, 1991.
10. Saiki, R.K., Gelfand, D.H., Stoffel, S., et al.: Primer directed enzymatic amplification of DNA with a thermostable DNA polymerase. Science, *239*:487, 1988.
11. Peter, J.B.: DNA amplification by polymerase chain reaction. Clin. Chem. News, 19/June, 1990.
12. Scriver, C.R., Beaudot, A.L., Sly, W.S., and Valle, D. (Eds.): The Metabolic Basis of Inherited Disease. 6th Edition. New York, McGraw-Hill, 1989.
13. Nyhan, W.L.: Abnormalities in Amino Acid Metabolism in Clinical Medicine. Norwalk, CT, Appleton-Century-Crofts, 1984.
14. Cohn, R.M., and Roth, K.S.: Metabolic Disease: A Guide to Early Recognition. Philadelphia, W.B. Saunders, 1983.
15. Hicks, J.M., and Boeckx, R.L.: Pediatric Clinical Chemistry. Philadelphia, W.B. Saunders, 1984.
16. Thomas, G.H., and Howell, R.R.: Selected Screening Tests for Genetic Metabolic Diseases. Chicago, Year Book Medical Publishers, 1973.
17. Perry, T.L., Hansen, S., and MacDougall, L.: Urinary screening tests in the prevention of mental deficiency. Can. Med. Assoc. J., *95*:89, 1966.
18. Berry, H.K., and Spinanger, J.: A paper spot useful in the study of Hurler's syndrome. J. Lab. Clin. Med., *55*:136, 1960.
19. Whitley, C.B., Ridnour, M.D., Draper, K.A., et al.: Diagnostic test for mucopolysaccharidosis. I. Direct method for quantifying excessive urinary glycosaminoglycan excretion. Clin. Chem., *35*:374, 1989.
20. Cohen, S.A.: The pico-tag method. Millipore Corp., Milford, MA.
21. Rutledge, J.C.: HPLC qualitative amino acid analysis in the clinical laboratory. Am. J. Clin. Pathol., *87*:614, 1987.
22. Roth, M.J., Hampai, A.: Column chromatography of amino acids with fluorescence detection. J. Chromatogr. *83*:353, 1973.
23. Hamilton, P.B.: Ion exchange chromatography of amino acids. A single column, high resolving, fully automatic procedure. Anal. Chem., *35*:2055, 1963.
24. Breningstall, G.N.: Carnitine deficiency syndromes. Pediatr. Neurol., *6*:75, 1990.
25. McGarry, J.D., and Foster, D.W.: An improved and simplified radioisotopic assay for the determination of free and esterified carnitine. J. Lipid Res., *17*:277, 1976.

REVIEW QUESTIONS

1. What are the strands of DNA composed of, and how do they bind together?
2. What are two methods one could use to assess the quality of DNA?
3. What is a DNA probe, and how is it utilized?
4. What is the polymerase chain reaction (PCR), and what is the main advantage to its use?
5. What is the defect in phenylketonuria (PKU), what are the clinical findings if untreated, and what compound in blood do state laboratories test for screening purposes?
6. What one laboratory finding is most suggestive of a urea cycle disorder?
7. What are three causes of a secondary carnitine deficiency?
8. Who should be screened for Tay-Sachs disease?

Appendices

APPENDIX 1*
REFERENCE VALUES†

TEST	SPECIMEN	REFERENCE RANGE	SI CONVERSION FACTOR‡	REFERENCE RANGE SI UNITS
Acetoacetic acid, qual	U	Negative		
Acetone, qual	U	Negative		
Adrenocorticotropic hormone (ACTH)	P	0800 hr: 25–50 pg/mL 2400 hr: ~5 pg/mL	1.0	25–50 ng/L ~5 ng/L
Alanine aminotransferase (ALT)	S	3–30 U/L		
Albumin	S	3.5–5.2 g/dL	10	35–52 g/L
	CSF	15–45 mg/dL	10	150–450 mg/L
	U	50–80 mg/day	—	50–80 mg/day
Aldosterone, upright	S	5–20 ng/dL	27.8	140–560 pmol/L
	or			
recumbent	P	2–10 ng/dL		56–280 pmol/L
Ammonia nitrogen	P	15–45 μg/dL	0.714	11–32 μmol/L
Amylase (saccharogenic methods)	S	25–135 U/L	—	Same
	U	2–19 U/hr		
Aspartate aminotransferase (AST)	S	6–25 U/L		
Bicarbonate (HCO_3^-)	S	Art, NB 17–24 mmol/L Art, Ch, and Ad 22–26 mmol/L		Same
Bilirubin, total	S	cord: <2.0 mg/dL 1–24 hr: <6 mg/dL 1–2 day: <8 mg/dL 3–5 day: <12 mg/dL 30 day and older: 0.2–1.0 mg/dL	17.1	<34 μmol/L <103 μmol/L <137 μmol/L <205 μmol/L 3.4–17 μmol/L
Esterified	S	30 day–60 yr: 0–0.2 mg/dL	17.1	0–3.4 μmol/L
qual	U	Negative	—	
Calcium, total (Ca_T)	S	8.5–10.4 mg/dL	0.25	2.13–2.60 mmol/L
ionized (Ca^{2+})	S, P or B	4.68–5.32 mg/dL		1.17–1.33 mmol/L
Catecholamines	U	<100 μg/day	0.0059	<0.59 μmol/day

APPENDIX 1*
REFERENCE VALUES† *Continued*

TEST	SPECIMEN	REFERENCE RANGE	SI CONVERSION FACTOR‡	REFERENCE RANGE SI UNITS
Chloride	S	98–108 mmol/L		Same
	CSF	115–132 mmol/L		
	U	110–250 mmol/L	—	
	Sw	5–40 mmol/L		
Cholesterol, total		mg/dL	0.0259	μmol/L
HDL	S	See Table 8.4 for recommended and high-risk plasma concentrations.		
LDL				
CO_2, total	S	Inf: 20–26 mmol/L	—	Same
		Ad: 22–30 mmol/L		
CO_2, partial pressure, (pCO_2)	P, Art.	Inf: 26–41 mm Hg		
	P, Art.	Ch and Ad: 35–45 mm Hg		
	P, Ven	Ch and Ad: 40–50 mm Hg	—	
Cortisol	S or P	5–22 μg/dL	27.6	138–607 nmol/L
Creatine kinase (CK)	S	10–100 U/L		
Creatine kinase isoenzyme, Fraction 2 (MB)	S	1–6 U/L		
Creatinine	S	Ch: 0.4–1.0 mg/dL	88.4	36–88 μmol/L
		Ad: 0.6–1.2 mg/dL		53–106 μmol/L
	U	Ad: ~1 mg/min		~88 mmol/L
	Af	>2.0 mg/dL taken as a sign of maturity	88.4	>177 μmol/L
Creatinine clearance	S and U	Ad: 90–140 mL/min		
		2 yr: 55 mL/min		
		4–8 yr: 64–72 mL/min		
		12 yr: 100–120 mL/min		
Estriol (E_3), free	S	5–40 ng/mL	3.47	17–139 nmol/L
Ferritin	S	Male: 30–250 ng/mL	1.0	30–250 μg/L
		Female: 12–125 ng/mL		12–125 μg/L
α-Fetoprotein	S	Pregnant female, 16th wk: 10–60 (mean, 34) ng/mL	1.0	10–60 μg/L
Gastric content (Residue)				
Volume	Gast	20–100 mL		
pH	Res	1.5–3.5		
Gastric secretory rate				
Basal acid output		1–3 mmol/hr		Same
Maximal acid output		8–40 mmol/hr		
Gastrin	P	46–140 pg/mL	1.0	46–140 ng/L

TEST	SPECIMEN	REFERENCE RANGE	SI CONVERSION FACTOR‡	REFERENCE RANGE SI UNITS
Glucose	S	65–100 mg/dL		3.6–5.6 mmol/L
	CSF	45–70 mg/dL	0.0556	2.5–3.9 mmol/L
	U	<6 mg/dL; negative by dipstick test		<0.33 mmol/L
γ-Glutamyl transferase	S	3–35 U/L		
Growth hormone (GH) (during daytime)	S or P	Ad: <5 ng/mL	1.0	<5 μg/L
Hemoglobin, glycated (HbA_{1c})	B	3–6% of total Hb		
Human chorionic gonadotropin, β-subunit (hCG)	S or P			
Nonpregnancy		<3.0 mU/mL	1.0	<3 U/L
Pregnancy		14 day: >10 mU/mL	1.0	>10U/L
17-hydroxycorticosteroids (17-OHCS)	U	Ad: 5–14 mg/day	2.76	14–39 μmol/day
Insulin	P	<860 pg/mL	1.0	<860 ng/L
Iron	S	NB: 120–240 μg/dL		21.5–43.0 μmol/L
		6 mo–12 yr: 20–123 μg/dL		3.6–22.0 μmol/L
		12 yr–18 yr: 40–140 μg/dL	0.179	7.2–25.0 μmol/L
		Ad: 45–165 μg/dL		8.1–29.5 μmol/L
Iron-binding capacity (TIBC)		NB: 59–175 μg/dL	0.179	10.6–31.3 μmol/L
Transferrin saturation	S	3 yr–Ad: 260–440 μg/dL		46.5–78.8 μmol/L
(% saturation)		20–50%		
17-Ketogenic steroids (17-KGS)	U	Ad Male: 8–25 mg/day	3.46	28–87 μmol/day
		Ad Female: 5–18 mg/day		17–62 μmol/day
		Ch: 2–4 mg/day		7–14 μmol/day
17-Ketosteroids (17-KS)	U	Ad Male: 8-20 mg/day	3.46	28–69 μmol/day
		Ad Female: 5-15 mg/day		17–52 μmol/day
		Ch: <2 mg/day		<7 μmol/day
Lactate dehydrogenase (LD)	S			
Total		125–290 U/L		
Isoenzymes: LD_1		100 U/L or 20–34% of total		
LD_2		115 U/L or 32–40% of total		
LD_3		65 U/L or 17–23% of total		
LD_4		40 U/L or 3–13% of total		
LD_5		35 U/L or 4–12% of total		

TEST	SPECIMEN	REFERENCE RANGE	SI CONVERSION FACTOR‡	REFERENCE RANGE SI UNITS
Lactic acid (lactate)	P	4.5–17.1 mg/dL	0.111	0.50–1.9 mmol/L
Lecithin/sphingomyelin (L/S) ratio	Af	>2.0 indicates maturity		
Lipase	S	28–200 U/L		
Magnesium	S	0.65–1.05 mmol/L		Same
Metanephrines	U	0.3–0.9 mg/day	5.1	1.5–4.6 μmol/day
5′-Nucleotidase (5NT)	S	1–7 U/L		
Ornithine carbamoyl transferase (OCT)	S	1–6 U/L		
Osmolality	S	278–305 mOsmol/kg		Same
	U	Random: 40–1350 mOsmol/kg		Same
		Maximal: 850–1350 mOsmol/kg		
Oxygen, content	B art	~20 mL/dL		
	B ven	~15 mL/dL		
Oxygen, partial pressure (pO_2)	art	NB: 65–76 mm Hg		
	B art	Ch and Ad: 85–105 mm Hg		
	B ven	Ch and Ad: 30–50 mm Hg		
Oxygen, saturation	B art	95–98%		
	B ven	34–50%		
pH	B art	7.35–7.45		
	B ven	7.32–7.42		
Phenylalanine	S	NB: 1.2–3.5 mg/dL	0.0605	0.07–0.21 mmol/L
		Ad: 0.6–1.8 mg/dL		0.04–0.11 mmol/L
Phosphatase, acid (ACP)				
Total	S	0.2–1.8 U/L		
Tartrate inhibitable		0.2–0.6 U/L		
Phosphatase, alkaline (ALP)	S	0–3 mo: 70–220 U/L		
		3 mo–10 yr: 50–260 U/L		
		10 yr–puberty: 60–295 U/L		
		Ad: 20–105 U/L		
Phosphate, inorganic (as P)	S	NB: 4.2–9.5 mg/dL		1.34–3.36 mmol/L
		Ch: 4.0–6.0 mg/dL	0.323	1.28–1.94 mmol/L
		Ad: 3.0–4.5 mg/dL		0.96–1.44 mmol/L
Potassium	S	NB: 5.0–7.5 mmol/L		
		Inf: 4.0–5.7 mmol/L		Same
		Ch + Ad: 3.5–5.5 mmol/L		
Prolactin (PRL)	S	Ad female: <30 ng/mL	1.0	<30 μg/L
		Ad male: <20 ng/mL		<20 μg/L
Protein, total	S	6.0–8.2 g/dL	10	60–82 g/L
	CSF	15–45 mg/dL		150–450 mg/L
	U	<100 mg/day		

APPENDIX 1*
REFERENCE VALUE† *Continued*

TEST	SPECIMEN	REFERENCE RANGE	SI CONVERSION FACTOR‡	REFERENCE RANGE SI UNITS
Electrophoresis	S			
Fractions:				
Albumin		3.5–5.2 g/dL		35–52 g/L
α_1-globulin		0.1–0.4 g/dL		1–4 g/L
α_2-globulin		0.5–1.0 g/dL	10	5–10 g/L
β-globulin		0.6–1.2 g/dL		6–12 g/L
γ-globulin		0.7–1.6 g/dL		7–16 g/L
Pyruvic acid (pyruvate)	B	0.3–0.9 mg/dL	0.114	0.03–0.10 mmol/L
Sodium	S	136–145 mmol/L		Same
Testosterone	P	Ad male: 350–850 ng/dL	0.0347	12.1–29.5 nmol/L
		Ad female: 20–80 ng/dL		0.8–2.8 nmol/L
Thyroid hormones				
Thyroxine (T_4), total	S	5.1–11.0 μg/dL	12.9	66–142 nmol/L
free		1.5–3.5 ng/dL		19–45 pmol/L
Triiodothyronine (T_3), total		70–200 ng/dL	0.0154	1.1–3.1 nmol/L
free		200–750 pg/dL		3.1–11.6 pmol/L
T_3 Resin uptake (T_3U)	S	25–35%		
Thyroid stimulating hormone (TSH)	P	0.4–6 μU/mL	1.0	0.4–6 mU/L
Triglycerides	S	35–200 mg/dL	0.0113	0.40–2.26 mmol/L
Urea nitrogen	S	6–18 mg/dL	0.357	2.1–6.4 mmol/L
Uric acid (urate)	S	Ch: 2.0–5.5 mg/dL	0.060	0.12–0.33 mmol/L
		Ad male: 3.5–7.5 mg/dL		0.21–0.45 mmol/L
		Ad female: 2.5–6.5 mg/dL		0.15–0.39 mmol/L
Urobilinogen	U	2 hr: 0.1–1.0 Ehrlich unit		
		24 hr: 0.5–4 Ehrlich units		
VMA (3-methoxy-4-hydroxy mandelic acid)	U	2–8 mg/day	5	10–40 μmol/day
Zinc protoporphyrin/heme ratio (ZPP/H)		30–80 μmol ZPP/mol heme		

*The following abbreviations are used:

Ad = adult	CSF = cerebrospinal fluid	S = serum
Af = amniotic fluid	Gast res = gastric residue	Sw = sweat
Art = arterial	Inf = infant	U = urine
B = whole blood	P = plasma	V = venous
Ch = children	qual = qualitative	

†The therapeutic and toxic serum concentrations of common drugs and toxins are listed in Table 14.2.
‡Multiply conventional units by this factor for conversion to SI.

APPENDIX 2
CONCENTRATIONS OF COMMON REAGENTS AND VOLUMES REQUIRED TO DILUTE TO 1 MOL/L

REAGENT*	CONCENTRATION OF CONCENTRATED SOLUTION MOL/L	PURITY %	ML CONCENTRATED REAGENT TO PREPARE 1 MOL/L SOLUTION
Acetic acid	17.4	99–100[†]	57
Hydrochloric acid	11.6	36[†]	86
Nitric acid	16.4	69[†]	61
Phosphoric acid	14.6	85[†]	68
Sulfuric acid	17.8	95[†]	56
Ammonium hydroxide	14.8	28[‡]	67

*Reagent grade chemicals.
[†]% (w/v) of the acid.
[‡]% (w/v) of NH_3.

APPENDIX 3
ATOMIC WEIGHTS OF THE COMMON ELEMENTS*

NAME	SYMBOL	INTERNATIONAL ATOMIC MASS	NAME	SYMBOL	INTERNATIONAL ATOMIC MASS
Aluminum	Al	26.98	Manganese	Mn	54.94
Antimony (stibium)	Sb	121.8	Mercury	Hg	200.6
Argon	Ar	39.95	Molybdenum	Mo	95.94
Arsenic	As	74.92	Neon	Ne	20.18
Barium	Ba	137.3	Nickel	Ni	58.71
Beryllium	Be	9.012	Nitrogen	N	14.01
Bismuth	Bi	209.0			
Boron	B	10.81	Oxygen	O	16.00
Bromine	Br	79.91	Palladium	Pd	106.4
Cadmium	Cd	112.4	Phosphorus	P	30.97
Calcium	Ca	40.08	Platinum	Pt	195.1
Carbon	C	12.01	Potassium (kalium)	K	39.10
Cerium	Ce	140.1	Selenium	Se	78.96
Cesium	Cs	132.9	Silicon	Si	28.09
Chlorine	Cl	35.45	Silver (argentum)	Ag	107.9
Chromium	Cr	52.00	Sodium (natrium)	Na	22.99
Cobalt	Co	58.93	Strontium	Sr	87.62
Copper	Cu	63.54	Sulfur	S	32.06
Fluorine	F	19.00	Tellurium	Te	127.6
Gold (aurum)	Au	197.0	Thallium	Tl	204.4
Helium	He	4.003	Thorium	Th	232.0
Hydrogen	H	1.008	Tin (stannum)	Sn	181.7
Iodine	I	126.9	Titanium	Ti	47.90
Iron (ferrum)	Fe	55.85	Tungsten (wolfram)	W	183.8
Lanthanum	La	138.9	Uranium	U	238.0
Lead (plumbum)	Pb	207.2	Vanadium	V	50.94
Lithium	Li	6.939	Xenon	Xe	131.3
Magnesium	Mg	24.31	Zinc	Zn	65.37

*Values as of 1963, based on carbon-12 and rounded off to 4 significant figures.

APPENDIX 4
MANUFACTURERS AND SUPPLIERS

Abbott Laboratories, Abbott Park, IL 60064
Advanced Instruments, Needham Heights, MA 02194
Aldrich Chemical Co., Milwaukee, WI 53201
American Association for Clinical Chemistry, Therapeutic Drug Therapy, Washington, D.C., 20006-1008
American Dade, Miami, FL 33152
American Optical Corp., Buffalo, NY 14240
Ames, Div. of Miles Laboratories, Tarrytown, NY 10591
Applied Biosystems, Foster City, CA 94404
J.T. Baker Chemical Co., Phillipsburg, NJ 08865
Baxter Dade Div., Baxter Healthcare Corp., Miami, FL 33172
Baxter Healthcare Corp., Hospital Supply, McGraw Park, IL 60085
Beckman Instruments, Brea, CA 92621
Becton Dickinson Vacutainer Systems, Rutherford, NJ 07070
Behring Diagnostics, Somerville, NJ 08876
Bio Rad Laboratories (Diagnostics), Hercules, CA 94547
Boehringer Mannheim Diagnostics, Indianapolis, IN 46250
Brinkmann Instruments, Westbury, NY 11590-0207
Burdick & Jackson Div., Baxter Healthcare Corp., Muskegon, MI 49442
Calbiochem Corp., La Jolla, CA 92037
Ciba-Corning Diagnostic Corp., Walpole, MA 02032
College of American Pathologists, Northfield, IL 60093-2570
Corning Inc., Science Products, Corning, NY 14831
DuPont Co., Biotechnology Systems, Wilmington, DE 19880-0016
Eastman Kodak Co., Rochester, NY 14652-3512
Eppendorf North America, Inc., Madison, WI 53711
Fisher Scientific, Pittsburgh, PA 15219
Gelman Sciences, Ann Arbor, MI 48106
Helena Labs, Beaumont, TX 77704–0752
Hewlett Packard Co., Analytical Products Group, Palo Alto, CA 94304
Hollister Inc., Libertyville, IL 60048
Hybritech, San Diego, CA 92126
Instrumentation Laboratory, Lexington, MA 02173
Isolab Inc., Akron, OH 44313
Laboratory Safety Systems, Janesville, WI 53545
Mallinckrodt, Inc., St. Louis, MO 63134
Mead, Johnson and Co., Evansville, IN 47721
Merck and Co., Rahway, NJ 07065
Mettler Instruments, Hightstown, NJ 08520-0071
Miles Inc., Diagnostics Division, Tarrytown, NY 10591
Millipore Corp., Bedford, MA 01730
National Bureau of Standards, Office of Standard Reference Materials, Washington, D.C. 20234
Nichols Institute of Diagnostics, San Juan Capistrano, CA 92690
Nova Biomedical Corp., Waltham, MA 02254
Novametrix Medical System, Wallingford, CT 06492
Orion Research, Cambridge, MA 02129
Oxford Labwares, Div. of Sherwood Medical, St. Louis, MO 63103
Perkin-Elmer, Norwalk, CT 06859-0177
Pfizer, New York, NY 10017
Pharmacia LKB Biotechnology, Piscataway, NJ 08855–1327
Radiometer America, Westlake, Oh 44145
Rainin Instrument Co., Inc., Emeryville, CA 94608-1097
Roche Diagnostic Systems, Branchburg, NJ 08876

Scientific Products Div., Baxter Healthcare Corp. McGraw Park, IL 60085
Sigma Diagnostics, St. Louis, MO 63178
SMI—see Baxter Healthcare Corp.
Syva Co., San Jose, CA 95161
Toxi-Lab, Inc., Irvine, CA 92718
Upjohn, Kalamazoo, MI 49001
Waters Chromatography Div., Millipore Corp., Milford, MA 01757
Waters Instruments, Inc., Rochester, MN 55901
Yang Laboratories, Inc., Bellevue, WA 98007

Index

Note: Page numbers in *italics* indicate illustrations; numbers followed by "t" indicate tables.